The Subnuclear Series • Volume 42

Proceedings of the International School of Subnuclear Physics

HOW AND WHERE TO GO BEYOND THE STANDARD MODEL

THE SUBNUCLEAR SERIES

Series Editor: ANTONINO ZICHICHI, *European Physical Society, Geneva, Switzerland*

1. 1963 STRONG, ELECTROMAGNETIC, AND WEAK INTERACTIONS
2. 1964 SYMMETRIES IN ELEMENTARY PARTICLE PHYSICS
3. 1965 RECENT DEVELOPMENTS IN PARTICLE SYMMETRIES
4. 1966 STRONG AND WEAK INTERACTIONS
5. 1967 HADRONS AND THEIR INTERACTIONS
6. 1968 THEORY AND PHENOMENOLOGY IN PARTICLE PHYSICS
7. 1969 SUBNUCLEAR PHENOMENA
8. 1970 ELEMENTARY PROCESSES AT HIGH ENERGY
9. 1971 PROPERTIES OF THE FUNDAMENTAL INTERACTIONS
10. 1972 HIGHLIGHTS IN PARTICLE PHYSICS
11. 1973 LAWS OF HADRONIC MATTER
12. 1974 LEPTON AND HADRON STRUCTURE
13. 1975 NEW PHENOMENA IN SUBNUCLEAR PHYSICS
14. 1976 UNDERSTANDING THE FUNDAMENTAL CONSTITUENTS OF MATTER
15. 1977 THE WHYS OF SUBNUCLEAR PHYSICS
16. 1978 THE NEW ASPECTS OF SUBNUCLEAR PHYSICS
17. 1979 POINTLIKE STRUCTURES INSIDE AND OUTSIDE HADRONS
18. 1980 THE HIGH-ENERGY LIMIT
19. 1981 THE UNITY OF THE FUNDAMENTAL INTERACTIONS
20. 1982 GAUGE INTERACTIONS: Theory and Experiment
21. 1983 HOW FAR ARE WE FROM THE GAUGE FORCES?
22. 1984 QUARKS, LEPTONS, AND THEIR CONSTITUENTS
23. 1985 OLD AND NEW FORCES OF NATURE
24. 1986 THE SUPERWORLD I
25. 1987 THE SUPERWORLD II
26. 1988 THE SUPERWORLD III
27. 1989 THE CHALLENGING QUESTIONS
28. 1990 PHYSICS UP TO 200 TeV
29. 1991 PHYSICS AT THE HIGHEST ENERGY AND LUMINOSITY: To Understand the Origin of Mass
30. 1992 FROM SUPERSTRINGS TO THE REAL SUPERWORLD
31. 1993 FROM SUPERSYMMETRY TO THE ORIGIN OF SPACE–TIME
32. 1994 FROM SUPERSTRING TO PRESENT-DAY PHYSICS
33. 1995 VACUUM AND VACUA: The Physics of Nothing
34. 1996 EFFECTIVE THEORIES AND FUNDAMENTAL INTERACTIONS
35. 1997 HIGHLIGHTS OF SUBNUCLEAR PHYSICS: 50 Years Later
36. 1998 FROM THE PLANCK LENGTH TO THE HUBBLE RADIUS
37. 1999 BASICS AND HIGHLIGHTS IN FUNDAMENTAL PHYSICS
38. 2000 THEORY AND EXPERIMENT HEADING FOR NEW PHYSICS
39. 2001 NEW FIELDS AND STRINGS IN SUBNUCLEAR PHYSICS
40. 2002 FROM QUARKS AND GLUONS TO QUANTUM GRAVITY
41. 2004 FROM QUARKS TO BLACK HOLES: PROGRESS IN UNDERSTANDING THE LOGIC OF NATURE
42. 2007 HOW AND WHERE TO GO BEYOND THE STANDARD MODEL

Volume 1 was published by W. A. Benjamin, Inc., New York; 2–8 and 11–12 by Academic Press, New York and London; 9–10 by Editrice Compositori, Bologna; 13–29 by Plenum Press, New York and London; 30–42 by World Scientific, Singapore.

The Subnuclear Series • Volume 42

Proceedings of the International School of Subnuclear Physics

HOW AND WHERE TO GO BEYOND THE STANDARD MODEL

Edited by

Antonino Zichichi

European Physical Society
Geneva, Switzerland

World Scientific

NEW JERSEY • LONDON • SINGAPORE • BEIJING • SHANGHAI • HONG KONG • TAIPEI • CHENNAI

Published by

World Scientific Publishing Co. Pte. Ltd.

5 Toh Tuck Link, Singapore 596224

USA office: 27 Warren Street, Suite 401-402, Hackensack, NJ 07601

UK office: 57 Shelton Street, Covent Garden, London WC2H 9HE

British Library Cataloguing-in-Publication Data
A catalogue record for this book is available from the British Library.

HOW AND WHERE TO GO BEYOND THE STANDARD MODEL

ISBN-13 978-981-256-750-5
ISBN-10 981-256-750-X

Printed in Singapore by World Scientific Printers (S) Pte Ltd

PREFACE

During August/September 2004, a group of 52 physicists from 29 laboratories in 14 countries met in Erice to participate in the 42nd Course of the International School of Subnuclear Physics. The countries represented by the participants were: Belarus, France, Germany, Italy, Mexico, Poland, Russia, Slovenia, Sweden, Turkey, The Netherlands, Ukraine, United Kingdom and the United States of America.

The School was sponsored by the Academies of Sciences of Estonia, Georgia, Lithuania, Russia and Ukraine; the Chinese Academy of Sciences; the Commission of the European Communities; the European Physical Society; the Italian Ministry of Education, University and Scientific Research; the Sicilian Regional Government; the Weizmann Institute of Science; the World Federation of Scientists and the World Laboratory.

The purpose of the School was to focus attention on the phenomenological and theoretical developments in Gauge Theories, as well as in global and local Supersymmetry, and in all the other sectors of Subnuclear Physics. Experimental highlights from the most relevant sources of new data were presented and discussed, as reported in the contents.

An original feature of the School, introduced in 1996, is a series of special sessions devoted to "New Talents". This is a serious problem in Experimental Physics where collaborations count several hundreds of participants and it is almost impossible for young fellows to be known. Even if with much less emphasis the problem exists also in Theoretical Physics. So we decided to offer the young fellows a possibility to let them be known. Twenty-six "new talents" were invited to present a paper, followed by a discussion. Four were given awards: one for an original theoretical work; one for an original experimental work; one for the best overview of a theoretical subject; and one for the best overview of an experimental subject. These special sessions devoted to New Talents represent the projection of Subnuclear Physics on the axis of the young generation.

As every year, the discussion sessions have been the focal point of the School's activity.

During the organization and the running of this year's Course, I enjoyed the collaboration of my colleague and friend, Gerardus 't Hooft, who shared with me the Directorship of the Course. I would like to thank him, together with the group of invited scientists and all the people who contributed to the success of this year's Course.

I hope the reader will enjoy the book as much as the students attending the lectures and discussion sessions. Thanks to the work of the Scientific Secretaries, the discussions have been reproduced as faithfully as possible. At various stages of my work I have enjoyed the collaboration of many friends whose contributions have been extremely important for the School and are highly appreciated. I thank them most warmly. A final acknowledgement to all those in Erice, Bologna and Geneva, who have helped me on so many occasions and to whom I feel very indebted.

Antonino Zichichi
Geneva, October 2004

CONTENTS

Lectures

New Talents

Present Status of Supersymmetry

Sergio Ferrara

CERN, Physics Department - Theory Division
CH - 1211 Geneva 23
and
University of California - Department of Physics and Astronomy
Los Angeles California - USA

1 Introductory remarks

The 42th course of the Erice International School of Subnuclear Physics happens to be in a particular year, which coincides with the following events:

Fifty years of Yang-Mills theories (1954) [1]

Fifty years of CERN (1954) [2, 3]

Thirty years of supersymmetric Yang-Mills theories (1974)

The latter one [4], which is for the time beeing an achievement of Mathematical physics and of modern quantum field theory, may actually become a fundamental physical theory if the Large Hadron Collider under completion at CERN will detect supersymmetric particles (sparticles).

There are already some theoretical hints of supersymmetry which can be summarized as follows:

1. Supersymmetry may be the solution of a double hierarchy problem; the first is the hierarchy of the weak versus the Planck scale ($m_W/m_P <<$

1), the second is the smallness of the cosmological constant [5]. Both problems may be related to the interconnection betwen ordinary electroweak and strong interactions and the gravitational force.

2. Quantum gravity: supersymmetry and its local extension (supergravity) suggests fundamental theories based on 'extended objects' (p-branes) [6]. these new objects are naturally incorporated in superstrings [7] and M-theory [8] which are formulated in ambient spaces with space-time dimensions $D = (9, 1)$ and $D = (10, 1)$.

3. Supersymmetric quantum theories have milder quantum corrections than ordinary quantum field theories and they satisfy some 'non-renormalizaic theorems' [11, 12, 13, 14] which allow to extrct some exact results in any given theory. These theorems are at the basis of the two previous motivations.

4. Holographic description [9] of strings or M-theory in some particular regimes. This in particular implies a relation between gauge field theories in D dimensions and gravitational theories in D+1 dimensions.

5. From a more phenomenological point of view, supersymmetry appears to be a natural candidate theory for the explanation of the unification of gauge couplings, a light Standard Model Higgs, dark matter and dark energy, deviations from the Standard Model predictions for the anomalous magnetic moment $(g-2)_\mu$ [10], large top quark mass, slow proton decay and some non perturbative aspects of QCD, such as glueballs, quark-gluon plasma, chiral symmetry breaking and confinement.

Supersymmetry appears also as a technical tool to understand some aspects and some mathematical properties of non-supersymmetric theories. One example is a proof of the positive energy theorem in gravity [15]. Another use is in the understanding of some particular properties of tree-level gravitons and gluons amplitudes in the pure theory of gravity and of gluons. Indeed, these tree-level amplitudes are the same as in $N = 1$ supergravity and $N = 1$ Super Yang-Mills theories, since fermions (gravitinos or gauginos) cannot be exchanged in tree-level graphs only involving external boson lines [16, 17]. As a consequence, these amplitudes satisfy the same properties as implied from the Ward identities of supersymmetry, even if the theories are not supersymmetric. Further higher loop properties of super Yang-Mills and supergravity theories have been thoroughly investigated in recent years [18].

Another interesting application of $N = 1$ supersymmetric gauge theories is obtained [19] in the large N limit of one-flavour QCD, as discussed by G. Veneziano in his contribution to this school.

2 The case for supersymmetry and its superworld

One of the most fascinating ideas in the quest for unity of the basic forces of nature is certainly the one that states that the physical laws that govern elementary particle processes have a higher, yet undiscovered, invariance called supersymmetry[20]–[21].

Supersymmetry is a symmetry that intertwines the basic interactions of the two classes of elementary particles out of which everything else is made, fermions and bosons.

Fermions and bosons obey two different kinds of statistics, Fermi–Dirac and Bose–Einstein respectively and, because of the Pauli exclusion principle, their roles in making our Universe as it is seem quite distinct. For example, their relativistic kinematical attributes, such as the spin, in Planck units is half-integer for fermions and integer for bosons, because of the PCT-spin statistics theorem. Also fermions appear as the basic constituents of nucleons, atoms and molecules built out of quarks, protons, neutrons and electrons. On the other hand, owing to the quantum nature of the subnuclear world, the forces underlying their mutual interactions are due to the exchange of particle quanta which are bosons, such as photons, gluons and weak vector bosons. Last but not least, the tiniest of these forces, gravity, when extended to the quantum domain, is believed to be due to an exchange of a spin two particle, the graviton, the basic carrier of the gravitational interaction.

The great importance of the supersymmetry hypothesis lies in the fact that such proposed invariance would not only predict the existence of new particles, as new symmetries often imply, but would unravel a drastic change of our notion of space and time. In fact, because of its intrinsic mathematical structure, it would imply the existence of extra-coordinates; rather than the usual kind of coordinates, described by complex numbers, these would be defined by anticommuting "numbers", known to mathematicians as Grassmann variables. Such space, named "superspace" by its discoverers (Salam, Strathdee [22]), carries D bosonic dimensions x_μ (in Minkowski space it has one time and $D-1$ space dimensions) ($\mu = 0, ..., D-1$), but also $2^{[D/2]}$ anticommuting dimensions θ_α ($[D/2]$ is the integral part of $D/2$). Indeed, while the bosonic coordinates have a "vector" index (a vector in D dimensions has D components), so that their component number grows linearly in D, the θ_α coordinates are spinors, which obey the mathematical theory of spinors mainly developed by Cartan and Weyl in the first half of the last century, and their number thus grows exponentially in D.

A particle moving in superspace is called a superparticle. When seen from ordinary space-time, it corresponds to a collection (or multiplet) of ordinary particles with spins differing by a half-unit. For instance, in ordinary four-dimensional space-time, there are four θ's and the simplest multiplet, carrying two bosons and a (Weyl) chiral fermion, is the so-called Wess–Zumino multiplet [23]. These authors were the first to propose an interacting Lagrangian for this multiplet in 1973 [24], by extending to $D = 4$ [25] a symmetry discovered two years before, in two-dimensional systems describing the world-sheet of strings, by Neveu, Schwarz, Ramond, Gervais and Sakita [26, 7]. In the same years, quite independently from one another, the Russian physicists Gol'fand, Likhtman [27] and Volkov, Akulov [28] proposed an algebraic structure, extending the Poincaré algebra, called graded Lie algebra by mathematicians (or superalgebra).

Superalgebras turn out to be the mathematical machinery [29] underlying supersymmetry, extending the concept of ordinary Lie algebras and Lie groups, the basic tools used to describe ordinary continuous symmetries in physics. As the group of motion of a point-particle in ordinary space-time is realized by the Poincaré group, a group of motion in "superspace" is induced by a supersymmetry transformation (super-Poincaré group). A Lagrangian in superspace corresponds to an ordinary Lagrangian for an entire "supermultiplet" of ordinary particles in ordinary space-time. This is due to properties of functions of "anticommuting" variables, and the so-called Berezin integration rule, which also underlies the path integral approach in quantum field theory when extended to fermionic variables [13].

As seen in superspace, the original Wess–Zumino Lagrangian appears as a "cubic" self-interaction of a "chiral" superfield [23], the simplest interaction that could be imagined in superspace. Soon later it was realized that such interactions not only are renormalizable, when extended to the quantum theory, but they have milder "ultraviolet" properties than ordinary renormalizable field theories [30, 31]. This property was not just a mathematical curiosity: it also opened the way to thinking that supersymmetric field theories may solve the problem of quantum gravity and the so-called hierarchy problem of Grand Unified Theories (GUTs) [12, 21]. In the latter, the separation of two different physical scales, such as the Fermi scale (250 GeV) and the Grand Unification scale (10^{16} GeV) is made quite unnatural when quantum corrections are taken into account. This is due to quadratic divergences occurring in the effective scalar potential, which controls the symmetry breaking in these theories [21].

Two pivotal developments were the unification of supersymmetry with ordinary gauge invariance, which led to super-Yang–Mills theories [4], and the unification with gravity, which led to supergravity [34, 35].

This unification made it possible to confront supersymmetry with an extension of the standard model of electroweak and strong interactions called the MSSM (Minimal Supersymmetric standard model) [32, 33]. The discovery of supergravity [34, 35, 36, 37, 38] made it possible to study quantum theories of gravity, where a new fundamental gauge particle of spin 3/2, the supersymmetric partner of the graviton, called the gravitino, plays a major role.

An amazing discovery made in studying fundamental interactions in superspace, partly due to W. Nahm [39] (who classified the Poincaré supermultiplets in arbitrary dimensions), is that supersymmetry gives an upper bound to the dimension of bosonic and fermionic coordinates. For example, supersymmetric interactions of only spin 0 and spin 1/2 particles exist only up to $D = 6$ (and 8 fermionic coordinates), super-Yang–Mills theories exist up to $D = 10$ dimensions (and 16 fermionic coordinates), supergravity theories exist up to $D = 11$ dimensions (and 32 fermionic coordinates). The maximal superspaces, with total dimension 26 and 43, play a major role in superstring [26, 7] and M-theory [40, 8]. These theories, which admit maximal supersymmetry, generalize point-like particle quantum field theories to extended objects of diferent space extension.

It is crucial, for the consistency of these theories, that two kinds of supersymmetry are at work, one present on the world-volume of a given extended object (called p-brane [6] if it has a p space extension), the other present on the "bulk", i.e. the D-dimensional ambient space. Such interplay of symmetries on the brane and bulk spaces and their dynamical role has made a major appearance in recent works on extra dimensions [41], realization of the so-called Randall–Sundrum [42] scenario, and the possibility of having two kinds of elementary particles, one living on the bulk and the other on the brane, something that makes their interactions quite distinct.

It is at present believed, especially by the pratictioners of superstring and M-theory, that supersymmetry should play a basic role at the Planck scale to make these theories quantum-mechanically consistent. However, since supersymmetry is not observed in our physical world, the case of supersymmetry requires an understanding of how this symmetry is broken and at what scale the symmetry breaking occurs [12, 13, 21].

After the discovery of supersymmetric Yang–Mills theories, in the mid 70's, P. Fayet made the proposal [43] of a minimal extension of the spectrum of the standard model, in which all ordinary particles, such as quarks, leptons, as well as the force carriers, are extended to supermultiplets. In doing this, it was noticed that, to realize the Higgs breaking, avoiding anomalies, two Higgs doublets were required, something that already goes beyond the minimal spectrum of the standard model. This implies, in particular, that in any supersymmetric extension of the standard model a new (adimensional) parameter arises, namely the ratio of the VEVs ($\tan\beta$) of the two Higgs doublet scalars[13, 21]. If supersymmetry is spontaneoulsy broken in these theories, the low-energy theorems can be studied and the gravitino acquires a mass through the supersymmetric version of the Higgs mechanism. The mass of the gravitino is closely related to the mass scale of supersymmetry breaking.

The compelling reason to invoke a low-energy supersymmetry for non-gravitational interactions essentially comes from the so-called hierarchy problem of GUTs [12, 13, 21]. A basic tree-level quadratic-mass sum rule implies that the effective potential of a supersymmetric gauge theory does not receive quadratic corrections above some cut-off scale Λ, at energies at which supersymmetry is effectively unbroken. So if the effective scale of supersymmetry breaking is E_0, quadratic corrections disappear for $\Lambda \gg E_0$. In particular, if the Fermi-scale is close to E_0, then it is stable with respect to a Grand Unification scale which occurs close to the string or Planck scale E_{P}, since $E_{\mathrm{P}} \gg E_0$.

It is of primary importance to disentangle the experimental constraints that make this scenario plausible: if it turns out to be correct, supersymmetry may indeed be discovered at present or future atom smashers such as the Tevatron in the US and the forthcoming LHC at CERN.

There are many indirect pieces of evidence for the introduction of supersymmetry in the basic laws of physics, both of theoretical and of "observational" nature. As previously alluded to, supersymmetry seems a basic ingredient for a theory that encompasses the gravitational force. The point-like particle limit, at low energies, of such a theory must be supergravity coupled to matter field multiplets, the latter describing the non-gravitational forces.

Supergravity is the gauge theory of supersymmetry as much as standard general relativity is the gauge theory of the Poincaré group, the space-time symmetry of relativistic systems. If gravity is extended at miscroscopic scales, such as the Planck scale, it is believed that the basic objects become extended objects [26, 7]. The ultimate theory based on such principles lives in $D = 11$ dimensions: it describes membranes and five-branes and reduces to all known string theories when some dimensions get compactified. This theory, called M-theory, can reproduce GUTs and supersymmetric extensions of the standard model when seven dimensions are compactified and the 32 original supersymmetries get reduced to 4. It is also believed that such theories should explain the present smallness of the cosmological constant and the inflationary evolution of our Universe.

In all these constructions, there is nothing that fixes the supersymmetry-breaking scale. However, such a scale becomes relevant when supersymmetry is advocated at low energies, in the TeV range, to solve the hierarchy problem and eventually to explain the origin of the Fermi scale.

3 The supersymmetric treshold and gauge coupling unification

Indirect signals of such low-energy supersymmetry are usually claimed to be the following:

1) The non-observation of proton decay as predicted by the minimal Georgi–Glashow SU(5) GUT;

2) The LEP precision measurements;

3) The unusually large top–Yukawa coupling;

3) Possible candidates, such as neutralinos for dark matter

Important ingredients in low-energy supersymmetry are the low-energy supersymmetry-breaking parameters [44], which are usually restricted by embedding supersymmetric gauge interactions in a supergravity low-energy effective theory [45, 46].

In the early 80's, the idea of supergravity, as a messenger of supersymmetry breaking to the observed elementary particles, such as quarks, leptons, Higgs and their superpartners, was elaborated [47]. A fundamental role is played by the gravitino mass $M_{3/2}$ which has its origin in the super-Higgs mechanism.

An appealing class of theories, which recently found place in superstring constructions, are the so-called no-scale supergravities [48]. In such theories the gravitino mass is a sliding scale at the tree level and it is dynamically fixed by radiative corrections, with the possibility of a hierarchical suppression with respect to the Planck or GUT scale[49].

Among the experimental implications of the supersymmetric extension of the standard model and of its GUT extensions, there is a prediction for the gauge-coupling unification and the supersymmetric threshold, namely the scale at which the supersymmetry breaking occurs [50], [51].

Although a naive analysis [52] preditcs this threshold to be at 1 TeV, for a unification of couplings at 10^{16} GeV, a more careful and detailed analysis, made by exploiting some work of Peterman and Zichichi in the late 70's, indicates that the supersymmetric threshold can be as low as the Fermi scale, making supersymmetry detectable in a wider energy range [53]. Peterman and Zichichi realized in 1979 that supersymmetric particles would imply a better convergence of gauge couplings because of the strong modifications on the β functions due to superpartners of the particle spectrum of the standard model[54].

Indeed, in 1974, it was already known that, in pure super-Yang–Mills theory (without matter), the one-loop β function is given by[4]

$$\beta_N(g) = \frac{A_N g^3}{16\pi^2} C_2(G),$$

where C_2 is the quadratic Casimir of G in the adjoint representation and

$$A_N = -\frac{11}{3}, -3, -2, 0$$

for $N = 0, 1, 2$ and 4 supersymmetry respectively. We therefore see that the β function is less and less negative as we have more and more supersymmetry, and that it vanishes in the maximally extended $N = 4$ Yang–Mills theory, which is conformal-invariant.

In a series of seminal papers, Peterman and Zichichi, in collaboration with Anselmo and Cifarelli [51, 53, 54], critically made a series of detailed studies on the basic problems connected with gauge-coupling unification. The goal of these studies was to have a deeper understanding on the lowest possible value of the supersymmetry threshold. It is shown in particular that the evolution of the gaugino masses, one of the supersymmetric soft-breaking terms, has a tendency to lowering the supersymmetric threshold. Two-loop effects were also included. A consequence of these studies was that the unification of gauge couplings, based on the initial data given by the LEP precision measurements, implies a wide range of values (from GeV to PeV) for the scale of supersymmetry breaking around the TeV energy.

A plausible and economic theoretical scenario for string unification has been given by Lopez, Nanopoulos and Zichichi [55, 56] in the context of a string-derived no-scale supergravity model based on a single parameter, a universal gaugino mass. The observable sector gauge group is $SU(5)\times U(1)$ [56].

An important element that deserves explanation and can be accommodated in the $SU(5) \times U(1)$ model is the gap between the GUT scale ($\simeq 10^{16}$ GeV) and the string scale (10^{18} GeV).

There are nowadays other possible scenarios to explain a unification scale below the string or Planck scale. One is to work in the strongly coupled heterotic string which is related to weakly coupled type I string or M-theory on a segment [57]. The other possibility is to invoke large extra dimensions and to have a different running depending on the energy scale with respect to the Kaluza–Klein masses [41].

4 The basic gauge interactions and their unification

Supersymmetric theories of gravitation appear as the low energy description of superstring and M-theory. Gravitational interactions are mediated by a spin two massless quantum whose gauge field is the metric symmetric tensor $g_{\mu\nu}$, $(\mu, \nu = 1, \cdots, D)$. The gravitational coupling in D dimensions k_D is related to the Newton constant G_D by the relation

$$k_D^2 = 4\pi G_D$$

Its dimensions in natural units are $[k_D^2] = 2 - D$ and the Newtonian force between two masses m_1, m_2 at distance r is given by the following expression

$$F = 4\pi G_D \frac{m_1 m_2}{r^{D-2}}$$

On the other hand, the gauge interactions in D dimensions, mediated by massless vector particles A_μ, have a coupling g_{YM}^D which has dimension $[(g_{YM}^D)^2] = 4 - D$. In string theory, where a new fundamental constant is present, $\alpha' = \ell_s^2$, where α' is the slope of Regge-trajectories in the Veneziano amplitude and ℓ_s is the string lenght, the gauge and gravitational coupligs get related

$$k^2 = (g_{YM})^2 \alpha' \qquad \text{(closed string)}$$

$$k = (g_{YM})/\alpha' \qquad \text{(open string)}$$

where these relations hold in $D = 10$. By defining a dimensionless string coupling g in closed strings, we further get

$$k^2 = g^2 \alpha'^4$$

$$g_{YM}^2 = g^2 \alpha'^3$$

For open strings, denoting by g_s the open string coupling, we instead have

$$k^2 = g_s^2 \alpha'^4$$

$$g_{YM}^2 = g_s \alpha'^3$$

The above relations indicate that the introduction of the new dimensionful couplings related to the string spatial extension allows to connect the gravitational coupling to the Yang-Mills coupling. This is the basic reason why extended objects allow to unify gauge and gravitational forces.

In recent years there have been new ideas which indicate that strings may be extended objects with an inverse size in the TeV region [58]. This is so because the origin of the weakness of the gravitational force may be related to the fact that gravity is spread in other dimensions which are large. Indeed, the relation between the gravitational couplings in four and in D dimensions is

$$k_4^2 = \frac{k_D^2}{R^{D-4}}$$

where R is the average size of the $D - 4$ extra dimensions.

Therefore if R is large k_D may be not too small [42]. In the old scenario both R and k_D where considered to be small. A consequence of this new scenario is that Kaluza-Klein states may show up at much lower energies than previously thought, and a rich phenomenology may emerge under this circonstancy.

To make this scenario realistic one may actually introduce p-dimensional extended objects and consider the simultaneous appearence of extra dimensions, $9 - p$ transverse to the brane (of average radius R_T) and $p - 3$ longitudinal to the brane (of average radius R_L). These extra dimensions play a different role in the relation between the Planck mass and the Yang-Mills coupling of the four dimensional theory which lives on the wrapped brane world-volume. The relation is [41]

$$M_{4P}^2 = \frac{1}{g_{YM}^4} M_s^2 (R_T M_s)^{9-p} (R_L M_s)^{3-p}, \qquad g_s = g_{YM}^2 (R_L M_s)^{p-3}.$$

This relation shows in particular that the relation between the Planck mass and the string scale can be arranged in such a way that the gauge coupling unification occurs at the string unification scale.

5 Super Yang-Mills and Supergravity theories

One of the most appealing features of supersymmetric field theories is the possible existence in a limited number of space-time dimensions. In fact the requirement that the spin 2 graviton and the spin 1 gauge fields be embedded into supermultiplets (as highest helicity components) gives an upper bound on the space-time dimension [39]. For spin one gauge fields $D \leq 10$ and for spin two gauge fields $D \leq 11$. The actual number of supersymmetric theories relies on the number of supercharges that can exist in a given dimension. This number is always bounded by 32 if one demands that the spin content of a multiplet does not exceed two, and is bounded by 16 if one demands not to exceed one.

There is just one Poincaré superalgebra in $D = 11$ and there are three in $D = 10$. The $D = 11$ superalgebra corresponds to eleven dimensional supergravity [59] and does not contain usual one-form gauge fields, but rather a three-form (three-index antisymmetric tensor) gauge field, other than the graviton. [1] Since a (p+1)-form gauge field naturally couples to a source which is a p-dimensionally extended object (a p-brane), eleven dimensional supergravity is the theory of membranes [8].

The magnetic dual extended object of a p-brane is a $(D-p-4)$-extended object since the magnetic dual gauge field of a $(p+1)$-form is a $(D-p-2)$-

[1]In mathematical language a p-form corresponds to a p-index totally antisymmetric tensor field.

form. Therefore the monopole analog of a membrane in $D = 11$ is a five-brane!

There are indeed two- and five-brane solitonic solution for the 11 dimensional Einstein equations coupled to the three-form gauge field. One may then say that 11 dimensional supergravity describes a world where the fundamental sources are two- and five-branes and the force fields are the gravitational and the rank three gauge fields.

In $D = 10$, we have two theories of supergravity called Type IIA and Type IIB theories. These theories differ for the chirality properties of the ten dimensional chiral supercharges. In type IIA the two supercharges are 16 component- Majorana-Weyl spinors of opposite chirality, while they have the same chirality in type IIB theory.

Since in the latter theory the fermions in the supergravity multiplet are of chiral nature, potential gravitational anomalies may exist. This is a phenomenon which only exists in $D = 10$ and $D = 6$ dimensions, namely gravity itself can have anomalies. In fact gravitaitonal anomalies are related to the possibility of having chiral fermions which are both Weyl and Mjorana (or pseudo-Majorana).

In the $D = 10$ type IIB theory this anomaly miraculously cancels [61]. The same problem appears in the $D = 10$, $N = 1$ superalgebra with a single Weyl charge which implies that N=1 supergravity coupled to N=1 super Yang-Mills theory in $D = 10$ is potentially anomalous.

The cancellation of gauge and gravitational anomalies in these theories led to the string revolution in 1984, since Green and Schwarz [60] showed that these theories are anomaly free for the partcular choices of gauge groups $E_8 \times E_8$ and $SO(32)$.

We can summarize all possible supersymmetric theories in $D = 4$ by giving the particle content for each theory [62].

We note that lower is the number of supercharges higher is the number of independent multiplets that may exist. The theories with maximal supersymmetry are N=8 supergravity (32 supercharges) and N=4 Yang-Mills theory (16 supercharges).

Not surprisingly these theories can be obtained by dimensional reduction

N	massless $\lambda_{MAX} = 2$ multiplet	massless $\lambda_{MAX} = 3/2$ multiplet
8	$[(2), 8(\frac{3}{2}), 28(1), 56(\frac{1}{2}), 70(0)]$	none
6	$[(2), 6(\frac{3}{2}), 16(1), 26(\frac{1}{2}), 30(0)]$	$[(\frac{3}{2}), 6(1), 15(\frac{1}{2}), 20(0)]$
5	$[(2), 5(\frac{3}{2}), 10(1), 11(\frac{1}{2}), 10(0)]$	$[(\frac{3}{2}), 6(1), 15(\frac{1}{2}), 20(0)]$
4	$[(2), 4(\frac{3}{2}), 6(1), 4(\frac{1}{2}), 2(0)]$	$[(\frac{3}{2}), 4(1), 7(\frac{1}{2}), 8(0)]$
3	$[(2), 3(\frac{3}{2}), 3(1), (\frac{1}{2})]$	$[(\frac{3}{2}), 3(1), 3(\frac{1}{2}), 2(0)]$
2	$[(2), 2(\frac{3}{2}), (1)]$	$[(\frac{3}{2}), 2(1), (\frac{1}{2})]$
1	$[(2), (\frac{3}{2})]$	$[(\frac{3}{2}), (1)]$

Table 1: Massless $\lambda_{MAX} = 2, 3/2$ multiplets.

from the theories with the same numbers of supercharges in D=10 and D=11 dimensions.

There are plenty of other possibilities where the naive dimensional reduction is replaced by a compactification on a given manifold. Depending on the particular properties of the manifolds, one can obtain theories with different number of supersymmetries in lower dimensions. It is important to observe that there is a class of supersymmetric field theories which exhibits an higher symmetry than just Poincaré supersymmetry.

These are the conformal field theories which, in the supersymmetric case, possess a higher spinorial invariance underlying a larger superalgebra called superconformal algebra. The superconformal algebra contains twice the number of supercharges of the Poincaré superalgebra.

A classification of superconformal algebras and their particle content, given by W. Nahm, has shown that in such a case the upper bound in the space-time dimension is $D = 6$. These results apply under the same assumptions of the so-called Coleman-Mandula theorem which states that the

N	massless $\lambda_{MAX} = 1$ multiplet	massless $\lambda_{MAX} = 1/2$ multiplet
8,6,5	none	none
4	$[(1), 4(\frac{1}{2}), 6(0)]$	none
3	$[(1), 4(\frac{1}{2}), 6(0)]$	none
2	$[(1), 2(\frac{1}{2}), 2(0)]$	$[2(\frac{1}{2}), 4(0)]$
1	$[(1), (\frac{1}{2})]$	$[(\frac{1}{2}), 2(0)]$

Table 2: Massless $\lambda_{MAX} = 1, 1/2$ multiplets.

bosonic part of a superalgebra must be a direct product of the space-time symmetry times an internal symmetry.

Conformally invariant Yang-Mills theories only exist in D=4. In D=5 the only conformal multiplet is the so-called hypermultiplet, while in D=6 one has the hypermultiplets and the so-called tensor multiplets.

These superconformal theories have the same superalgebra symmetry of theories of gravity in D+1 dimensional Anti de Sitter space. In fact, the conformal group in D dimensions $SO(D,2)$, is also the isometry group of AdS space in D+1 dimensions: $SO(D,2)/SO(D,1)$. More precisely, the group $SO(D,2)$ acts as the group of conformal transformations on the D dimensional boundary of AdS space.

This general property is at the basis of the strong dynamical conjecture, originally due to Maldacena [63], called the AdS/CFT correspondence, which postulates a duality between a theory of gravity (strings) in AdS space and a quantum field theory on its boundary [64].

The most popular example is the connection between N=4 Yang-Mills at D=4 and type IIB supergravity on $AdS_5 \times S^5$. The latter is the space-time geometry at the horizon of a D3-brane whose world-volume degrees of freedom are the N=4 super Yang-Mills multiplet.

The AdS/CFT correspondence is a particular case of the holographic conjecture and does open a new arena of investigation in which supersymmetric gauge theories play a role in connection with a possible dual description of strings or M-theory.

There is in particular another limit of the gravitational metric, called the pp-wave limit [65], in which the dual gauge theory also captures genuine stringy effects and not just the supergravity approximation of superstrings [66].

6 String revolutions and beyond

The first string revolution, in 1984, is marked by the Green-Schwarz discovery of the quantum consistency of N=1 supergravity in D=10 coupled to Yang-Mills theory [60]. This result applies to open and closed strings at D=10 and it has led to the subsequent discovery of heterotic theory by Gross, Harvey, Martinec and Rohm [67].

Soon after, in the seminal paper by Candelas, Horowitz, Strominger, Witten [68], the role of Calabi-Yau manifolds i.e. Ricci-flat Kähler manifolds with reduced holonomy made contact with chiral gauge theories in four dimensions. This result opened the way to phenomenological applications and model building constructions for stringy-inspired Grand Unified Theories.

The second string revolution came ten years later, in 1995 [69], with the understanding of the important role of duality symmetries in quantum field theories, superstrings and M-theory [70]. These dualities make connections among apparently unrelated theories.

In the low energy effective supergravity lagrangians, dualities manifest themselves as a generalization of the electric-magnetic duality of the Maxwell equations [71], which further imply the possible existence of monopoles and dyons [72]. Black holes and more in general p-branes are important quantum states where the duality conjecture plays an important role.

In the same year, Kallosh, Strominger and the author [73] realized that the dynamics of BPS extremal black holes is governed by an 'attractor mechanism' when the black hole follows its evolution towards the horizon. The black hole entropy, given by the Bekenstein-Hawking area-entropy formula [74], is given by a duality invariant expression in terms of the electric and magnetic charges of the black hole, and does not depend on the initial (moduli) data at infinity. Indeed the moduli fields follow an attractor equation which fixes their value in terms of the asymptotic electric and magnetic charges.

For a p=0 brane (a black hole) in D=4, the horizon geometry is conformal and it is given by the so-called Bertotti-Robinson geometry [75] $AdS_2 \times S^2$. For BPS p-branes in D dimensions, Gibbons and Townsend [76] proved that the horizon geometry is $AdS_{p+2} \times S^{D-p-2}$. This result was a key ingredient in the Maldacena conjecture of 1997, anticipated in the previous paragraph, between theories of supergravity in those backgrounds and conformal field theories in p+1 dimensions [64].

In the AdS/CFT correspondence, the theories with maximal supersymmetry are the most predictive and they correspond to p=2 and p=5 branes in 11 dimensions and p=3 in D=10. The holografic conformal field theories live in three, six and four dimensions respectively and their superalgebra symmetries correspond to the classical simple superalgebra denoted by $OSp(8|4)$, $PSU(2,2|4)$ in the mathematical dictionary [77]. The conformal fields on the brane world volumes are known in the mathematical-physics literature are singletons [78] and in this setting they are the basic constituents of all the states.

7 M-theory: Mother, Mystery or Magic?

The maximal supergravity theory lives in D=11 dimensions and its content is the metric tensor $g_{\mu\nu}$, a three-index antisymmetric tensor $C_{\mu\nu\rho}$ and a spin-3/2 Rarita-Schwinger (Majorana) field $\psi_{\mu\alpha}$. The (on-shell) physical degrees of freedom of these fields are respectively given by $44 + 84 = 128$ bosonic states and 128 fermionic states making an overall number of $2^8 = 256$ states. This is actually the number of states of any theory with 32 supersymmetries in any dimensions.

As anticipated in the previous paragraph, the sources of the three-form C and of its (six-form) magnetic dual are two-branes and five-branes [79].

These sources may coexist provided a Dirac-Schwinger quantization condition is met, analogous to the electric and magnetic charge quantization in four dimensions.

M-theory is the quantum theory underlying eleven dimensional supergravity. This theory has no dimensionless coupling constant and the only physical scale is the Planck mass. Its non-perturbative definition is still obscure, and only a weak coupling expansion (in the space-time curvature) is known. In particular, the lowest order curvature term is D=11 supergravity.

The M-theory super-Poincaré algebra in D=11 [80, 81] contains a 32 component spinorial charge and 528 (commuting) bosonic charges. They are

- P_μ (11) the space-time vielbein
- $Z_{\mu\nu}$ (55) the membrane charge
- $Z_{\mu\nu\rho\sigma\lambda}$ (462) the five-brane charge

The presence of these charges in the supersymmetry algebra allows to have one-half BPS membrane and five-brane configurations, preserving 16 supersymmetries. These are indeed the number of Poincaré supersymmetries of the corresponding field theories on the world-volumes of two- and five-branes, which are respectively an $N = 8$ three dimensional theory and a $(2,0)$ six dimensional theory. We are now ready to explain the three differnt meanings of M given in the title of this section. 'Mother' resides in its connection to string theory. Indeed, this theory hides in the eleven-th dimension the string coupling constant. In fact, by dimensionally reducing 11 dimensional supergravity on a circle S^1 of radius R, one precisely gets type IIA supergravity, which is the effective theory of type IIA superstring theory. Comparison of the two theories gives the identification

$$g_s^2 \sim \left(\frac{R}{\ell_P}\right)^3$$

This formula suggests that the decompactification limit $R \to \infty$ of M-theory is the strongly coupled type IIA string! Another indication of this argument is that the mass of the D0-brane becomes the Kaluza-Klein mass of M-theory [69]

$$M_{KK} \sim \frac{1}{R} \sim g_s^{-2/3}$$

which diverges for g_s going to zero and is small for g_s large.

Horava and Witten [82] in 1996, have further shown that by compactification of M-theory on S^1/Z_2 (a segment) anomaly cancellation introduces new states confined at the 10 dimensional end-points of the segment, each carrying an E_8 gauge group. Moreover, the Z_2 modding only preserves a half of the supersymmetries, then giving an N=1 theory on the ten dimensional walls. This is the M-theory interpretation of the $E_8 \times E_8$ heterotic string theory in 10 dimesnions.

The second word, Mistery, is somewhat related to the lack of a perturbative definition of M-theory. There are many subtleties in M-theory, mostly connected to the Chern-Simons term for the three-form C present in the lagrangian [83], and also to its coupling to five-branes, whose world-volume admits chiral fermions [84]. There is also some evidence that M-theory may admit a topological string analog that is called Z-theory [85].

The last word, Magic, may depend on some magic property of M-theory, like its chain of duality symmetries when the theory is compactified on a d-dimensional torus T^d. Indeed, the duality groups in $11 - d$ space-time dimensions are given by the exceptional series $E_{d(d)}$ [86] which are 'classical' for $d < 6$, exceptional for $d = 6, 7, 8$ and infinite for $d = 9, 10, 11$. The latter infinite groups are related to cosmological billiards [87].

The groups E_6, E_7, for example, classify the black-hole charges in $d = 5$ and 4 maximal supergravities and the Bekenstein-Hawking entropy is expressed in terms of the cubic and quartic invariants of the fundamental representations of these groups [88, 89].

8 Supersymmetry breaking: no-scale supergravity, Scherk-Schwarz breaking and flux compactifications

8.1 No-scale supergravity

An important feature of a generic supergravity theory is the possibility of undergoing spontaneous supersymmetry breaking without a cosmological constant. By studying the universal coupling of a Goldstone fermion to supergravity, one can see that in a spontaneously broken supergravity theory the contributions to the vacuum energy could in principle cancel [90]. The first concrete example, based on a field theory lagrangian, was given by Polony [91, 92]. He considered $N = 1$ supergravity coupled to a single chiral multiplet with canonical kinetic term and linear superpotential and showed that it is possible to fine tune the parameters (α and β) of the superpotential $W = \alpha z + \beta$ in such a way that the potential stabilizes the scalar fields with vanishing vacuum energy. The scalar field masses satisfy the sum rule $m_A^2 + m_B^2 = 4m_{3/2}^2$ [92]. Polony type superpotentials were used in the first phenomenological studies of broken supergravity. They generate the soft breaking terms of the observable sector of standard (electroweak and strong) interactions in the supersymmetric extension of the standard model and of grand unified theories (For a review, see Ref. [93]).

The Polony classical potential is rather unnatural because it requires an ad hoc superpotential. Subsequent studies of the superHiggs sector of supergravity models lead to the introduction of a more appealing class of theories, the so called no-scale supergravities [48, 49]. In these models, the vanishing vacuum energy of the classical potential is obtained without stabilizing the scalar superpartner of the Goldstino. Instead, there is an exact cancellation, prior to minimization, of the positive Goldstino contribution against the negative gravitino contribution to the vacuum energy without the need of fine-tuning the parameters. The no-scale structure of these models poses further constraints on the soft-breaking terms which enter in the phenomenological Lagrangians [49].

The first construction of an extended supergravity exhibiting a no-scale structure was in the context of $N = 2$ supergravity coupled to abelian vector multiplets in presence of a Fayet-Iliopoulos term [94].

The no-scale structure for $N > 2$ extended supergravity is encountered in the context of eleven dimensional supergravity with Scherk-Schwarz generalized dimensional reduction. This produces spontaneously broken supergravity theories in four dimensions [95, 96]. The four dimensional interpretation of these theories [97] is an $N = 8$ gauged supergravity whose gauge algebra (a "flat" algebra according to Ref. [95]) is a 28 dimensional Lie subalgebra of $E_{7,7}$., the U-duality group of (ungauged) D=4 supergravity.

More recent examples of no-scale extended supergravity are type IIB $N = 4$ and $N = 2$ orientifolds on tori and Calabi-Yau manifolds.

A simple example is the T^6/Z^2 orientifold model in presence of n D3-branes with three-form fluxes turned on [98]. This theory corresponds to a gauged supergravity with gauge group the direct product $T_{12} \times G(n)$ ($G(n)$ is a gauge group of rank n) which are a particular set of isometries of the $N = 4$ scalar sigma model. In the superstring interpretation, six of the vector multiplets come from the supergravity fields on the bulk and the rest comes from a non abelian D3-brane Born-Infeld action coupled to supergravity.

8.2 Scalar potential in N-extended supergravity: vacua without cosmological constant

We consider an N-extended supergravity theory in $D = 4$. We will denote by $\psi_{\mu A}$, $A = 1, \ldots N$ the spin 3/2 gravitino fields and by λ^I the spin 1/2 fields (the spinor indices are not shown explicitly). They are all taken to be left handed, and the right handed counterparts are denoted by ψ_μ^A and λ_I. The scalar fields will be denoted by q^u, and are coordinates on a Riemann manifold $\mathcal{M}$. Supersymmetry requires that $\mathcal{M}$ has a restricted holonomy group $H = H_R \times H_M$, with H_R being U(N) or SU(N) (U(N) being the R-symmetry group) and H_M varying according to the different matter multiplet species. It also requires that on $\mathcal{M}$ there is an H_R-bundle with a connection whose curvature is related to the geometric structure of $\mathcal{M}$ [99].

For $N = 1$ supergravity coupled to n chiral multiplets we have a Kähler-Hodge manifold of complex dimension n, with $H_M = \mathrm{SU}(n)$ and $H_R = \mathrm{U}(1)$. On $\mathcal{M}$ there is a U(1) bundle whose Chern class is equal to the Kähler class.

For $N = 2$ coupled to n vector multiplets we have a special Kähler-Hodge manifold. If we have n_h hypermultiplets, then $\mathcal{M}$ is a quaternionic manifold of quaternionic dimension n_h. The holonomy is $H = H_R \times H_M$ with $H_R = \mathrm{SU}(2)$ and $H_M = \mathrm{USp}(2n_h)$. On $\mathcal{M}$ there is an SU(2)-bundle with curvature equal to the triplet of hyperKähler forms on $\mathcal{M}$.

For $N > 2$ the manifolds of the scalars are maximally symmetric spaces $\mathcal{M} = G/H$ with $H = H_R + H_M$. Then there is also an H_R-bundle on $\mathcal{M}$ whose connection is the H_R part of the spin connection.

For $N = 3$ with n_v vector multiplets $H_M = \mathrm{SU}(n_v)$ and $H_R = \mathrm{U}(3)$. For $N \geq 4$ the supergravity multiplet itself contains scalars. For $N = 4$ with n_v vector multiplets $H_M = \mathrm{SO}(n_v)$ and $H_R = \mathrm{SU}(4) \times \mathrm{U}(1)$. For $N > 4$ there are no matter multiplets and $H = H_R = \mathrm{U}(N)$ except for $N = 8$ where $H = \mathrm{SU}(8)$ [89].

The above considerations imply that the covariant derivative of the supersymmetry parameter, $\mathcal{D}_\mu \epsilon_A$, contains, in presence of scalar fields, an H_R connection in addition to the spacetime spin connection.

The supersymmetry variations of the fermions in a generic supergravity theory can be expressed as [100]

$$\delta\psi_{A\mu} = \mathcal{D}_\mu \epsilon_A + \frac{1}{2} S_{AB} \gamma_\mu \epsilon^B + \cdots \tag{1}$$

$$\delta\lambda^I = i P^I_{\mu A} \gamma^\mu \epsilon^A + N^{IA} \epsilon_A + \cdots , \tag{2}$$

where $S_{AB} = S_{BA}$, and N^{IA} are sections of H_R bundles on $\mathcal{M}$ which depend on the specific model under consideration. The dots stand for terms which contain vector fields. $P^I_{\mu A} dx^\mu$ is pullback into spacetime of $P^I_{uA} dq^u$, the vielbein one-form on $\mathcal{M}$, so

$$P^I_{\mu A} = P^I_{uA} \partial_\mu q^u .$$

The variation of the scalars is then given by

$$\delta q^u P^I_{uA} = \bar{\lambda}^I \epsilon_A .$$

The supergravity lagrangian contains the following terms

$$\frac{1}{\sqrt{-g}} \mathcal{L} = \cdots + S_{AB} \bar{\psi}^A_\mu \sigma^{\mu\nu} \psi^B_\nu + i N^{IA} \bar{\lambda}_I \gamma^\mu \psi_{\mu A} + M^{IJ} \bar{\lambda}_I \lambda_J \quad + \mathrm{c.c.} - V(q)$$

where M^{IJ} is the mass matrix of the spin 1/2 fields and $V(q)$ is the potential of the scalar fields. The potential must be such that the supersymmetry variation of all these terms cancel. This implies [101, 102]

$$\begin{aligned} \delta^A_B V(q) &= -3S^{AC}S_{BC} + N^{IA}N_{IB} \\ \frac{\partial V}{\partial q^u} P^I_{uA} &= 2iN^{IB}S_{BA} + 2M^{IJ}N_{JA}, \end{aligned} \tag{3}$$

where $N_{IA} = (N^{IA})^*$ and $S^{AC} = S^*_{AC}$.

Flat space requires that on the extremes $\partial V/\partial q^u = 0$ the potential vanishes, so

$$3\sum_C S^{AC}S_{CA} = \sum_I N^{IA}N_{IA}, \qquad \forall A.$$

The first term in the potential (3) is the square of the gravitino mass matrix. It is hermitian, so it can be diagonalized by a unitary transformation. Assume that it is already diagonal, then the eigenvalue in the entry (A_0, A_0) is non zero if and only if $N^{IA_0} \neq 0$ for some I. On the other hand, if the gravitino mass matrix vanishes then N^{IA} must be zero.

For no-scale models, there is a subset of fields $\lambda^{I'}$ for which

$$3\sum_C S^{AC}S_{CA} = \sum_{I'} N^{I'A}N_{I'A}, \qquad \forall A \tag{4}$$

in all $\mathcal{M}$. This implies that the potential is given by

$$V(q) = \sum_{I \neq I'} N^{IA}N_{IA},$$

and it is manifestly positive definite. Zero vacuum energy on a point of $\mathcal{M}$ implies that $N^{IA} = 0$, $I \neq I'$ at that point. This happens independently of the number of unbroken supersymmetries, which is controlled by $N^{I'A}$ (gravitino mass matrix).

In N extended supergravities, the axion couplings to the gauge fields

$$\mathcal{D}_\mu a^i = \partial_\mu a^i - g^i_\Lambda A^\Lambda_\mu$$

are related to the gravitino mass matrix S_{AB} through the existence, for each pair of indices i, Λ of a section $X^\Lambda_{i,AB}$ of an H_R bundle over $\mathcal{M}$ such that

$$S_{AB} = g^i_\Lambda X^\Lambda_{i,AB}, \qquad X^\Lambda_{i,AB} = X^\Lambda_{i,BA}.$$

In the next section we will give the particular form of S_{AB} in $N = 8$ and $N = 4$.

8.3 $N = 8$ Scherk-Schwarz spontaneously broken supergravity

In $N = 8$ spontaneously broken supergravity à la Scherk-Schwarz, the R-symmetry that is manifest is USp(8) $\subset$ SU(8). The spin 3/2 gravitinos are in the fundamental representation of USp(8) (**8**), while the spin 1/2 fermions are in the **8** and **48** (threefold Ω-traceless antisymmetric representation). We will denote them as $\psi_{\mu a}$, λ_a and λ_{abc}.

From a dimensional reduction point of view, the scalar potential is originated by the five dimensional σ-model kinetic energy term

$$\sqrt{-g} g^{\mu\nu} P_\mu^{abcd} P_{\nu abcd}, \qquad \text{for} \quad \mu = \nu = 5,$$

where P_μ^{abcd} is the pullback on spacetime of the vielbein one form of the coset $E_{6,6}/\text{USp}(8)$.

From the generalized dimensional reduction, the four dimensional scalar potential is

$$V = \frac{1}{8} e^{-6\phi} P_5^{abcd} P_{5abcd}, \tag{5}$$

where ϕ is the radion field. This term would not appear in a standard dimensional reduction.

The five dimensional supersymmetry variations are

$$\begin{aligned} \delta\psi_{\mu a} &= \mathcal{D}_\mu \epsilon_a + \cdots \\ \delta\lambda_{abc} &= P_{\mu abcd} \gamma^\mu \epsilon^d + \cdots . \end{aligned}$$

We denote by $Q_{\mu ab}$ the USp(8) connection in five dimensions. The functions S_{AB} and N^{IA} of the previous section (2) are then

$$S_{ab} = \frac{1}{\sqrt{3}} e^{-3\phi} Q_{5ab}, \quad N^{ab} = e^{-3\phi} Q_5^{ab}, \quad N^{abcd} = e^{-3\phi} P_5^{abcd}$$

(the indices can be raised or lowered with the antisymmetric metric Ω_{ab}). P_5^{abcd} satisfies the identity [103]

$$P_5^{abcd} P_{5\,ebcd} = \frac{1}{8} \delta_e^a P_5^{fbcd} P_{5\,fbcd},$$

which is crucial to have (3)

In the computation of the scalar potential using (3) there is an exact cancellation between the gravitino and the spin 1/2 fermions in the **8** as in (4),

$$3|S_{ab}|^2 = |N^{ab}|^2,$$

so that

$$V = \frac{1}{8}|N^{abcd}|^2.$$

This explains formula (5) from a four dimensional point of view. Note that at a linearized level (near the origin of the coset, where the exponential coordinates ϕ^{abcd} are small),

$$\begin{aligned} P_5^{abcd} &= M_{a'}^{[a}\phi^{a'bcd]} \quad - \quad \Omega\text{–traces} + \mathcal{O}(\phi^{abcd})^2, \\ Q_{5ab}\Omega^{ba'} &= M_a^{a'} + \mathcal{O}(\phi^{abcd}), \end{aligned}$$

where $M_a^{a'}$ is the Scherk-Schwarz phase of the 5d gravitino.

The vacua with zero potential correspond to $P_5^{abcd} = 0$, while the supersymmetry breaking depends on the vanishing eigenvalues of the matrix Q_{5ab}. When all the eignevalues m_i of $M_a^{a'}$ are different from zero, the requirement $P_5^{abcd} = 0$ determines all but two coordinates which are the two scalars which are neutral with respect to the CSA of USp(8). Together with the radion, they are the flat directions of the potential. There are three additional massless scalars, the three axions in the 27 of USp(8) which are neutral under the CSA. All together, they form the moduli space of the Scherk-Schwarz compactification.

8.4 $N = 4$ supergravity and type IIB orientifolds with fluxes

We now consider no-scale $N = 4$ spontaneously broken supergravity. This theory is the low energy limit of type IIB 10 dimensional supergravity compactified on orientifolds in presence of three form fluxes and n D3 branes with non commutative coordinates [98, 104].

The six $N = 4$ vector multiplets coming from the bulk lagrangian contain 36 scalars, 21 of which are the metric deformation of the 6-torus T^6 g_{IJ},

$I, J = 1, \dots 6$, and 15 scalars coming from the four form gauge field $C_{\mu\nu\rho\lambda}$, whose components along the 6-torus are dual to a two form

$$B^{IJ} = {}^*C^{IJ}, \qquad I, J = 1, \dots 6.$$

Turning on the three form fluxes corresponds in the effective theory to gauge particular isometries of the coset SO(6,6)/SO(6) × SO(6) [105]. More explicitly, the gauge isometries are twelve of the fifteen translational isometries T_{12} inside $SO(6,6)$.

In the case when Yang-Mills $N = 4$ multiplets are added (describing the D3 brane degrees of freedom), the gauge group is $T_{12} \times G(n)$. This theory gives rise to a no-scale supergravity with four arbitrary parameters for the gravitino masses [106].

The SU(4) (R-symmetry) representations of the bulk fermions are as follows:

spin 3/2 (gravitinos) in the **4**,

spin 1/2 (dilatinos) in the $\bar{\mathbf{4}}$,

spin 1/2 (gauginos, from the 6 vector multiplets) in the $\mathbf{20}+\bar{\mathbf{4}}$.

The fermions in the brane (gauginos) form $n^2 = \dim G(n)$ copies of the representation **4** of SU(4).

Computing the potential (3), the subset of fields $\lambda^{I'}$ (4) are the bulk gauginos in the $\bar{\mathbf{4}}$. The condition for vanishing potential [107] is then $N^{IA} = 0$ for $I \neq I'$. For the bulk fermions it fixes the complex dilaton, 18 radial moduli and 12 axions. For the brane gauginos it fixes all the scalars but the ones in the CSA of $G(n)$.

9 Quantum corrections to the cosmological constant

In the present models, giving partial super-Higgs around Minkowski vacuum, we can make some discussion on the one-loop corrections to the cosmological constant.

Let us remind that the quartic, quadratic and logarithmic divergent parts, in any field theory, are respectively controlled by the following coefficients [108, 109]

$$a_k = \sum_J (-1)^{2J}(2J+1)m_J^{2k}\,, \quad (k=0,1,2). \tag{6}$$

While $a_0 = 0$ in any spontaneously broken supersymmetric theory (quartic divergence), the vanishing of a_1, a_2 is model-dependent [108]

However, under some mild assumption, in any spontaneously broken N-extended theory where the partial super-Higgs $N \to N-1$ is permitted, the following main formulae are true

$$a_k = \sum_J (-1)^{2J}(2J+1)m_J^{2k} = 0\,, \quad 0 \leq k < N. \tag{7}$$

This relation follows from the fact that a non vanishing contribution to the vacuum energy must be proportional to $\Pi_{i=1}^{N} \frac{m_i^2}{M_P^2} M_P^4$.

This is the leading behavior in the variable $X = \Pi_{i=1}^{N} \frac{m_i^2}{M_P^2}$, at least if we assume analyticity in this variable.

From the above we conclude that there is only a finite correction in $N = 3, 4$ models, while there is a logarithmic correcton to the $N = 2$ models and a quadratic correction to the $N = 1$ models.

Note however that in the Scherk–Schwarz $N = 8$ models [95] the same formulae were true but for $0 \leq k < \frac{N}{2}$. This is because in that case the gravitino masses were pairwise degenerate so that the hypotesis of partial breaking $N \to N-1$ was invalid.

Acknowledgements

I would like to thank Dr. Laura Andrianopoli, from Centro Enrico Fermi, Rome, for substantial help in the preparation of this contribution.

References

[1] C.N. Yang and R. Mills, *Phys. Rev.* **95** (1954) 631 and *Phys. Rev.* **96** (1954) 191.

[2] "Infiniment CERN, Témoins de cinquante ans de recherches", Sciences et Histoires (Ed. Suzanne Hurter).

[3] "The Roots of LEP and LHC", on the occasion of the 50th Anniversary of CERN, Edited by L. Cifarelli CERN/DG-2004-306/0, September 2004.

[4] S. Ferrara and B. Zumino, "Supergauge invariant Yang-Mills theories", *Nucl. Phy.***B79** (1974) 413. A. Salam and J. Strathdee, "Supersymmetry and nonabelian gauges," *Phys. Lett.* **B51** (1974) 353.

[5] S. Weinberg, "The Cosmological Constant Problem," Rev. Mod. Phys. **61** (1989) 1, "The cosmological constant problems," arXiv:astro-ph/0005265..

[6] M. J. Duff, "Anti-de Sitter space, branes, singletons, superconformal field theories and all that", *Int. J. Mod. Phys.* **A14** (1999) 815 [arXiv:hep-th/9808100].

[7] M. Green, J, Schwarz and E. Witten, "Superstring theory" (Cambridge University Press, 1987).

[8] M. J. Duff, "M theory (the theory formerly known as strings)", *Int. J. Mod. Phys.***A11** (1996) 5623 [arXiv:hep-th/9608117].

[9] G. 't Hooft, "Dimensional reduction in quantum gravity," arXiv:gr-qc/9310026. L. Susskind, "The World as a hologram," J. Math. Phys. **36** (1995) 6377 [arXiv:hep-th/9409089].

[10] W. J. Marciano, "Anomalous magnetic moments," Int. J. Mod. Phys. A **19S1** (2004) 77.

[11] P. Fayet and S. Ferrara, "Supersymmetry", *Phys. Rep.* **32** (1977) 249.

[12] "Supersymmetry", vols. I, II, Reprints volume (ed. S. Ferrara, North Holland and World Scientific, 1987).

[13] "Supersymmetry and supergravity", A reprint volume of *Phys. Rep.* , (ed. M.Jacob, North Holland, World Scientific, 1986).

[14] P. West, Introduction to supersymmetry and supergravity", World Scientific, 1986.

[15] E. Witten, "A Simple Proof Of The Positive Energy Theorem," Commun. Math. Phys. **80** (1981) 381.

[16] M. L. Mangano and S. J. Parke, "Multiparton Amplitudes In Gauge Theories," Phys. Rept. **200** (1991) 301.

[17] S. J. Parke and T. R. Taylor, "An Amplitude For N Gluon Scattering," Phys. Rev. Lett. **56** (1986) 2459.

[18] Z. Bern, L. J. Dixon, D. C. Dunbar, M. Perelstein and J. S. Rozowsky, "On the relationship between Yang-Mills theory and gravity and its implication for ultraviolet divergences," Nucl. Phys. B **530** (1998) 401 [arXiv:hep-th/9802162]. Z. Bern, L. J. Dixon, M. Perelstein, D. C. Dunbar and J. S. Rozowsky, "Perturbative relations between gravity and gauge theory," Class. Quant. Grav. **17** (2000) 979 [arXiv:hep-th/9911194].

[19] A. Armoni, M. Shifman and G. Veneziano, "SUSY relics in one-flavor QCD from a new 1/N expansion," Phys. Rev. Lett. **91** (2003) 191601 [arXiv:hep-th/0307097].

[20] G.R. Farrar, S. Ferrara, A. Zichichi, contributions to "The new aspects of subnuclear physics", vol. 16 of Subnuclear Series, Erice 1978.
E. Witten, contribution to "The unity of the fundamental interactions", vol. 19 of Subnuclear Series, Erice 1981 (ed. A. Zichichi).

[21] S. Weinberg, "The quantum theory of fields", vol. 3: "Supersymmetry", (Cambridge University Press, 2000).

[22] A. Salam and J. Strathdee, "Supergauge transformations", *Nucl. Phys.* **B76** (1974) 477.

[23] J. Bagger and J. Wess, "Supersymmetry and supergravity", JHU-TIPAC-9009 (Princeton University Press, 1992).

[24] J. Wess and B. Zumino, "A Lagrangian model invariant under supergauge transformations," *Phys. Lett.* **B49** (1974) 52.

[25] J. Wess and B. Zumino, "Supergauge transformations in four dimensions," *Nucl. Phys.* **B70** (1974) 39.

[26] "Superstrings, the first 15 years of superstring theory", vols. I, II, (ed. J. Schwarz), Reprints volumes (World Scientific, 1985).

[27] Y. A. Gol'fand and E. P. Likhtman, "Extension of the algebra of Poincaré group generators and violation of P", *JETP Lett.* **13** (1971) 323 [*Pisma Zh. Eksp. Teor. Fiz.* **13** (1971) 452].

[28] D. V. Volkov and V. P. Akulov, "Is the neutrino a Goldstone particle?," *Phys. Lett.* **B46** (1973) 109.

[29] V.S. Varadarajan, "Supersymmetry for Mathematicians: An Introduction", Courant lecture notes; 11 (2004), Library of Congress Cataloging-in-Publication Data.

[30] J. Iliopoulos and B. Zumino, "Broken Supergauge Symmetry And Renormalization," Nucl. Phys. B **76** (1974) 310. S. Ferrara, J. Iliopoulos and B. Zumino, "Supergauge Invariance And The Gell-Mann - Low Eigenvalue," Nucl. Phys. B **77** (1974) 413.

[31] M. T. Grisaru and W. Siegel, "Supergraphity. 2. Manifestly Covariant Rules And Higher Loop Finiteness," Nucl. Phys. B **201** (1982) 292 [Erratum-ibid. B **206** (1982) 496].

[32] P. Fayet, "Spontaneously Broken Supersymmetric Theories Of Weak, Electromagnetic And Strong Interactions," Phys. Lett. B **69** (1977) 489.

[33] S. Dimopoulos and H. Georgi, Nucl. Phys. B **193** (1981) 150.

[34] D. Z. Freedman, P. van Nieuwenhuizen and S. Ferrara, "Progress toward a theory of supergravity", *Phys. Rev.* **D13** (1976) 3214.

[35] S. Deser and B. Zumino, "Consistent supergravity," *Phys. Lett.* **B62** (1976) 335.

[36] D. Freedman and P. van Nieuwenhuizen, "Supergravity and the unification of the laws of physics", Scientific American, No. 238 (1978).

[37] P. Van Nieuwenhuizen, "Supergravity," Phys. Rept. **68** (1981) 189.

[38] L. Castellani, R. D'Auria and P. Fre, "Supergravity and superstrings: A Geometric perspective. 3 Vol.s", Singapore: World Scientific (1991)

[39] W. Nahm, "Supersymmetries And Their Representations", Nucl. Phys. B **135** (1978) 149.

[40] M.B. Green, contribution to "Highlights: 50 years later" in vol. 35 of Subnuclear Series, Erice 1995 .

[41] I. Antoniadis, "Physics with large extra dimensions," CERN-TH/2001-318, lecture given at the "2001 European School on HEP", Beatenberg, Switzerland, 2001.

[42] L. Randall and R. Sundrum, "A large mass hierarchy from a small extra dimension," *Phys. Rev. Lett.* **83** (1999) 3370 [arXiv:hep-ph/9905221]. L. Randall and R. Sundrum, "An alternative to compactification", *Phys. Rev. Lett.* **83** (1999) 4690 [arXiv:hep-th/9906064].

[43] P. Fayet in refs. [11, 12, 13].

[44] L. Girardello and M. T. Grisaru, "Soft Breaking Of Supersymmetry," Nucl. Phys. B **194** (1982) 65.

[45] R. Arnowitt, contribution to "From supersymmetry to the origin of space-time" in vol. 31 of Subnuclear Series, Erice 1993.

[46] R. Barbieri, contribution to "The superworld III" in vol. 26 of Subnuclear Series, Erice 1988.

[47] R. Barbieri, S. Ferrara and C. A. Savoy, "Gauge models with spontaneously broken local supersymmetry", *Phys. Lett.* **B119** (1982) 343. A. H. Chamseddine, R. Arnowitt and P. Nath, "Locally supersymmetric grand unification," *Phys. Rev. Lett.* **49** (1982) 970. L. J. Hall, J. Lykken and S. Weinberg, "Supergravity as the messenger of supersymmetry breaking", *Phys. Rev.* **D27** (1983) 2359.

[48] E. Cremmer, S. Ferrara, C. Kounnas and D.V. Nanopoulos, "Naturally vanishing cosmological constant in N=1 supergravity", *Phys. Lett.* **B133** (1983) 61.

[49] J.R. Ellis, A.B. Lahanas, D.V. Nanopoulos and K. Tamvakis, "No-scale supersymmetric standard model", *Phys. Lett.* **134** (1984) 429. J. Ellis, C. Kounnas and D.V. Nanopoulos, "No scale supersymmetric Guts", *Nucl.Phys.* **B247** (1984) 373. For a review see A. B. Lahanas and D. V. Nanopoulos, "The road to no scale supergravity", *Phys. Rep.* **145** (1987) 1.

[50] A. Zichichi, "New developments in elementary particle physics," *Riv. Nuovo Cim.* **2N14** (1979) 1.

[51] A. Zichichi, contributions to "The superworld I", in vol. 24 of Subnuclear Series, Erice 1986, to "From superstrings to the real super superworld", in vol. 30, Erice 1992, and to "From supersymmetry to the origin of space-time", Erice 1993.

[52] U. Amaldi, W. de Boer and H. Furstenau, "Comparison of grand unified theories with electroweak and strong coupling constants measured at LEP," *Phys. Lett.* **B260** (1991) 447.

[53] F. Anselmo, L. Cifarelli, A. Peterman and A. Zichichi, "The effective experimental constraints on M (susy) and M (gut)", *Nuovo Cim.* **A104** (1991) 1817 and "The simultaneous evolution of masses and couplings: consequences on supersymmetry spectra and thresholds", *Nuovo Cim.* **A105** (1992) 1179.

[54] A. Zichichi, "Subnuclear physics, the first 50 years: highlights from Erice to ELN", vol. 24 (World Scientific Series in 20th Century Physics, 1999).

[55] J. L. Lopez, D. V. Nanopoulos and A. Zichichi, "A string no scale supergravity model and its experimental consequences", *Phys. Rev.* **D52** (1995) 4178 [arXiv:hep-ph/9502414].

[56] J. L. Lopez, D. V. Nanopoulos and A. Zichichi, contribution to "From superstring to the real superworld", vol. 30 of Subnuclear Series, Erice 1992.

[57] E. Witten, "Strong coupling expansion of Calabi-Yau compactification", *Nucl. Phys.* **B 471** (1996) 135 [arXiv:hep-th/9602070].

[58] I. Antoniadis, N. Arkani-Hamed, S. Dimopoulos and G. R. Dvali, "New dimensions at a millimeter to a Fermi and superstrings at a TeV," Phys. Lett. B **436** (1998) 257 [arXiv:hep-ph/9804398].

[59] E. Cremmer, B. Julia and J. Scherk, "Supergravity Theory In 11 Dimensions," Phys. Lett. B **76** (1978) 409.

[60] M. B. Green and J. H. Schwarz, "Anomaly Cancellation In Supersymmetric D=10 Gauge Theory And Superstring Theory," Phys. Lett. B **149** (1984) 117.

[61] L. Alvarez-Gaume and E. Witten, "Gravitational Anomalies," Nucl. Phys. B **234** (1984) 269.

[62] J. Strathdee, "Extended Poincare Supersymmetry," Int. J. Mod. Phys. A **2** (1987) 273.

[63] J. M. Maldacena, "The large N limit of superconformal field theories and supergravity," Adv. Theor. Math. Phys. **2** (1998) 231 [Int. J. Theor. Phys. **38** (1999) 1113] [arXiv:hep-th/9711200].

[64] O. Aharony, S. S. Gubser, J. M. Maldacena, H. Ooguri and Y. Oz, "Large N field theories, string theory and gravity," Phys. Rept. **323** (2000) 183 [arXiv:hep-th/9905111].

[65] R. Penrose, "Any space-time has a plane wave as a limit", Differential geometry and Relativity, Reidel, Dordrecht, 1976, pp. 271-275.

[66] D. Berenstein, J. M. Maldacena and H. Nastase, "Strings in flat space and pp waves from N = 4 super Yang Mills," JHEP **0204** (2002) 013 [arXiv:hep-th/0202021].

[67] D. J. Gross, J. A. Harvey, E. J. Martinec and R. Rohm, "The Heterotic String," Phys. Rev. Lett. **54** (1985) 502.

[68] P. Candelas, G. T. Horowitz, A. Strominger and E. Witten, "Vacuum Configurations For Superstrings," Nucl. Phys. B **258** (1985) 46.

[69] E. Witten, "String theory dynamics in various dimensions," Nucl. Phys. B **443** (1995) 85 [arXiv:hep-th/9503124].

[70] C. M. Hull and P. K. Townsend, "Unity of superstring dualities," Nucl. Phys. B **438** (1995) 109 [arXiv:hep-th/9410167].

[71] S. Ferrara, J. Scherk and B. Zumino, "Algebraic Properties Of Extended Supergravity Theories," Nucl. Phys. B **121** (1977) 393.

[72] N. Seiberg and E. Witten, "Electric - magnetic duality, monopole condensation, and confinement in N=2 supersymmetric Yang-Mills theory," Nucl. Phys. B **426** (1994) 19 [Erratum-ibid. B **430** (1994) 485] [arXiv:hep-th/9407087].

[73] S. Ferrara, R. Kallosh and A. Strominger, "N=2 extremal black holes," Phys. Rev. D **52** (1995) 5412 [arXiv:hep-th/9508072].

[74] J. Bekenstein, Lett. Nuov. Cim. **4** (1972) 737; Phys. Rev. **D 7** (1973) 2333; Phys. Rev. **D 9** (1974) 3292;
S.W. Hawking, Nature **248** (1974) 30; Comm. Math. Phys. **43** (1975) 199

[75] T. Levi-Civita, R.C. Acad. Lincei **26**, 519 (1917); B. Bertotti, Phys. Rev. **116**, 1331 (1959); I. Robertson, Bull. Acad. Polon. **7**, 351 (1959).

[76] G. W. Gibbons and P. K. Townsend, "Vacuum interpolation in supergravity via super p-branes," Phys. Rev. Lett. **71** (1993) 3754 [arXiv:hep-th/9307049].

[77] I. Bars and M. Gunaydin, "Unitary Representations Of Noncompact Supergroups," Commun. Math. Phys. **91** (1983) 31.

[78] M. Flato and C. Fronsdal, "Interacting singletons," Lett. Math. Phys. **44** (1998) 249 [arXiv:hep-th/9803013].

[79] M. J. Duff, R. R. Khuri and J. X. Lu, "String solitons," Phys. Rept. **259** (1995) 213 [arXiv:hep-th/9412184].

[80] R. D'Auria and P. Fre, "Geometric Supergravity In D = 11 And Its Hidden Supergroup," Nucl. Phys. B **201** (1982) 101 [Erratum-ibid. B **206** (1982) 496].

[81] P. K. Townsend, "P-brane democracy," arXiv:hep-th/9507048.

[82] P. Horava and E. Witten, "Heterotic and type I string dynamics from eleven dimensions," Nucl. Phys. B **460** (1996) 506 [arXiv:hep-th/9510209].

[83] G. W. Moore, "Anomalies, Gauss laws, and page charges in M-theory," arXiv:hep-th/0409158.

[84] E. Witten, "Five-brane effective action in M-theory," J. Geom. Phys. **22** (1997) 103 [arXiv:hep-th/9610234].

[85] N. Nekrasov, "A la recherche de la m-theorie perdue. Z theory: Casing m/f theory," arXiv:hep-th/0412021.

[86] . E. Cremmer in "Superspace And Supergravity. Proceedings, Nuffield Workshop, Cambridge, Uk, June 16 - July 12, 1980," S. W. Hawking and M. Rocek, eds. Cambridge, UK. Univ. Press (1981). B. Julia in "Superspace And Supergravity. Proceedings, Nuffield Workshop, Cambridge, Uk, June 16 - July 12, 1980," S. W. Hawking and M. Rocek, eds. Cambridge, UK. Univ. Press (1981).

[87] T. Damour, M. Henneaux and H. Nicolai, "Cosmological billiards," Class. Quant. Grav. **20** (2003) R145 [arXiv:hep-th/0212256].

[88] S. Ferrara and R. Kallosh, "Universality of Supersymmetric Attractors," Phys. Rev. D **54** (1996) 1525 [arXiv:hep-th/9603090].

[89] L. Andrianopoli, R. D'Auria and S. Ferrara, "U-duality and Central Charges in Various Dimensions Revisited". *Int, J. Mod. Phys.* **A13** 431 (1998).

[90] S. Deser and B. Zumino, "Broken Supersymmetry and Supergravity". *Phys. Rev. Lett.* **38** 1433 (1997).

[91] J. Polony, Budapest Preprint KFKI-1977-93 (1997) unpublished.

[92] E. Cremmer, B. Julia, J. Scherk, S. Ferrara, L. Girardello, P. van Nieuwenhuizen, "Spontaneous Symmetry Breaking and Higgs Effect in Supergravity without Cosmological Constant". *Nucl. Phys.* **147** 105 (1979).

[93] H. P. Nilles, "Supersymmetry, Supergravity and Particle Physics". *Phys. Rept.* **110** 11 (1984).

[94] E. Cremmer, C. Kounnas, A. Van Proeyen, J. P. Derendinger, S. Ferrara, B. de Wit and L. Girardello, "Vector Multiplets Coupled to N=2 Supergravity: SuperHiggs Effect, Flat Potentials and Geometric Structure ", *Nucl. Phys. B* **250** 385 (1985).

[95] J. Scherk and J. H. Schwarz, "How To Get Masses From Extra Dimensions". *Nucl. Phys. B* **153**, 61 (1979).

[96] E. Cremmer, J. Scherk and J. H. Schwarz, "Spontaneously Broken N=8 Supergravity". *Phys. Lett. B* **84**, 83 (1979).

[97] L. Andrianopoli, R. D'Auria, S. Ferrara and M. A. Lledó, "Gauging of Flat Groups in Four Dimensional Supergravity". *JHEP* **0207** 010 (2002).

[98] A. R. Frey and J. Polchinski, "N = 3 Warped Compactifications". *Phys. Rev.* **65** 126009 (2002).

[99] L. Andrianopoli, M. Bertolini, A. Ceresole, R. D'Auria, S. Ferrara, P. Fre and T. Magri, "$N = 2$ Supergravity and $N = 2$ Super Yang-Mills on General Scalar Manifolds:Symplectic Covariance, gaugings and Momentum Map". *J. Geom. Phys.* **23** 111 (1997).

[100] R. D'Auria and S. Ferrara, "On Fermion Masses, Gradient Flows and Potential in Supersymmetric Field Theories". *JHEP* **0105**, 034 (2001).

[101] S. Ferrara and L. Maiani, "An Introduction to Supersymmetry Breaking in Extended Supergravity". Proceedings of SILARG V. Baricloche, Argentina. O. Bressan, M. Castagnino and V. Hamity editors. World Scientific (1985).

[102] S. Cecotti, L. Girardello, M. Porrati, "Constraints in Partial Super-Higgs". *Nucl. Phys.* **B268**, 295 (1986).

[103] E. Sezgin and P. van Nieuwenhuizen, "Renormalizability Properties of Spontaneously Broken $N = 8$ Supergravity". *Nucl. Phys. B* **195** 325 (1982).

[104] S. Kachru, M. Schulz and S. Trivedi, "Moduli Stabilization from Fluxes in a Simple IIB Orientifold". hep-th/0201028.

[105] L. Andrianopoli, R. D'Auria, S. Ferrara and M. A. Lledó, "Duality and Spontaneously Broken Supergravity in Flat Backgrounds". *Nucl. Phys.* **B640**, 63 (2002).

[106] V. A. Tsokur and Y. M. Zinovev, "Spontaneous Supersymmetry Breaking in N = 4 Supergravity with Matter," *Phys. Atom. Nucl.* **59**, 2192 (1996);"Spontaneous Supersymmetry Breaking in N = 3 Supergravity with Matter". *Phys. Atom. Nucl.* **59** 2185 (1996).

[107] R. D'Auria, S. Ferrara and S. Vaulá, "$N = 4$ Gauged Supergravity and IIB Orientifold with Fluxes". *New J.Phys* **4** 71 (2002); R. D'Auria, S. Ferrara, M. A. Lledó and S. Vaulá,"No-scale $N = 4$ Supergravity Coupled to Yang-Mills: the Scalar Potential and superHiggs Effect". hep-th/0211027.

[108] B. Zumino, *Nucl. Phys. B* **89** (1975) 535.

[109] B.S. De Witt, "Dynamical Theory of Groups and Fields", (Gordon and Breach, New York, 1965) pgg. 231-233

CHAIRMAN: S. FERRARA

Scientific Secreteries: G. De Risi, W. Schleifenbaum

DISCUSSION I

Wagner :

How high in energy we have to go and not find SUSY, to say it does not solve the hierarchy problem ?

Ferrara:

Let me show you, before answering your question, the introduction of the vol. 3 of "The Quantum Theory Of Fields" by Weinberg. He writes that supersymmetry is a symmetry that unites particles of integer and half-integer spin and that this theory offers a possibility of solving the hierarchy problem, the mystery of the enormous ratio of the Planck mass to the 300 GeV energy scale of electroweak symmetry breaking. Supersymmetry also has the quality of uniqueness that we search for in any fundamental physical theory. There is an infinite number of Lie groups, that can be used to combine particles of the same spin in ordinary symmetry multiplets, but there are only eight kinds of supersymmetry in four space-time dimensions of which only one, the simplest, could be directly relevant to observed particles.

This has to do with what I was saying this morning, that there are only a finite number of supersymmetric theories. So the nice thing about supersymmetry is really its uniqueness and the fact that, whether it solves the hierarchy problem or not, it can be disproved.

So the answer to your question depends on how much fine-tuning you are ready to accept. There is a question of fine-tuning which essentially defines the hierarchy problem. I think if supersymmetry is not found in the 1 TeV range it will not solve the hierarchy problem. But there are now papers on this; for instance Dimopulos and Arkani-Hamed recently have written a paper called "High energy supersymmetry", where they introduce a fine-tuning which allow them to push the scalar supersymmetric particle high in energy scales, something like the 10-100 TeV range, and the light supersymmetric particle is still at the Fermi scale. So in that model you would find, for instance, incomplete multiplets. But I would certainly say that if supersymmetry is not found in the one TeV range, then it's hard to believe it is connected to the hierarchy problem. This does not mean supersymmetry is not relevant in field theory, because we believe that uniqueness of supersymmetry probably has to do something with the Plank scale. For instance, the construction of superstring theories which need supersymmetry for their consistency at the quantum level is an indication of that, but the fact that supersymmetry is relevant for the hierarchy problem is something that we will probably know the answer to in this decade.

Of course, supersymmetry is very much connected to the Higgs. If the Higgs particle is found, whose lower bound for the mass is at about 115 GeV, it will turn out to be lower than 150 GeV. This small energy range is a prediction of supersymmetry which is very much connected to the existence of fundamental scalar particles. I strongly believe that there are fundamental scalar particles in nature and the Higgs would be the first one to be observed. Before the work of ‘t Hooft no theory with a spontaneously broken gauge symmetry was known, because the strong forces and electromagnetic forces are all based on exact symmetries. So the fact that theories with spontaneously broken gauge symmetries show up, points to the existence of a fundamental scalar particle. In the limit of unbroken gauge symmetry of a supersymmetric standard model the Higgs particle is in the same multiplet as the vector boson. So there is a fundamental reason why the Higgs particle exists in supersymmetry. But of course, only experiment will tell what is the answer.

Bourjaily:

If the hierarchy problem is addressed with finely-tuned, high-energy supersymmetry then we have not solved the problem: we have traded one fine-tuning problem for another. Could you quantify how this could be a quantitative problem?

Ferrara:

The fine-tuning problem of the Higgs mass must be compared to the fine-tuning of the cosmological constant, which is much more severe. This relates to the comparison of complexity and reductionism that Zichichi talked about. Weinberg argues that the smallness of the cosmological constant must originate from an antropic principle, others say it must be some kind of symmetry. Zumino proved that the cosmological constant is zero if there is an unbroken supersymmetry. Obviously, suprsymmetry is broken, so the cosmological constant is suppressed by the gravitino mass. But the suppression is not enough to reach observed values for the cosmological constant if the supersymmetry scale is too low. For this reason Dimopulos et al. are ready to accept a larger fine-tuning for the Higgs mass than one would accept in constructing low-energy supersymmetry because this is anyway much smaller than the fine-tuning for the cosmological constant.

Zichichi:

But if it is valid to link the two fine-tuning problems, then we are stuck. This means that supersymmetry is broken at the Planck scale.

Ferrara:

I agree. It would be nice if supersymetry solved this problem, but I cannot cheat. Whether supersymmetry is going to be real or not, you need fine-tuning in the standard model to avoid quadratic divergencies beyond the weak scale. This is something we will probably know within this decade.

Supersymmetry, in contrary to composite models, is well-defined and it can be disproved. A theory that cannot be disproved is not a true physical theory in my belief. This is a problem of string theory, for instance. String theories are beautiful, but at the moment they have very little prediction. For supersymmetry, on the other hand, we are close to showing if we are near to a new revolution in physics, or whether this idea will turn out to be irrelevant.

Kamenik:

Have there been any attempts to analyze the gauge groups, or better the generator algebra, in a way analog to the SUSY enlargment of the Poincaré group?

Ferrara:

Internal symmetries cannot change the spin of a particle, so this would be again a violation of the Coleman-Mandula theorem. It is possible to have a violation of the Coleman-Mandula theorem: if you have a generator *X* that commutes with the spacetime generators, this not necessarily means that the symmetry group is the product of the two groups, because *X* could be in the center of the spacetime symmetry group. In extended supersymmetric theories this implies that one has a "central charge", which has to do with the masses of the so called "BPS states". For example, in the supersymmetric version of the Georgi-Glashow model, BPS states are 't Hooft monopoles, whose masses appears in the right-hand side of the supersymmetry algebra, and similar things happen in gravitational theories with the so called "extremal black holes", which again are BPS states whose masses and charges are related by the supersymmetry algebra.

Ferrara:

By the way, coming back to the question about the hierarchy problem, I would like to quote a sentence by Albert Einstein, that Schwinger reports in his paper on the unity of the fundamental interaction, which could be provocative: "The axiomatic basis of theoretical physics cannot be extracted from experiment, but must be freely invented"

Zichichi:

This is simply not true!

Ferrara:

This is what Einstein said.

Zichichi:

Who cares? History of physics proves that it is not true!

Ferrara:

The amazing thing is that this is written in the paper by Schwinger that was fundamental in constructing the standard model.

Zichichi:

The most recent proof that this is not true is the discovery of the non-Abelian forces. You need experimental results to construct theories.

Ferrara:

But we have other examples in which Einstein was right. As you said, Yang-Mills theories were formulated for a completely different problem.

Korthals-Altes:

Perhaps string theory is an example in which Einstein was right.

Ferrara:

String theory was formulated as a theory of hadrons.

Zichichi:

Sting theory starts by the attempt by Veneziano to explain strong resonances.

Ferrara:

On the other hand experience tells us that if we have nice mathematical structures which are consistent with rules of quantum field theory, then these structures are often correct. I don't know if you have a counterexample about this. Supersymmetry would be the first counterexample.

Zichichi:

That is why I am convinced that supersymmetry does exist. You know what disturbs me. It is the Higgs field. Theorists should invent something less metaphysical. It is like the Dirac sea. It cannot exists.

't Hooft:

Dirac would say "It's your problem if you don't like Higgs field. We have to wait and see."

Zichichi:

I make the following statement: sooner or later it will be discovered that the Higgs field does not exist. It's too crazy! It's a great invention, as the Dirac sea, but it cannot be real. The point is that all this was stimulated by two hundred years of experiments in electricity, magnetism and optics, and led to the Maxwell unification and Lorentz invariance. Without this, how could Dirac have invented his equation?

De Risi:

You have said that SUSY is a physical theory because it could be falsified, differently from what happens in string theory. But don't you think that, since SUSY and SUGRA are necessary for the consistency of string theory, if one takes seriously SUSY one has to trust in string theory?

Ferrara:

The previous question dealt with how supersymmetry could address the hierarchy problem, and this problem is completely separated with the problems dealing with quantum gravity, that are at the Planck scale. If you believe that string theory is the solution to the quantum gravity problem, then supersymmetry must be there, because it is the fundamental symmetry that makes quantum gravity consistent. But it could be that supersymmetry has nothing to do with the Fermi scale.

De Risi:

But could it be the contrary? I mean, if you believe that supersymmetry could solve the hierarchy problem at the Fermi scale, you are then led to believe in string theory.

Ferrara:

I agree with you, but in the sense that this will suggest that the low energy effective theory of the theory that would describe quantum gravity at the Plank scale must have those symmetries. Nowadays only string theory and M theory have this property, but there could be another theory different from superstring not yet constructed that could share the same symmetries.

Salur:

What is the most important next step if the SUSY partners are observed at LHC?

Ferrara:

Of course, you might as well find new light particles which have nothing to do with supersymmetry. Most importantly, you have to find the Higgs particle with a mass lower than 150 GeV implied by supersymmetry. To find this particle would be a proof of supersymmetry. Then, of course, you have to do the spectroscopy of MSSM in order to fix the ratio of the masses of supersymmetric particles, and to determine some of the soft-breaking terms of the MSSM. The spectrum one finds must show up a set of parameters that is compatible with the allowed region of the parameter space predicted by supersymmetric models and which are consistent with experimental bounds.
Another hint to supersymmetry is dark matter. One of the most natural candidates for dark matter is the neutralino. Evidence for supersymmetry might come from outer space and not in an accelerator.

Bourjaily: Even if the neutralino is discovered tomorrow in dark matter experiments we will not be able to identify it until it is seen, and measured, at colliders. Specifically, this means that no conclusive demonstration of supersymmetry can come from dark matter experiments alone.

Alwall:

What did Weinberg mean when he said that low-energy SUSY is unthinkable without supergravity?

Ferrara:

The point is, supersymmetry is a space-time symmetry, and gravity exists, so this space-time symmetry must be a local symmetry. This implies the existence of a spontaneous symmetry breaking mechanism. From the phenomenological point of view, this might be irrelevant, because the supergravity sector could be completely decoupled, e.g. in the gauge mediated supersymmetry breaking scenarios. But there are some scenarios, such as the gravity mediated supersymmetry breaking, in which the supergravity interactions are responsible for the supersymmetry breaking and thus for the soft terms that appear in MSSM. Whether or not supergravity is responsible for supersymmetry breaking is, however, a model dependent problem. But from a fundamental point of view, supergravity must be there.

Loktik:

You briefly mentioned the connection between gravity and gauge theory. Could you elaborate on that?

Ferrara:

This has to do with the celebrated AdS/CFT correspondence. According to the works of Maldacena, Witten and Gubser, Klebanov and Polyakov, there is a duality relation between supersymmetric gauge theories and supergravity theories, which, in the most simplified case, relate conformal supersymmetric theories with supergravity in an Anti-de Sitter space. AdS space is a maximally symmetric spacetime with constant curvature, essentially an hyperboloid. For instance in 4 dimensions, AdS is SO(3,2)/SO(3,1). This is the only maximally symmetric space, together with the de Sitter space, which has the same number of isometries as the Poincaré group.

This space-time is important because it allows unbroken supersymmetries, like Minkowski space, and also deSitter spaces are also important in constructing inflationary models, because they have a non-zero cosmological constant.

What is intriguing about it is that d-dimensional AdS has a (d-1)-dimensional boundary, and the isometry group of the AdS space actually acts as a conformal group on the boundary. For example, 5-dimensional AdS space has the ordinary 4-dimensional Minkowski space as a boundary. The idea is that when you have gravitational theories with branes, the degrees of freedom of those branes are described by SYM theories. These branes are black branes, so when they move towards their horizon, the geometry that results is exactly an AdS space, and so there is a correspondence between SYM theories and gravitational theories in AdS spaces. The most important example is the duality between type IIB supergravity compactified on $AdS_5 \times S_5$ and the N=4 four dimensional SYM theory that describes 3-branes. Because of this duality one can make a correspondence between correlator functions evaluated in the two different theories. For example, the stress-energy tensor correlator is related to the graviton correlator of the five-dimensional theory.

What was proved by Maldacena is that the weakly coupled regime of the supergravity theory corresponds to the strongly coupled regime of the SYM theory, more precisely the limit with large N and large 't Hooft coupling with the 't Hooft coupling being $\lambda=g^2N$. So you could in principle evaluate correlators of a strongly coupled theory by using results from a weakly coupled theory. This becomes very important for the calculation of the glueball masses and for the quark-gluon plasma discussed in another lecture. What is unrealistic is that this theory has 4 supersymmetries, and so it is hard to understand why it should be a correct description of QCD.

This enables, for the first time, calculations on a strongly coupled gauge theory different from lattice calculations.

CHAIRMAN: Sergio Ferrara

Scientific Secretaries: Jacob Bourjaily and Benjamin Labonne

DISCUSSION II

— *Gabriele Veneziano:*

In view of the dependence of spinors on the number of spacetime dimensions, how can we understand the way fermions transform as the result of compactifications?

— *Sergio Ferrara:*

The state of chirality of compactification is a very interesting problem. The answer depends very much on the manifold of compactification. For instance, if you start with a theory in ten- or eleven-dimensions—which is either chiral or non-chiral—and compactify on a torus, the theory for the massless states in lower dimensions is always non-chiral. So you cannot get chiral fermions when you compactify on a torus starting from a higher dimensional theory. This is the main problem of finding realistic Kaluza-Klein theories using standard compactification techniques.

In fact, you can get chiral fermions only if the compactification is done in a particular way. This can occur in the famous compactifiction of heterotic string theory in a ten-dimensional spacetime which is the product of Minkowski spacetime with a Calabi-Yau three-fold. In this case, you start from ten-dimensions with pure Yang-Mills theory and all but four spacetime dimensions are compactified. The original gauge group of ten-dimensions is then broken into a four-dimensional group. This group has the property that it is essentially a subgroup of E_8 which commutes with the holonomy group of the Calabi-Yau space. In the heterotic string, the gauge group is E_8xE_8 and one of its E_8 factor group is decomposed into SU(3)x E_6. Because the Calabi-Yau tree-fold is a complex Kähler manifold which is Ricci flat, SU(3) is its holonomy group. The commutant of E_8 with the holonomy group SU(3) is E_6. Now for the E_6 gauge group—which is the unbroken gauge group that you're looking for—there is an index theorem which tells you the number of chiral fermions. Specifically, the number of chiral fermions is related to the difference between the number of 27 and 27bar representations. If a theory is chiral, the numbers of chiral fermions in the 27 and of antichiral fermions in the 27bar representations are not equal. This difference is given by the Euler characteristic of the SU(3) Calabi-Yau manifold. In this case, it is simply given by the integral of the trace of the square of the Riemann tensor.

If the theory in ten-dimensions would have been pure gravity, without Yang-Mills fields, then the kind of compactification above would not have existed. It essentially realizes an anti-Kaluza-Klein idea: rather than starting with a theory of only gravity in ten-dimensions, consider adding a gauge symmetry. The theory you get in four-dimensions has a smaller gauge symmetry. It has an E_8xE_6 instead of the original E_8xE_8 gauge symmetry. It took almost thirty years to find this way out of standard Kaluza-Klein compactification. If you start with the standard compactification where the gauge symmetry comes only from the isometry of the internal manifold, you will never find chiral fermions in lower dimensions. This is the answer if a smooth geometry such as Calabi-Yau manifold is used.

There is another way to get chiral fermions which comes from string theory. This is what happens when you use singular manifolds like orientifolds or orbifolds. For example, consider the compactification manifold of an N-dimensional torus modded by a discrete symmetry group. The famous example is in the connection between M-theory and heterotic string theory due to Horava and Witten: consider an eleven-dimensional supergravity compactified on $S^1/\mathbb{Z}_2$. The circle becomes, after $\mathbb{Z}_2$-modding, half a circle: a segment. When you do so, what you discover is that at the end points of the segment, there are ten-dimensional slices and the theory acquires new degrees of freedom on those slices which come from anomaly cancellation. In the case of eleven-dimensional theory compactified on a ten-dimensional segment, those additional degrees of freedom are precisely the ten-dimensional Yang-Mills, $\mathcal{N}$=1 multiplets of the gauge group E_8 at $x^{11}=0$ and another E_8 at $x^{11}=\pi$, which are the end points of the segment. When this is done, you have a chiral theory because in ten-dimensions, $\mathcal{N}=1$ supersymmetry is chiral.

This is an example where you start with a non-chiral theory in eleven-dimensions and you get a chiral theory in ten-dimensions which is the heterotic string theory. The way to get chiral fermions living in a ten-dimensional theory from an eleven-dimensional theory without chirality is as following: start from a theory of pure gravity in eleven-dimensions, and you will get a theory of gravity in ten-dimensions. Because of the singular parts of the compactified manifold, new degrees of freedom emerge!

Lastly, there is a third way to obtain chiral string theories, which uses branes. Even if the theory is non-chiral in the bulk you can have a chiral theory on the branes. If the branes invade the entire spacetime, then chiral fermions on the branes become those of our world.

These three examples especially exhaust the methods to produce chiral fermions from higher dimensional theories.

— *Christian Korthals-Altes:*

You showed a table with spinors on columns and the number of supersymmetries on rows. In $\mathcal{N}=4$, for example, there is one vector, four spin-½ fermions, and six scalars. Is this multiplicity representative of any internal symmetry?

— *Sergio Ferrara:*

Yes, the multiplicity is nothing but the dimension of the antisymmetric representations of SU(N). The reason why you get six scalars is that in $\mathcal{N}=4$ SYM, you have an SU(4) R-symmetry and the scalars are in the two-fold antisymmetric representations of SU(4).

— *Luisa Cifarelli:*

We know from LEP that supersymmetry may allow for grand unification. Could you tell us more about the scales involved? In particular, how do the coupling unification, string, and Planck scales arise?

— *Sergio Ferrara:*

I should say that, as you have shown with Antonino Zichichi and others, the low energy coupling extrapolated to high energies does not uniquely fix the supersymmetry breaking scale. This is interesting because today there is a model by Dimopoulos and collaborators that has a high supersymmetry scale which can allow at least an order of magnitude higher for the supersymmetry breaking scale relative to that expected by grand unification.

It was pointed out that the grand unification scale seems to be two orders of magnitude lower than the string scale. This is because it was assumed that the string scale is close to the Planck scale. However, I should point out that this conclusion is not general. Rather, it is only true in the models which were known at the time when these calculations were done—namely, perturbative, heterotic string theory. This is a theory where you naturally have a gauge theory in four dimensions. However, there are models which show completely different dependence of the string scale in terms of the Planck scale. This will be discussed in detail in my third lecture. For now, allow me to summarize the results: the relation between the gravitational constant, the gauge coupling, and the string coupling is different for open and closed strings.

Closed and open strings acquire lower dimensional gauge groups in different ways. Closed strings get gauge groups *i.e.* from the compactification on a Calabi-Yau manifold. For open strings, however, it arises from the singularities at the endpoints of the string. Furthermore, for open strings the ten-dimensional gauge group in the modern interpretation of open strings is actually the gauge group which lives on a 9-brane. A 9-brane is a ten-dimensional space and so you can think of the ten-dimensional space of open strings as a 9-brane which fills the whole space. For the open string, however, there are not only 9-branes, but also p-branes for $p<9$. When you have these branes, they can be compactified either longitudinally or transversely. For a 9-brane, all directions of compactification are longitudinal because there are no transverse directions. The opposite extreme of this situation is when you have a 3-brane. Because a 3-brane lives in ten-dimensions, all of the six compactified directions must be transverse to the brane. In this case, the 3-brane must fill the four-dimensional spacetime after the compactification. If you want to compactify ten-dimensions to four, then all the coordinates of the 3-brane must be on spacetime and all the compactified directions will be transverse to the brane.

It is an interesting fact that if you compactify a theory with a brane that has both longitudinal and transverse compactified directions, the string coupling is found to be lower than that expected in the heterotic, closed string case. In particular, the string coupling can be made to coincide with the grand unification scale. So it may not be a problem that the gauge coupling unification occurs at energy much lower than the Planck scale. The fact that the gauge coupling unification occurs early may be an indication that the perturbative heterotic string is not the correct way to make predictions from string theory. Rather, one should consider scenarios where the standard model arises from branes. It is natural to use the fact that the heterotic string is not the only possibility to obtain grand unification. Essentially, the string scenario which looks the most promising is that of branes in which the particles of the standard model correspond to the degrees of freedom on the branes.

— *Antonino Zichichi:*

You are saying that brane technology can be used to reduce the string scale by three orders of magnitude from the Planck scale. These three orders of magnitude must correspond to some scale. Where does this come from?

— *Sergio Ferrara:*

When the compactification takes place, new scales are always introduced corresponding to the radius of compactification. In the case of branes, these scales are associated with the compactification of the directions longitudinal and transverse to the brane.

— *Antonino Zichichi:*

But these scales are completely arbitrary. How can they know where the gauge coupling unification occurs?

— *Sergio Ferrara:*

Remember, that the string scale is *a priori* arbitrary. There are always some input scales. The question is whether or not you arrange the numbers to give you the other scales you need.

— *Antonino Zichichi:*

We know that the gauge couplings unify around 10^{16} GeV, three orders of magnitude below the Planck scale. *A priori*, there appears to be no reason why this unification could not take place three orders of magnitude above the Planck scale. Do the brane calculations uniquely determine which scale is larger?

— *Sergio Ferrara:*

We are talking about possibly very large scales: the Planck scale, the string scale, and the unification scale. So we are naturally talking about scales where we can assume that supersymmetry is unbroken. From this general framework, the interesting fact is that when you have these branes, you generate new fundamental scales because these branes need compactification. When you have a brane, the dimensions you compactify can be either longitudinal or transverse to the brane which enter very differently to the low energy couplings. In fact, there is a general formula relating the Planck mass, the Yang-Mills coupling constant, the string scale, and the orthogonal and longitudinal radii of compactification to four-dimensions. So there is explicit dependence between the Yang-Mills coupling and the Planck scale. This implies that you must have a large dimension transverse to the brane to compensate the two orders of magnitude between the grand unification scale suggested by LEP data and the naïve string scale.

This is a relatively natural explanation because no *ad hoc* assumptions are required. If you have a theory where the gauge interactions do not come from the bulk but come from some particular brane, then there is dependence on both longitudinal and transverse compactification scales. Of course, this is only one possibility but it is generally true that compactification introduces new scales.

— *Gabriele Veneziano:*

Although it is nice to have an extra parameter to play with the relation between M_{Pl}, M_S, and g_{YM}, the problem remains of what can fix that extra parameter. Since there is a scalar field associated with (R_T, M_S) it looks as if the parameter should be dynamically fixed when the scalar field will acquire a mass.

— *Sergio Ferrara:*

In the tree-level theory, this is a free parameter because it is related to the compactification radius which is a free parameter of the theory. However, we expect that this free parameter, which originaly corresponds to a massless scalar, will be stabilized by quantum corrections. So it may be that these theories undergo a spontaneous compactification where quantum corrections fix the radius. Whether these corrections fix the radius to be that required for the desired reduction in string scale is an open question. It would depend on the details and I don't know the answer. The nice thing about these arguments is that with large extra dimensions, you can lower the Planck scale. Whether or not extra dimensions are large, however, it is believed to be a dynamical question.

— *Johan Alwall:*

What is meant by 'spontaneous compactification' of extra dimensions? Has there been any attempt to achieve this through string dynamics?

— *Sergio Ferrara:*

The term 'spontaneous compactification' is somewhat misleading. It is often used outside of the context of string theory. In particular, it is used to describe the situation where a higher dimensional solution to the Einstein equations of general relativity has a geometry that is the product of a four-dimensional Minkowski space with an internal manifold. This is called spontaneous compactification because it is analogous to choosing a particular vacuum of a gauge theory. It means simply that there is a choice of the vacuum of the field equations, which breaks the original higher dimensional Lorentz symmetry.

In string theory, there appear to be an enormous number of vacua that are completely degenerate at the classical level and it is not known which would be preferred. This is a problem that must be solved at the quantum level of the theory. There is today no way of choosing one over another. However, there have been recent attempts to analyze these many vacua statistically; the collection of degenerate vacua is nowadays called the 'landscape.'

In many ways, the problems facing string theory today are orthogonal to those of the past. Once, it was believed that string theory would be the unique 'theory of everything.' Nowadays, we find that string theories have an enormous number of vacua and some of them may be consistent with our universe. Therefore, the central question of string theory is how to select the vacuum.

CHAIRMAN: S. FERRARA

Scientific secretaries: Johan Alwall, Jakub Wagner

DISCUSSION III

- *Miklos Guylassy:*

You described different compactification paths from D=11 down. Are all possible compactifications down to D=4 dual to each other with the same spectrum?

- *Sergio Ferrara:*

The idea is that, the more you go down in dimension, the more theories are you able to construct. But because of the duality symmetries, some of these on the surface different theories actually look the same. For instance, if you stay in ten dimensions, if you believe in this unification with superstrings, there are five seemingly different theories, type IIA, type IIB, heterotic and type I, and then there is the M theory, so there are six theories (considering that the heterotic string can exist in two different gauge groups). However, already in ten dimensions, type I with gauge group SO(32) and the heterotic string with the same group are claimed to be the same theory in two different regimes. These theories are related by the S duality, meaning that one theory in the weak coupling is the strong coupling of the other theory. Since these theories live in ten dimensions, the weakly coupled and the strongly coupled theories have the same massless spectrum, because the D=10 theory is unique. So the type I and the heterotic theory with gauge group SO(32) are already dual in ten dimensions.

Also, in ten dimensions the type IIA theory is actually related to the D=11 theory in the sense that in the strong coupling it becomes the D=11 theory, because in the M theory interpretation, the radius of compactification of the 11th dimension, if you compactify it on a circle, is the coupling constant of the D=10 theory. If you instead compactify the D=11 theory on a segment, the theory is dual to the heterotic theory with the other gauge group E8xE8 at strong coupling.

What are not equivalent in D=10 are the type IIA and type IIB, since they actually have different massless spectra; type IIB has two chiral gravitini, while type IIA has one chiral and one anti-chiral gravitino. Also the number of supersymmetries are different, in type IIA the supersymmetry algebra is called (1,1) and in type IIB it is (2,0). Dual symmetries must relate theories with the same supersymmetry algebra.

However, if you go to D=9, i.e. you just compactify further from D=10 to D=9, these theories (type IIA and type IIB) actually become dual to each other; type IIA compactified on a circle of radius R is dual to type IIB compactified on a circle of radius 1/R. You can actually prove that thanks to the previously mentioned dualities all five theories are linked, by a web of dualities. So really, already in nine dimensions all these theories are linked to each other. When you go further down in dimension, there are different manifolds of compactifications which are all possible solutions of string theory, but you still have duality in lower dimensions. Of course, the number of duality symmetries in lower dimensions increase enormously, because there are extra dualities related to the internal manifolds. Using these bigger dualities, you can prove that there are also complicated theories in lower dimension which have the same number of supersymmetries that are dual to each other. What classifies the theories are the amount of supersymmetries plus these duality relations. Since the theories in D=9 are all dual to each other, you can prove that essentially all theories in lower dimensions with the same number of supersymmetries fall in equivalence classes, meaning that they are essentially dual.

- *Antonino Zichichi:*

So this means that really 9 is the magic number, i.e. 9+1+1?

- *Ferrara:*

Yes, since in D=9 all the theories are dual, and when you go from D=11 to D=9 it corresponds to the M theory compactified on a torus.

- *Guylassy:*

Why then are there so many possible models with different cosmological constant?

- *Ferrara:*

If the theories preserve supersymmetry, then there are only two possibilities. Either the cosmological constant is zero because you are in a Minkowsky vacuum, or the cosmological constant is negative because you are in an anti-de Sitter vacuum. If supersymmetry is broken, the cosmological constant can be rather arbitrary, but then it is much harder to make these duality connections.

- Zichichi:

But in real life, supersymmetry is broken.

- Ferrara:

In supergravity theories, unlike global supersymmetry, there is actually a false intuition, which is extrapolated from global supersymmetry. That is that in supersymmetric theories the potential is positive and then when you have broken supersymmetry, there is an induced cosmological constant. This, however, is not correct. In supergravity theories, the scalar potential has a more general form and is actually given by supersymmetry Ward identities which tell us that a scalar potential has a positive term and a negative term. You see that if you have gravity, there is a negative gravitino mass contribution which does not exist in global supersymmetry. Therefore in local supersymmetry the terms can cancel each other and give zero cosmological constant. This happens in what is called the no-scale supergravity. It means that you can have broken supersymmetry in Minkowski space with a zero cosmological constant, at least in the tree-level analysis.

- Guiseppe De Risi:

Is it possible to relate some supersymmetry breaking scale to the so-called brane-world scenarios, such as the Randall-Sundrum scenario? I mean, if something could happen in the bulk, that induce supersymmetry breaking on the brane?

- Ferrara:

To have these vacua with zero cosmological constant, so called flux vacua, you need a certain number of branes. In the case of type IIB orientifold what you need are D3 branes. You have essentially a Yang-Mills theory on the D3 brane. These flux vacua actually break supersymmetry but at the tree level, the D3 brane degrees of freedom don't feel this breaking. There is no splitting in the gauginos and scalars which are the superpartners of the D3 brane gauge field. There is only a splitting in the hidden sector. This realises the situation when the hidden sector of the usual no-scale theory is made by the bulk fields. The observable sector corresponds to the degrees of freedom of the brane. We expect that the radiative corrections will transmit breaking to the observable sector and hopefully stabilise the gravitino mass.

- *Sevil Salur:*

Could you please give an example of a direct application of M theory to the physical problem without lowering down the 11 dimensions?

- *Ferrara:*

The phenomenological application of M theory is to reproduce a model in D=4. I don't know what would be the phenomenology which would distinguish in D=4 between M-theory and string theory.

- *Zichichi:*

The question could be generalized by asking if there is any experimental verification to establish?

- *Ferrara:*

A string theorist might tell you that the discovery of supersymmetry is a signal that string theory is correct. The argument is that supersymmetry includes supergravity, and supergravity is consistent only if you have a quantum theory of gravity. And the only known quantum theory of gravity we have is superstring theory. That is the bottom-up argument. But it could be that there is another theory at the Planck scale which is neither string theory nor the M-theory, and still supersymmetry would be a good symmetry in low energy, if this theory extends the low energy supersymmetry to the Planck scale. I believe that the only possible signal from experiment is the signal for supersymmetry.

- *Gabriele Veneziano:*

You said before that M-theory, seen as 11-D supergravity, is a low-energy effective theory. Is it known up to what energy scale the effective theory is valid? Is it the 11-D Plank mass, is it the string scale, or some combination of the two?

- *Ferrara:*

I think that it is the Planck scale. The 11-D theory has no stringy parameter, only the Planck scale. Also, we know that it cannot be the usual gravity, in fact the gravity corrections can only be $(1/M_{Pl})^p$ times some higher order curvature terms. So, in D=11, it's low energy in respect to the Planck scale.

- *Veneziano:*

How can you be sure that the string scale cannot reappear in the M theory as an effective cutoff?

- *Ferrara:*

That is a good question. I think that the answer is known, because in M theory there are higher order correction terms to the Einstein action, which are proportional to powers of the curvature of the gravitational field. Some of these are known from its connection to string theory, but I don't remember the answer. This will answer your question.

- *Jacob Bourjaily:*

You have mentioned the enormous number of degenerate vacua, or consistent string theories. Specifically, Douglas and collaborators estimated that of the $\sim 10^{1000}$ possible string vacua roughly 10^{100}-10^{200} theories could reproduce all the observables in the Standard model. Could you please comment on what this implies for the uniqueness of string theory?

- *Ferrara:*

Today the situation is reversed compared to the early days of string theory, when they thought there were very few, perhaps a unique, vacua of string theory. Today, because of all these possible compactification schemes, it seems that there are millions of string vacua. In the flux models that I mentioned before, you can actually count the number of vacua since they are discreet, and finite. This is called the "landscape of vacua", and there are some people, such as M. Douglas et al., who try to make a statistical analysis of which of the vacua are preferred. These vacua have a certain distribution, for example some of them has a zero cosmological constant and some have a non-zero cosmological constant. Since most of these vacua have zero cosmological constant, you could say that the cosmological constant is zero. This reasoning is similar to the anthropic principle of Weinberg. Of course, which vacuum is preferred is a dynamical question; the full quantum corrections of this theory would remove the degeneracy of these vacua. Similarly, you could count the vacua which have broken supersymmetry and compare to the ones that have unbroken supersymmetry, to make some argument.

- *Miklos Guylassy:*

Your answer to ms. Salur was that she would have no way to measure in the lab which compactification and topological classification is the one we live in. But doesn't the observable spectrum in our D=4 constrain the many scenarios (i.e. experimental input from LHC and beyond will be necessary to select which compactification and which physical vacuum we live in)?

- *Ferrara:*

If you had a realistic model for all these vacua, you could possibly compute the spectra of the real world, for example the W mass.

- *Zichichi:*

Are you saying that you could compute the W mass without any knowledge of the Fermi coupling?

- *Ferrara:*

Computation means that from these models you could get some model for the soft breaking terms of the MSSM. You have to see how many of the possible vacua are close to the real world. The only input is the model, although of course you must put in the string scale, and a gauge group. The low-energy scales are generated dynamically, through gaugino condensate or some mechanism which stabilizes the gravitino mass against the quantum corrections.

- *Giovanni Imponente:*

Could we expect any phenomenology to look for to rule out or confirm the extra dimensions?

- *Ferrara:*

If you are referring to low energy strings or extra dimensions, then there are a lot of experimental consequences. This is a class of models that I have not covered in my lectures, so called Randall-Sundrum scenarios. In my opinion these models with large extra dimensions are rather artificial from the point of view of string theory. They are constructed to solve the hierarchy problem but they get enormous quantum corrections and it is not known how to stabilise them, becouse they have to deal with a non-renormalizable theory like gravity. Therefore you can consider them as an effective theory.

- *Jernej Kamenik:*

If you are using low energy phenomenology and known experimental data to select between the numerous degenerate string vacua, where is the progress in comparison to fitting the low energy parameters of SM or MSSM ?

- *Ferrara:*

The progress is that you are assured that the supersymmetry breaking parameters can be derived from first principles, from some theory at the Planck scale. It is the possibility of making a connection between the physics at the low energy and the physics at the Planck scale. It is of course important if you want to have this theory unified with gravity.

Testing the Standard Model at HERA and first results from HERA II

A. Geiser
DESY

Only the discussion is available for this lecture

CHAIRMAN: A. GEISER

Scientific secretaries: Iwona Grabowska-Bold, Tatsiana Klimkovich

DISCUSSION

- *Luisa Cifarelli:*

Could you clarify what is the situation concerning the observation of strange pentaquarks at HERA among the various experiments? Does HERA-B see a signal?

- *Achim Geiser:*

HERA-B has not seen a θ^+ signal. And H1 has not yet made a statement whether they see a signal or not. H1 has a little bit less statistics than ZEUS in inclusive DIS events because their trigger was less efficient. That might be the reason why they could not yet make a significant statement. Concerning HERA-B I can show one plot from the ZEUS measurements in this context. The ZEUS pentaquark signal appears mainly at high Q^2. If we go to lower Q^2, which approaches the HERA-B situation of hadronic interactions, then the signal starts to disappear in the background. Our conclusion on that for the time being is that for different Q^2 cuts the signal is consistent with scaling with the cross section which you would expect. The background goes up more strongly with decreasing Q^2 than the signal, such that for low Q^2 the signal just disappears in the background. And of course we cannot make a straight extrapolation to HERA-B. Therefore there is not necessarily a contradiction between the ZEUS signal and the non-observation of such a signal by HERA-B. It is very hard to make any conclusive statement. More statistics and more checks will have to show whether the ZEUS signal is actually a pentaquark or not.

- *Sergio Ferrara:*

Is SU(5) grand unification ruled out from experiment? Even in presence of the supersymmetry?

- *Achim Geiser:*

No, I did not say that. I did not make any definite statement. The only point which I was trying to make is that it is interesting to reduce the error on the measurement of α_s, such that this kind of studies can be done better.

- *Sergio Ferrara:*

But I think SU(5) grand unification is ruled out by the proton decay.

- *Achim Geiser:*

Yes, but I do not see the connection to…

- *Sergio Ferrara:*

Well I believe these people use SU(5) as a grand unified group.

- *Achim Geiser:*

But this has nothing to do with the measurements we are making.

- *Antonino Zichichi:*

To improve α_3 is the most important.

- *Achim Geiser:*

That is the point which I wanted to make. That we are not just measuring α_s for the purpose of measuring it, but that the result can be used to derive other conclusions.

- *Sevil Salur:*

What is the antiproton/proton ratio at ZEUS?

- *Achim Geiser:*

In the central rapidity region of the ZEUS, which is actually the backward region if you would go to the center of momentum because of the asymmetric collisions, indeed the ratio of protons to antiprotons is essentially one to one up to small corrections due to detector efficiencies and other systematic uncertainties. We are not really sensitive to the valence quarks because the valence quarks, even if they do interact, go either down the beam pipe or to the region where we have calorimetry but not tracking.

- *Sevil Salur:*

What are the peaks next to the θ^+ signal (at 1.47 and 1.58 GeV)?

- *Achim Geiser:*

Yes of course we have worried about these peaks. But this analysis was specifically about looking for states, that have been seen before by other people in a specific mass region. The fact that we maybe see bumps also in other regions is an issue from the point of view of whether we describe the background curve correctly. I showed you a plot where we have a double peak fit. We also did single peak and triple peak fits and the conclusion from that is that the peak of interest is the one at 1521 MeV which could correspond to the pentaquark signal. Its parameters do not critically depend on what assumptions you make for the background shape for these potential neighbouring bumps. The conclusion is that the impact of these potential bumps on the signal is very small.

- *Sevil Salur:*

How do you describe the background?

- *Achim Geiser:*

It is an empirical parametrization of a threshold function plus up to two Gaussians for the neighbouring peaks that you were asking me about.

- *Jacob Bourjaily:*

You showed a plot of the ep scattering cross section as a function of the longitudinal polarization of the electrons (or positrons). That data ranged from approximately –40 to +40% polarization. What are the fundamental limiting factors preventing the use of 100% polarized beams?

- *Achim Geiser:*

There is a competition between natural build up of polarization and its destruction due to beam-beam interactions, and interactions of the beam with various beam elements. And that is actually what limits the polarization. The spin rotators are a potential additional source of destroying the polarization because they need to compensate each other very precisely. Since the beams are circulating and going through these elements many times per second a tiny perturbation would immediately destroy the polarization. There is a certain competition between these technical effects which destroy the polarization and the physical process which builds it up. This yields an equilibrium at a certain point. Right now we are happy with 40%, and 50% is the final goal.

- *Miklos Gyulassy:*

H1 reported several years ago a Λ/Λbar > 1 excess in 6-7 units of rapidity from the proton target. The data of that time were really interesting. Kopelevich made an interesting analysis suggesting that this may yield valuable estimate on baryon number transport dynamics and the determination of the Regge intercept of the Veneziano-Rossi function. What is the status of this study? There was a 10% asymmetry - excess of the positive baryon number lambda over antilambda.

- *Achim Geiser:*

I am not aware of any more recent measurements which show a significant difference between baryons and antibaryons in the central rapidity region. In the forward region of course, there would be. But not in the central rapidity region.

On the level of the statistics which we have been discussing here for the pentaquark signal, a 10% difference would not be a very significant difference. From the ZEUS detector point of view, there are differences in acceptance between negative and positive particles of the order of 10-20%. If one wants to make a statement beyond this level of precision a very detailed study of systematics is needed. This was partially done for this analysis but only partially because a remaining difference of 10% would not significantly affect the result.

Chapter 1: Highlights from Virgo.
Chapter 2: Advanced interferometers for gravitational wave detection: Quantum non-demolition techniques.

Adalberto Giazotto –INFN-Pisa

Chapter 1: Hilights from Virgo

Gravitational Waves (GW) give foundamental informations on the Universe. By comparing the four foundamental nature's interaction constants we realize that gravitational interaction is 34 order of magnitude smaller than the other:

Strong	Electromagnetic	Weak	Gravitational
$\alpha_s=1$	$e^2=1/137$	$G_F M^2=10^{-5}$	$G_N M^2=10^{-39}$

where M is the proton mass and we have set Planck's constant and light speed equal to 1.
This smallness has as a consequence that gravitons escape undisturbed when other particles interact many times; this happens in supernovae collaps where neutrinos undergoes about 1000 interactions before escaping and gravitons none.
This peculiarity makes that gravitons decouples from Big-Bang hot matter after 10^{-43}s , the Planck's time, corresponding to $\sim 10^{18}$ TeV. Graviton detection seems the only way to reach times closed to 10^{-23}s (because we are sensitive only to the "acustic" band of frequencies) for studying Big-Bang.

The existence of Gravitational Waves

Einstein's General Relativity Theory[1] (GTR) has as a consequence the existence[2] of GW; this has been demonstrated by Taylor ed Hulse[3] by observing for about 20 years the neutron stars binary system PSR 1913+16. They demonstrated that the binary system energy loss follows at 1/1000 level the GTR predictions:

$$\boxed{\dot{P}_{MEAS} / \dot{P}_{GTR} = 1.0023 \pm .0047}$$

Where P_{MEAS} and P_{GTR} are the revolution periods measured and predicted by GTR respectively. This measure give a non contradicted evidence of GW existance.

GW-Mass interaction, polarization effects and the GW transmitter

Gravitational Waves are a consequence of Einstein's General Relativity; they are shown to be ripples in the space -time curvature traveling at light speed. In curved space-time the invariant distance ds between two neighbouring events is described by Metric Tensor $g_{\mu\nu}$:

$$\begin{cases} ds^2 = g_{\mu\nu} dx^\mu dx^\nu \\ g_{\mu\nu} g^{\nu\lambda} = \delta^\lambda_\mu \end{cases}$$

In Flat Space

$$g_{\mu\nu} \Rightarrow \eta_{\mu\nu} = \begin{pmatrix} 1 & 0 & 0 & 0 \\ 0 & -1 & 0 & 0 \\ 0 & 0 & -1 & 0 \\ 0 & 0 & 0 & -1 \end{pmatrix}$$

The evolution of the metric tensor is given by Einstein's equations:

$$\begin{cases} g_i^k R_{k\mu\nu}^i = R_{\mu\nu} = \dfrac{8\pi G}{c^4}\left(T_{\mu\nu} - \dfrac{1}{2} g_{\mu\nu} g_i^l T_l^i\right) & \textbf{Einstein Eq.} \\ R_{klm}^i = \dfrac{\partial \Gamma_{km}^i}{\partial x^l} - \dfrac{\partial \Gamma_{kl}^i}{\partial x^m} + \Gamma_{nl}^i \Gamma_{km}^n - \Gamma_{nm}^i \Gamma_{kl}^n & \textbf{Riemann Tensor} \\ \Gamma_{kl}^i = \dfrac{1}{2} g^{im}\left(\dfrac{\partial g_{mk}}{\partial x^l} + \dfrac{\partial g_{ml}}{\partial x^k} - \dfrac{\partial g_{kl}}{\partial x^m}\right) & \textbf{Christoffel Symbols} \end{cases}$$

Where $T_{\mu\nu}$ is the energy-momentum tensor and G is Newton's constant .
The presence of Momentum-Energy tensor $T_{\mu\nu} \neq 0$ changes $\eta_{\mu\nu}$ in $g_{\mu\nu}$ according to Einstein's equations. When gravity is "small" we may write $g_{\mu\nu} = \eta_{\mu\nu} + h_{\mu\nu}$, where $h_{\mu\nu}$ is infinitesimal; to first order in $h_{\mu\nu}$ Einstein equation becomes:

$$\Box \Psi_{\mu\nu} = (8\pi G/c^4)\, \tau_{\mu\nu} \qquad \Psi_{\mu\nu} = h_{\mu\nu} - \frac{1}{2}\delta_{\mu\nu} h_\lambda^\lambda$$

and $\tau_{\mu\nu}$ is $T_{\mu\nu}$ expanded to first order in $h_{\mu\nu}$. From momentum-energy conservation

$$\partial_\mu \tau_{\eta\nu} = 0$$

and considering that $\tau_{00} = \rho c^2$, where ρ is the matter density, it follows:

$$\Psi_{\alpha\beta} = -\frac{2G}{c^4 R_o}\left(\frac{\partial^2}{\partial t^2}\int \rho x_\alpha x_\beta dV\right)_{t-R_0/c}$$

where R_0 is the source distance and ρ source mass density.
$\Psi_{\mu\nu}$ has 16 el. , Symmetry reduces to 10, Gauge trasf. x'=x+ϵ ($\Box\epsilon=0$) reduces to 6, Transversality and Tracelessnes (TTsystem)reduces to 2:

$$h_{\mu\nu}^{TT} = h_{11}^{TT}\begin{pmatrix} 0 & 0 & 0 & 0 \\ 0 & 1 & 0 & 0 \\ 0 & 0 & -1 & 0 \\ 0 & 0 & 0 & 0 \end{pmatrix} + h_{12}^{TT}\begin{pmatrix} 0 & 0 & 0 & 0 \\ 0 & 0 & 1 & 0 \\ 0 & 1 & 0 & 0 \\ 0 & 0 & 0 & 0 \end{pmatrix} = h^+ e_{ik}^+ + h^\times e_{ik}^\times$$

GW along X_3

It is easy now to show that GW have spin 2 because a 45 degree rotation exchange h^+ with $h^\times$ and viceversa.
The Riemann tensor in the TT gauge reaches its simplest form:

$$R_{iklm} \Rightarrow R_{0\alpha 0\beta} = -\frac{1}{2}\ddot{h}_{\alpha\beta}^{TT} \qquad \textit{where} \quad \ddot{h}_{\alpha\beta}^{TT} = -\frac{2G}{c^4 R_0}\left(\frac{\partial^2}{\partial t^2}\int \rho (x_\alpha x_\beta)^{TT} dV\right)_{t-R_0/c}$$

i.e. in the limit of weak gravity, GW amplitude is proportional to the second time derivative of the mass quadrupole moment. This is the GW Transmitter.

The GW receiver

A particle moving freely under the action of a gravitational field has its x^μ coordinate satisfying the geodesic equation:

$$\frac{d^2x^\mu}{d\tau^2}=\Gamma^\mu_{\nu\lambda}\frac{dx^\nu}{d\tau}\frac{dx^\lambda}{d\tau}$$

Where τ is the proper time; It is always possible to find a space-time trajectory in which $\Gamma^\mu_{\nu\lambda}=0$ at any time; along this trajectory a particle is freely falling. Let's consider two freely falling particles A and B (see Fig. 1), their separation $\xi^\alpha=(x_A-x_B)^\alpha$ satisfies the geodesic deviation equation:

$$\frac{D^2\xi^\alpha}{d\tau^2}+R^\alpha_{\beta\gamma\delta}\xi^\gamma\frac{dx^\beta}{d\tau}\frac{dx^\delta}{d\tau}=0$$

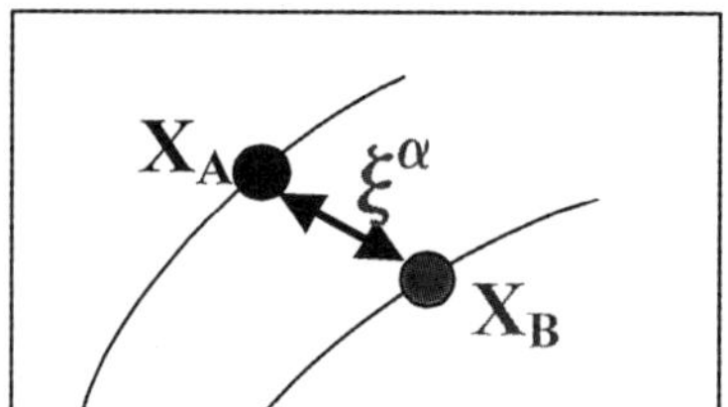

Fig.1

D^2 is the second covariant derivative:

With the purpose of evaluating ξ^α let's put $x^A=0$ in the A particle CMS, $x^A{}_0$ equal to proper time

$$\frac{D^2\xi^\alpha}{d\tau^2}=\frac{d^2\xi^\alpha}{d\tau^2}+\frac{d\Gamma^\alpha_{\beta\mu}}{d\tau}\xi^\beta\frac{dx^\mu}{d\tau}+\Gamma^\alpha_{\beta\mu}\frac{d}{d\tau}(\xi^\beta\frac{dx^\mu}{d\tau})+\Gamma^\alpha_{\beta\mu}(\frac{d\xi^\beta}{d\tau}+\Gamma^\beta_{\varepsilon\nu}\xi^\varepsilon\frac{dx^\nu}{d\tau})\frac{dx^\mu}{d\tau}$$

τ and the coordinate axis connected to gyroscopes carried by A. At $x^A=0$ since A is freely falling along the geodesic line, we can obtain $(\Gamma^\mu_{\nu\lambda})_{x=0}=0$, $(d\Gamma^\mu_{\nu\lambda}/d\tau)_{x=0}=0$, consequently:

$$\frac{D^2\xi^\alpha}{d\tau^2}=\frac{d^2\xi^\alpha}{d\tau^2}=-R^\alpha_{\beta\gamma\delta}\xi^\gamma\frac{dx^\beta}{d\tau}\frac{dx^\delta}{d\tau}$$

To first order in $h^{TT}_{\mu\nu}$, since the only surviving terms are for $\beta=\delta=0$ and $\tau=t$, and remembering that

$$R_{0\alpha 0\beta}=-\frac{1}{2}\ddot{h}^{TT}_{\alpha\beta}$$

it follows that the relative acceleration of B with respect A is:

$$\frac{d^2\xi_\alpha}{d\tau^2}=\frac{1}{2}\ddot{h}^{TT}_{\alpha\beta}\xi^\beta$$

and the Riemann force is

$$F_\alpha = \frac{1}{2} M \ddot{h}^{TT}_{\alpha\beta} \xi^\beta$$

Consequently the receiver is a device measuring space-time curvature i.e. the relative acceleration of two freely falling masses or their relative displacement.
GW produce real tidal accelerations on the bodies embedded in space-time; considering the reference system shown in fig.2, the two masses M will undergoes a relative acceleration A_β and the Riemann force F.

$$A_\beta = -\frac{1}{2} L_\alpha \frac{d^2}{dt^2} h^\alpha_\beta \qquad F_\beta = MA_\beta \qquad (1)$$

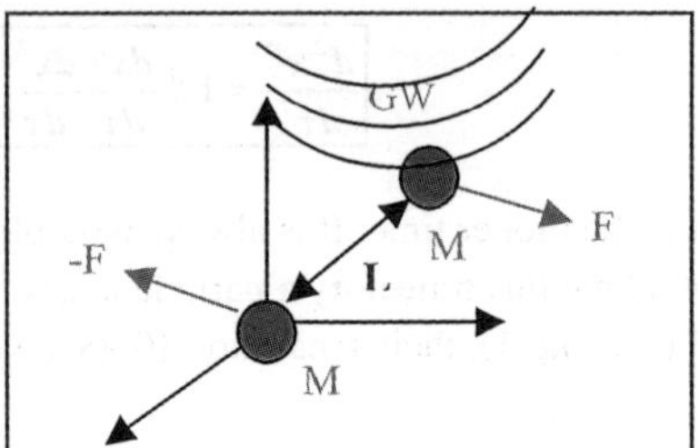

Fig.2

where L_α is the mass separation. This equation is valid in the limit $\lambda >> L$ and in this limit we note that Riemann force increases with L.
GW have two state of polarization h_+ and h_x; the effect on a massive ring whose plain is perpendicular to wave propagation direction, is to produce a deformation as shown in fig.3.

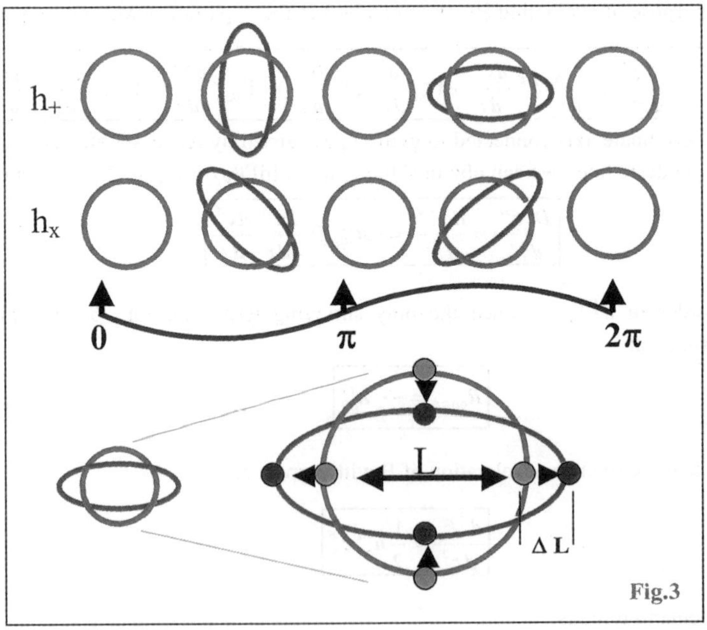

Fig.3

Short outline about GW sources

Coalescing binaries

These systems are formed by two compact stars, neutron stars and black-holes, rotating one around the other. In fig.4 the reference system for a coalescing binaries system having with respect to the equatorial system (X_0,Y_0,Z_0), direction (α,δ) and angles (i,ϑ) of the orbit plane normal with respect to the direction (α,δ).

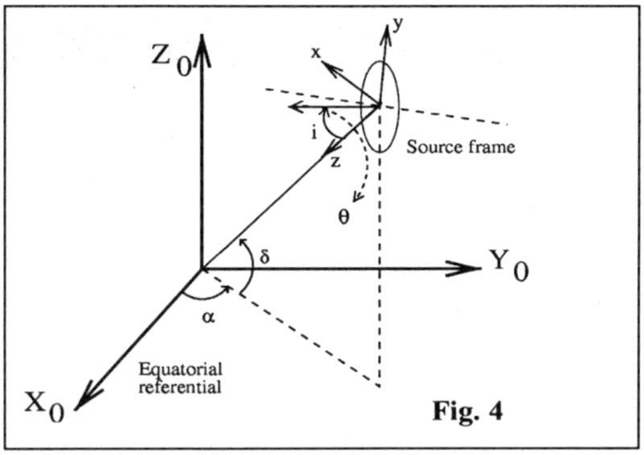

Fig. 4

The detector responds to the GW action in the following way:

$$h(t) = h_+(t)\,F_+(\alpha,\delta) + h_\times(t)\,F_\times(\alpha,\delta) \qquad (2)$$

Where F_+ and F_x dipend from detector orientation and h_+ ed h_x are given by:

$$\begin{pmatrix} h_+ \\ h_\times \end{pmatrix} = 2.6\times 10^{-21}\left[\frac{1Mpc}{D}\right]\left[\frac{\mu^{3/5}M^{2/5}}{M_\oplus}\right]^{5/3}\left[\frac{\nu(t)}{100\ Hz}\right]^{2/3}\times$$
$$\begin{pmatrix} \cos\Phi(t)\dfrac{\cos^2 i+1}{2}\cos(2\theta)+\sin\Phi(t)\cos i\sin 2\theta \\ \sin\Phi(t)\cos i\cos(2\theta)-\cos\Phi(t)\dfrac{\cos^2 i+1}{2}\sin(2\theta) \end{pmatrix} \qquad (3)$$

Where μ=M_1M_2/(M_1+ M_2) is the reduced mass and M= M_1+ M_2 is the total mass, $M_\oplus$ la the Sun mass and Φ is:

$$\Phi(t) = \Phi_0 + \int_{t_0}^{t} \nu(\tau)\,d\tau \qquad (4)$$

The frequency as a time function is given by:

$$\nu(t) = \frac{1}{\pi}\left(\frac{5}{256}\frac{1}{K^{5/3}}\frac{1}{(t-t_0)}\right)^{3/8} \qquad (5)$$

Where $K=(\mu M^{2/3})^{3/5}$ is the Chirp Mass and t_0 the coalescence time.

The typical received signal will have a shape similar to the one shown in fig.5; on the last ms strong GRT nonlinearities will occur.

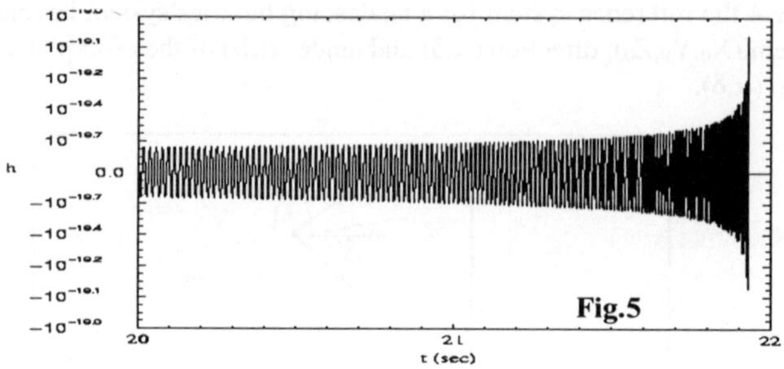

Fig.5

The signal Fourier transform h(ν) is given by:

$$h(\nu) \propto \nu^{-7/6} e^{i\Phi(K,...)} \tag{6}$$

The signal analysis consists in filtering the detector signal with a series of templates in which the unknown parameters M, μ, the initial phase.... are cyclically explored by means of Wiener integrals:

$$\frac{s}{n} = \int d\nu \frac{\mathrm{Re}\,(S+h)V_n^*}{S^2} e^{i\omega\tau} \tag{7}$$

Where S is the spectral noise and S+h is the detector signal+ noise, V_n is the n-th template and τ the arrival time.

The expected rate is some event per year in a 50Mpc sphere.

Supernovae (SN)

The theoretical evaluation of GW emission from SN explosions has a large uncertainty since the fraction of energy converted in GW is uncertain too; it is assumed a fraction between 10^{-6} and 10^{-3}. The expected rate is about 1/30 years in our Galaxy with $h\sim10^{-20}$ and about 1/year in the Virgo cluster with $h\sim10^{-23}$.

Periodic Sources

In our Galaxy are about 10^9-10^{10} Neutron Stars (NS) which may emit because of some mass distribution asymmetry both at the rotation and twice the rotation frequency ν.

Since Earth rotates both around the Sun and arount his polar axis, the detector receives a non-monochromatic signal because of Doppler shift:

$$e^{i\omega t} \Rightarrow e^{i\omega(t-\vec{n}\bullet\vec{R}/c)} \tag{8}$$

where $\bar{n}$ is the NS direction, $\omega=2\pi\nu$ and $\bar{R}$ is the vector connecting the detector to the cms of the Sun-Earth system. The Doppler shift widens the relative natural frequency width $\Delta\nu=1/T$, where T is the observation time, to $\Delta\nu/\nu{\sim}10^{-4}$. For eliminating this effect we need to know ω and $\bar{n}$ and this happens only for known Pulsars.
The known Pulsars are about 1400 and those in the Virgo bandwidth, above 4 Hz, about 700.
If ω and $\bar{n}$ are unknown the optimal analysis method it the "blind search" consisting in dividing the space (ω , $\bar{n}$) in cells and evaluate (ω_i , $\bar{n}_i$) in each cell and compute the FFT

of the data multiplied by $e^{+i\omega_i \vec{n}_i \bullet \vec{R}/c}$ the FFT will be repeated in every cell and the

sky will be explored till a high S/N ratio is found.
The payed price is enormous because for obtaining the maximal frequency resolution $\Delta\nu=1/T$ we need about 10^{22} sky cells; consequently the computing power rquired becomes impossibly large hence hierarchical methods are under study.

Cosmological BKG[4,5]
Perhaps the most important signal to search for; it gives informations on the Big-Bang at time close to Planck time. For being detected it requires coincidence between two close high sensitivity interferometers.

Mass displacement due to GW action.

As shown in Fig. 6, GW produce a mass displacement; this displacement, Δ L, is extremely small and may be optimistically evaluated to be in the order $\Delta L \sim 10^{-22}L$.
Thermal noise, created by mirror and mirror suspension wires atom, is the dominant noise at room temperature. To detect GW the thermal stockastic force[13] should be smaller than the one produced by GW:

$$\frac{1}{2}M\omega^2 L_\alpha h_\beta^\alpha > \sqrt{\frac{4kTM\omega_0^2}{Q\omega}}\frac{N}{\sqrt{Hz}}$$

Where M is the oscillator's mass, Q his quality factor, $\nu_0=\omega_0/2\pi$ the resonance frequency and T the temperature.

By defining the spectral GW amplitude as $\tilde{h}_\beta^\alpha (1/\sqrt{Hz})$ and defining, $L_\alpha \tilde{h}_\beta^\alpha = L\tilde{h}$

The measurability condition becomes $$\tilde{h} > \frac{2}{L\omega^2}\sqrt{\frac{4kT\omega_0^2}{MQ\omega}}\frac{1}{\sqrt{Hz}}$$

From this condition detector topology is following:

-**If L is small** (some meter) and the source has Fourier components around 1000Hz, the **Bar crygenic**[6,7,8,9,10] detectors working at very low teperature (1K) are needed.

-If L is large (some km) it is possible to have high sensitivity at low frequency too (some Hz) and the detector may work at room temperature (but at low temperature seems even better); this is the working condition for **Interferometric detectors** [11].

In the following we will concentrate our attention on the Virgo interferometric detector.

Interferometers as GW detector

Eq. 1 gives the relative acceleration, due to GW interaction, of two freely falling objects separated by the distance L.
We should find out the effect of interferometric measurement of masses distance. For doing this evaluation let's observe test masses displacement in a system in which they are at rest.
From General Relativity Theory it follows that if we change reference system by ε_μ also $h_{\mu\nu}$ changes according to a transformation, which leaves invariant ds^2:

$$\begin{cases} x'_\mu = x_\mu + \varepsilon_\mu \\ h'_{\mu\nu} = h_{\mu\nu} - \dfrac{\partial \varepsilon_\mu}{\partial x_\nu} - \dfrac{\partial \varepsilon_\nu}{\partial x_\mu} \end{cases}$$

It is then possible to find the appropriate ε_μ in such a way that mirror separation does not varies; in this new reference system the perturbed photon ds^2 is:

$$ds^2 = c^2 dt^2 - (1 + h_{11}^{TT})dx^2 - (1 + h_{22}^{TT})dy^2 = 0$$

where $h^{TT}_{11} = - h^{TT}_{22} = h(t) = h\cos(\Omega_g t + \phi)$; integrating we obtain, for a ray traveling along x axis and scattered back by a mirror at distance x=L, the round trip retarded time t_r:

$$cdt(1 - \frac{1}{2} h \cos(\Omega_g t + \varphi)) = dx$$

$$c(t_r - t) - \frac{1}{2} h \frac{\sin(\Omega_g t_r + \varphi) - \sin(\Omega_g t + \varphi)}{\Omega_g} = -2L$$

To first order in h we obtain:

$$c(t_r - t) - \frac{1}{2} h_g \frac{\sin(\Omega_g (t - \frac{2L}{c}) + \varphi) - \sin(\Omega_g t + \varphi)}{\Omega_g} = -2L$$

$$t_r = t - \frac{2L}{c} - \varepsilon h_g \frac{L}{c} \frac{\sin \eta}{\eta} \cos(\Phi - \eta)$$

Where $\eta = \Omega_g L/c$ and $\epsilon = \pm 1$ if the photon is traveling along x or y axis respectively.
The time dependent part of the EM fields along the trajectory, after a round trip 2L long, is:

$$A(t) = A_0 e^{-i\omega\left(t - \frac{2L}{c} - \varepsilon h \frac{L}{c}\frac{\sin\eta}{\eta}\cos(\Phi-\eta)\right)} \sim A_0 e^{-i\omega\left(t-\frac{2L}{c}\right)}\left(1 + i\omega\varepsilon h \frac{L}{c}\frac{\sin\eta}{\eta}\cos(\Phi - \eta)\right)$$

For F round trips in a Fabry-Perot cavity we obtain[11] the phase shift produced by GW:

$$\varphi(t) = \frac{\omega h F L}{c\sqrt{1 + \left(F\Omega_g L / c\right)^2}} \qquad \textit{cutoff freq.} \quad \nu_g = \frac{c}{2\pi F L}$$

We see that the cavity has a storage time a FL/c acts as a low pass filter ; above that cutoff frequency the cavity starts loosing sensitivity. A tipical value ν_g=1kHz is set for being maximally sensitive to supernova explosions events whose characteristic GW emission time is ≅1 ms.

The Virgo Collaboration

Virgo is a french-italian collaboration between CNRS and INFN. The following groups are involved: Lapp (IN2P3 Annecy), OCA (CNRS Nice), SMA (IN2P3 Lyon), Orsay(IN2P3 Orsay), ESPCI (CNRS Paris) on the french side and the italian INFN groups of Firenze-Urbino, LNF, Perugia, Pisa, Roma1, Napoli.
The collaboration is composed by about 100 physicists and a similar number of technicians. Virgo is built at S. Stefano a Macerata locality, about 15 km from Pisa.

The typical diagram of Virgo interferometric detector

In Fig.7 the typical scheme for the interferometric read out system of test mass displacement is shown [12]; since the GW changes L_A and L_B in different way, the measurement of $\Delta L = L_A - L_B \neq 0$ gives a GW amplitude measurement.

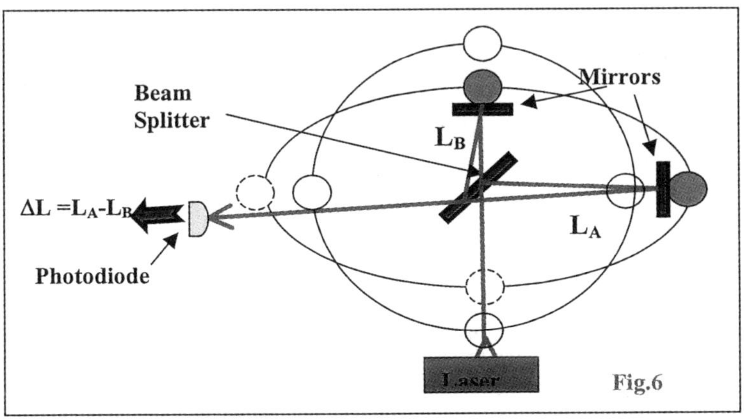

Fig.6

The interferometric measurement sensitivity may reach $\sim 10^{-19}$-10^{-20} m/Hz$^{1/2}$ hence for measuring $\Delta L/L \sim 10^{-22}$ at room temperature, L_A and L_B should be Km long. The large interferomer arm length and frequency band are enumerated in table 1.

Table 1

Italy	VIRGO	Cascina	3km	4-10000 Hz
USA	LIGO	Hanford	4km	70-10000 Hz
USA	LIGO	Hanford	2km	70-10000 Hz
USA	LIGO	Louisiana	4km	70-10000 Hz
Ge-GB	GEO	Hannover	600m	70-10000 Hz
Japan	TAMA	Tokyo	300m	100-10000 Hz

In fig. 8 the typical optical scheme of Virgo and LIGO like detectors is shown: in Virgo a laser NdYAG having ~ 20 w is prestabilized by a Fabry-Perot (FP) triangular rigid cavity 30 cm long to spectral frequency width $\sim 10^{-2}$ Hz/Hz$^{1/2}$ @10 Hz. The prestabilized laser is then injected in a FP triangular cavity 144 m long in such a way of filtering out all the transvers modes and leaving only the TEMoo. The beam then traverse the Recycling Mirror (RM) and enters through the Beam Splitter (BS) mirror which split it into Nord and West beams. The two beams then enter the two FP cavities where about 50 reflections occur for a total of 150 km. Subsequently the photons reach the BS mirror and the phase can be varied in such a way that all the power flows toward RM and the sidebands, containing the signal,flows through a mode cleaner, to the photodiodes. The beam sent to RM is then sent toward BS by RM and the power at BS increased by a factor ~50 with the purpose of reducing shot noise. The RM effect is not only to reduce shot noise but also to stabilize laser frequency at level of 10^{-6} Hz/Hz$^{1/2}$ @10 Hz. This spectral frequency width reduction is vital since FP arms might have storage time asymmetries as large as 10%.

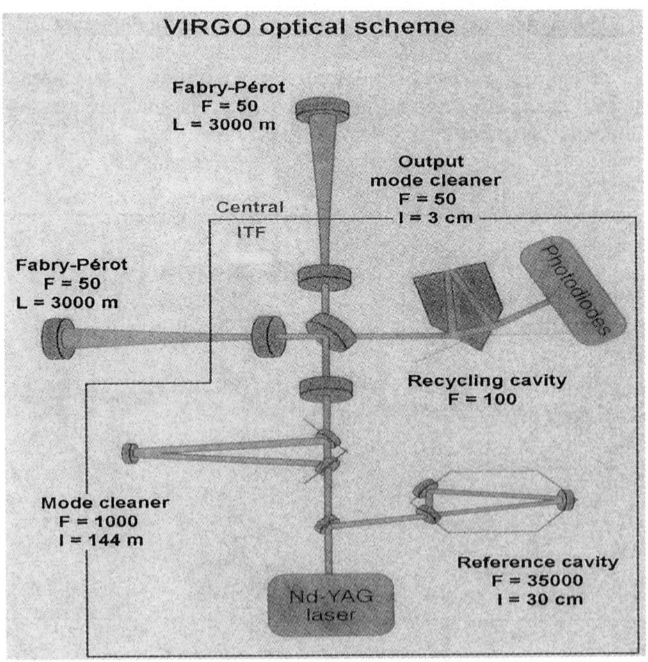

Fig.7

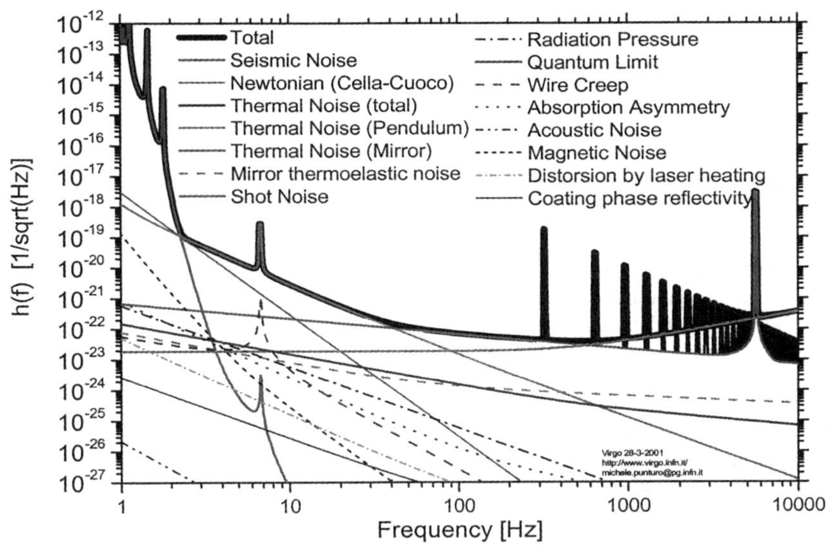

Fig. 8

The interferometric detector dominant noises.

Main contributions are shown in fig. 8 :

Seismic noise [13]: It is due to the continuous agitation of Earth crust due both to man and natural activity. If not attenuated it may give contributions as large as $\sim 10^{10}$ than GW signal.

Thermal noise[14]: Is created by the thermal agitation of atoms composing mirror substrate and his wire suspension. The thermal stockastic force is created by the materials internal friction, hence for having low thermal noise monocristal mirror and mirror suspension wires are needed.

Shot noise:Is due to the granular nature of photons who is creating a phase fluctuation $\Delta\varphi = N^{-1/2}$, where N is photon number. This noise is reduced increasing laser power.

Radiation Pressure noise: In FP cavities about 100 KW are stored; the photon numbert fluctuations produce a fluctuting force on the mirrors $\Delta f = Fd(N^{1/2} h\nu_L/c)/dt$, (where F is the cavity finesse and ν_L the laser frequency); this force produces a mirror position fluctuation.

Quantum Noise: represents the precision limit for mass M position measurement in a time t at the oserving frequency ν; it is obtained by minimizing the equivalent displacements due both to shot noise and radiation pressure noise. From this operation it follows the optimal laser power as function of M and ν.

Newtonian noise: Is produced by newtonian fluctuating force due to the seismic movement of Earth masses close to the test masses. Also wind is producing Newtonian noise. The Earth Newtonian noise can be reduced by putting a grid of seismometers around the detector and by reconstructing the produced noise.

The Laser and the frequency stabilization requirements

The laser system (see Fig. 9) is a 20 W ring with cristal Nd: YVO_4; this power unit is Injection-Locked to a 1W Monolitic YAG master prestabilized to a 30 cm triangular Ultra Low Expansion reference cavity. The prestabilization can reach 10^{-4} $Hz/Hz^{0.5}$@50 Hz.
Due to the unavoidable storage time mismatch of the two 3km cavities, for reducing laser frequency noise effects, it is necessary to stabilize the laser by using as a frequency reference the 3 km cavities common mode. This allows to reach the frequency fluctuations specifications (10^{-6} $Hz/Hz^{0.5}$@10 Hz), needed for reaching the design sensitivity

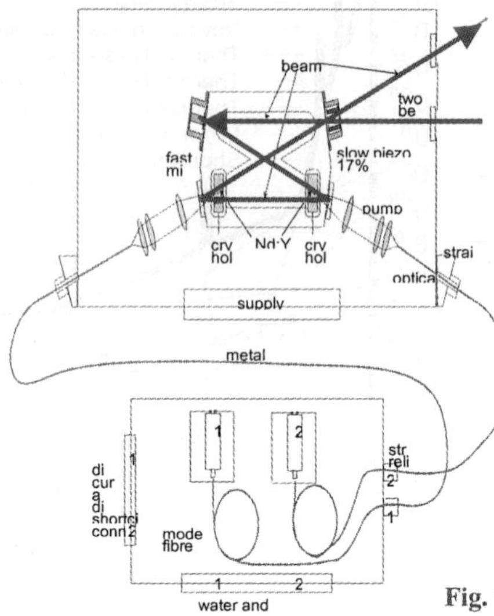

Fig. 9

The Virgo Superattenuators and the Ierarchical control

It is very important to underline the extremely high level of technology for Virgo design and specifications; Virgo mirror suspensions, called Superattenuators (SA), are unique and kill seismic noise in such a way to cross thermal noise at 4 Hz. The seismic performances are so good because every mechanical filter of SA isolate the rigid body in 6 dimensions, 3 translations and 3 rotations; this is the key to extreme seismic isolation i.e. for every mechanical filter there are not un-isolated degrees of freedom who may leave noise to go through.
In Fig. 10 the required specifications and in Fig. 11 the schematic diagram are shown. The inverted pendulum supports the inertial table actively isolated by means of 4 accelerometers, 3 on the horizontal plane and one in the vertical direction. This table supports the chain of mechanical filters together with marionette and mirror.
We have measured the upper limit of SA isolation at 4 Hz ; the value(see Fig. 12) perfectly stands on the attenuation transfer function(TF) obtained by measuring SA attenuation stage by stage and then multiplying the TF so obtained. SA structure allows to implement a very efficient mirror control; infact we may keep un-interrupted locking even in presence of large tidal SA displacement by distributing locking control signal at different levels of SA, as shown in Fig. 13.
With by the jerarchical control we obtain an optimal dinamicswise system; infact with SA the contro lsystem should not introduce noise down to 4 Hz.

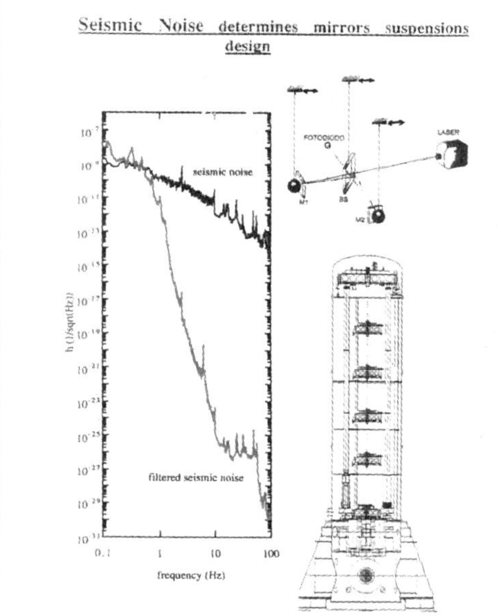

Fig.10

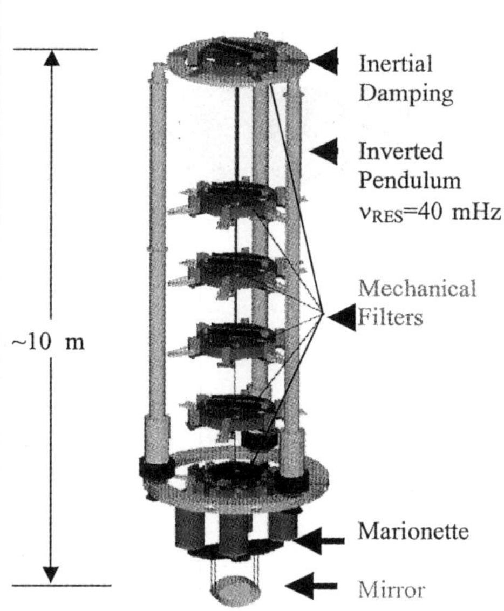

Fig.11

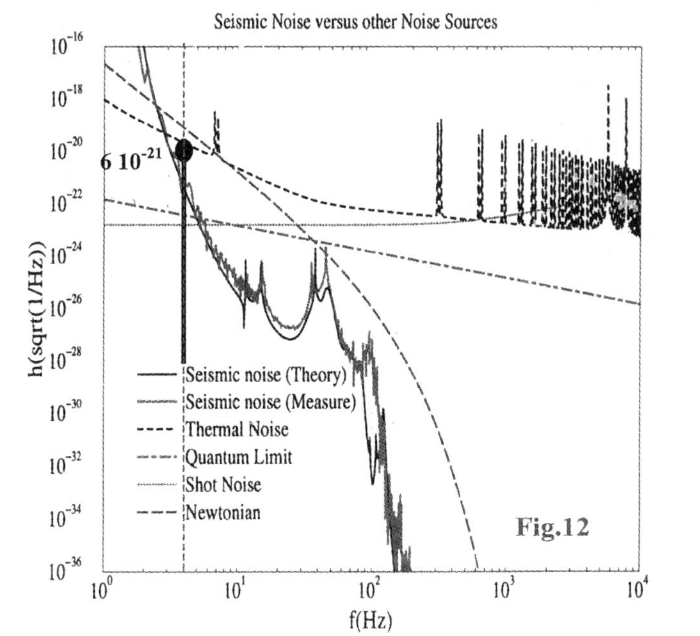

Fig.12

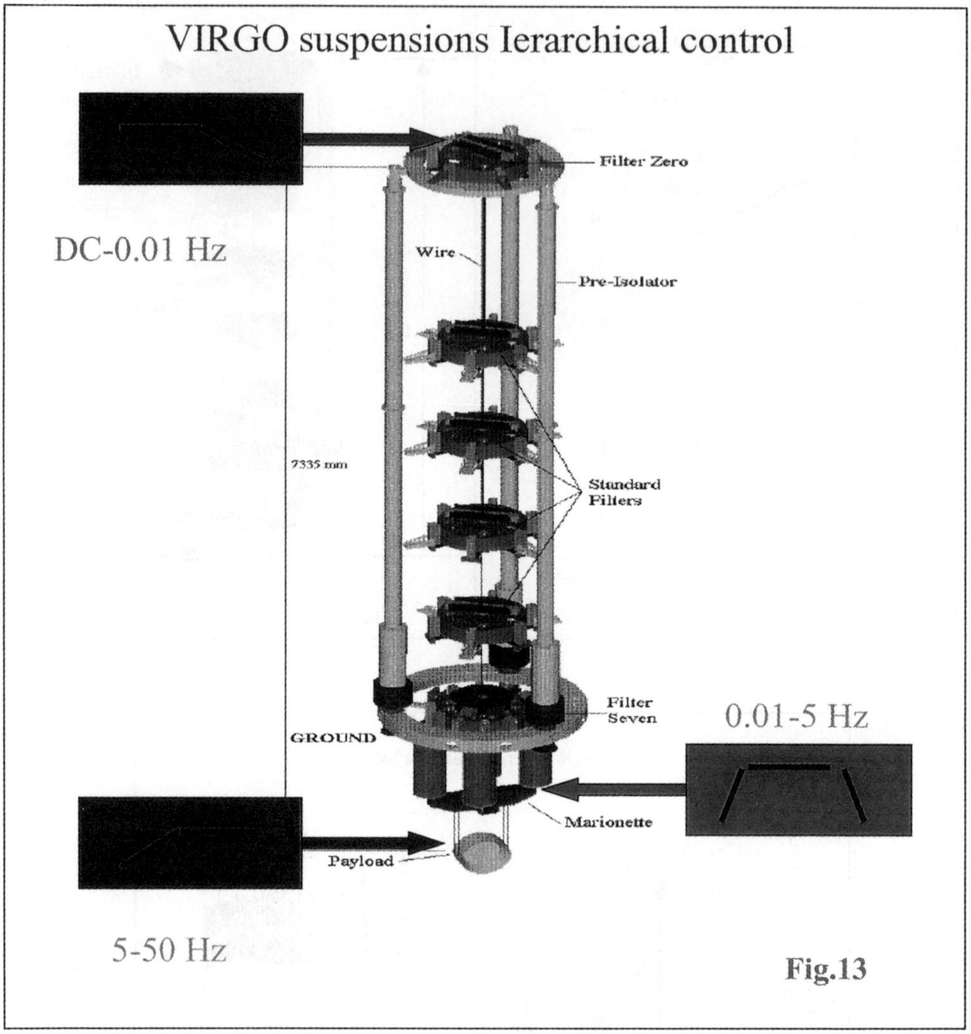

Fig.13

Results from the Central Interferometer (CITF) Commissioning

In the CITF operation the north and west input mirrors were used to create a short Michelson interferometer (see interferometer inside green part of Fig.7): in the latest runs also the recycling mirror was used.
Results from the CITF operation are shown in Fig. 14, and expressed in meter/Hz$^{-1/2}$.
Central Interferometer operation ended succesfully in july 2002 and the milestone required by INFN and CNRS was fulfilled.

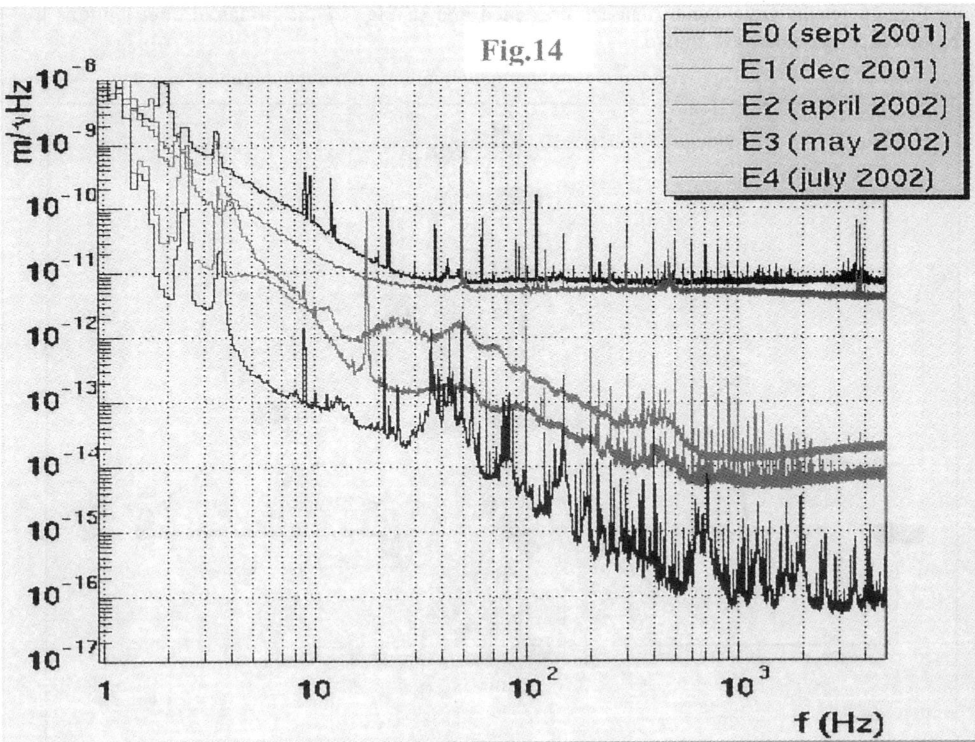

<u>Virgo Upgrading to 3km</u>

Starting on september 2002 till october 2003 the following upgradings to Virgo were performed:
1) Completion of vacuum.
2) Mounting Nord and West end Superattenuators.
3) Mounting of high quality mirrors.
4) New Mode-Cleaner Mirror suspension.
5) More performant local controls.
6) More performant DSP for suspension control.

In July 2003 the upgradings ended and the 23 July 2003 the Virgo inauguration was performed. In september 2003 Virgo 3km Commissioning Started and the recombined 3+3 km cavities were operated (see diagram in fig. 15 and note the misaligned power recycling mirror).

During these runs, C2 and C3, the longer was uninterrupted 28 hours, we had the following running conditions:
1)Laser frequency stabilized on the 3 km cavities common mode.
2)Cavities mirror were automatically aligned by using the (0.1) or (1,0) modes excitation.
3)The output mode cleaner, on the detection bench, needed for increasing black fringe contrast, was in operation.

In Fig. 16 results from these runs are presented and in Fig. 17 the principal contributions to recombined ITF noise are shown.

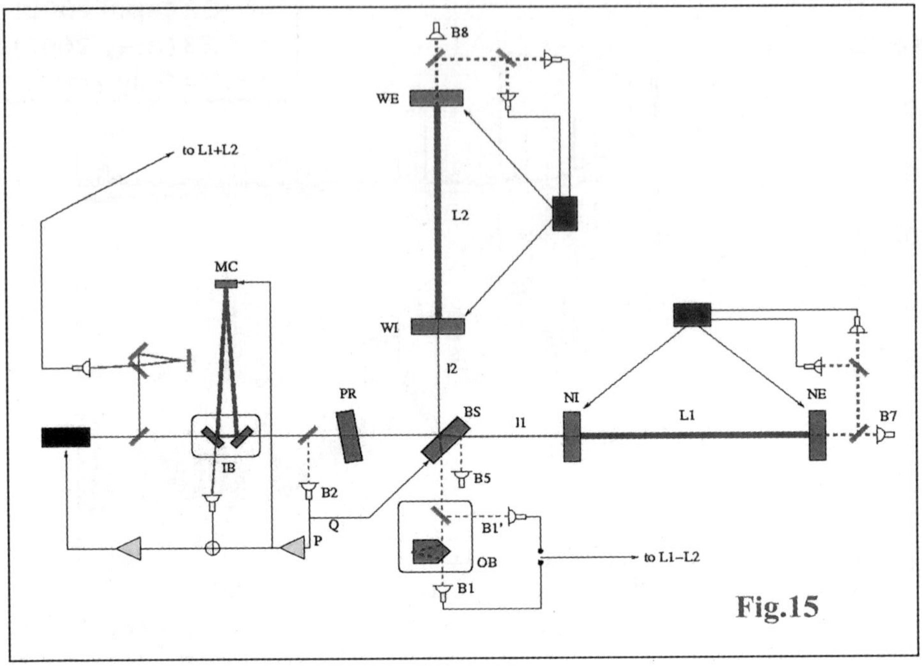

Fig.15

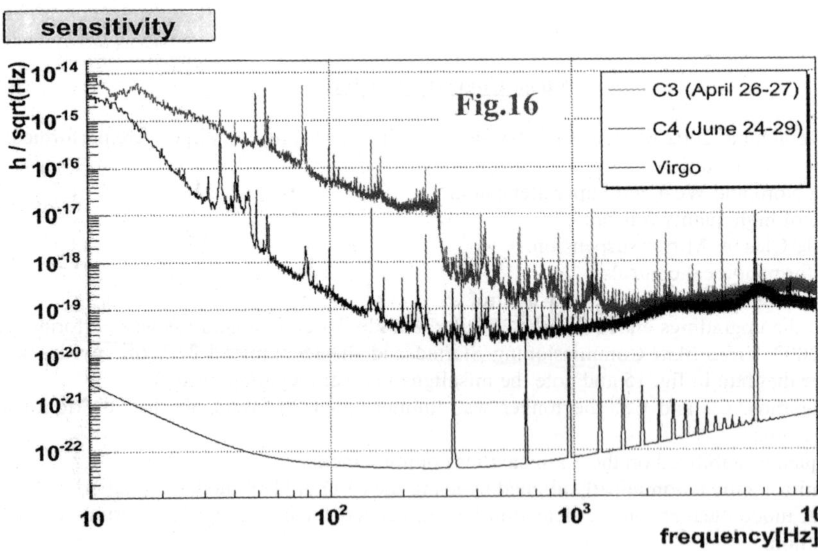

Fig.16

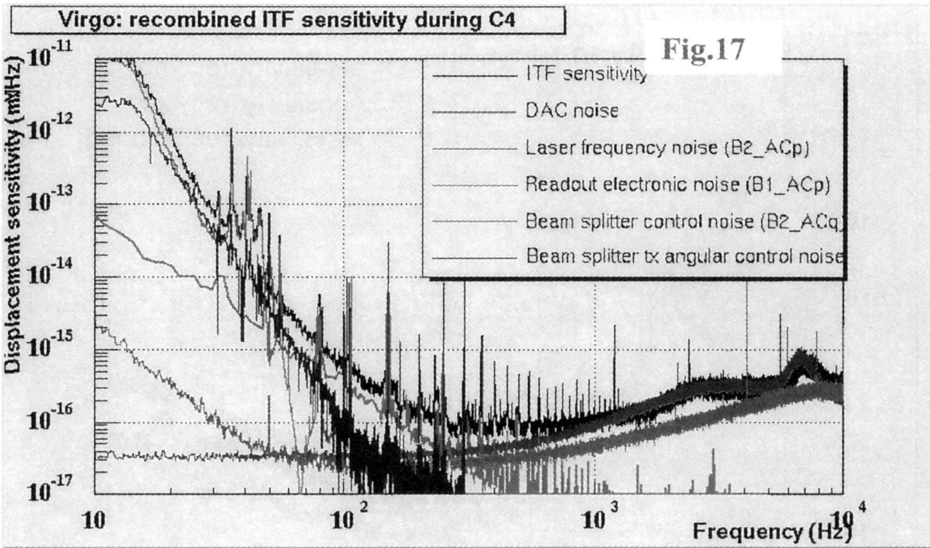

Fig.17

From the operation of recombined ITF and from the first attempt of recycling locking, we realized that the need of a Faraday isolator in the injection system
In September 2004 we started the recycled operation of Virgo; the Power Recycling mirror was aligned and the mirrors were aligned only with the local controls and not with the automatic alignment system (see Fig. 18), being the latter not yet ready for the recycled ITF. In Fig. 19 the sensitivities of C4 and of C5 both recombined and recycled, are shown.

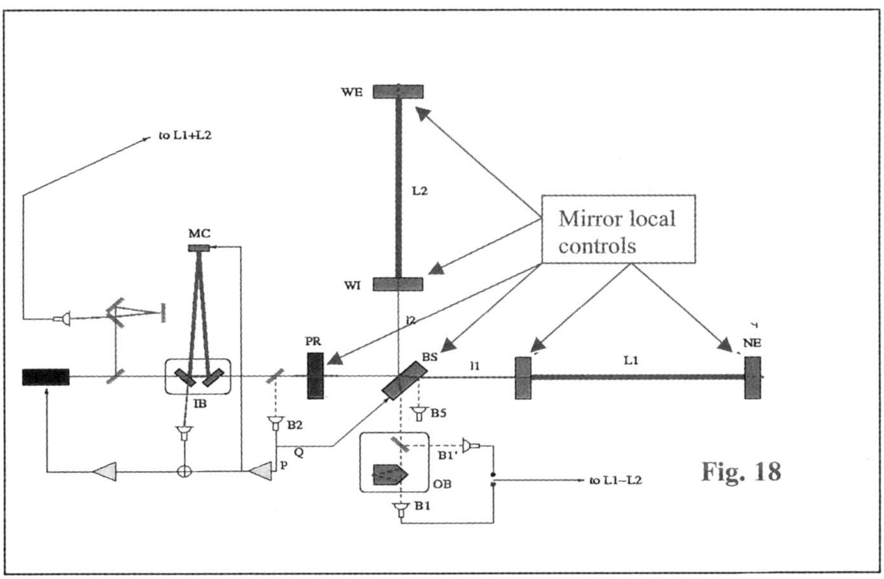

Fig. 18

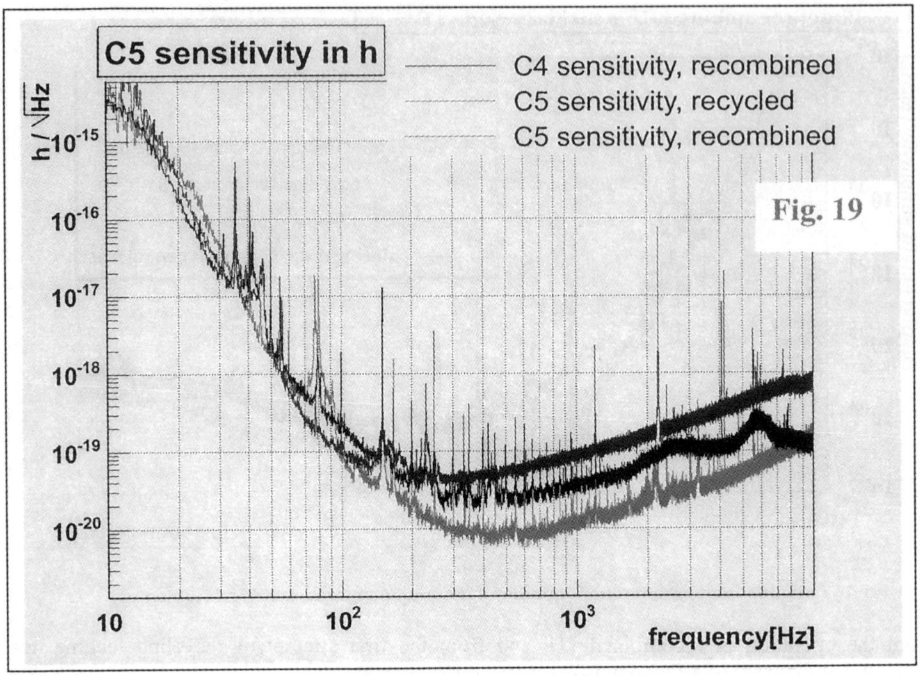

Fig. 19

Close future program

The main changes we have to perform during the first half 2005 are:

1)New injection bench equipped with Faraday isolator.

2)Change of mode cleaner mirror, being it too contaminated.

3)Substitution of power recycling curve-flat mirror with a flat-flat mirror.

After this changes are performed we can start to dig deeply into the noise hoping to reach the design sensitivity before end 2005.

References

(1) -A.Einstein, Sitz. Ber. Kon.Preus. **Ak. Wiss**, 688(1916); id. 154(1918)

(2) -C.W.Misner, K.S.Thorne, J.Wheeler, 1973, "*Gravitation*" (W.H.Freeman Eds.), S.Francisco.

-K.S.Thorne, 1987, "Gravitational radiation" in "*300 years of gravitation*" (Edited by S.W.Hawking and W.Israel, Cambridge University Press), New York (9) 330-458.

(3) -R.A.Hulse and J.H. Taylor, **Astrophys. J. Lett**. 195, 151 (1975)-H.V.Shapiro, S.A.Teukolsky, 1984, "*The physics of compact objects*" (J.Wiley and - -Sons, Eds.), New York.

(4)-Danilo Babusci (Frascati), Massimo Giovannini (Lausanne U.) "SENSITIVITY OF A VIRGO PAIR TO RELIC GW BACKGROUNDS". Aug 2000. 25pp. **Class.Quant.Grav**.17:2621-2633,2000
-Michele Maggiore "GRAVITATIONAL WAVE EXPERIMENTS AND EARLY UNIVERSE COSMOLOGY " **Phys.Rept**.331:283-367,2000
(5)-J. Weber, **Phys. Rev**.117,306 (1960)
-J. Weber, **Phys. Rev**. Lett. 22,1302(1969)
(6) -E. Coccia, 1995 "Resonant-mass gravitational wave detectors" in "*14th International Conference on General Relativity and Gravitational Physics*, Florence, August 6-12, 1995"
(7) -P. Astone et al. "Recent results of Nautilus" in *Gravitational Waves*, third Amaldi Conference, Pasadena Ca. 12-16 july 1999 , AIP Conference Proceedings # 523, Melville, New York).
(8) -M. Cerdonio et al. "Status Report of the Gravitational Wave Detector AURIGA" in " *Gravitational Waves*, third Amaldi Conference, Pasadena Ca. 12-16 july 1999 , AIP Conference Proceedings # 523, Melville, New York).
(9) -D. Blair et al. "Niobe: improved noise temperature and Background Noise Suppression" in *Gravitational Waves,* third Amaldi Conference, Pasadena Ca. 12-16 july 1999 , AIP Conference Proceedings # 523, Melville, New York).
(10) - J.J. Johnson et al. in *Gravitational Wave Experiments*, Proceedings of the First Edoardo Amaldi Conference, Frascati 1994,(World Scientific, Singapore, 1995)"
(11) - A. Giazotto "Interferometric Detection of of Gravitational Waves", **Physics Rept**, 182, 6, 1989.
(12) - J.Y.Vinet, B.Meers, C.N.Man, A.Brillet**, Phys. Rev**. *D* 38 (1988) 433.
- A.Brillet, *Ann. de Phys.* 10 (1985) 219.
(13) - R. Del Fabbro et al., 1988,Phys. Lett. A132,237
-S.Braccini et al., 1997, "*Design of the superattenuators for the VIRGO construction"* VIRGO Internal Report (VIR-TRE-PIS-4600-134).
(14) - H.B.Callen, R.F.Greene, **Phys. Rev**. 86 (1952) 702.

Chapter 2: Advanced interferometers for gravitational wave detection: Quantum non-demolition techniques.

Some considerations on 2 very important quantum noises: Shot noise and Radiation pressure

1)Shot Noise: Uncertainty principle gives to a photon beam $\Delta\varphi\Delta N>1$

The phase of a coherent light beam fluctuates as

$$\varphi \geq \frac{1}{\sqrt{N}} = \sqrt{\frac{h\nu}{Wt}}$$

The phase produced by GW signal is:

$$\varphi_{sig}(t) = \frac{\omega hFL}{c\sqrt{1+\left(F\Omega_g L/c\right)^2}}$$

Where F is the cavity finesse

The measurability condition is $\tilde{\varphi}_{sig} > \tilde{\varphi}$

$$\tilde{h} > \frac{\lambda}{4\pi}\sqrt{\frac{h\nu}{W}}\,\frac{\sqrt{1+\left(\frac{\Omega FL}{c}\right)^2}}{FL}$$

i.e shot noise decreases by increasing $W^{1/2}F$

2)Radiation Pressure Noise

The photon number fluctuations create a fluctuating momentum on the mirrors of the FP cavities :

$$\delta P \approx F\frac{h\nu}{c}\sqrt{N} = F\frac{h\nu}{c}\sqrt{\frac{Wt}{h\nu}}$$

The spectral force on the mirrors is:

$$\frac{\delta\tilde{P}}{\delta t} \approx F\frac{h\nu}{c}\sqrt{\frac{W}{h\nu}} = \frac{F}{c}\sqrt{h\nu W}$$

For the measurability condition this force should be smaller than Riemann force:

$$F_{Riem.} = \frac{1}{2}M\tilde{h}\Omega^2 L > \frac{F}{c}\sqrt{h\nu W} \qquad \tilde{h} > \frac{2F}{cM\Omega^2 L}\sqrt{h\nu W}$$

The measurability cond. for Shot noise and Radiation Pressure noise is:

$$\tilde{h}^2 > \left(\frac{2}{LM\Omega^2 c}\right)^2 h\nu WF^2 + \frac{\lambda^2}{16\pi^2}\frac{h\nu}{WF^2}\,\frac{1+\left(\frac{\Omega FL}{c}\right)^2}{L^2}$$

The term WF^2 produces dramatic effects on the sensitivity; minimizing h with respect to W we obtain the Standard Quantum limit:

$$\begin{cases} W = \dfrac{\lambda M\Omega^2 c}{16\pi F^2} \\ \tilde{h}_{\substack{\text{Standar}\\\text{Quantum}\\\text{Limit}}} \approx \dfrac{1}{L\Omega}\sqrt{\dfrac{\hbar}{M}} \end{cases}$$

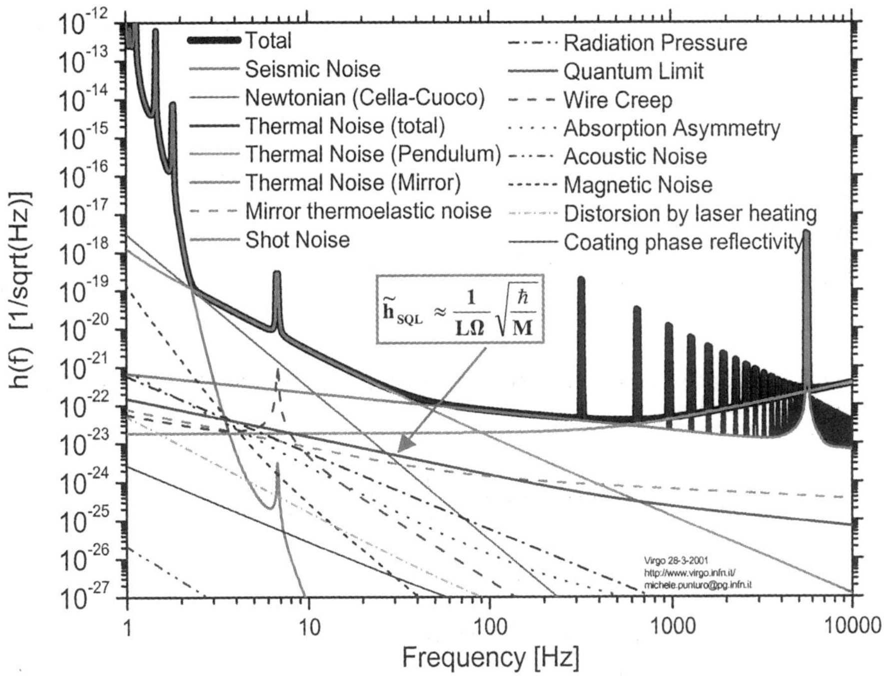

SQL can not be reduced in conventional interferometers; in the existing interferometers we measure the sum of Shot noise and Radiation pressure fluctuation i.e. We need to find a clever way to measure only phase shifts without pollution from Photon number fluctuations.
In the following I'll give some ideas on how SQL can be perhaps circumvented.
In Quantum Electrodynamics it is shown that Electric field E_x and magnetic field H_y of a z-propagating em wave do not commute and satisfy the following commutation relation:

$$\left[E_x(r_1,t), B_y(r_2,t)\right] = \frac{\hbar}{i\varepsilon_0}\frac{\partial}{\partial z_1}\delta^3(r_1 - r_2)$$

We may expand the vector potential and the Hamiltonian as a sum of creation and destruction operators:

$$A(r,t) \propto \sum_{k,s} a_{k,s}\varepsilon_{k,s}e^{i(kr-\omega t)} + h.c.$$

From E&H commutation relations we obtain

$$\left[a_{k,s}, a^{+}_{q,j}\right] = \delta_{k,q}\delta_{i,j}$$

In interferometric detectors, GW produce sidebands at frequency $\omega_0 \pm \Omega$, where $2\pi.10<\Omega<2\pi.10000$ is the acoustical band; this process involves the emission of two correlated photons of frequency $\omega_0 \pm \Omega$:

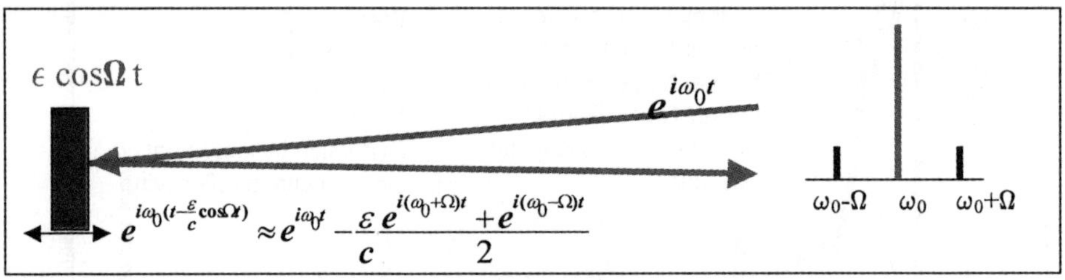

Consequently it is appropriate to deal with two photon processes. The positive frequency of the fluctuating electric field is:

$$E^{(+)}_{in} = Ce^{-i\omega_0 t}\int_0^\infty (a_+e^{-i\Omega t} + a_-e^{i\Omega t})\frac{d\Omega}{2\pi} \qquad C = \sqrt{\frac{2\pi\hbar\omega_0}{Ac}}$$

Where $a_\pm = a_{\omega\pm\Omega}$ satisfy the following commutation relations:

$$\left[a_+, a^{'+}_+\right] = \left[a_-, a^{'+}_-\right] = 2\pi\delta(\Omega - \Omega')$$

the prime means that a is evaluated at $\omega\pm\Omega'$. The total fluctuating electric field is:

$$E_{in} = E^{(+)}_{in} + \left(E^{(+)}_{in}\right)^+ = C\left[e^{-i\omega_0 t}\int_0^\infty (a_+e^{-i\Omega t} + a_-e^{i\Omega t})\frac{d\Omega}{2\pi} + h.c.\right] =$$
$$= C\left[a_0(t)\cos\omega_0 t + a_{\pi/2}(t)\sin\omega_0 t\right]$$

where

$$a_0(t) = \int_0^\infty \left[\frac{a_+ + a^+_-}{\sqrt{2}}e^{-i\Omega t} + \frac{a_- + a^+_+}{\sqrt{2}}e^{i\Omega t}\right]\frac{d\Omega}{2\pi}$$

$$a_{\pi/2}(t) = \int_0^\infty \left[\frac{a_+ - a^+_-}{i\sqrt{2}}e^{-i\Omega t} + \frac{a_- - a^+_+}{i\sqrt{2}}e^{i\Omega t}\right]\frac{d\Omega}{2\pi}$$

If we include the carrier, the total field is defined:

$$E_{in} = C\left[(\alpha + a_{0,in}(t))\cos\omega_0 t + a_{\pi/2,in}(t)\sin\omega_0 t\right]$$

Where $\alpha=(2W/h\nu_0)^{1/2}$ is the amplitude of the carrier in a coherent state and W is the laser power entering the cavity.See Fig.1.

We may rewrite E_{in} in more self explicative way (only first order terms in $a_{0,\ \pi/2}$ are kept):

$$E_{in}(t) \approx \alpha(1+\frac{a_{0,in}}{\alpha})\cos(\omega_0 t+\frac{a_{\pi/2,in}}{\alpha})$$

From eq.1 it is evident the role of $a_{0,in}$ and a $_{\pi/2,in}$ in creating radiation pressure and phase fluctuations respectively..

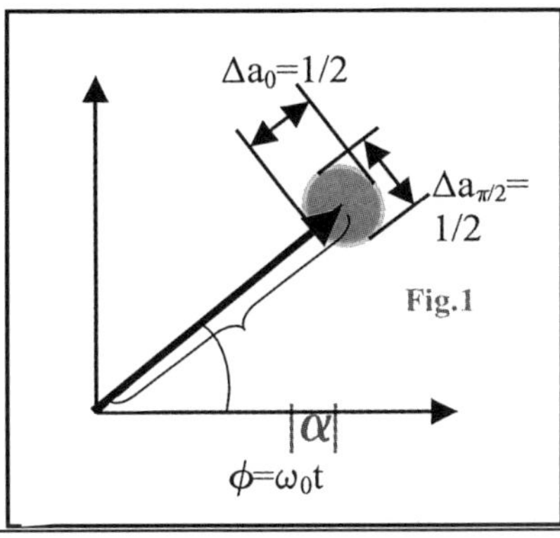

Fig.1

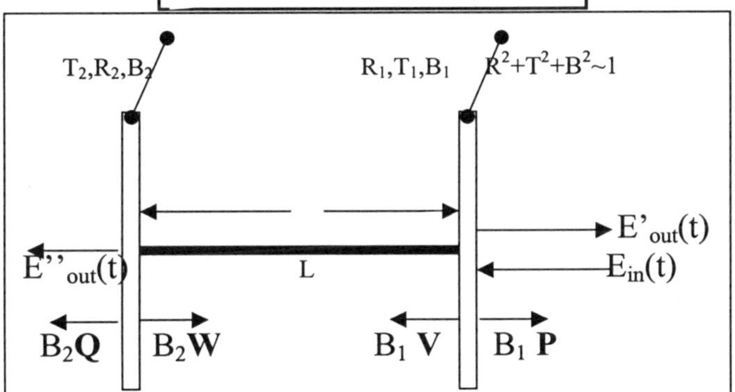

Fig.2

It is important to describe the scattering of light by a Fabry-Perot cavity with suspended mirrors as shown in Fig.2 B_1 and B_2 are the losses of mirror 1 and 2 respectively; W,V, P and Q are the fluctuating vacuum fields entering and leaving the cavity trough cavity losses.We may now define E_{in},, the field entering the FP cavity, and E'_{out} the reflected field:

$$E_{in} = C[(\alpha + a_{0,in}(t))\cos\omega_0 t + a_{\pi/2,in}(t)\sin\omega_0 t]$$
$$E'_{out} = C[(\alpha_0 + \Gamma_0 + a_{0,out}(t))\cos\omega_0 t + (\alpha_{\pi/2} + \Gamma_{\pi/2} + a_{\pi/2,out}(t))\sin\omega_0 t] \qquad \textbf{Eq. 2}$$

Where α_0 and $\alpha_{\pi/2}$ are carrier's quadratures and Γ_0 and $\Gamma_{\pi/2}$ are the classical contribution to the phase shift due to mirror motion.
With reference to Fig. 2 and to Eq.2, we obtain for E'_{out}:

$$\begin{pmatrix}\alpha_0\\ \alpha_{\pi/2}\end{pmatrix} = \begin{pmatrix}\frac{1}{2}\Lambda(\omega_0,0,R_1,R_2,L) & -\frac{R_2T_1^2}{2i}Y(\omega_0,0,R_1,R_2,L,-)\\ \frac{R_2T_1^2}{2i}Y(\omega_0,0,R_1,R_2,L,-) & \frac{1}{2}\Lambda(\omega_0,0,R_1,R_2,L)\end{pmatrix}\begin{pmatrix}\alpha\\ 0\end{pmatrix} \qquad \textbf{Eq. 3}$$

$$\begin{pmatrix}\Gamma_0\\ \Gamma_{\pi/2}\end{pmatrix} = \begin{pmatrix}0 & -R_2T_1^2\frac{\alpha\omega_0}{ic}Z(\omega_0,\Omega,R_1,R_2,L,-)\\ 0 & -R_2T_1^2\frac{\alpha\omega_0}{c}Z(\omega_0,\Omega,R_1,R_2,L,+)\end{pmatrix}\begin{pmatrix}0\\ e^{-i\Omega\frac{L}{c}}\delta\tilde{X}_1 - \delta\tilde{X}_2\end{pmatrix}_{NQ} - \begin{pmatrix}0\\ R_1\frac{2\alpha\omega_0}{c}\delta\tilde{X}_1\end{pmatrix}_{NQ}$$

$$\begin{pmatrix}a_{0,out}\\ a_{\pi/2,out}\end{pmatrix} = \begin{pmatrix}\frac{1}{2}\Lambda(\omega_0,\Omega,R_1,R_2,L) & -\frac{R_2T_1^2}{2i}Y(\omega_0,\Omega,R_1,R_2,L,-)\\ \frac{R_2T_1^2}{2i}Y(\omega_0,\Omega,R_1,R_2,L,-) & \frac{1}{2}\Lambda(\omega_0,\Omega,R_1,R_2,L)\end{pmatrix}\begin{pmatrix}a_{0,in}\\ a_{\pi/2,in}\end{pmatrix} +$$

$$+\begin{pmatrix}0 & -R_2T_1^2\frac{\alpha\omega_0}{ic}Z(\omega_0,\Omega,R_1,R_2,L,-)\\ 0 & -R_2T_1^2\frac{\alpha\omega_0}{c}Z(\omega_0,\Omega,R_1,R_2,L,+)\end{pmatrix}\begin{pmatrix}0\\ e^{-i\Omega\frac{L}{c}}\delta\tilde{X}_1 - \delta\tilde{X}_2\end{pmatrix}_{Quant} - \begin{pmatrix}0\\ R_1\frac{2\alpha\omega_0}{c}\delta\tilde{X}_1\end{pmatrix}_{Quant}$$

$$+T_1R_2\begin{pmatrix}\frac{1}{2}Y(\omega_0,\Omega,R_1,R_2,L,+) & -\frac{1}{2i}Y(\omega_0,\Omega,R_1,R_2,L,-)\\ \frac{1}{2i}Y(\omega_0,\Omega,R_1,R_2,L,-) & \frac{1}{2}Y(\omega_0,\Omega,R_1,R_2,L,+)\end{pmatrix}\left\{B_1\begin{pmatrix}V_0\\ V_{\pi/2}\end{pmatrix} - R_1B_2e^{-i\Omega\frac{L}{c}}\begin{pmatrix}W_0\\ W_{\pi/2}\end{pmatrix}\right\} + B_1\begin{pmatrix}P_0\\ P_{\pi/2}\end{pmatrix}$$

Where NQ means Non-Quantic i.e. that part of Ã0 and Ãδ/2 not containing to first order both a0 and aδ/2, while the functions Y,Ë and Z are defined as follows:

$$Y(\omega_1,\Omega,R_1,R_2,L,\pm)=\frac{e^{2i(\omega_1-\Omega)\frac{L}{c}}}{1-R_1R_2e^{2i(\omega_1-\Omega)\frac{L}{c}}}\pm\frac{e^{-2i(\omega_1+\Omega)\frac{L}{c}}}{1-R_1R_2e^{-2i(\omega_1+\Omega)\frac{L}{c}}}$$

$$\Lambda(\omega_1,\Omega,R_1,R_2,L)=T_1^2R_2Y(\omega_1,\Omega,R_1,R_2,L,+)-2R_1=$$

$$=\frac{(T_1^2+R_1^2)R_2e^{2i(\omega_1-\Omega)\frac{L}{c}}-R_1}{1-R_1R_2e^{2i(\omega_1-\Omega)\frac{L}{c}}}+\frac{(T_1^2+R_1^2)R_2e^{-2i(\omega_1+\Omega)\frac{L}{c}}-R_1}{1-R_1R_2e^{-2i(\omega_1+\Omega)\frac{L}{c}}}$$

$$Z(\omega_1,\Omega,R_1,R_2,L,\pm)=\frac{e^{i(2\omega_1-\Omega)\frac{L}{c}}}{(1-R_1R_2e^{2i\omega_1\frac{L}{c}})(1-R_1R_2e^{2i(\omega_1-\Omega)\frac{L}{c}})}\pm$$

$$\pm\frac{e^{i(-2\omega_1-\Omega)\frac{L}{c}}}{(1-R_1R_2e^{-2i\omega_1\frac{L}{c}})(1-R_1R_2e^{-2i(\omega_1+\Omega)\frac{L}{c}})}$$

The Origin of Ponderomotive Squeezing

Putting ω_0 the resonance frequency ($e^{2i\omega_0 L/c}=1$) , ω the detuning frequency, we obtain for the Intra Cavity (IC) amplitudes:

$$\begin{pmatrix}\alpha_0\\ \alpha_{\pi/2}\end{pmatrix}_{IC}=\frac{R_2T_1}{2}\begin{pmatrix}Y(\omega_0-\omega,0,R_1,R_2,L,+) & -\frac{1}{i}Y(\omega_0-\omega,0,R_1,R_2,L,-)\\ \frac{1}{i}Y(\omega_0-\omega,0,R_1,R_2,L,-) & Y(\omega_0-\omega,0,R_1,R_2,L,+)\end{pmatrix}\begin{pmatrix}\alpha\\ 0\end{pmatrix}$$

$$\begin{pmatrix}a_0\\ a_{\pi/2}\end{pmatrix}_{IC}=\frac{R_2T_1}{2}\begin{pmatrix}Y(\omega_0-\omega,\Omega,R_1,R_2,L,+) & -\frac{1}{i}Y(\omega_0-\omega,\Omega,R_1,R_2,L,-)\\ \frac{1}{i}Y(\omega_0-\omega,\Omega,R_1,R_2,L,-) & Y(\omega_0-\omega,\Omega,R_1,R_2,L,+)\end{pmatrix}\begin{pmatrix}U_{0,in}\\ U_{\pi/2,in}\end{pmatrix}$$

Where

$$Y(\omega_0-\omega,\Omega,R_1,R_2,L,\pm)=\frac{e^{2i(\omega_0-\omega-\Omega)\frac{L}{c}}}{1-R_1R_2e^{2i(\omega_0-\omega-\Omega))\frac{L}{c}}}\pm\frac{e^{-2i(\omega_0-\omega+\Omega)\frac{L}{c}}}{1-R_1R_2e^{-2i(\omega_0-\omega+\Omega)\frac{L}{c}}}$$

$$\begin{pmatrix}U_{0,in}\\ U_{\pi/2,in}\end{pmatrix}=\left\{\begin{pmatrix}a_{0,in}\\ a_{\pi/2,in}\end{pmatrix}+\frac{B_1}{T_1}\begin{pmatrix}V_0\\ V_{\pi/2}\end{pmatrix}+\frac{R_1B_2}{T_1}e^{-2i\Omega\frac{L}{c}}\begin{pmatrix}W_0\\ W_{\pi/2}\end{pmatrix}\right\}\qquad\textbf{Eq. 4}$$

Since we are interested in reflective FP, putting $R_2 \sim 1$, $T_2=0$ $T=T_1$ and $R_1=R$, from eq.s 2, 3, 4 and Fig.2 and considering that $T^2+R^2 \approx 1$ and $1+R \sim 2$, it follows that at resonance ($e^{2iù0L/c}=1$, ù=0) the IC power is :

$$W_{IC} = \frac{h\nu_0 T^2}{2}\left[\frac{\alpha}{1-R} + \left(a_{0,in} + \frac{B_1 V_0 - B_2 R W_0}{T}\right)\frac{e^{-2i\Omega L/c}}{1-\mathrm{Re}^{-2i\Omega L/c}}\right]^2$$

$$\delta\widetilde{W}_{IC} = h\nu_0\left[T^2\,\frac{\alpha\left(a_{0,in} + B_1\dfrac{BV_0 - B_2 R W_0}{T}\right)}{1-R}\,\frac{e^{-2i\Omega L/c}}{1-\mathrm{Re}^{-2i\Omega L/c}}\right] \approx \qquad \textbf{Eq. 4.1}$$

$$\sim \sqrt{2(1+R)^2 W h\nu_0}\left(a_{0,in} + \frac{B_1 V_0 - B_2 R W_0}{T}\right)\frac{e^{-2i\Omega L/c}}{1-\mathrm{Re}^{-2i\Omega L/c}}$$

Since RP fluctuations are producing displacements fluctuations δX_{1RP} and δX_{2RP} we define:

$$\begin{aligned}
\delta\widetilde{X}_1 &= \frac{\delta W_{IC}}{Mc\Omega^2}(1+e^{2i\Omega L/c}) + \left(\delta\widetilde{X}_1\right)_{NQ} \\
\delta\widetilde{X}_2 &= -\frac{\delta W_{IC}}{Mc\Omega^2}2e^{i\Omega L/c} + \left(\delta\widetilde{X}_2\right)_{NQ} \qquad \textbf{Eq. 5} \\
\delta\widetilde{X}_1 e^{-i\Omega L/c} - \delta\widetilde{X}_2 &= \frac{\delta W_{IC}}{Mc\Omega^2}\left(e^{-i\Omega L/c} + 3e^{i\Omega L/c}\right) + \left(e^{-i\Omega L/c}\delta\widetilde{X}_1 - \delta\widetilde{X}_2\right)_{NQ}
\end{aligned}$$

Where M is the mirror mass and we have assumed the observation frequency Ω to be much larger than mirror suspension mode frequencies. For the reflected field E'_{out} , at resonance, from eqs.2, 4, 5 we obtain:

K

$$\begin{cases}
a_{0,out} = \left[a_{0,in} + T(B_1 V_0 - B_2 R W_0)\dfrac{e^{-2i\Omega L/c}}{e^{-2i\Omega L/c} - R}\right]e^{2i\beta} + B_1 P_0 \\
a_{\pi/2,out} \approx \left[a_{\pi/2,in} + T(B_1 V_{\pi/2} - B_2 R W_{\pi/2})\dfrac{e^{-2i\Omega L/c}}{e^{-2i\Omega L/c} - R}\right]e^{2i\beta} + B_1 P_{\pi/2} - \\
-\left[a_{0,in} + T(B_1 V_0 - B_2 R W_0)\dfrac{e^{-2i\Omega L/c}}{e^{-2i\Omega L/c} - R}\right]\dfrac{16W\omega_0\left(e^{-i\Omega L/c} + 3e^{i\Omega L/c}\right)}{Mc^2\Omega^2(1-R)^2}\,\dfrac{e^{2i\beta}}{(1+\dfrac{4\Omega^2 L^2}{c^2(1-R)^2})}
\end{cases}$$

$$\begin{pmatrix}\Gamma_0 \\ \Gamma_{\pi/2}\end{pmatrix} = -\frac{2\pi}{\lambda}\sqrt{\frac{32W}{h\nu_0}}\,\frac{e^{i\Omega L/c}}{1-\mathrm{Re}^{-2i\Omega L/c}}\left(e^{i\Omega L/c}\delta\widetilde{X}_1 - \delta\widetilde{X}_2\right)_{NQ}\begin{pmatrix}0 \\ 1\end{pmatrix} \qquad \textbf{Eq. 6}$$

Because of the strong $1/\Omega^2$ dependence favoring low Ω we neglect in eq.6 the term $4\,\Omega^2 L^2/c^2$ with respect $(1\text{-}R)^2$; we finally obtain in matrix form :

where

$$K = \frac{64W\omega_0}{Mc^2\Omega^2(1-R)^2} \qquad \textbf{Eq. 7}$$

Let us suppose to have W~10 w, M~2.10^1 kg, ω_0=2 10^{15} Hz, 1-R=10^{-5} ;with these parameters K becomes:

$$K = \frac{32.10.2.10^{15}}{2.10^1 9.10^{16}\Omega^2 10^{-10}} \approx \frac{4.10^9}{\Omega^2}$$

i.e. K is very large.
Due to ponderomotive squeezing the state out has different uncertainty contour than state in. See Fig.3.

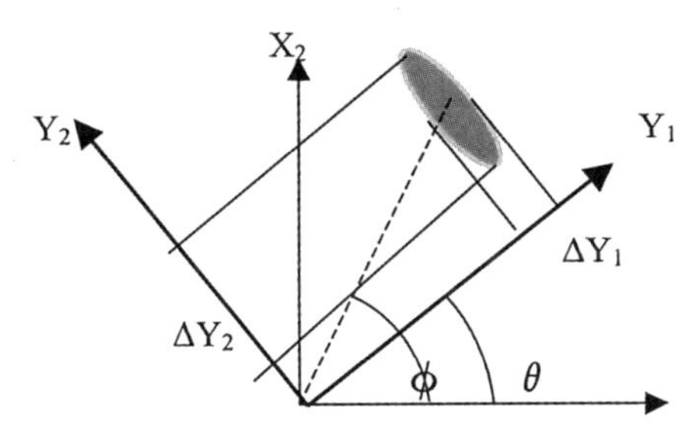

Fig.3

Radiation Pressure and Shot noise in a Michelson Interferometer

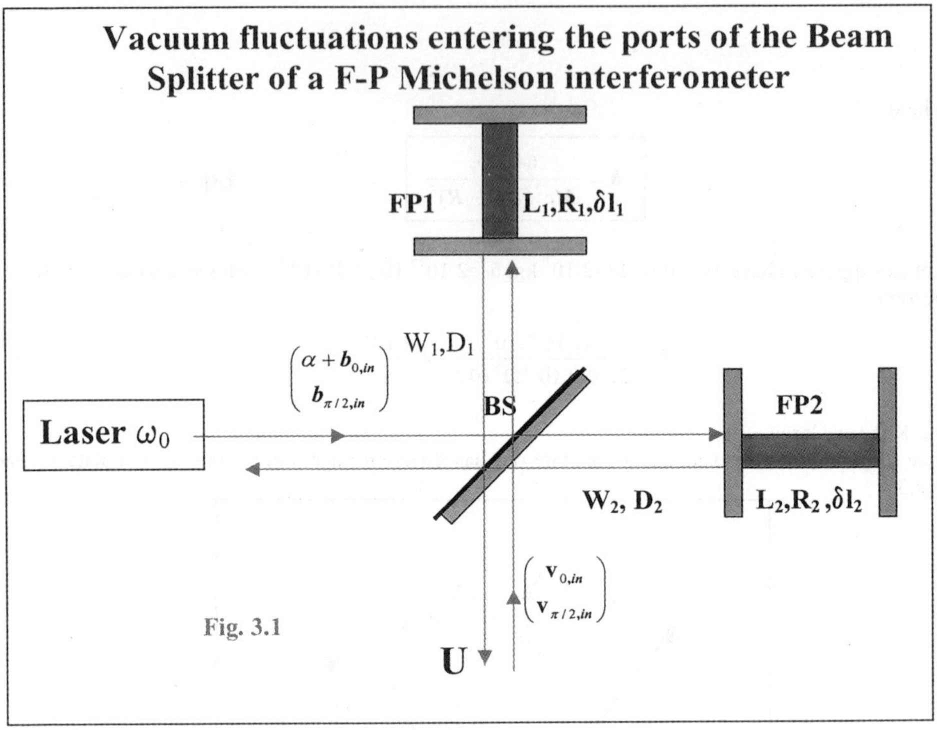

Fig. 3.1

With reference to Fig. 3.1, the FP cavities have length $L_{1,2}$ and entrance mirror reflectivity $R_{1,2}$ respectively. The far mirrors have $R_2= 1$ and the two entrance mirrors have distance $D_{1,2}$ from BS respectively. The Laser has power W and circular frequency ω_0 and $W_1=W_2=W/2$ are the powers entering the FP cavities; the quadrature operators $b_{0,\pi/2}$ and $v_{0,\pi/2}$ are vacuum fluctuations entering the laser line and the unused BS ports respectively. With reference to eq.s 7 we obtain for the output field U:

$$U=\frac{1}{2}e^{2i\omega_0 D_1/c}\begin{pmatrix} b_{0,in}+v_{0,in} \\ b_{\pi/2,in}+v_{\pi/2,in}-(b_{0,in}+v_{0,in})K_1-Q_1\delta\ell_1 \end{pmatrix}e^{2i\beta_1}+$$

$$+\frac{1}{2}e^{2i\omega_0 D_2/c}\begin{pmatrix} -b_{0,in}+v_{0,in} \\ -b_{\pi/2,in}+v_{\pi/2,in}-(-b_{0,in}+v_{0,in})K_2+Q_2\delta\ell_2 \end{pmatrix}e^{2i\beta_2} \qquad \textbf{Eq. 7.1}$$

$$J = 1,2$$

$$\delta\ell_i = e^{-i\Omega\frac{L_i}{c}}\delta X_i^1-\delta X_i^2 \quad i=1,2$$

$$\beta_J = 2\Omega L_J/(c(1-R_J))$$

$$K_J=\frac{32W_J\omega_0}{Mc^2\Omega^2(1-R_J)^2}$$

$$Q_J=\frac{2\pi}{\lambda}\sqrt{\frac{32W_J}{h\nu_0}}\frac{e^{i\Omega L_J/c}}{1-R_Je^{-2i\Omega L_J/c}}$$

When $D_1=D_2+n\lambda/2$ and the two FP cavities are equal, we obtain:

$$U = e^{2i\omega_0 D_1/c}\begin{pmatrix} v_{0,in} \\ v_{\pi/2,in} - v_{0,in}K_1 - Q_1 hL \end{pmatrix} e^{2i\beta_1} \quad \textbf{Eq. 7.3}$$

In the configuration in which U contains the GW signal ($\delta X_1 - \delta X_2 = hL$), we obtain:

$$U = e^{2i\omega_0 D_1/c}\begin{pmatrix} v_{0,in} \\ v_{\pi/2,in} - v_{0,in}K_1 - Q_1(\delta\tilde{X}_1 - \delta\tilde{X}_2) \end{pmatrix} e^{2i\beta_1} \quad \textbf{Eq. 7.2}$$

Eqs.7.2,7.3 clearly show that the only contribution to the noise is given by the vacuum quadrature operators $v_{0,\pi/2}$ entering the BS from the dark fringe port; in this symmetric configuration there is no noise contribution from $b_{0,\pi/2}$. <u>This is a foundamental result showing that we may reduce shot noise by injecting in the black port of the BS a squeezed vacuum.</u>

The Squeezed State Ellipse Contour

We know that the coherent two photon state annihilation operators $a_{0,in}$ and $a_{\pi/2,in}$ have fluctuations $\Delta a_{0,in} = \Delta a_{\pi/2,in} = 1/2$, where Δa means square root of a variance.For evaluating the fluctuations, omitting the signal δX_2^{ext} , the unessential (for this demonstration) $e^{2i\beta}$ factor, and the external vacuums, with ref. to eq.7, we obtain:

$$\begin{cases} \Delta a_{0,out} = \Delta a_{0,in} \\ \Delta a_{\pi/2,out} = \Delta a_{\pi/2,in} - K\Delta a_{0,in} \end{cases}$$

Let us define

$$\begin{cases} \Delta a_{\pi/2,in} = \rho\sin\varphi \\ \Delta a_{0,in} = \rho\cos\varphi \end{cases} \quad \textbf{Eq. 8}$$

and evaluate the fluctuation contour for $a_{0,out}$ and $a_{\pi/2,out}$ as a function of ϕ:

$$\begin{cases} R^2(\varphi) = (\Delta a_{\pi/2,out})^2 + (\Delta a_{0,out})^2 = \rho^2\left(1 + K^2/2 + \sqrt{\dfrac{K^4}{4} + K^2}\cos(2\varphi + \psi)\right) \\ tg\psi = \dfrac{2}{K} \end{cases}$$

From eq.8 it follows $\rho^2 = 1/2$; we may then evaluate the two ellipse radii to be:

$$\begin{cases} R^2_{\min}(\varphi) = \lim\limits_{K\to\infty} \rho^2\left(1 + K^2/2 - \sqrt{\dfrac{K^4}{4} + K^2}\right) \sim \dfrac{1}{2K^2} = \dfrac{1}{2}e^{-2r} \\ R^2_{\max}(\varphi) = \lim\limits_{K\to\infty} \rho^2\left(1 + K^2/2 + \sqrt{\dfrac{K^4}{4} + K^2}\right) \sim \dfrac{K^2}{2} = \dfrac{1}{2}e^{2r} \end{cases} \quad \textbf{Eq. 9}$$

Where $e^{\pm 2r}$ is the squeezing factor, and the two radii eq.s 9 correspond to the angles j=(π+arctg(2/K))/2 and j=(arctg(2/K))/2 respectively. We evaluate the angles of the ellipse minor axis in the out system of reference i.e. $\partial R^2_{min}/\partial\varphi=0$:

$$tg\theta = \frac{\Delta a_{\pi/2,out}}{\Delta a_{0,out}} = \frac{\Delta a_{\pi/2,in} - K\Delta a_{0,in}}{\Delta a_{0,in}} = tg\varphi - K \qquad tg\theta_{R_{min}} = \sqrt{\frac{\sqrt{K^2+4}+K}{\sqrt{K^2+4}-K}} - K \qquad \lim_{K\to\infty} tg\theta_{R_{min}} = \frac{1}{K} \qquad \lim_{K\to 0} tg\theta_{R_{min}} = \frac{\pi}{4}$$

Eq.s 10

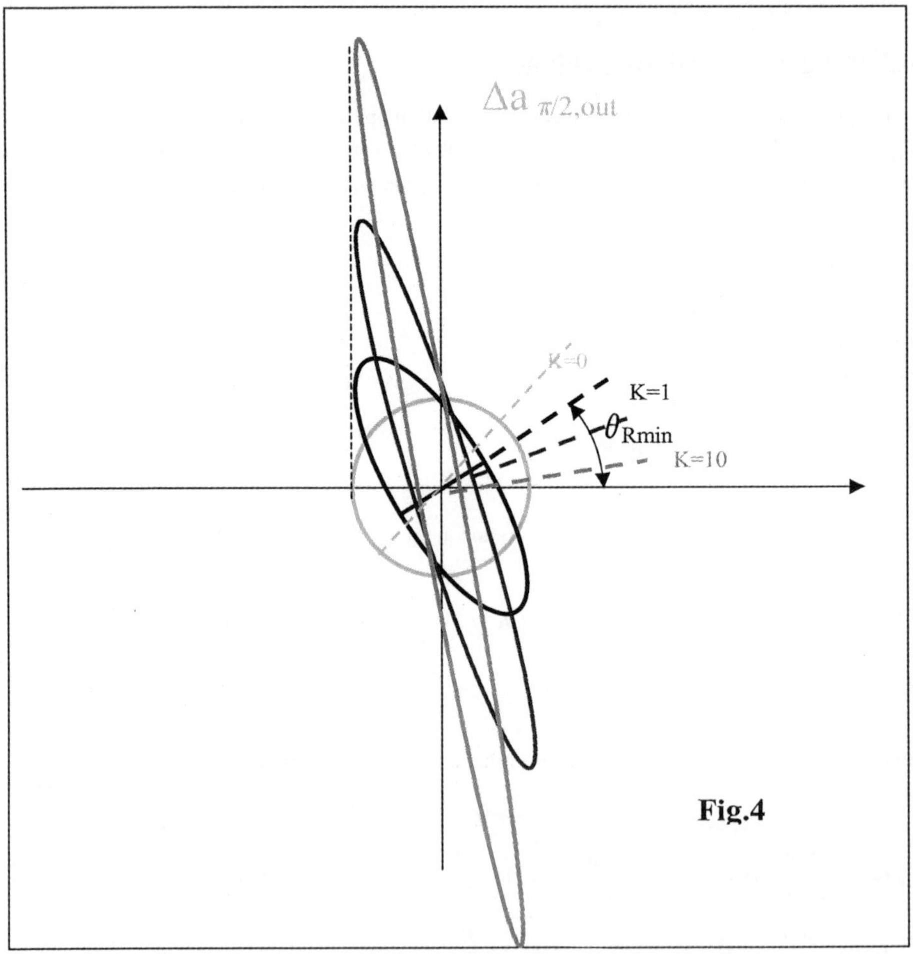

Fig.4

We may rotate the ellipse by an angle $\Phi=\pi/2-\theta_{Rmin}$, in such a way to coincide the minor axis direction with the $a_{\pi/2,out}$ axis (see fig.4); this quadrature corresponds to a "squeezed" shot noise. Since K depends upon Ω, so will do θ_{Rmin}, consequently for keeping the minor axis direction coinciding with the $a_{\pi/2,out}$ axis for all Ω, we need a more complex optical structure filtering the out beam.

For obtaining R_{min} aligned with the $a_{\pi/2}$ axis we need to perform in the quadrature space a rotation of an angle $\Phi=\pi/2-\theta_{Rmin}$ i.e

$$tg\Phi = \cot g\theta_{R_{min}} = \left(\sqrt{\frac{\sqrt{K^2+4}+K}{\sqrt{K^2+4}-K}} - K \right)^{-1} = \tilde{K} \qquad \textbf{Eq. 11}$$

In the limit K>>1, see eq.s 10, 11, we obtain:

$$tg\Phi=\cot g\theta_{R_{min}}=K\approx\frac{4.10^9}{\Omega^2}$$

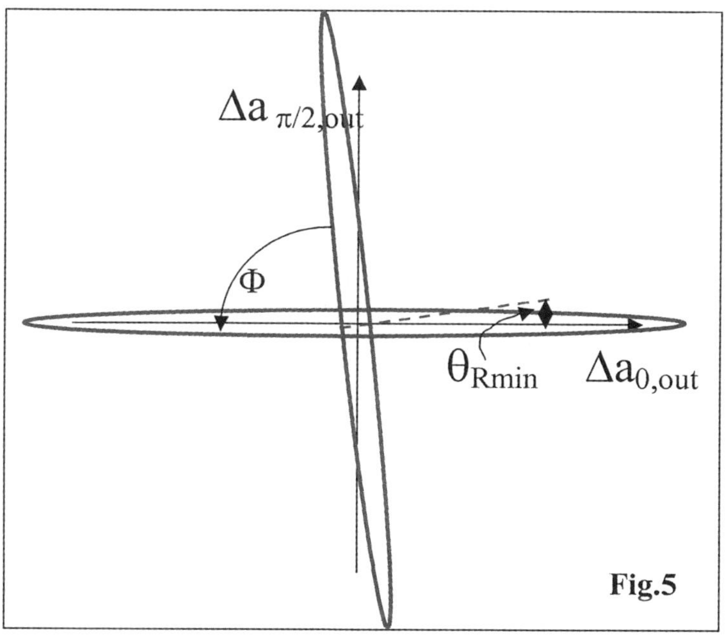

Fig.5

The Detuned Cavity (DC)

For obtaining such a rotation in the quadrature space let us filter the out fields of eq.7 with a detuned Fabry-Perot cavity (see fig. 4). The action of the cavity on the fields is the following:

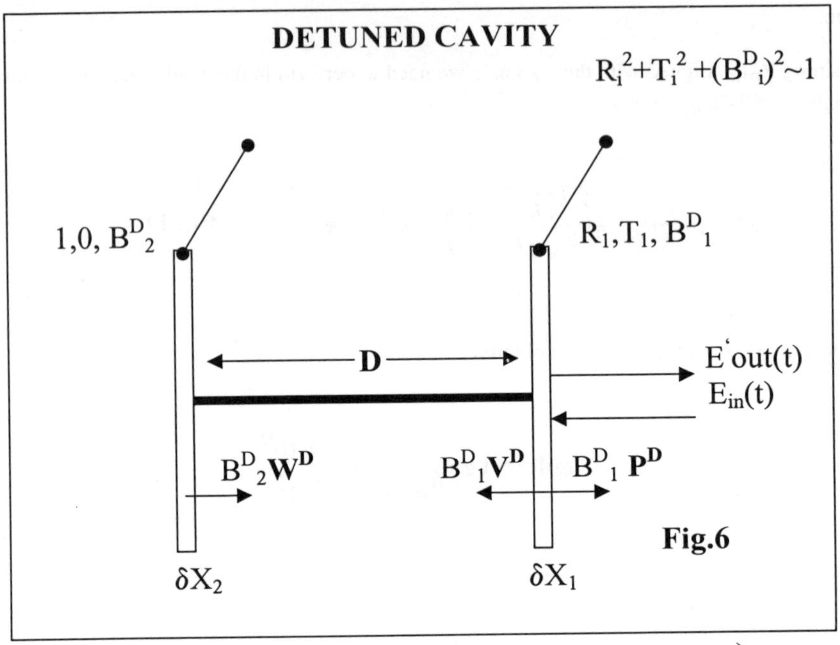

Fig.6

$$\begin{pmatrix} a_{0,out} \\ a_{\pi/2,out}\end{pmatrix}' = \begin{pmatrix} \cos\Phi \equiv \frac{1}{2}\Lambda(\omega_0-\omega,\Omega,R_1,1,D) & -\sin\Phi \equiv -\frac{T_1^2}{2i}Y(\omega_0-\omega,\Omega,R_1,1,D,-) \\ \sin\Phi \equiv \frac{T_1^2}{2i}Y(\omega_0-\omega,\Omega,R_1,1,D,-) & \cos\Phi\equiv\frac{1}{2}\Lambda(\omega_0-\omega,\Omega,R_1,1,D)\end{pmatrix}\begin{pmatrix}a_{0,in}\\ a_{\pi/2,in}\end{pmatrix}+$$

$$+\left(e^{-i\Omega\frac{D}{c}}\delta\tilde{X}_1-\delta\tilde{X}_2\right)\begin{pmatrix}-T_1^2\frac{\alpha\omega_0}{ic}Z(\omega_0-\omega,\Omega,R_1,1,D,-)\\ T_1^2\frac{\alpha\omega_0}{c}Z(\omega_0-\omega,\Omega,R_1,1,D,+)\end{pmatrix}-\begin{pmatrix}0\\ R_1\frac{2\alpha\omega_0}{c}\delta\tilde{X}_1\end{pmatrix}_{Quant}$$

$$+T_1\begin{pmatrix}\frac{1}{2}Y(\omega_0-\omega,\Omega,R_1,1,D,+) & -\frac{1}{2i}Y(\omega_0-\omega,\Omega,R_1,1,D,-)\\ \frac{1}{2i}Y(\omega_0-\omega,\Omega,R_1,1,D,-) & \frac{1}{2}Y(\omega_0-\omega,\Omega,R_1,1,D,+)\end{pmatrix}\left\{B_1^D\begin{pmatrix}V_0^D\\ V_{\pi/2}^D\end{pmatrix}-R_1B_2^De^{-i\Omega\frac{D}{c}}\begin{pmatrix}W_0^D\\ W_{\pi/2}^D\end{pmatrix}\right\}+B_1^D\begin{pmatrix}P_0^D\\ P_{\pi/2}^D\end{pmatrix}$$

$$\begin{pmatrix}\alpha_0\\ \alpha_{\pi/2}\end{pmatrix}=\begin{pmatrix}\frac{1}{2}\Lambda(\omega_0-\omega,0,R_1,R_2,L) & -\frac{R_2T_1^2}{2i}Y(\omega_0-\omega,0,R_1,R_2,L,-)\\ \frac{R_2T_1^2}{2i}Y(\omega_0-\omega,0,R_1,R_2,L,-) & \frac{1}{2}\Lambda(\omega_0-\omega,0,R_1,R_2,L)\end{pmatrix}\begin{pmatrix}\alpha\\ 0\end{pmatrix}$$

$$\begin{pmatrix}\Gamma_0\\ \Gamma_{\pi/2}\end{pmatrix}=\begin{pmatrix}0 & -R_2T_1^2\dfrac{\alpha\omega_0}{ic}Z(\omega_0-\omega,\Omega,R_1,R_2,L,-)\\ 0 & -R_2T_1^2\dfrac{\alpha\omega_0}{c}Z(\omega_0-\omega_0,\Omega,R_1,R_2,L,+)\end{pmatrix}\begin{pmatrix}0\\ e^{-i\Omega\frac{L}{c}}\delta\tilde{X}_1-\delta\tilde{X}_2\end{pmatrix}_{NQ}-\begin{pmatrix}0\\ R_1\dfrac{2\alpha\omega_0}{c}\delta\tilde{X}_1\end{pmatrix}_{NQ}$$

$$\frac{T_1^2}{2}Y(\omega_0-\omega,\Omega,R_1,1,D,+)\approx\frac{\left(1-R_1^2\right)\left[\left(1-R_1-\dfrac{\Delta^2-R_1\Sigma^2}{2}\right)^2+R_1^2\Sigma^2\right]^{1/2}}{\Pi}e^{i(\Sigma+arctgQ+arctgI)}$$

$$T_1^2\frac{\alpha\omega_0}{ic}Z(\omega_0-\omega,\Omega,R_1,1,D,-)\approx\frac{2\alpha\omega_0}{c}\frac{(1-R_1^2)\Delta(1-R_1^2)}{(1-R_1)^2\left[\left((1-R_1)^2+2R_1\Delta^2+\Sigma^2R_1(1-2R_1)\right)^2+4R_1^2(1-R_1)^2\Sigma^2\right]^{1/2}}\bullet$$

$$\bullet e^{i\left(\frac{\Sigma}{2}+arctgQ+arctgH-arctg\frac{R_1^2\Sigma}{1-R_1^2}\right)}$$

$$T_1^2\frac{\alpha\omega_0}{c}Z(\omega_0-\omega,\Omega,R_1,1,D,+)\approx\frac{2\alpha\omega_0}{c}\frac{\left(1-R_1^2\right)\left[\left((1-R_1)^2-\dfrac{(R_1^2+1)}{2}\Delta^2+\dfrac{R_1}{2}(1-R_1)\Sigma^2\right)^2+\Sigma^2R_1^2(1-R_1)^2\right]^{1/2}}{(1-R_1)^2\left[\left((1-R_1)^2+2R_1\Delta^2+\Sigma^2R_1(1-2R_1)\right)^2+4R_1^2(1-R_1)^2\Sigma^2\right]^{1/2}}\bullet$$

$$\bullet e^{i\left(\frac{\Sigma}{2}+arctgQ+arctgH-arctg\frac{R_1G}{1-R_1}\right)}$$

Where ω is the detuning frequency, ω_0 the resonance frequency ($e^{2i\omega_0L/c}=1$). In the following we put $\Delta=2D\omega/c$ and $\Sigma=-2D\Omega/c$:

$$\Pi\approx\sqrt{\left((1-R_1)^2+R_1\Delta^2+R_1(1-2R_1)\Sigma^2\right)^2+4R_1^2(1-R_1)^2\Sigma^2}$$

$$Q=\frac{2R_1\Sigma(1-R_1)}{(1-R_1)^2+R_1\Delta^2R_1+R_1(1-2R_1)\Sigma^2}\qquad H=\frac{2R_1(1-R_1)\Sigma}{(1-R_1)^2+2R_1\Delta^2+\Sigma^2R_1(1-R_1)^2(1-2R_1)}$$

$$G=\frac{(1-R_1)^2\Sigma}{(1-R_1)^2-\dfrac{(R_1^2+1)}{2}\Delta^2+\Sigma^2\left(R_1-\dfrac{(1-R_1)^2}{2}\right)}\qquad I=\frac{R_1\Sigma}{1-R_1-(\Delta^2-R_1\Sigma^2)/2}$$

$$H=A\cos\Phi=\frac{1}{2}\Lambda(\omega_0-\omega,\Omega,R_1,1,D)\approx\frac{\left(\left((1-R_1)^2-\dfrac{(R_1^2+1)}{2}\Delta^2+\Sigma^2\left(R_1-\dfrac{(1-R_1)^2}{2}\right)\right)^2+(1-R_1)^4\Sigma^2\right)^{1/2}}{\Pi}\bullet$$

$$\bullet e^{i(\Sigma+arctgQ+arctgG)}$$

Eq.13

$$\Xi=A\sin\Phi=\frac{T_1^2}{2i}Y(\omega_0-\omega,\Omega,R_1,1,D,-)\approx\frac{(1-R_1^2)\Delta}{\Pi}e^{i(\Sigma+arctgQ)}$$

If we detune the cavity i.e. putting

$$(1-R_1)^2-\frac{(R_1^2+1)}{2}\Delta^2=0$$

and keeping terms up to second order in Δ and Σ, we obtain:

$$\Pi\approx\sqrt{\left(\frac{(R_1+1)^2}{(R_1^2+1)}(1-R_1)^2+R_1(1-2R_1)\Sigma^2\right)^2+4R_1^2(1-R_1)^2\Sigma^2}\approx$$

$$\approx(1-R_1)^2\frac{(R_1+1)^2}{(R_1^2+1)}\left(1+\frac{2(R_1^2+1)(3R_1+1)}{(R_1+1)^4}\frac{\Sigma^2}{(1-R_1)}\right)^{1/2} \qquad \textbf{Eq.14}$$

$$Q=\frac{2R_1\Sigma(R_1^2+1)}{(R_1+1)^2(1-R_1)}\qquad G=\frac{(1-R_1)^2}{\Sigma\left(R_1-\frac{(1-R_1)^2}{2}\right)}$$

$$H=A\cos\Phi=\frac{1}{2}\Lambda(\omega_0-\omega,\Omega,R_1,1,D)\approx\frac{\Sigma^2\left(\left(R_1-\frac{(1-R_1)^2}{2}\right)^2+\frac{(1-R_1)^4}{\Sigma^2}\right)^{1/2}}{\Pi}\bullet$$
$$\bullet e^{i(\Sigma+arctgQ+arctgG)}$$

$$\Xi=A\sin\Phi=\frac{T_1^2}{2i}Y(\omega_0-\omega,\Omega,R_1,1,D,-)\approx\frac{(1-R_1^2)(1-R_1)}{\Pi}\sqrt{\frac{2}{(1+R_1^2)}}e^{i(\Sigma+arctgQ)}$$

$$\frac{T_1^2}{2}Y(\omega_0-\omega,\Omega,R_1,1,D,+)\approx\frac{(1-R_1^2)(1-R_1)R_1\left[\frac{(1+2R_1)^2}{(1+R_1^2)^2}+\frac{\Sigma^2}{(1-R_1)^2}\right]^{1/2}}{\Pi}e^{i\left(\Sigma+arctgQ-arctg\frac{R_1\Sigma}{1-R_1-\Delta^2/2}\right)}$$

$$T_1^2\frac{\alpha\omega_0}{ic}Z(\omega_0-\omega,\Omega,R_1,1,D,-)\approx\frac{2\alpha\omega_0}{c}\frac{\Delta(1+R_1)^2\bullet e^{i\left(\frac{\Sigma}{2}+arctgQ-arctg\frac{R_1^2\Sigma}{1-R_1^2}\right)}}{(1-R_1)^2\left(1+\frac{4R_1}{(1+R_1^2)}\right)\left[1+\frac{\Sigma^2}{(1-R_1)^2}\frac{2R_1}{\left(1+\frac{4R_1}{(1+R_1^2)}\right)}\left(1-2R_1+\frac{2R_1}{\left(1+\frac{4R_1}{(1+R_1^2)}\right)}\right)\right]^{1/2}}$$

$$T_1^2\frac{\alpha\omega_0}{c}Z(\omega_0-\omega,\Omega,R_1,1,D,+)\approx\frac{2\alpha\omega_0}{c}\frac{\Sigma(1+R_1)R_1}{(1-R_1)^2\left(1+\frac{4R_1}{(1+R_1^2)}\right)}\bullet e^{i\left(\frac{\Sigma}{2}+arctgQ-arctg\frac{R_1G}{1-R_1}\right)}$$

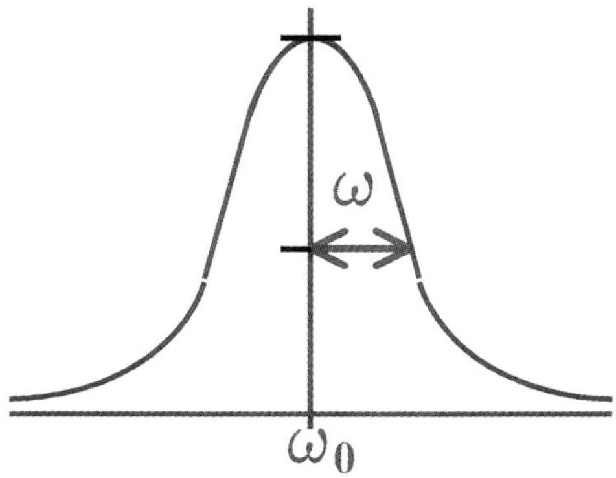

It is now very important to evaluate ponderomotive effects in the DC; to this purpose we evaluate from eqs.13 the following ratios up to first order in Σ.

$$\frac{Z(\omega_0-\omega,\Omega,R_1,1,D,+)}{Z(\omega_0,\Omega,R_1,1,D,+)} \approx \frac{R_1\Sigma}{(1-R_1)\left(1+\frac{4R_1}{(1+R_1^2)}\right)}$$

$$\frac{Z(\omega_0-\omega,\Omega,R_1,1,D,+)}{Z(\omega_0-\omega,\Omega,R_1,1,D,-)} = \frac{R_1\Sigma}{(1-R_1)(1+R_1)}\sqrt{\frac{1+R_1^2}{2}} \qquad \textbf{Eq.15}$$

Eq. 15 is showing that in a DC the coupling to mirror motion $\delta X_1 - \delta X_2$ i.e the function $Z(+, \Delta=(1-R)(2/(1+R^2))^{1/2})$, is negligible with respect to the same coupling in a tuned cavity if

$\Sigma \ll 1-R_1$. Since we want a negligible ponderomotive effect in the $\pi/2$ quadrature component coming out from the DC we impose the condition:

$$\boxed{\Sigma \ll 1-R_1} \qquad \textbf{Eq.16}$$

A second condition is coming from eq.11, infact from eq. 14 it follows:

$$tg\Phi \approx \frac{(1+R_1)(1-R_1)^2\sqrt{\frac{2}{(1+R_1^2)}}\,e^{-iarctgG}}{\Sigma^2\left(\left(R_1-\frac{(1-R_1)^2}{2}\right)^2+\frac{(1-R_1)^4}{\Sigma^2}\right)^{1/2}} = \tilde{K} \cong \frac{32W\omega_0}{Mc^2\Omega^2(1-R)^2} \qquad \textbf{Eq.17}$$

For having an Ω-independent squeezing, eq. 17 should be proportional to Σ^{-2} , consequently the term $(1-R_1)^4/\Sigma^2$ must be <<1. With this condition G $\to$0 and eq. 17 becomes real. From eq.16 it follows:

$$\boxed{(1-R_1)^2 \ll \Sigma \ll 1-R_1} \qquad \textbf{Eq.18}$$

For checking inequalicy eq.18 as an example let us fix K(1000 Hz)=10:

$$\frac{(1+R_1)(1-R_1)^2}{\Sigma^2 R_1}\sqrt{\frac{2}{(1+R_1^2)}} = \tilde{K} \cong \frac{32W\omega_0}{Mc^2\Omega^2(1-R)^2} = \frac{y}{\Sigma^2} = \frac{U}{\nu^2}$$

$$K(1000Hz) = 10 = \frac{U}{\nu^2} = \frac{10^7}{\nu^2} \qquad U = \frac{8W\omega_0}{\pi^2 Mc^2(1-R)^2} \qquad \textbf{Eq.19}$$

$$y = \frac{128W\omega_0 D^2}{Mc^4(1-R)^2} = \frac{\Sigma^2 U}{\nu^2} = \frac{16\pi^2 UD^2}{c^2}$$

$$\begin{cases} \Sigma > (1-R_1)^2 \cong \dfrac{R_1\Sigma^2 K}{(1+R_1)} = \dfrac{R_1}{(1+R_1)}y = \dfrac{R_1}{(1+R_1)}\dfrac{16\pi^2 UD^2}{c^2} \\ \Sigma < (1-R_1) = \sqrt{\dfrac{R_1 U}{(1+R_1)}}\dfrac{4\pi D}{c} \end{cases}$$

$$(1-R_1) = \sqrt{\frac{R_1 U}{(1+R_1)}}\frac{4\pi D}{c} \sim 810^{-5} D \qquad (1-R) = \sqrt{\frac{8W\omega_0}{\pi^2 Mc^2 U}} \approx \sqrt{\frac{W}{M}}410^{-5}$$

$$2\cdot 10^{-1} D(Hz) \sim \frac{4\pi D}{c}\frac{R_1 U}{(1+R_1)} \le V \le \sqrt{\frac{R_1 U}{(1+R_1)}} \sim 2\cdot 10^3 (Hz)$$

Note that while the PC finesse 1/(1-R) is independent from length L, the finesse of the DC $1/(1-R_1)$ is inversely proportional to D.

The DC Quadrature Rotation Matrix

From eqs.7, 14 and 17 it follows the total output from the DC, where PC means Ponderomotive Cavity :

$$\begin{pmatrix} a_{0,out} \\ a_{\pi/2,out} \end{pmatrix} + \begin{pmatrix} \Gamma_0 \\ \Gamma_{\pi/2} \end{pmatrix}_{NQ}^{PC\to DC} + \begin{pmatrix} \Gamma_0 \\ \Gamma_{\pi/2} \end{pmatrix}_{NQ}^{DC} \cong D(\widetilde{K}) = \frac{\widetilde{K}}{\sqrt{1+\widetilde{K}^2}} \begin{pmatrix} 1/\widetilde{K} & -1 \\ 1 & 1/\widetilde{K} \end{pmatrix} \left\{ \begin{pmatrix} a_{0,in} \\ a_{\pi/2,in} - \widetilde{K} a_{0,in} \end{pmatrix} + \begin{pmatrix} \Gamma_0 \\ \Gamma_{\pi/2} \end{pmatrix}_{NQ}^{DC} \right\} +$$

$$+ \frac{1}{(1+R_1)\sqrt{1-R_1^2}} \begin{pmatrix} R_1(1+2R_1) & -\sqrt{2(1+R_1^2)} \\ \sqrt{2(1+R_1^2)} & R_1(1+2R_1) \end{pmatrix} \left\{ B_1^D \begin{pmatrix} V_0^D \\ V_{\pi/2}^D \end{pmatrix} - R_1 B_2^D e^{-i\Omega\frac{D}{c}} \begin{pmatrix} W_0^D \\ W_{\pi/2}^D \end{pmatrix} \right\} + B_1^D \begin{pmatrix} P_0^D \\ P_{\pi/2}^D \end{pmatrix}$$

$$\begin{pmatrix} (1-R_1^2)\sqrt{2(R_1^2+1)} \\ \Sigma(1+R_1^2)R_1 \end{pmatrix} \frac{16 W_D \omega_0}{Mc^2\Omega^2} \frac{R_2^2\left(R_1^2(1+2R_1)^2 + (1+R_1)(1+R_1^2)\right)}{(1+R_1)^2(1-R_1)^3\left(1+R_1^2+4R_1\right)} U_{0,in} + \begin{pmatrix} \Gamma_0 \\ \Gamma_{\pi/2} \end{pmatrix}_{NQ}^{DC}$$

$$\begin{pmatrix} \Gamma_0 \\ \Gamma_{\pi/2} \end{pmatrix}_{NQ}^{DC} = \begin{pmatrix} (1-R_1^2)\sqrt{2(R_1^2+1)} \\ \Sigma(1+R_1^2)R_1 \end{pmatrix} \sqrt{\frac{2W_D}{h\nu_0}} \frac{4\pi(1+R_1)}{\lambda(1-R_1)^2\left(1+R_1^2+4R_1\right)} \left(\delta\widetilde{X}_1 - \delta\widetilde{X}_2\right)_{NQ}^{DC}$$

$$\begin{pmatrix} \Gamma_0 \\ \Gamma_{\pi/2} \end{pmatrix}_{NQ}^{PC} = -\frac{2\pi}{\lambda}\sqrt{\frac{32W}{h\nu_0}} \frac{e^{i\Omega L/c}}{1-\mathrm{Re}^{-2i\Omega L/c}} \left(e^{i\Omega L/c}\delta\widetilde{X}_1 - \delta\widetilde{X}_2\right)_{NQ}^{PC} \begin{pmatrix} 0 \\ 1 \end{pmatrix} \qquad \textbf{Eq. 20}$$

It is very interesting now to see how this technique creates squeezed photon states; eq.7 shows how a F-P cavity with suspended mirrors, due to radiation pressure, adds a_0 to the $a_{\pi/2}$ line. From eq.s 21, by keeping only the first term, we obtain:

Eq.21

$$\begin{pmatrix} a_{0,out} \\ a_{\pi/2,out} \end{pmatrix} + \begin{pmatrix} \Gamma_0 \\ \Gamma_{\pi/2} \end{pmatrix}_{NC}^{PC\to DC} \approx \begin{pmatrix} 1/K & -1 \\ 1 & 1/K \end{pmatrix} \left\{ e^{2i\beta} \begin{pmatrix} a_{0,in} \\ a_{\pi/2,in} - a_{0,in}K \end{pmatrix} + \right.$$

$$\left. -\frac{2\pi}{\lambda}\sqrt{\frac{32W}{h\nu_0}} \frac{e^{i\Omega L/c}}{1-\mathrm{Re}^{-2i\Omega L/c}} \left(e^{i\Omega L/c}\delta\widetilde{X}_1^{PC} - \delta\widetilde{X}_2^{PC}\right)_{NQ} \begin{pmatrix} 0 \\ 1 \end{pmatrix} \right\} =$$

$$= e^{2i\beta} \begin{pmatrix} \frac{a_{0,in}}{K} - \left(a_{\pi/2,in} - a_{0,in}K\right) \\ \frac{a_{\pi/2,in}}{K} \end{pmatrix} - \frac{2\pi}{\lambda}\sqrt{\frac{32W}{h\nu_0}} \frac{e^{i\Omega L/c}}{1-\mathrm{Re}^{-2i\Omega L/c}} \left(e^{i\Omega L/c}\delta\widetilde{X}_1^{PC} - \delta\widetilde{X}_2^{PC}\right)_{NQ} \begin{pmatrix} 1 \\ 1/K \end{pmatrix}$$

In the limit of K>>1, eq.16 becomes :

$$\begin{pmatrix} a_{0,out} \\ a_{\pi/2,out} \end{pmatrix} + \begin{pmatrix} \Gamma_0 \\ \Gamma_{\pi/2} \end{pmatrix}_{NC}^{PC\to DC} \approx e^{2i\beta} \begin{pmatrix} a_{0,in} K \\ \dfrac{a_{\pi/2,in}}{K} \end{pmatrix} - \frac{2\pi}{\lambda} \sqrt{\frac{32W}{h\nu_0}} \frac{e^{i\Omega L/c}}{1-\mathbf{R}e^{-2i\Omega L/c}} \left(e^{i\Omega L/c} \delta\tilde{X}_1^{PC} - \delta\tilde{X}_2^{PC} \right)_{NQ} \begin{pmatrix} 1 \\ 1/K \end{pmatrix}$$

Eq.22

From eq. 17 it follows that while the shot noise term together with $\delta X^{ext}{}_{12}$ are reduced by a factor

K, the amplitude noise is increased by the same factor i.e. we obtain a squeezed state. Finally we write eq. 28 in matrix form:

Detuned F-P Cavity ↓ **Ponderomotive F-P Cavity** ↓

Eq.24

$$\begin{pmatrix} a_{0,out} \\ a_{\pi/2,out} \end{pmatrix} + \begin{pmatrix} \Gamma_0 \\ \Gamma_{\pi/2} \end{pmatrix}_{NC}^{PC\to DC} \approx \begin{pmatrix} 1/K & -1 \\ 1 & 1/K \end{pmatrix} \begin{pmatrix} 1 & 0 \\ -K & 1 \end{pmatrix} \left\{ \begin{pmatrix} a_{0,in} \\ a_{\pi/2,in} \end{pmatrix} e^{2i\beta} - \frac{2\pi}{\lambda} \sqrt{\frac{32W}{h\nu_0}} \frac{e^{i\Omega L/c}}{1-\mathbf{R}e^{-2i\Omega L/c}} \left(e^{i\Omega L/c} \delta\tilde{X}_1^{PC} - \delta\tilde{X}_2^{PC} \right)_{NQ} \begin{pmatrix} 0 \\ 1 \end{pmatrix} \right\}$$

Conclusions

From Eq. 24 it seems that it should be possible to create experimentally light squeezed states efficiently; these states could the be injected in the dark port of a gravitational wave interferometric detector; the result will be to reduce shot noise by a factor K which means that we can obtain the same shot noise by reducing the power by a factor K^2. This will also allows to operate detectors free from mirror lensing.
These techniques will deserve thorough experimentation in the close future.

Bibliography

Two books give a lot of important informations in this research field; they are:

1) Marlon O. Scully and M. Suhail Zubary-Quantum Optics
   Cambridge University Press (Sept. 4,1997)

2) S. M. Barnett and P. M. Radmore-Methods in theoretical quantum optics
   Oxford University Press (Nov. 1,1997)

The following papers are instrumental:

C. M. Caves, B.L. Schmaker:-New formalism for two-photon quantum optics.I. Quadrature phases and squeezed states.- Phys. Rev. A 31,5, may 1985, 3086.

C. M. Caves, B.L. Schmaker:-New formalism for two-photon quantum optics.II. Mathematical foundations and compact notations- Phys. Rev. A 31,5, may 1985, 3093

J.M. Courty, A. Heidmann and M. Pinard- Back action cancellation in interferometers by quantum locking.-gr-cq/0301068 v1, jan 2003

CHAIRMAN: A. GIAZOTTO

Scientific Secretaries: D. Baldassari, V. Bindi

DISCUSSION I

- *G. Veneziano*:

I wonder whether an easier way to defend our field would not be to point at the spin-offs of the research. By pushing the research for fundamental physics to its limits one is also developing various techniques to those limits. Examples can easily be found in cryogenics, in computing, in the way seismic noise is handled, etc.

It is like developing Formula 1 cars with the idea that it will eventually bring about improvement for anyone's car.

- *A.Giazotto*:

We think that what we aim for is studying some kind of new astronomy so I guess this could be a real extremely important bonus, more than anything else we could extract from the technological point of view. To me this target is really so big, so important that justifies also the enormous technological effort that we are making; we have to carry on this line as strong as we can because it will be rewarding, in my opinion, for the overall universe knowledge.

I agree with you, the by-products of what we are doing are numerous, we have a lot of improvement on the laser technology, and also in the technology dealing with these quantum things that I'm about to give some ideas tomorrow morning. These are things that will change the way in which some kind of physics will be done in the future. Per example with ours superattenuators you may suspend 200 kg and have only thermal as remnant noise without any seismic above 4 Hz; we didn't find any real direct application for them, but we guess that someone will find it soon.

Moreover the investigation in the direction of unexpected things could be interesting, like movement of the earth core; a very important fact is that we have the multiple superattenuator pendula suspending the mirrors, pointing toward the Earth center and with the sensitivity we have, we can really try to detect movement in the vertical direction of the earth core. It is a completely new branch of things that can come out from this very sophisticated and nasty technology.

- *A. Zichichi* added:

We have to convince people that the discoveries in other fields like medicine, chemistry, and so on, are spin-offs of physics research, and to convince them that modern technology has come from physics discoveries and we can prove this.

People classify us as philosophers and claim there is a medical technology, but the truth is that not even modern chemistry could exist without quantum mechanics.

I have been at the opening session of the world conference of radiology and in the opening lecture the new great field was the T-rays. I had to convince these guys that the medical technology does not exist, it comes from us, the Tera-hertz-rays are a physics discovery, not a medical invention.

A very non trivial task is that we also have to defend our field against the attitude of politicians.

- *Christian Klein-Bosing:*

How did you determine the different contributions to the noise and can't you just subtracted them?

- *A.Giazotto*:

We determine it by measuring the noise of the different contributions and the different parts. We can measure, for example, the noise of the power amplifier which drives the coils to push the mirrors in the longitudinal direction, just taking the power amplifier out and measuring how many nanovolts or millivolts of noise it produces. Then we scale them to what we see because we know how many amperes coming out produce how many newtons, and so we can understand how much is the displacement noise in the mirror. All these things can be measured indipendently.

You cannot subtract the noise contributions at the end, you have to kill the noise at the origin. When you subtract you can make a factor of ten better, not more than that. You really have to go down by a factor of more than one million in some kind of noise.

- *M. Gyulassy*:

Continuing the line of questions about how to show that fundamental "pure" physics, such as research for gravitational waves (GW), can lead to major advances useful to society, I feel that there must be tremendous amount of very usefull information in the 20 orders of magnitude "noise" that your highly sensitive detecter try to subtract away to see GW. Have you contacted geologists and seismic engineers to see how they could perhaps utilize your "noise" to help predict earthquakes and other geological effects on human society? Your detector technology may be useful to develop new early warning systems.

- *A. Giazotto:*

We have a group in Moscow, lead by Valentino Rudenko from the Moscow University, claiming, since long time, the possibility of measuring the earth core by means of these devices. When the core goes up, we can measure the increasing in the angle between the two verticals so we can really try to have the informations on the vertical motion of the core. To have a tridimentional study, the vertical motion can be measure in different positions on the earth surface.

At the moment, the real problem we are facing is the lack of forces and we'd like to start a collaboration with LIGO people and other communities to solve this problem.

- *C. Aidala:*

In what ways are the various gravitation wave experiments around the world coordinated?

- *A. Giazotto*:

At the moment the coordination is some way connected with the status of every detector at the moment in wich is running, because duty cycles are not very much complete over the years at the moment. So there will be a coverage of the data taking time in such a way to make coincidence in the common analysis. As an example, machine upgradings should be coordinated in such a way to avoid that all the machines are off at same time.

- *S. Salur:*

Can you reproduce the noise with simulations? What is the signal signature that we expect to see if we detect GW?

- *A. Giazotto*:

Of course, we can reproduce the noise by simulations. Simulation is at the basis of everything, we have a lot of simulation power: we simulate mechanics, optics, almost everything. At the moment we are still in the phase in which the time to simulate everythings takes too long, anyway we have heavily using simulation for understanding in particular the optics behaviour.

Concerning the second question, suppose to take one between the thousand pulsar known, of which you know direction, frequency and slowing down time, then you can apply the correction for the Earth doppler shift and reconstruct the monocromatic behavior. Then you run the system with your Fourier device and you get a peak at the frequency you expect. This peak should be reproducible and should not fluctuate in time and should be visible by VIRGO, LIGO and other spectrometers.

In the more complex cases of coalescing bynaries, when you don't know all the parameters, but you know the wave-form family, you can build up a family of these curves, called templates. Then you make a linear analysis taking the output of the interferometer and introducing it in a Wiener integral where one of the component is the template, and you make the integral of the product between the template and this signal weighted with the square of the detector noise. Wiener integral gives a signal proportional to signal/noise. For selecting the right template we try every millisecond of data something like 10^5 templates each tested in the Wiener integral with 300 s of data. When the statistical significance is high, i.e. if the template copy the wave-form, there will be higher output from the Wiener integral and this is a possible candidate for GW detection. This analysis requires at least a 300 Gflop machine continuously running during data taking.

- *O. Loktik:*

Is there a quantum limit on measuring signals? Wouldn't the Heisenberg uncertainty principle prohibit 10^{-20}m measurement?

- *A. Giazotto*:

Yes, uncertainty principle puts some limitations on the measurement but there are ways to take it into account. You can measure only one parameter, the phase, not the number of photons so you can measure the phase as well as you want, making much larger the fluctuations in the number of photons. I will speak about tomorrow.

- *J. Boursaily:*

The detection of pulsar sources requires spherical asymmetry of the star. Surely this asymmetry is not static. How does this evolve - if known – and how can this affect the monocromaticity/amplitude of the signal?

- *A. Giazotto*:

There is no knowledge about that, we don't know what glitches are exactly, we don't know if there is a change of mass distribution, and these GW can give out other informations on top of the radio telescope, but for the moment we are still back in this kind of sophisticated problem.

-G. Imponente:

How is the thermal noise generated for the mirror oscillations and the related plots at high frequencies? Is it in principle possible to discriminate the GW background of cosmological origin from the superposition of many (also distant) astrophysical objects and eventually distinguish different cosmological models?

- A. Giazotto:

The high frequencies are due to the thermally excited violin modes of the wires suspending the mirrors. The mirrors are suspended by two wires going around it in such a way that changing the position with the marionette, you can produce movement of the mirror. These wires are working at half of the breaking point, so they are very much under tension, and all the modes they have are excited by thermal noise. The wires being attached to the mirror produce mirror displacement at the harmonic frequencies; what is shown on the sensitivity plot are the equivalent h produced by those displacements.

At the moment, I think that the cosmological background is very low, unless there is some particular theory turning out to be the effective one governing nature and giving intense GW production. The background produced by the overall system from the infinity of supernovae explosion, coalescing binaries etc… which produce also a background, has a signature which is quite similar to the other one. I would say that these two things are still very much to be understood.

- G. Veneziano:

Following the question by G. Imponente and answer from Prof.Giazotto: I think that the stochastic background from astrophysical searches tends to be important at rather low frequency. This could help distingushing it from a cosmological background if the latter is peacked at higher frequencies.

- A. Giazotto:

Yes, but the shape of the cosmological background is very model dependent.

-G. De Risi :

Apart from the work by Taylor and Hulse, we don't have any direct evidence of the existence of GW. If both LIGO and VIRGO fail to prove experimentally the existence of GW, could it be a reason to believe that they do not exist at all?

- A. Giazotto:

The problem could depend on the sensitivity of the experimental apparatus. I think that one proof will comes from the measurement of the standard candle. Unfortunately, in the frequency range for the terrestrial gravitational detector, there is no standard candle. You don't have a source which you know to have a precise GW amplitude in our frequency range. This happens in the much lower frequency range, per example the AM-CVN Cataclismic binary star system has a period of about 500 seconds and it is locate in a well precise point of our galaxy. This source can be seen, instead, by the LISA projec; its frequency band includes several galactic standard candles.

- *J. Bourjaily:*

A related question to the above one: null results in physics can have enormous implications on physics. What would be the implications of a null results from all GW detectors in the next 15 years?

- *A. Giazotto*:

I think that one has to examine, first of all, if the experimental apparatus is ok. If there is nothing wrong, then you can try to understand what doesn't work, for example, in the Taylor PSR-1913; if there is some other kind of phenomenon which is really slowing down the system by simulating a GW emission?. Of course it is very difficult to copy General Relativity theory result with the experimental data in such a precise way. So it would be quite a challanging thing.

-*G. De Risi:*

Do you think we will be able to see GW in the forthcoming ten years?

- *A. Giazotto*:

I hope so, otherwise I wouldn't do all this!

- *M. Clarke:*

Why is the blind analysis so computationally expensive?

- *A. Giazotto*:

Well, to extract the numbers of 10^{13} Tflops is easy, you have about 10^{30} sets to explore, some 10^{10} in frequency, about10^{7} in slowing down time and 10^{14} in angle, because you don't know where the source is, and the angle cell has to be small, in such a way that the signal should come out monocromatic. If you are wrong in locating the angle cell, once you have fixed one frequency for a given source, you don't cancel exactly the Doppler shift and then you loose sensitivity.

There is another techniqe which is, in my opinion, very interesting, it is the analysis done with two interferometers colocated and then making the product of the two outputs. If the source is phase modulated, not amplitude modulated, there is the complete cancellation of the Doppler shift for every source in the sky. This cancellation can vastly reduce the computational time but you need two colocated interferometers.

- *J. Alwall:*

How is the superattenuator connected with the fixed stars?

- *A. Giazotto*:

The superattenuator is formed by three legs about 7 m long, they form an inverted pendulum with 40 millihertz proper frequency, supporting an inertial platform on which three accelerometers, sensitive to horizontal motion and put in trigonal symmetry plus two in the vertical direction, by means of a feed-back loop minimize the movement of the platform with respect to the fixed stars. The platform then supports the long chain of 6 six dimensional filters which finally supports the mirror.

- *A. Zichichi:*

When I visited the spectacular VIRGO, I suggested that the spectrometer technology they have been able to implement in this laboratory should focus some attention to establish if the earth core is moving. This would have an enormous interest in the field of seismo-physics and of all related effects.

- *A. Giazotto*:

We have a group in Moscow university which is now preparing for coming to Virgo and to analyze the data for extracting this kind of signal. Actually we could see the angle movement between the two superattenuator vertical when the Earth center of mass changes position. The thing that is still unknown is how precise this measurement could be.

CHAIRMAN: A. GIAZOTTO

Scientific Secretaries: L. Quadrani, R. Silvestri

DISCUSSION II

- *M. Clark*:

After gravitational waves are proven to exist (perhaps by the VIRGO project) what will the next target project be?

- *A.Giazotto*:

It depends from what you want to see. You have to go on with increasing the sensitivity, and this was the reason of my talk (this morning talk), how to try to circonvent one limit which, until same years ago, was considered to be uncirconventable, and now it seem that we start understanding how to do it. It is clear that we have to increase the sensivity, and infact at VIRGO and at Ligo we are designing already the advanced one, which will be done in the next three or four years from now.

The target is the gravitational wave astronomy; this is the only really relevant thing that we are looking for.

Demonstration of the GW existence we think that has already been done.

- *J. Bourjaily* :

The use of gravitational waves interferometers to measure the core of the Earth surely requires collaboration between detectors located in very different places on the Earth. Have such collaborations been established and are there groups studying this now?

- *A.Giazotto*:

There is no bilateral agreement of any kind yet. We still have to demonstrate that the sensitivity in the single interferometer is good enough, because, at the end, this is the only relevant thing. If the Earth core moves, it is the angle variation of the two plumb lines pointing to the Earth CMS which is changing, and we have to demonstrate that we have enough sensitivity for measuring it, and once we have demonstrated that this method is really working we will start with bilateral agreements.

- *J. Bourjaily* :

How high is the priority? and other people are working on it right now?

- *A.Giazotto*:

The group of Valentin Rudenko at Moskow University is working on it and they will come on Virgo before the end of the year for starting measuring signal to noise ratio in the milli Hz frequency range.

- *O. Loktik*:

Do you need some special techniques in order to make the sqeezed elliptic light to interfere?

- *A.Giazotto*:

Yes, we need special very complicated cavities. Special cavities which need to have very little losses. The Ponderomotive cavities is the one in which you inject a laser beam and you take out a state, which is mixing both the two quadratures, one responsible for the shot noise fluctuations and the other responsible of the radiation pressure fluctuations i.e. a_0 and $a_{\pi/2}$. The power fluctuations created by a_0 move the mirrors so they creates phase shifts. Consequently the reflected beam coming out after this action contains the two quadratures summed with the certain coefficient; then the detuned cavity performs a rotation in the quadrature space. It depends what you want to measure. If you want to measure the phase noise only without any pollution from the radiation pressure, you can find the way by tuning your detuned cavities with a frequency detuning equal to the half width half maximum of the cavity resonance curve. This is the key point. Here, it opens another very complex point: to control these cavities in such way that the first one, producing Ponderomotive sqeezing, should stay in in resonance as well as the detuned one. And for staying in resonance you need a feedback keeping the mirrors in the right position; but the feed-back is acted by a photodiode detecting light coming from detuned cavity, and diodes are lossy elements. And these losses can be compared to the mirror losses, and so the vacuum fluctuations reenter and pollute the squeezed vacuum. It is a very complex matter.

- *J. Alwall*:

Are there any knewn astronomical objects that can produce gravitational waves with the frequency in the 1-1000 Hz region?

- *A.Giazotto*:

Yes, per example two solar masses neutron star making coalescence. At merging point, they have the final frequency of about 1 KHz. So we start from the lowest frequency limit we have, which is about 4 Hz, and we observe for about 2000 s before to reach the merging frequency of about 1kHz.

- *L. Cifarelli*:

Do you already have an estimate on the sensitivity improvement achievable with the sqeezing procedure? Would this improvement help in the detection of signals from Supernove explosions?

- *A.Giazotto*:

The importance of the squeezing is that you can with much smaller power obtain the same sensitivity. It goes with the inverse of the square root of the power, so, now, you must pump a lot power for a modest sensitivity increase. If we are able to have a factor ten in squeezing, we could have the same sensitivity with hundred time less power. And if you have a detector in which you may use less power you have a lot of advantages because power in the mirror produces lensing, and so the mirrors heat up and deform; squeezing is the way to get high sensitivity with low power. This is the reason why we are trying very much to understand experimentally how to make this kind of squeezing. At the moment there are no experiments obtaining large squeezing factors; the Ponderomotive squeezing seem to be the way to get large squeezing factors but there are also enormous problems for the light detection and the control of these special cavities because the lossy photodiodes spoil squeezing. For this reason we are thinking to change from photodiodes to bolometers, devices which may, in principle have photon detection efficiency much closer to one. But anyway it is right, if we can go in this direction this could increase the sensitivity for supernova explosions detection.

- *C. Aidala*:

Could you give a quick, intuitive explanation of Ponderomotive sqeezing? Do you know where the name comes from?

- *A.Giazotto*:

Intuitive explanations is that we are able to ascribe the origin of quantum noise to two operators, a_0 and $a_{\pi/2}$. a_0 affects the intensity noise fluctuations and $a_{\pi/2}$ affects the phase fluctuations of the em field. The cavity suspended mirrors under the action of the a_0 term starts to fluctuate in position because a_0 is responsible for the intensity fluctuation. Their position changes and this produces a phase shift which adds to the one produced by $a_{\pi/2}$, the ponderomotive squeezing has transformed the confusion circle in an ellipse with the same area; the ellipse major axis reflects the fact that there was a noise addition. If we are able to rotate the ellipse in such a way to have the ellipse minor axis in the wanted direction, in this direction we will have reduced noise. The detuned cavity performs such a rotation.

- *C. Aidala*:

Do you know where the name comes from?

- *A. Giazotto*:

Something that it was invented some years ago, and expresses the fact that there is some kind of action connected with weight movement, the mirrors in this case. I don't know the exact origin.

- *J. Bourjaily* :

A stable aspherically rotating neutron star may oscillate about a particular shape producing amplitude-modulated gravitational waves. It appears as if gravitational waves detectors would be insensitive to this because of the need to Fourier-transform $O(10^2)$ days worth of data. Is there any way for this type of information be recovered? Surely, this could have implications for our understanding of extremely-dense matter (i.e. neutron stars).

- *A.Giazotto*:

Well, you have the usual ellipticity which is responsible for the GW emission. And you will see modulation of the continuos signal. If the modulation is non-periodic we can fail detection.

- *J.F. Sandin*:

As far as I know, in theory, gravitational waves are described as a perturbation in a flat space-time background. I wonder, is the space-time in our Galaxy sufficiently flat for gravitational wave signals to be detected?

- *A.Giazotto*:

Well, it is detectable, and the level of such a curvature is not very important for us, for the time being. It will become important when GW astronomy will begin.

GW passing through enormous masses very concentrated, can have some lensing, the Einstein lensing. But close to here, I don't think there are such a concentration of masses that can really perturbe GW motion, in a detectable way. Someone has proposed to use Sun for focussing GW, but the focus is a bit far away in to space.

- *G. Veneziano*:

Is the upgrade that you described today specific to VIRGO or would it be applicable to other interferometers, such as LIGO?

- *A.Giazotto*:

I have no evidence of any experiment, but we are thinking very much a little bit togeter on all these things. So, it is a common target also for GEO group in Hannover. It is somethings that requires big efforts. In gravity waves, in particular in Europe, we are very few scientist doing this kind of job, compared to those who make high energy physics. And we would really like an improvement in the number of people, also for performing tests on Ponderomotive squeezing on a bench table and other very important tests; we are just delaying these important studies because we have too many things to do on the machine before to start making this kind of R&D.

- *G. Veneziano:*

There is something which is very specific to VIRGO that makes this sqeezing work for you and not for instance in LIGO?

- *A. Giazotto*:

No I don't think so, it is a general request. In the same way as the laser is common to everybody, more or less, squeezing should be an upgrade of every detector.

- *J. Kamenik*:

What are the estimates of the furthest sources that can still be detected?

If those can be even outside our Galaxy their signals could be disturbed by the strong gravitational field (curvature) of the center of our Galaxy?

- *A.Giazotto*:

It depends very much on the source. The periodic sources, are so weak that you can only detect galactic ones, like pulsars or rotating neutron stars. Even if you integrate for one year and you are so clever to point the right direction in such way to get rid of the doppler shift, the signal will be excruciately small. Instead the cohalescing binaries have a very large conversion efficiency in GW, they are between the most efficient sources and can be detected from 20-30 Mpc far away.

- *S.Ferrara*:

Modern theories of gravitation, such as superstrings or M theories, predict the possible existence of light gravitationals coupled scales. Could such particles have some influence on your experiments?

- *A.Giazotto*:

In the case in which the waves are not coming along the perpendicular to Virgo plane we may have a net phase shift difference between the two arms and the GW can be detected. So it is something that has to be looked for.

- *S.Ferrara*:

There is today even in a new cosmology, which is call modular cosmology. We just do with the fact that in this theories like string or M theories you predict many scalar which come from geometry of space-time. That would be very interesting to have a test coming from gravitational…

- *A.Giazotto*:

This is also in the package of the things we want to observe.

How and Where to go *Within* the Standard Model?: sQGP/CGC @ RHIC/LHC

Miklos Gyulassy[a] and Larry McLerran[b]

[a]*Physics Department, Columbia UniversityNew York, NY USA*

[b]*Physics Department POB 5000 Brookhaven National Laboratory Upton, NY 11973 USA*

Abstract

Three lectures are presented on two new forms of QCD matter discovered in ultra-relativistic nuclear collisions at RHIC. A large body of data now provides compelling evidence that an especially strongly interacting form of Quark Gluon Plasma (sQGP) is created in $AuAu$ collisions in the $\sqrt{s} = 100 - 200$ AGeV energy range. There is also a growing body of evidence that the sQGP originates from another novel form of matter called the CGC.

1 Introduction

Since the Relativistic Heavy Ion Collider began operations at the Brookhaven National Laboratory in 2000, a very large amount of new experimental information has become available about the properties of matter at extremely high energy densities. In May 2004, a special RBRC workshop was organized[1] to try to assess the new physics implications of the first few hundred experimental publications. The three lectures presented in this school are based on the report written in collaboration with Larry McLerran in Ref.[2] and updated here for these proceedings. Other relevant references for this material can be found in refs.[3]-[8]. Even more general extensive reviews of this field can be found in Ref.[10].

While the $\sim$ 1000 experimentalists at RHIC remain cautious about their discoveries and remain unsure of what to name the new states of matter [11]-[14], most theorists have adopted the abbreviated name *s*QGP (*Strongly Interacting* Quark-Gluon Plasma), that was suggested by T.D. Lee[3]. To me the "s" abbreviation also stands for *"surprising"* , *"special"*, and even, *"spectacular"* and *"startling"* in many respects. As terabytes of new data continue to stream in at a furious baud rate, the mystery of the true nature of the sQGP continues to deepen. At the same time a second, entirely different, form of nuclear matter may have also been discovered at RHIC. This has been named CGC, Color Glass Condensate, by Larry McLerran.

The aim of these Erice lectures is to introduce both the data and the evolving theoretical frameworks that provide the complelling case that both the sQGP and the CGC have been sighted. The relentless stream of new data from RHIC is now driving the evolution of both concepts and points to a very promising set of future directions on the question of "How and Where to go *Within* the Standard Model?" even as the new directions discussed by other speakers materialize *Beyond* the Standard Model.

1.1 Historical Background

Thirty years ago[15], T.D. Lee suggested that it would be interesting to explore new phenomena "by distributing high energy or high nucleon density over a relatively large volume." In this way one could temporarily restore broken symmetries of the physical vacuum and possibly create novel abnormal dense states of nuclear matter[16]. W. Greiner and collaborators pointed out that the required high densities could be achieved via relativistic heavy ion collisions[17]. Concurrently, Collins and Perry and others[18] realized that the asymptotic freedom property of Quantum Chromodynamics (QCD) implies the existence of an ultra-dense form of matter with deconfined quarks and gluons, called later the Quark-Gluon Plasma (QGP)[19]. In 1982 J.D. Bjorken developed a relativistic hydrodynamic theory[20] of the novel "inside-out" evolution of the central low baryon density regions of ultra-relativistic heavy ion collisions. While many signatures of QGP formation were proposed, he suggested that "if the quark-gluon plasma is produced, it will manifest itself in experimental signatures as yet unforeseen. The system is after all a complicated relativistic fluid subject to highly nonlinear forces. ... If very interesting and novel phenomena will be seen in ion-ion collisions, theory must develop the capability to interpret them, if not to predict them".

With this theoretical background, the 1983 DOE/NSAC Long Range Plan set in motion plans that led to the construction of the Relativistic Heavy Ion Collider (RHIC) [21] to explore properties of ultra-dense matter above the deconfinement transition point. Now, after its first three years of operation, a vast data base[22] on $p+p, D+Au$, and $Au+Au$ at $\sqrt{s}=20-200$ AGeV has been harvested from RHIC. The published data are available through 22 (4 PRL) publications from BRAMHS[23], 92 (15 PRL) from PHENIX[24], 34 (6 PRL) from PHOBOS[25], and 127 (21 PRL) from STAR[26]. This body of data extends and builds upon the knowledge gained about dense hadronic matter measured at the SPS/CERN (publications include 108 NA49/35, 69 NA50/38, 26 CERES/NA45, 79 WA98/80, 32 na57/wa97) at $\sqrt{s}\leq 20$ AGeV The SPS heavy ion data already displayed several signatures that hinted at the onset of QGP formation[27]. Based on the SPS data and theoretical predictions, RHIC with its factor of 10 increase in the center of mass energy to 200 AGeV was assured to create matter well above the deconfinement transition point. The fourth year run of RHIC with Au+Au at 200 AGeV has just concluded with a gain of about another factor ~ 20 integrated luminosity ($\sim 1400/\mu$b). These and future RHIC data with significantly upgraded detectors will provide even higher resolution measurements of the detailed properties of the new forms of matter discovered at RHIC.

While there naturally remain many open questions and unsolved puzzles, a striking set of new phenomena have been conclusively discovered at RHIC. Those phenomena furthermore can now be readily interpreted and predicted from significant advances in theory and from the empirical information gained from the past 30 years from the Bevalac, SIS, AGS, and SPS experiments.

In this report, we discuss[28, 29] the evidence that at least one and possibly two new forms of QCD matter have been discovered at RHIC. We consider the Quark Gluon Plasma, which is a form of matter characterized by a thermal equilibrium density matrix of a system of quarks and gluons. We also consider the Color Glass Condensate (CGC), which is a form of matter characterized by a universal initial density matrix which describes high energy strongly interacting particles - including nuclei. The QGP is the incoherent thermal limit of QCD matter at high temperatures while the CGC is the coherent limit of QCD at high energies. Since the QGP has to be created at RHIC from the interaction of initial nuclear enhanced coherent chromo electric magnetic fields, both limiting forms of QCD matter need to be considered at RHIC.

The first 300+ published experimental papers from RHIC have of course only barely scratched the surface of the new physics of these forms of matter, but the data are so striking and decisive that several strong physics conclusions can already be drawn. They establish empirically that a special form of strongly coupled QGP (sQGP) exists with remarkable properties. In addition, there is growing evidence that its source is well described by a saturated gluon CGC initial state. These RHIC discoveries and those at the SPS/CERN pave a clear path for future systematic studies of these new forms of matter in the laboratory.

We begin in section 2 by describing the basic ingredients of the QGP hypothesis. The hypothesis concerning the existence and properties of this form of matter has a firm basis in QCD. Much is known about its theoretical properties on the basis of numerical computation within QCD (lattice gauge theory). In section 3 we review the CGC hypothesis, which is newer but is also based firmly in QCD. The CGC hypothesis can be tested in a wide range of experimental environments (HERA, RHIC,LHC,eRHIC) not restricted to heavy ion collisions. It is newer, and because of this, somewhat more tenuous than is the QGP hypothesis.

The scientific method is based on the paradigm that theories are tested by falsification. This is an important concept, since simple models with many adjustable parameters are often used to "fit" heavy ion data. However, one of the most compelling motivations for extending studies of heavy ion physics into the ultra-relativistic RHIC energy frontier $\sqrt{s} \sim 20 - 200$ AGeV, is that predictions, based on the QCD theory itself, of new physics under extreme conditions of temperature and energy can be tested. For RHIC and higher energies controlled QCD theoretic approximations and methods have been developed that are applicable to a wide class of observables.

At lower energies, BEVALAC, SIS, AGS, and SPS, the physics of nuclear collisions is now known to be dominated by the non-equilibrium dynamics of the confined *intermediate* phase of QCD, known as hadronic resonance matter[30]. Unfortunately, even lattice QCD methods are not yet powerful enough to predict the dynamics or thermodynamics of this intermediate form of matter. For low temperature nuclear matter, effective quantum hadro-dynamic theories (QHD, Chiral Perturbation Theory) have successfully been constructed incorporating the known (Lorentz and Chiral) symmetries of QCD. However, at moderate temperatures where matter is in the hadron resonance excitation region (below the deconfinement temperature) no quantitative theory yet exists. This handicaps the interpretation of data at such lower energies in terms of fundamental QCD properties. Many phenomenological hadronic dynamical models have been advanced to help interpret data at lower energies, but as yet they have not evolved into a consistent effective theory. The existing huge data base from lower energies provides valuable information for further development of such an effective theory. However, RHIC energies provide for the first time the possibility of exploring a qualitatively new kinematic regime where the uncertainties due to our as yet incomplete understanding of the intermediate hadronic resonance phase of QCD may be minimized.

The case that we present in this report, based on the published data from RHIC, about the Quark Gluon Plasma and the Color Glass Condensate is predicated by the firm root of these concepts in first principles in QCD. As such, these concepts must be tested thoroughly through a wide array of observables to see whether they are consistent with available measurements, and that they are robust in their predictions. It was a priori not at all obvious whether any observables exist at RHIC that can be described quantitatively by the QGP or CGC concepts. This is where a healthy bit of experimental "luck" was essential in order to find the "needles in the haystack" that are least distorted by uncertain non-equilibrium hadronic final state dynamics. We make the case in this report that a few sharp needles have been found through three convergent lines of empirical evidence that point to the discovery of a new strongly coupled QGP and its source, the saturated CGC.

It is also important to understand when an approximation to the QCD theory breaks down. Only special limits of QCD can be quantitatively predicted at this time: (1) long wavelength QCD thermodynamics, (2) very short wavelength pQCD, and (3) very high energy CGC coherence. New quark coalescence techniques are being developed to address the intermediate wavelength observables. Any breakdown of a given approximation must be understood for solid reasons. One case in point is applying the QGP hypothesis to observables (such as radial flow) which are strongly influenced by the uncertain late time dynamics of hadronic matter in heavy ion collisions. Another case is at the earliest formation times, when the description changes between the initial coherent CGC and the produced thermal QGP. In the latter case, one needs to determine directly from careful control experiments, which approximation to QCD is better suited to the phenomena under study. In this report we will discuss both the strengths and limitations of the present theoretical understanding of the RHIC discoveries.

2 What is the Quark Gluon Plasma?

Quantum Chromodynamics (QCD) predicts the existence of a deconfined form of matter called a Quark Gluon Plasma (QGP), in which the quark and gluon degrees of freedom normally confined within hadrons are mostly liberated[18]. This transition occurs when the energy density of matter is of the order of that of matter inside a proton. This density is about an order of magnitude larger than the energy density inside of atomic nuclei, that is about $1 - 10\ GeV/fm^3$.

To get a more accurate determination of the energy of this transition to a Quark Gluon Plasma, QCD can be numerically studied using numerical methods[31]-[34]. Such computations show that there is a rapid rise of the energy density, $\epsilon(T)$, of matter when the temperature reaches $T \approx T_c \sim 160$ MeV. The energy density changes by about an order of magnitude in a narrow range of temperatures $\Delta T \sim 10 - 20\ MeV$ as can be seen from Fig. 1.

One can understand this transition as a change in the number of degrees of freedom of the system. Far below T_c, the active hadronic degrees of freedom are limited to three, corresponding to a dilute gas of 3 charge states of pions. Above T_c, 8 colors of gluons times two helicity degrees of freedom become activated. In addition, there are $N_f(T) \approx 2 - 3$ active light flavors of quarks (for T not far above T_c). Each flavor has equal number of quarks and antiquarks when the chemical potentials vanish, and each have two spin states and three colors. Therefore, there are $g_{q\bar{q}}(T) \approx 24 - 36$ quark degrees of freedom in a QGP. In the quark gluon plasma phase, there are then about 40 - 50 internal degrees of freedom in the temperature range $(1 - 3)T_c$, while the low temperature and vanishing chemical potentials the pion gas has 3. Since the energy density, pressure and entropy are all roughly proportional to the number of degrees of freedom, one understands this rapid change in the energy density over a narrow range of temperature as a change in the degrees of freedom between the confined and deconfined worlds.

The system above T_c is called a plasma because the degrees of freedom carry the non-Abelian analog of charge as in ordinary plasmas. Just as there are different regimes of ordinary plasmas, the QGP plasma also has weakly coupled and strongly coupled limits. At extremely high temperatures, the asymptotic freedom property of QCD predicts that it will be weakly coupled[18]. For

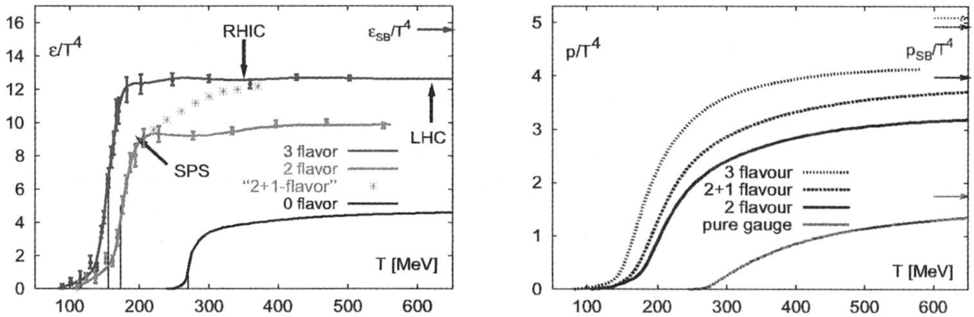

Figure 1: (a) The energy density as a function of temperature scaled by T^4 from lattice QCD [31]. Various number of species of quarks are considered. The realistic case is for $2+1$ flavors. An estimate of the typical temperature reached at SPS and RHIC, and estimated for LHC is included in the figure.(b) The pressure as a function of temperature scaled by T^4. Note that the pressure is continuous in the region where there is a sharp change in the energy density.

moderate temperatures $(1-3)T_c$ accessible at RHIC, on the other hand, the plasma is predicted by nonperturbative lattice techniques to be strongly coupled - even though the thermodynamic variables are near the ideal Stefan Boltzmann limit. Thus, nonperturbative correlations beyond dielectric phenomena persist in the QGP well beyond T_c.

The transition between the confined hadronic world and the deconfined QGP world may or may not be a phase transition in the strict statistical mechanical sense. Stricly speaking, a phase transition requires a mathematical discontinuity in the energy density or one of its derivatives in the infinite volume limit. The QGP transition may in fact be a "crossover", or rapid change, as is suggested numerical computation and a number of theoretical arguments. Nevertheless, the change as measured in numerical computation is very abrupt as seen in Fig. 1.

We shall refer to matter in the transition region as a "mixed phase". If there were a strict statistical mechanical first order phase transition, then matter in this region would be a mixture of hadronic gas phase and quark gluon plasma domains, as is the case when there is phase coexistence between water and ice. If there is a rapid crossover, then many of the bulk properties of the system, which are determined by the relation between energy density and pressure, are to a good approximation similar to those when there is a strict phase transition.

In the transition region, the energy density changes by roughly an order of magnitude, but the pressure is continuous and varies slowly. The sound velocity, $c_S^2 = dP/d\epsilon$ must therefore become very small in this range of energy densities[32, 33]. It is very difficult to generate pressure gradients and do mechanical work in the mixed phase region, since as we vary the energy density we generate little change in pressure. We will call a relation between energy density and pressure (an equation of state) stiff when the sound velocity is big (c_s of order the speed of light) and soft when it is small. The quark gluon plasma equation of state is stiff at high temperature, but becomes soft near T_c. The sound velocity as a function of temperature as determined by numerical computation is shown in Fig. 2a. It is expected that hadronic matter again becomes stiffer below T_c, and eventually softens as the temperature tends to zero. Unfortunately, lattice QCD is still not powerful enough to predict accurately the thermodynamic properties of hadronic matter below T_c. Preliminary results[35] (still with rather large pion mass ~ 700 MeV) are roughly consistent with a heavy hadronic resonance gas equation of state, but the thermodynamics of the confined phase of QCD remains an open problem. This fact again underscores the necessity of concentrating on those (few) observables that are least influenced by the poorly known hadronic phase dynamics.

The relationship between the energy density and pressure determines the evolution of matter from a given initial condition *IF* local equilibrium is maintained. This then can determine experimentally by measuring "barometric" observables. The softening of the QGP equation of state near T_c is the key feature that can be looked for in the collective hydrodynamic flow patterns produced when the plasma expands.

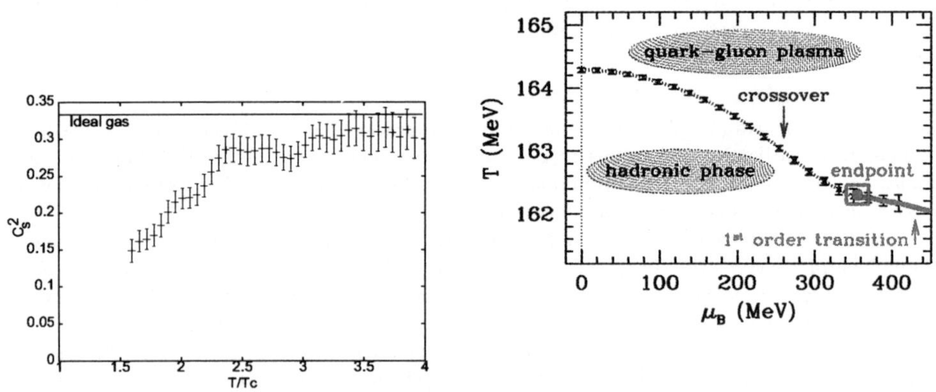

Figure 2: Important features of the QGP equation of state. (a) The speed of sound[32, 33] $c_s^2 = d\epsilon/dP$ drops below 1/3 for $T < 2T_c \approx 300$ MeV. (b) Right panel shows a current estimate of the location of the tri-critical point at finite baryon density[34]

Another distinctive feature of the QGP phase diagram is shown in the right panel in Fig.(2). Recent numerical calculations[34] have begun to show evidence that the QGP may have a phase transition in the strict statistical mechanical sense, if one also allows the system to have a high baryon number density as well as high temperature. [36, 37] In the plot on the right hand side of Fig. 2, the possible phases of QCD are shown as a function of both temperature and a measure of the baryon number density, μ_B. At low baryon number density, there is probably a rapid crossover from the low density hadronic world to that of the quark gluon plasma. At higher baryon number density, there is quite likely a first order phase transition. The computations on this issue are still work in progress and extrapolation to realistic quark masses are very difficult, so the position of the endpoint of first order phase transitions is still rather uncertain. There is an ongoing program at the SPS concentrating on lower energies to search for possible signatures of high baryon density quark plasma transition.

The quark gluon plasma has the property that quarks and gluons are no longer confined inside of hadrons. This is unlike zero temperature and baryon number density QCD. The transition between the confined and deconfined world may provide us insights into the nature of confinement.

In the QCD equations of motion the small mass of the "light" u,d quarks can be neglected and the system possesses a special chiral symmetry. This chiral symmetry prevails at high temperature, in spite of the fact that quarks acquire a collective dynamical mass $\sim gT$ as do gluons. Below T_c, not only do quarks and gluons become confined into hadrons, but the chiral symmetry of QCD is also broken as manifested by the appearance of the 3 light pion degrees of freedom while the quarks are bound in heavy nucleons.

The Quark Gluon Plasma is important to search for at RHIC because:

- It is the ultimate, primordial form of QCD matter at high temperature or baryon number density (at least up to the electro weak scale at about $T_{EW} \sim 10^3 T_c$).
- It was present during the first few microseconds of the Big Bang according to current cosmology.
- It may occur naturally in supernovae, gamma ray bursts and neutron stars as matter at high baryon number density and relatively low temperature.
- It provides an example of phase transitions which may occur at a variety of higher temperature scales in the early universe.
- It may provide us important information concerning the origin of mass for matter, and how quarks are confined into hadrons.

3 What is the Color Glass Condensate?

The ideas for the Color Glass Condensate are motivated by HERA data on the gluon distribution function shown in Fig. 3(a) [38]. The gluon density , $xG(x, Q^2)$, rises rapidly as a function of decreasing fractional momentum, x, or increasing resolution, Q. This rise in the gluon density ultimately owes its origin to the non-Abelian nature of QCD and that the gluons carry color charge. At higher and higher energies, smaller x and larger Q become kinematically accessible. The rapid rise with $\log(1/x)$ was expected in a variety of theoretical works[39]-[41]. Due to the intrinsic non-linearity of QCD, gluon showers generate more gluon showers - producing an exponential avalanche toward small x. The physical consequence of this exponential growth is that the density of gluons per unit area per unit rapidity of any hadron including nuclei must increase rapidly as x decreases [42]. This follows because the transverse size, as seen via the total cross sections, rise more slowly at high energies than the number of gluons. This is illustrated in Fig. 3(b). The non-linearity of gluon interactions, however, led to the conjecture that the transverse density of gluons measured at some fixed Q^2 resolution scale should eventually become limited , that is, there is gluon saturation at sufficiently high energies. [39]-[40], [42]

The low x gluons are closely packed together in the transverse direction, therefore, as in the thermal case at extreme temperatures or chemical potentials, the strong interaction strength must

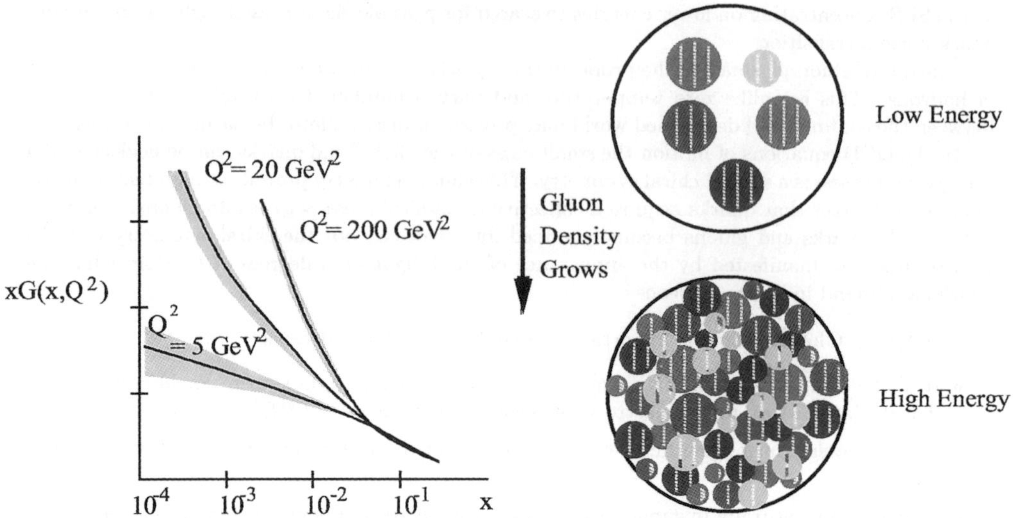

Figure 3: (a)The HERA data for the gluon distribution function as a function of x for various values of Q^2. (b) A physical picture of the low x gluon density inside a hadron as a function of energy

become weak, $\alpha_S \ll 1$. Weakly coupled systems should be possible to understand from first principles in QCD [42].

This dense but weakly coupled system is called a Color Glass Condensate for reasons we now enumerate:[43]

- **Color** The gluons which make up this matter are colored.
- **Glass** The gluons at small x are generated from gluons at larger values of x. In the infinite momentum frame, these larger momentum gluons travel very fast and their natural time scales are Lorentz time dilated. This time dilated scale is transferred to the low x degrees of freedom which therefore evolve very slowly compared to natural time scales. This is the property of a glass.
- **Condensate** The dimensionless transverse phase space density

$$\rho = \frac{1}{\pi R^2} \frac{dN}{dy d^2 p_T} \tag{1}$$

saturates at a value $\rho = c/\alpha_S(Q_s)$, where c is a constant of order unity. Saturation is due to the competition between extra gluon radiation originating from the source current, $\propto \rho$, and non-linear gluon fusion, $\propto -\alpha_S \rho^2$, that reduces the number of gluons at high density. Because $\alpha_S << 1$, this means that the quantum mechanical states of the system associated with the condensate are multiply occupied. They are highly coherent, and share some properties of Bose condensates. The gluon occupation factor is very high, of order $1/\alpha_S$, but it is only slowly (logarithmically) increasing when further increasing the energy, or decreasing the transverse momentum. This provides saturation and cures the infrared problem of the traditional BFKL approach [44].

There is a critical momentum scale, $Q_s(x, A)$, which controls the occupation number through $1/\alpha_S(Q_s)$. This is called the CGC saturation scale. The transverse phase space density, ρ, is constant only up to $p_T < Q_s(x, A)$. For wavelengths, $1/p_T$, much smaller than $1/Q_s$, the coupling becomes even weaker and perturbative QCD predicts that $\rho \propto \alpha(p_T)/(R^2 p_T^2)$.

Note that Q_s plays a role in the CGC form of matter similar to what T_c plays in the QGP form of matter. Both delineate two different phases of matter. However, in the QGP case, the critical scale, $T_c \sim \Lambda_{QCD}$, is small and the matter on both sides of the transition remain in the strong coupling sector of QCD! In the CGC case, on the other hand, it is possible to find kinematic regimes of x that depend on A, where the CGC can be weakly coupled but in nonlinear regime due to high occupation numbers.

The crux of the search for the CGC therefore is to locate those kinematic regimes where the corrections to the weak coupling methods can be controlled. The non-linear dependence of Q_s on x and A will be discussed further in section 6. We note here only that both as x becomes small and A becomes large, Q_s grows. The saturation momentum itself does not saturate. The gluon distribution function for resolution scale $Q \leq Q_s$ grows slowly and saturates, while the gluon distribution function for $Q \geq Q_s$ grows rapidly. The physics is easy to understand: As more gluons are added to the hadron, they have to go to the unsaturated region since the saturated region is already densely packed.

The Color Glass Condensate is important to search for at RHIC because:

- It represents the universal form of high energy QCD wavefunctions at small x, not only of hadrons but heavy nuclei as well.
- It is a new form of matter because the gluons inside the hadron are separated by distance scales small compared to the hadron size, they evolve on time scales long compared to microphysics time scales, and the CGC itself is specified by a special new density matrix that encodes the non-linear virtual fields of QCD in the high energy limit.
- It already begins to become important at $x \sim 10^{-2}$ in protons at HERA, and and should begin to become important at comparable values of x at RHIC.
- It provides a rigorous QCD theoretical description of the initial state in $A + A$ from which the QGP must evolve.
- It can be tested directly at RHIC via $p + A$ or $D + A$, where no final state interactions and hence no QGP formation can occur.

4 The Space-Time Picture of Ultra-Relativistic Nuclear Collisions

Heavy ion collisions at ultrarelativistic energies are visualized in Fig. 4 as the collision of two sheets of colored glass.[45] The nuclei appear as sheets at ultrarelativistic energies because of Lorentz contraction. The CGC gluons are shown as vectors which represent the polarization of the gluons, and by colors corresponding to the various colors of gluons.

At ultrarelativistic energies, these sheets pass through one another. In their wake is left melting colored glass, which eventually materializes as quarks and gluons. These quarks and gluons would naturally form in their rest frame on some natural microphysics time scale. For the saturated color glass, this proper formation time scale, τ_0, is of order the inverse saturation momentum. At RHIC, $1/Q_s \sim 0.2$ fm/c $\sim 5 \times 10^{-25}$ sec. Note that 0.2 fm/c is also comparable to the natural crossing time of two 10 fm nuclei, each contracted by a gamma factor 100, in the center of mass frame at RHIC. For particles with a large momentum or rapidity along the beam axis, this time scale is Lorentz dilated. This means that the slow (smaller rapidity) particles are produced first towards the center of the collision regions and the fast (larger rapidity) particles are produced later further away from the collision region.

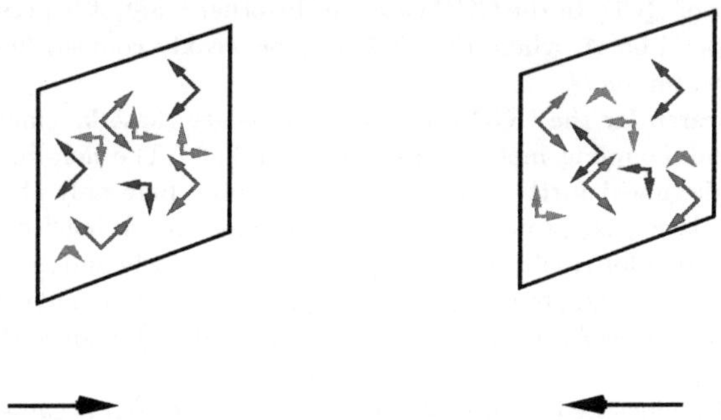

Figure 4: The collision of two sheets of colored glass. The arrows represent the polarization vectors for the gluons which live on the sheets, and their colors correspond to the different colors of gluons

This Bjorken "inside-out" correlation[20] between space and momentum is similar to what happens to matter in Hubble expansion in cosmology. The stars which are further away have larger outward velocities. This means that the matter produced at RHIC, like the universe in cosmology is born expanding. One important difference is that the "mini-bang" at RHIC is born with one dimensional Hubble flow along the collision axis, while the Big-Bang is three dimensional. This is shown in Fig. 5

As this system expands, it cools. On some time scale $\tau_{eq} > \tau_0$ the produced quarks and gluons may thermalize. It is by no means obvious that local equilibrium can be reached on the short time scale $\sim (1-2)R/c \sim 10$ fm/c available in nuclear collisions. *IF* local equilibrium is reached early with $\tau_{eq} < 1$ fm/c , then the QGP can develop collective flow according to the laws of hydrodynamics. In any case, when the energy density decreases due to expansion below about 1 GeV/fm^3, the QGP must begin to hadronize through some mixture of hadrons and quarks and gluons. Eventually, all the quarks and gluons must become confined into hadrons that may still interact as hadronic matter before being detected.

The particle multiplicity as a function of energy has been measured at RHIC[46], as shown in Fig. 6. Combining the multiplicity data together with the measurements of transverse energy or of typical particle transverse momenta, one can determine the energy density of the matter when it decouples.[47] One can then extrapolate backwards in time using 1 dimensional (Bjorken)

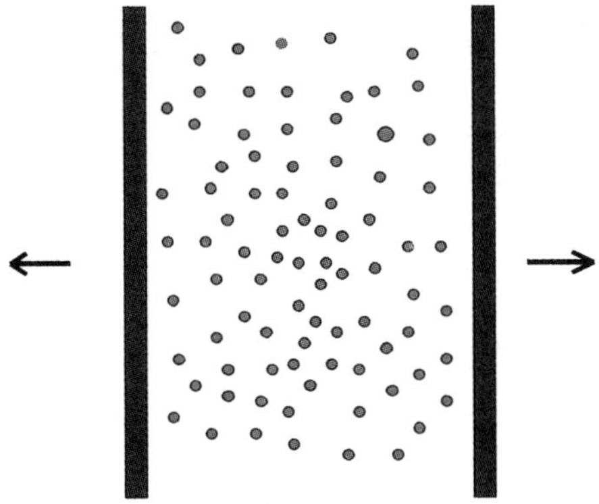

Figure 5: Particles being produced after the collision of two nuclei.

expansion, since decoupling occurs soon after the matter begins to expands three dimensionally. We can extrapolate backwards only until τ_0, when the matter is formed from the "shattered" Color Glass.

To do this extrapolation we use that the proper density of particles falls as $N/V \sim 1/\tau$ during 1 dimensional expansion. If the particles expand without interaction and work, then the energy per particle remains constant (and $E/V(\tau_0) \approx (dE_T/dy)/(\tau_0 \pi R^2)$ in terms of the final observed transverse energy per unit rapidity). If the particles thermalize, then $E/N(\tau) \sim 3T(\tau)$, and the *entropy* rather than the energy per particle remains constant. For a massless gas, the temperature then falls as $T \sim \tau^{-1/3}$. For a gas which is not massless or not in perfect equilibrium, the temperature falls somewhere in the range $T_o > T(\tau) > T_o(\tau_{eq}/\tau)^{1/3}$ This 1 dimensional expansion continues until the system begins to feel the effects of finite size in the transverse direction, and then rapidly cools through three dimensional expansion. We shall take a conservative overestimate of this time to be of order $t_0 \sim 0.3$ fm/c The extrapolation of the energy density backwards is bounded by $\epsilon_f(t_f/t) < \epsilon(t) < \epsilon_f(t_f/t)^{4/3}$. The lower bound is that assuming that the particles do not thermalize and their typical energy is frozen. The upper bound assumes that the system thermalizes as an ideal massless gas. These bounds on the energy density is shown in Fig. 7. On the left axis is the energy density and on the bottom axis is time. The system begins as a coherent Color Glass Condensate, then melts to Quark Gluon Matter which may eventually thermalize to a Quark Gluon Plasma. At a time $\sim 3fm/c$, the plasma becomes a mixture of quarks, gluons and hadrons which further expand together.

At a time of about 10 fm/c, the system falls apart and decouples. At a time of $t \sim 1$ fm/c, the estimate we make is identical to the Bjorken energy density estimate, and this provides a lower bound on the energy density achieved in the collision. (All estimates agree that by a time of order 1 fm/c, matter has been formed.) The upper bound corresponds to assuming that the system expands as a massless thermal gas from a melting time of .3 fm/c. (If the time was reduced, the upper bound would be increased yet further.) The bounds on the initial energy density are

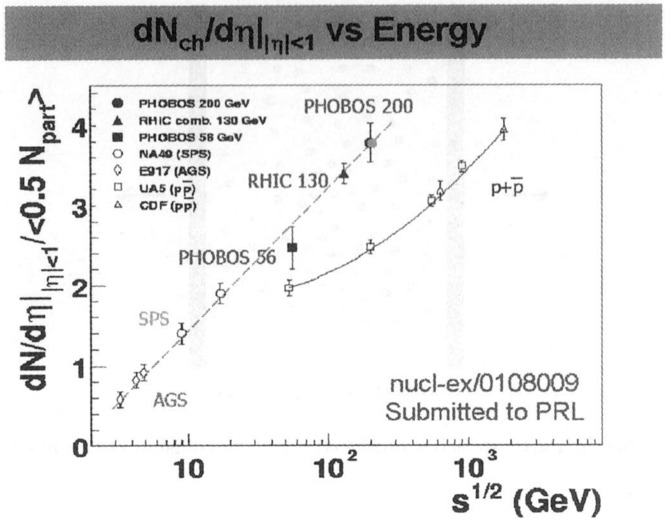

Figure 6: The particle multiplicity as a function of c.m. energy, $\sqrt{s}$, per nucleon pair at RHIC[46] and lower AGS and SPS energies.

therefore

$$2 - 3\ GeV/fm^3 \lesssim \epsilon \lesssim 20 - 30\ GeV/fm^3 \tag{2}$$

where we included a greater range of uncertainty in the upper limit because of the uncertainty associated with the formation time. The energy density of nuclear matter is about 0.15 GeV/fm^3, and even the lowest energy densities in these collisions is in excess of this. At late times, the energy density is about that of the cores of neutron stars, $\epsilon \sim 1\ GeV/fm^3$.

We conclude that based on the observed multiplicity at RHIC alone, the initial energy densities achieved in RHIC collisions can be high enough to produce a quark gluon plasma.

5 Empirical Evidence for QGP at RHIC

In this section, we discuss RHIC data which show that the matter produced in central collisions in RHIC becomes well thermalized, and behaving in a way consistent with theoretical expectations from QCD.

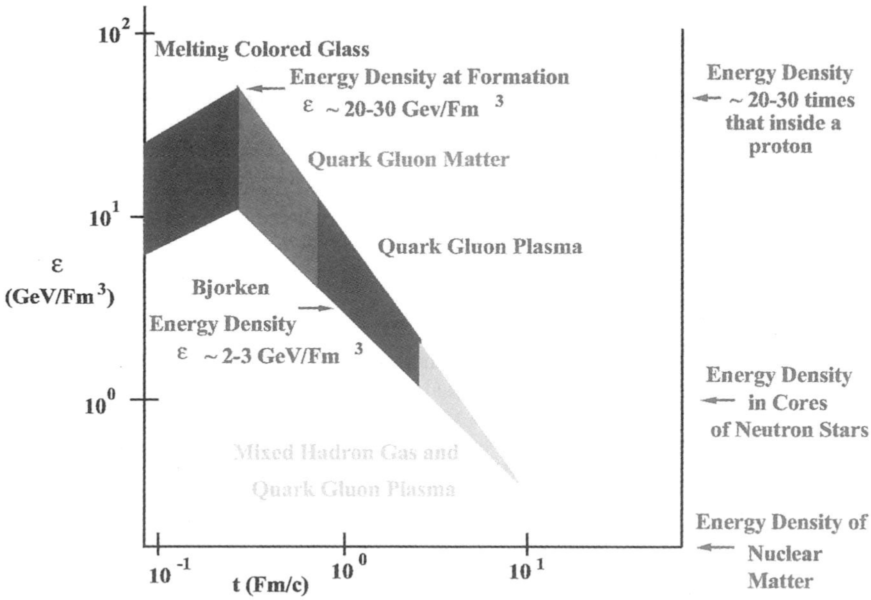

Figure 7: Bounds on the energy density as a function of time in heavy ion collisions.

5.1 Collective Flow

The identification of a new form of "bulk matter" requires the observation of novel and uniquely different collective properties from ones seen before. In heavy ion reactions the flow pattern of thousands of produced hadrons is the primary observable used to look for novel collective phenomena[17], [48]- [51]. The collective flow properties test two of the conditions necessary for the validity of the QGP hypothesis.

The first is the degree of thermalization. Nothing is yet known from lattice QCD about far off equilibrium dynamics of a QGP. However, the evolution of matter from some initial condition can be computed via the equations of viscous relativistic hydrodynamics if local equilibrium is maintained. These equations can be further approximated by perfect (Euler) fluid equations when the corrections due to viscosity can be neglected. Such viscous corrections can be neglected when scattering mean free paths are small compared to the scale of spatial gradients of the fluid.

The second condition is the validity of the numerically determined equation of state or relationship between energy density and pressure. The required input for perfect fluid hydrodynamical equations is the equation of state. With a specific initial boundary condition, the future evolution of the matter can be then predicted. We shall show that the data on elliptic flow confirms the idea that to a very good approximation, local thermal equilibrium is reached at RHIC energy and that the flow pattern is entirely consistent with numerical determinations of the equation of state from QCD.

The different types of collective flows are conveniently quantified in terms of the first few Fourier components of the azimuthal angle (angle around the beam axis for the collision) distribution. [52, 53], $v_n(y, p_T, N_p, h)$, of the centrality selected triple differential inclusive distribution of hadrons, h. The centrality or impact parameter range is usually specified by a range of associated multiplicities, from which the average number of participating nucleons, N_p, can be deduced. The azimuthal angle of the hadrons are measured relative to a globally determined estimate for the collision reaction plane angle $\Phi(M)$. The "directed" v_1 and "elliptic" v_2 flow components [51]-[53], [54]-[61] are readily identified from azimuthal dependence

$$\begin{aligned} \frac{dN_h(N_p)}{dy dp_T^2 d\phi} = \frac{dN_h(N_p)}{dy dp_T^2}\frac{1}{2\pi}(1 &+ 2v_1(y, p_T, N_p, h)\cos\phi \\ &+ 2v_2(y, p_T, N_p, h)\cos 2\phi + \cdots) \ . \end{aligned} \tag{3}$$

The first term in the above equation also contains information about flow. Produced particles should have their momentum spectrum broadened in heavy ion collisions relative to the case for proton-proton collisions. Because flow is due to a collective velocity, the flow effects should be largest for the most massive particles, and therefore the mass dependence is a powerful diagnostic tool.

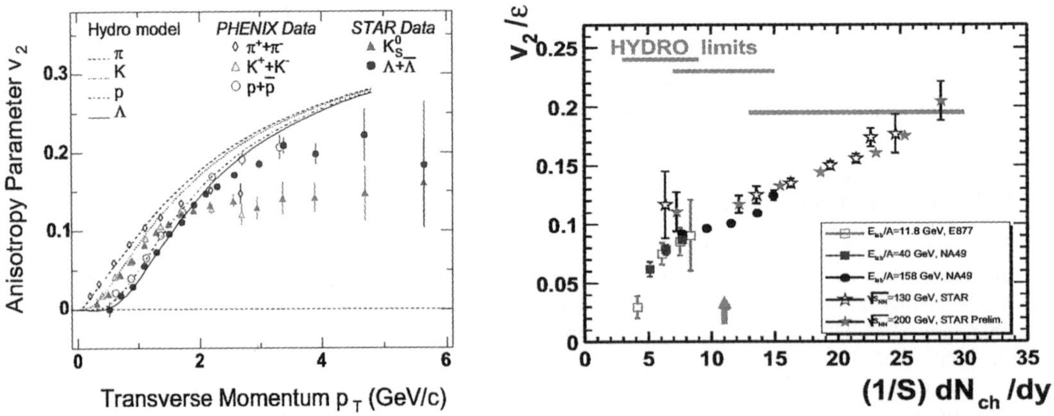

Figure 8: First line of evidence: Bulk collective flow is the barometric signature of QGP production. Left figure combines STAR [57]-[60] and PHENIX [61] measurements of the azimuthal elliptic flow ($v_2(p_T)$) of π, K, p, Λ in Au+Au at 200 AGeV. The predicted hydrodynamic flow pattern from [62]-[66] agrees well with observations in the bulk $p_T < 1$ GeV domain. Right figure from [55] shows v_2 scaled to the initial elliptic spatial anisotropy, ϵ, as a function of the charge particle density per unit transverse area. The bulk hydrodynamic limit is only attained at RHIC.

Figure (8) shows the striking bulk collectivity elliptic flow signature of QGP formation at RHIC. Unlike at SPS and lower energies, the observed large elliptic deformation $((1 + 2v_2)/(1 -$

$2v_2) \sim 1.5$) of the final transverse momentum distribution agrees for the first time with non-viscous hydrodynamic predictions [62]-[74] at least up to about $p_T \sim 1$ GeV/c. However, the right panel shows that when the local rapidity density per unit area [54, 55] drops below the values achieved at RHIC $\sim 30/\text{fm}^2$, then the elliptic flow (scaled by the initial spatial ellipticity, $\epsilon = \langle (y^2 - x^2)/(y^2 + x^2) \rangle \rangle$) falls below the perfect fluid hydrodynamic predictions. We will discuss in more detail the origin of the large discrepancy at SPS energies in the next section.

The most impressive feature in Fig.(8) is the agreement of the observed hadron mass dependence of the elliptic flow pattern for all hadron species, π, K, p, Λ, with the hydrodynamic predictions below 1 GeV/c. This is the QGP fingerprint that shows that there is a common bulk collective azimuthally asymmetric flow velocity field, $u^\mu(\tau, r, \phi)$. Such good agreement with the hadron mass dependence of the $v_2(p_T, m_h)$ data is furthermore only found when the input equation of state has the characteristic "softest point" near T_c as predicted by lattice QCD [62]-[74]. When equations of state without the predicted drop of speed of sound near T_c were used as input, the flow velocity field, especially that of the heavy baryon, was over estimated.

The flow velocity and temperature fields of a perfect (non-viscous) fluid obeys the hydrodynamic equations:

$$\partial_\mu \left\{ [\epsilon_{QCD}(T(x)) + P_{QCD}(T(x))] u^\mu(x) u^\nu(x) - g^{\mu\nu} P_{QCD}(T(x)) \right\} = 0 \,, \tag{4}$$

where $T(x)$ is the local temperature field, $P_{QCD}(T)$ is the QGP equation of state, and $\epsilon_{QCD}(T) = (TdP/dT - P)_{QCD}$ is the local proper energy density. The above equations apply in the rapidity window $|y| < 1$, where the baryon chemical potential can be neglected. Eq.(4) provides the barometric connection between the observed flow velocity and the theoretical properties of the QGP.

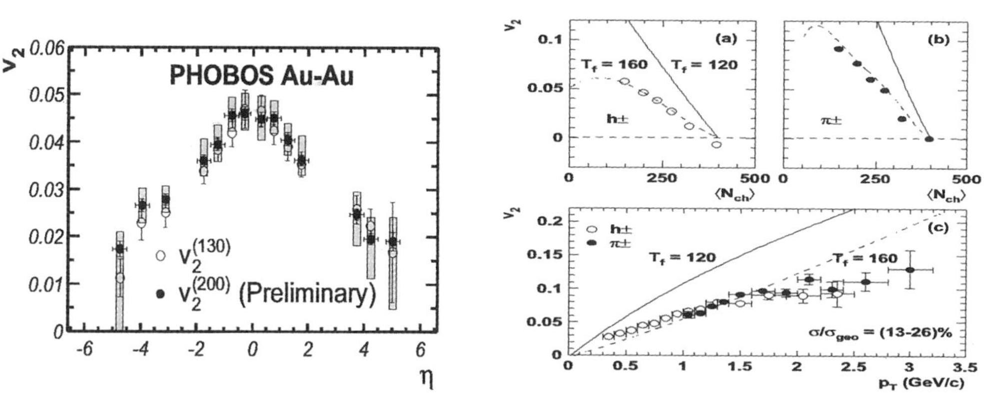

Figure 9: Left figure shows the pseudo rapidity dependence of elliptic from PHOBOS [76]. Right figure is CERES[77] data on elliptic flow at SPS. It is well below hydrodynamic predictions with freeze-out $T_f = 120$ MeV required to reproduce the single inclusive radial flow. Early freeze-out with $T_f = 160$ MeV, simulating effects of dissipation, is needed to reproduce the data.

We note that in any hydrodynamic treatment of collective flow there is a tradeoff between the combined effects due to the initial state boundary condition, the equation of state of the matter, and dissipation effects. In order to use the flow pattern to constrain the equation of state the initial condition must be constrained from other measurements and dissipation must be negligible. Here is where the measurements of the global multiplicity systematics[46, 47] are so important. The remarkably weak energy and centrality dependence of the bulk entropy observed via the $dNdy$ plays a pivotal role as perhaps the most convincing test of the CGC initial condition hypothesis at RHIC[75]. Without such an experimental and theoretical constraint on the initial condition no meaningful constraint on the QGP equation of state could have been found.

The study of the interplay between the equation of state and dissipative phenomena is more difficult and can only be untangled through detailed systematics of the flow pattern as a function of beam energy, centrality, and rapidity dependence. Here the detailed systematics from AGS and SPS data have played a pivotal role in helping sorting out the different viscous effects in hadronic and QGP matter as we discuss in the next section.

Why is v_2 more emphasized than v_1 or radial flow as a signature of QGP formation? The primary reason is that elliptic flow is generated mainly during the highest density phase of the evolution before the initial geometric spatial asymmetry of the plasma disappears. It comes from the azimuthal dependence of the pressure gradients, which can be studied by varying the centrality of the events [53]. Detailed parton transport [78] and hydrodynamic [68] calculations show that most of the v_2 at RHIC is produced before 3 fm/c and that elliptic flow is relatively insensitive to the late stage dissipative expansion of the hadronic phase. The reason for the generation of v_2 at relatively early times is that it is very difficult to convert the spatial anisotropy of the matter distribution into a momentum space anisotropy once the system cools into the mixed phase, since in the mixed phase pressure gradients cannot be set up. Actually, it was a surprise how well the observed collective flow agrees with perfect fluid hydrodynamic predictions. Ideal fluid flow requires very strong interactions of the quarks and gluons in the plasma at early times $\tau \gtrsim \tau_{eq} \approx 0.6$ fm/c.

In contrast, radial flow has been observed at all energies [79] and has been shown to be mainly sensitive to late time "pion wind" radial pressure gradients [30, 80], which continue to blow long after the QGP condenses into hadronic resonances.

5.2 The Breakdown of Bulk Collectivity

It is important to point out that no detailed 3+1D hydrodynamic calculation [70]-[74] has yet been able to reproduce the rapid decrease of $v_2(|\eta| > 1)$ observed by PHOBOS in Fig.(9). This is most likely due to the increasing role *hadronic* dissipation effects in the "corona" when the comoving density decreases with increasing y. The volume of the QGP shrinks while the hadronic corona thickens as the rapidity density dN/dy is reduced within a fixed nuclear geometry away from midrapidity. From the right panel of Fig.(8), we see that a decrease of the local transverse density from midrapidity RHIC conditions leads to an increasing deviation from the perfect fluid limit. The initial density was also observed to decrease at RHIC as $|y|$ increases [81]. Therefore, from the known SPS data, we should expect deviations from the perfect fluid limit away from the midrapidity region.

Another set of RHIC data that show deviations from perfect fluid hydrodynamic predictions is the centrality dependence of v_2. The observed $v_2(b)$ decreases relative to hydrodynamic predictions also when the impact parameter increases toward the more peripheral collisions. This is again due to the fact that the produced multiplicity, $dN/dy \propto N_p(b)$, decreases with increasing b. The hadronization time decreases with b since the QGP is formed with smaller initial density and the hadronic fluid is less efficient in transferring pressure into collective flow.

To elaborate further on this important point, Fig.9 shows CERES data [77] on $v_2(p_T)$ at SPS energy $\sqrt{s} = 17$ AGeV. In agreement with the NA49 data shown in the right panel of Fig.(8), the CERES data falls well below the hydrodynamic predictions. At even lower energies, AGS and BEVALAC, the v_2 even becomes *negative* and this "squeeze out" of plane [50] is now well understood in terms of low energy non-equilibrium nuclear transport theory [56, 82].

In order to account for the smallness of v_2 at SPS, hydrodynamics has to be frozen out at unphysically high densities and temperatures, $T_f \approx T_c$. However, the observed radial flow rules out this simple fix. The reduction of v_2 while maintaining radial flow can be approximately understood in approaches [68] that combined perfect fluid QGP hydrodynamics with dissipative final state hadronic evolution.

In light of the above discussion on the breakdown of collectivity due to hadronic dissipation at high rapidity and large impact parameters at RHIC and even at midrapidity at SPS and lower energies, the smallness of dissipative corrections in the central regions of RHIC is even more surprising. At mid-rapidities, the lack of substantial dissipation in the QGP phase is in itself a remarkable and unexpected discovery [83] at RHIC. Calculations based on parton transport theory [78] predicted large deviations from the ideal non-viscous hydrodynamic limit even in a QGP. Instead, the data show that the QGP at RHIC is almost a perfect fluid. A Navier Stokes analysis [69] of the RHIC data also indicates that the viscosity of the QGP must be about ten times smaller than expected if the QGP were a weakly interactive conventional Debye screened plasma. This unexpected feature of the QGP must be due to strong coupling QCD physics that persists to at least $3T_c$. (See [28, 83, 84] and refs therein for further discussion).

Summarizing this section: Elliptic flow measurements confirm that the quark-gluon matter produced at RHIC is to a very good approximation in local thermal equilibrium up to about 3 fm/c. In addition, the final hadron mass dependence of the flow pattern is remarkably consistent with numerical QCD computations of the equation of state. Viscous corrections furthermore appear to be surprisingly small during this early evolution. The produced Quark Gluon Plasma must therefore be very strongly interacting. Such behavior was not seen at lower energy because the highly dissipative hadronic fluid component masked the QGP flow signals. The perfect fluid behavior is also masked at RHIC at higher rapidities and in more peripheral reactions again due to the increased role of the dissipative hadronic "corona".

5.3 Perturbative QCD and Jet Quenching

In addition to the breakdown of perfect fluid collectivity at high rapidity seen in Fig.(9), Fig.(8) clearly shows that hydrodynamics also breaks down at very short wavelengths and high transverse momenta, $p_T > 2$ GeV. Instead of continuing to rise with p_T, the elliptic asymmetry stops growing and the difference between baryon vs meson v_2 even reverses sign. Between $2 < p_T < 5$ GeV the baryon $v_2^B(p_T)$ exceeds the meson $v_2^M(p_T)$ by approximately 3/2. For such short wavelength components of the QGP, local equilibrium simply cannot be maintained due the fundamental asymptotic freedom property of QCD, i.e. the coupling strength becomes too weak.

In this section, we concentrate on the $p_T > 2$ GeV meson observables that can be readily understood in terms of QGP modified perturbative QCD (pQCD) dynamics [85, 86]. (Baryons at intermediate $2\ GeV \leq p_T \leq 5\ GeV$ are outside the range of a perturbative treatment and several competing mechanisms have been proposed and are under theoretical development[87, 88, 9].)

The quantitative study of short wavelength partonic pQCD dynamics focuses on the rare high p_T power law tails that extend far beyond the typical (long wavelength) scales $p < 3T \sim 1$ GeV of the bulk QGP. The second major discovery at RHIC is that the non-equilibrium power law high p_T jet distributions remain power law like but are strongly quenched [89]-[97]. Furthermore, the quenching pattern has a distinct centrality, p_T, azimuthal angle, and hadron flavor dependence that can be used to test the underlying dynamics in many independent ways.

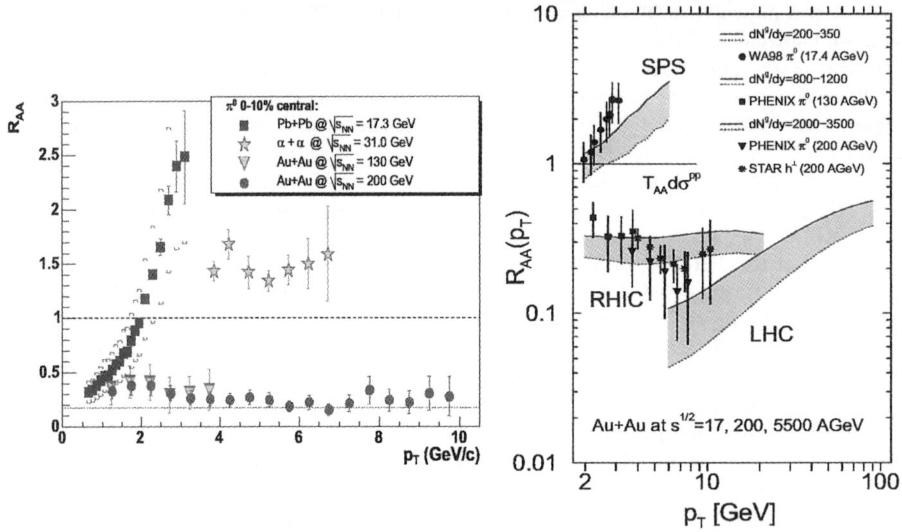

Figure 10: Jet Quenching at RHIC. Left [106] shows the jet quenching pattern of π^0 discovered by PHENIX [89, 90] at RHIC compared to previous observation of high p_T enhancement at ISR and SPS energies The nuclear modification factor $R_{AA} = dN_{AA}/T_{AA}(b)d\sigma_{pp}$ measures the deviation of AA spectra from factorized pQCD. Right shows predictions [104] of the $\sqrt{s}$ and p_T dependence from SPS, RHIC, LHC based on the GLV theory [105] of radiative energy loss.

Below RHIC energies, there is an enhancement of moderately high p_T tails that was observed in central $Pb+Pb$ reactions at the SPS. (Very recent reanalysis of the WA98 data shows a somewhat weaker enhancement at SPS [98]) This enhancement was expected as a consequence of the Cronin enhancement: now understood as an initial state effect[99]t which is also seen in pA collisions. Since the Cronin enhancement is an effect of the initial state nuclear wavefunction, it plays a role in the Color Glass Condensate, but we wish to isolate final state effects for our study of the Quark Gluon Plasma. In contrast, at RHIC a large suppression, by a factor of 4-5, was discovered in central $Au + Au$ that extends beyond 10 GeV for π^0.

Jet quenching in $A+A$ was proposed in [100, 101] as a way to study the dense matter produced at RHIC energies. As noted before, the pQCD jet production rates finally become large enough to measure yields up to high $p_T > 10$ GeV. Order of magnitude suppression effects were predicted based on simple estimates of induced gluon radiative energy loss. Ordinary, elastic energy loss [102] was known by that time to be too small to lead to significant attenuation.

As reviewed in [85, 86] refinements in the theory since then have opened the possibility of using the observed jet quenching pattern as a tomographic tool [103] to probe the parton densities in a QGP. The right panel of Fig.10 shows a recent jet tomographic analysis [104] of the PHENIX π^0 data [89, 90] based on the GLV opacity formalism [105]. This analysis concludes that the initial gluon rapidity density required to account for the observed jet quenching pattern must be $dN_g/dy \sim 1000 \pm 200$.

This jet tomographic measure of the initial dN_g/dy is in remarkable agreement with three other independent sources: (1) the initial entropy deduced via the Bjorken formula from the measured multiplicity, (2) the initial condition of the QGP required in hydrodynamics to produce the observed elliptic flow, and (3) the estimate of the maximum gluon rapidity density bound from the CGC gluon saturated initial condition, (which will be described later).

These four independent measures make it possible to estimate the maximal initial energy density in central collisions

$$\epsilon_0 = \epsilon(\tau \sim 1/p_0) \approx \frac{p_0^2}{\pi R^2}\frac{dN_g}{dy} \approx 20\frac{\text{GeV}}{\text{fm}^3} \sim 100 \times \epsilon_A \tag{5}$$

where $p_0 \approx Q_{sat} \approx 1.0 - 1.4$ GeV is the mean transverse momentum of the initial produced gluons from the incident saturated virtual nuclear CGC fields[29, 75, 107]. This scale controls the formation time $\hbar/p_0 \approx 0.2$ fm/c of the initially out-of-equilibrium (mostly gluonic) QGP. The success of the hydrodynamics requires that local equilibrium be achieved on a fast proper time scale $\tau_{eq} \approx (1-3)/p_0 < 0.6$ fm/c. The temperature at that time is $T(\tau_{eq}) \approx (\epsilon_0/(1-3) \times 12)^{1/4} \approx 2T_c$.

In HIJING model[108], the mini-jet cutoff is $p_0 = 2 - 2.2$ GeV limits the number of mini-jets to well below 1000. The inferred opacity of the QGP is observed to be much higher and consistent with the CGC[75] and EKRT[107] estimates.

5.3.1 I_{AA} and Di-Jet Tomography

Measurements of near side and away side azimuthal angle correlations of di-jet fragments provide the opportunity to probe the evolution of the matter produced at RHIC in even more detail. Fig.(11) show the discovery [94, 95, 96] of mono-jet production [100] in central collisions at RHIC. In peripheral collisions, the distribution $dN/d\Delta\phi$ of the azimuthal distribution of $p_T \sim 2$ GeV hadrons relative to a tagged $p_T \sim 4$ GeV leading jet fragment shows the same near side and away side back-to-back jetty correlations as measured in $p+p$. This is strong evidence that the kinematic range studied tests the physics of pQCD binary parton collision processes. For central collisions, on the other hand, away side jet correlations are almost completely suppressed.

The published data are as yet limited to $y_1 \approx y_2 \approx 0$, broad p_T cuts: $p_{T1} > 4$ GeV and $p_T \sim 2$ GeV, two bins of $\phi_1 - \phi_2$, and of course averaged over Φ_b. The measured modification of di-jet correlations is obtained by subtracting out the correlations due to bulk elliptic flow, and this

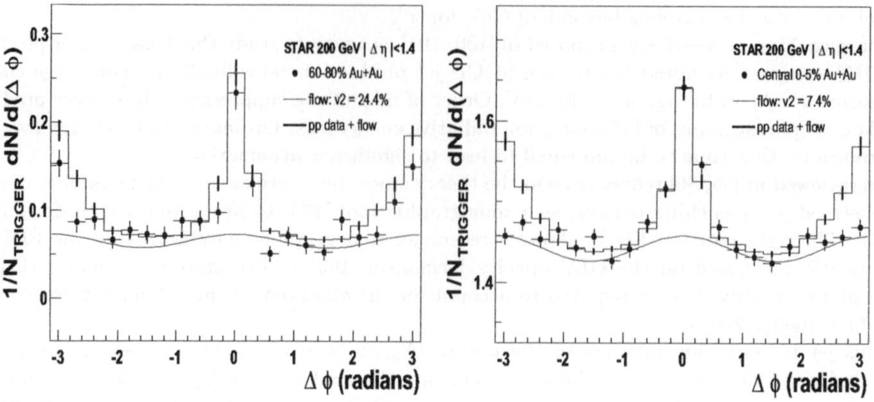

Figure 11: Monojets at RHIC from STAR [95, 96, 94]. Strongly correlated back-to-back di-jet production in pp and peripheral $AuAu$ left side is compared to mono-jet production discovered in central $AuAu$.

introduces some uncertainty. Analysis of present and future data at higher transverse momenta for a variety of rapidities will allow better tests of the underlying perturbative QCD dynamics.

Only one year ago [113] the interpretation of high p_T suppression was under intense debate because it was not yet clear how much of the quenching was due to initial state saturation (shadowing) of the gluon distributions and how much due to jet quenching discussed in the previous section. There was only one way to find out - eliminate the QGP final state interactions by substituting a Deuterium beam for one of the two heavy nuclei. In fact, it was long ago anticipated [101] that such a control test would be needed to isolate the unknown nuclear gluon shadowing contribution to the A+A quench pattern. In addition $D+Au$ was required to test predictions of possible initial state Cronin multiple interactions [99, 114, 115, 116, 117]. In contrast, one model of CGC [118] predicted a substantial suppression in $D+Au$ collisions. The data [109, 110, 111, 112] conclusively rule out large initial shadowing as the cause of the $x_{BJ} > 0.01$ quenching in Au+Au.

The I_{DAu} measurement from STAR [110] shows clearly how the suppression disappears in $D + Au$ collisions. The return of back-to-back jet correlation in $D + Au$ to the level observed in pp is seen in Fig.13. The data appear to be entirely consistent with jet quenching as a final state effect in $AuAu$ with little initial state effect in $D + Au$. These $D + Au$ data support the conclusion [119, 120] that the observed jet quenching in $AuAu$ is due to parton energy loss.

Theoretical analysics of jet quenching confirm the energy density estimates determined from measurements of particle multiplicity. They give large energy losses for jets propagating through the matter produced at RHIC, and strengthen the case for multliple strong interactions of the quark and gluon constituents of the matter made at RHIC.

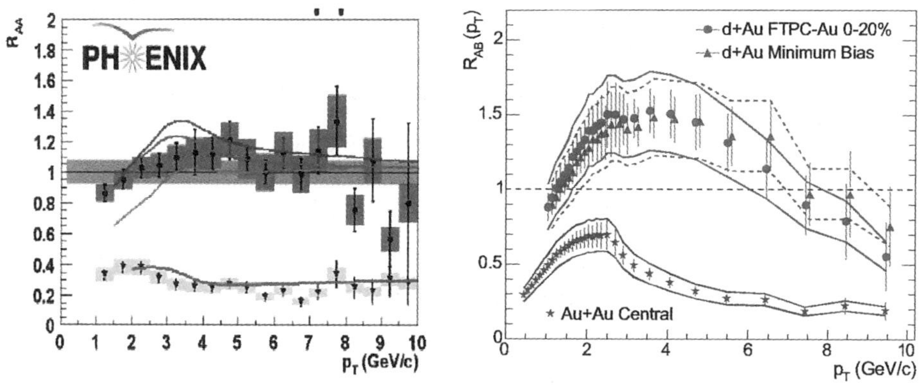

Figure 12: The **dA** control: PHENIX [109] π^0 and STAR [110] $h^{\pm}$ data compare R_{DAu} to R_{AuAu}. These and BRAHMS [111] and PHOBOS [112] data prove that jet quenching in $Au+Au$ must be due to final state interactions. Curves for π^0 show predictions from [104] for $AuAu$ and from [116] DAu. The curves for DAu show the interplay between different gluon shadow parameterizations (EKS, none, HIJING) and Cronin enhancement and are similar to predictions in [114, 115, 116]. In lower panel, the unquenching of charged hadrons is also seen in $D+Au$ relative to $Au+Au$ at high p_T.

6 Empirical Evidence for the Color Glass Condensate

In this section, the accumulated evidence for the Color Glass Condensate hypothesis is discussed[29]. The evidence rests in a variety of measurements done at different accelerators with different type of particles scattering. The discussion of the results from RHIC will be emphasized here, but it is first important to briefly review the results from HERA involving electron proton scattering.

6.1 Results from Electron-Hadron Scattering

Electron-hadron scattering provide information about the wavefunction of a hadron. The Color Glass Condensate describes the contribution to this wavefunction which have very many gluons in them. These pieces of the wavefunction control the physics at very small x, typically $x \leq 10^{-2}$. The various pieces of experimental information which support the CGC hypothesis come largely from *ep* scattering experiments at HERA:

- Geometrical Scaling

  Geometrical scaling is the observation[121]-[122] that the deep inelastic cross section for virtual photon scattering as a function of Q^2 and x is really only a function of

$$\sigma^{\gamma^* p} \sim F(Q^2/Q^2_{sat}) \tag{6}$$

  where the saturation momentum increases as the fractional momentum, x, of the gluon tends

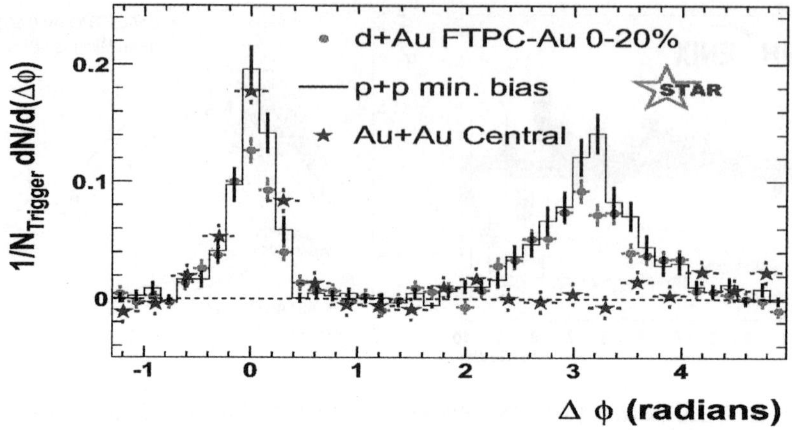

Figure 13: The **dA** "Return of the Jeti": Dijet fragment azimuthal correlations from STAR [110] in DAu are unquenched relative to the mono jet correlation observed in central $AuAu$.

to zero as

$$Q_{sat}^2(x) \sim (x_0/x)^\lambda \ 1GeV^2 \tag{7}$$

with $\lambda \sim 0.3$ and $x_0 \sim 10^{-4}$. This scaling with $\tau = Q^2/Q_{sat}^2$ works for $x \leq 10^{-2}$ and over the available Q^2 range at HERA as shown in Fig. 14.

It is straightforward to understand why this scaling works for the small $Q^2 \leq Q_{sat}^2$. This is the region of the CGC, and there is only one dimensionful scale which characterizes the system: the saturation momentum.[123] The surprise is that there is an extended scaling window for $Q_{sat}^2 \leq Q^2 \leq Q_{sat}^4/\Lambda_{QCD}^2$.[124] This can be proven analytically. As well, one now has reliable computation of the dependence on x of the saturation momentum, that is, one knows the exponent λ to about 15% accuracy, and it agrees with what is seen from the geometrical scaling curve.[125] What is not determined from the theory of the CGC is the scale x_0, and this must be found by experiment. This comes from the boundary conditions for the renormalization group equations.

- The Structure Function F_2

  Using the dipole description of the virtual photon wavefunction, the structure function F_2 can be related to the gluon distribution function which arises from the CGC. The results for the description of the data are remarkably good for $x \leq 10^{-2}$ and $Q^2 \leq 45\ GeV^2$. One should note that this description includes both the high and low Q^2 data. Descriptions based on DGLAP evolution can describe the large Q^2 points. The CGC description is very economical in the number of parameters which are used [126].

- Diffraction and Quasi-Elastic Processes

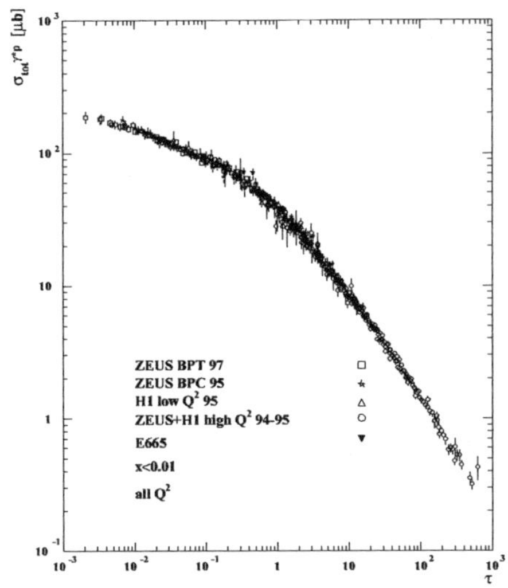

Figure 14: The cross section $\sigma^{\gamma^* p}$ as a function of the scaling variable $\tau = Q^2/Q^2_{sat}$.[122]

The CGC provides a description of the underlying structure of gluonic matter inside a hadron. As such, it should be sensitive to probes of the transverse extent of this matter, which can be experimentally studied in diffraction and related quasi-elastic particle production.[127]-[131] Diffractive scattering can be computed in the CGC description for small inclusive masses of produces particles. The CGC agrees both with generic features of the data, and provides a reasonably good quantitative description.

There are additional computations of quasi-elastic ρ meson production and J/Ψ production[132]-[133]. Again, up to uncertainties associated with the overall normalization (which arises from imprecise knowledge of hadronic wavefunctions), the CGC hypothesis provides a reasonably good quantitative description.

6.2 Heavy Ion Collisions

The collision of two ultrarelativistic heavy ions can be visualized as the scattering of two sheets of colored glass, as shown in Fig. 4. [134]-[137]

At very early times after the collision the matter is at very high energy density and in the form of a CGC. As time goes on, the matter expands. As it expands the density of gluons decreases, and gluons begin to propagate with little interaction. At later times, the interaction strength increases and there is sufficient time for the matter to thermalize and form a Quark Gluon Plasma. This scenario is shown in Fig. 7, with realistic estimates for energy density and time scales appropriate for the RHIC heavy ion accelerator.

6.3 The Multiplicity

The CGC allows for a direct computation of the particle multiplicity in hadronic collisions. If one naively tries to compute jet production, the total multiplicity is infrared divergent. This follows because of the $1/p_T^4$ nature of the perturbative formula for gluon production

$$\frac{1}{\pi R^2} \frac{dN}{dyd^2p_T} \sim \frac{1}{\alpha_S} \frac{Q_{sat}^4}{p_T^4} \tag{8}$$

In the CGC, when $p_T \leq Q_{sat}$, this divergence is cutoff and the total gluon multiplicity goes as

$$\frac{1}{\pi R^2} \frac{dN}{dy} \sim \frac{1}{\alpha_S} Q_{sat}^2 \tag{9}$$

One can compute the proportionality constant and before the RHIC data appeared, predictions were made for the gluon multiplicity. In Fig. 15 from Ref.[138], the predictions for 200 AGeV to extrapolations from the first RHIC run at 130 AGeV are shown. The CGC correctly predicted

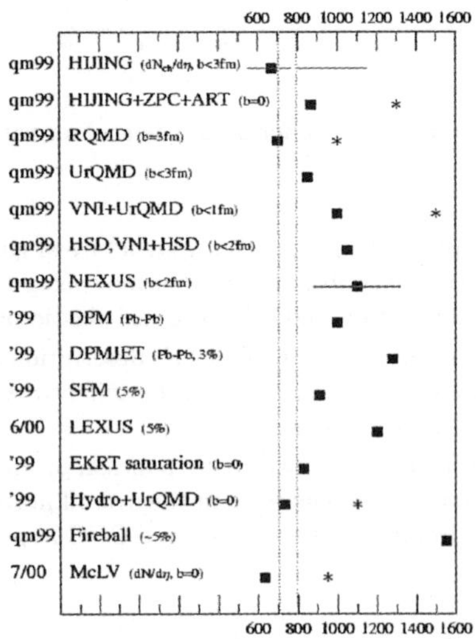

Figure 15: Predictions for the charged hadron rapidity density for central $Au + Au$ at $\sqrt{s} = 200$ AGeV compiled by Eskola [138] compared to data band (740 ± 50) measured later by all four experiments at RHIC. The CGC prediction is marked McLV.

the surprizingly low multiplicity.

Also, the dependence of the multiplicity on the number of participants can also be computed, realizing that the saturation momentum squared should be (for not too small x) proportional

$N_{part}^{1/3}$. This leads to

$$\frac{1}{N_{part}}\frac{dN}{dy} \sim \frac{1}{\alpha_S(Q_{sat}^2)} \sim \log N_{part} \tag{10}$$

so that a very slow logarithmic dependence on the number of participants is predicted in agreement with experiment, as shown in Fig. 16.[139]-[142]

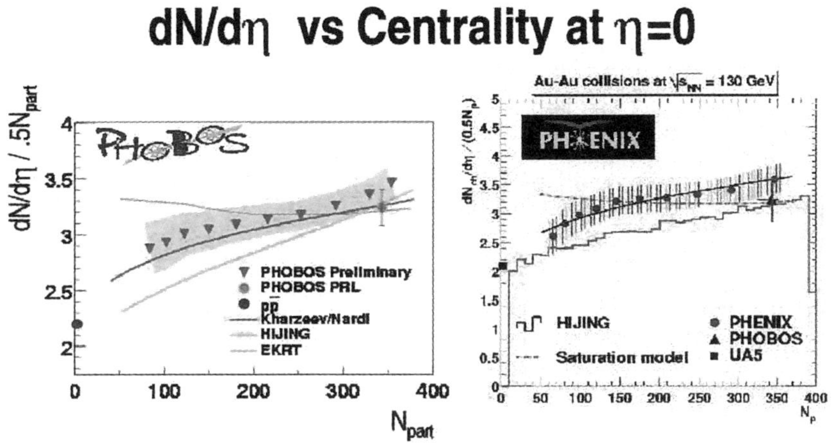

Figure 16: The total multiplicity as a function of the number of participants as measured by Phobos and Phenix. [76, 143, 144]

One can go much further[73] and compute the dependence of the multiplicity on rapidity and centrality, and the transverse momentum distribution of produced hadrons by using CGC initial conditions matched together with a hydrodynamic calculation which evolves the matter through the Quark Gluon Plasma[74]. At high $p_T \gtrsim 2$ GeV, jet quenching of the pQCD power law tails is also taken into account using [105]. This combined hydrodynamic calculation of Nara and Hirano accounts very well for many important features of the data at RHIC. [89],[143]

We emphasize the importance of constraining the QGP hydrodynamics with a QCD theoretic initial condition, the CGC. At SPS energies, the uncertainty about initial conditions together with failure of hydrodynamics to account for the highly dissipative hadronic flow are in remarkable contrast to situation at RHIC. If the initial condition were not well constrained at RHIC, then the conclusion that a QGP was formed could not be sustained. Hydrodynamics is a dynamical mapping of a given given initial condition to final spectra that depends on the equation of state. Only with a known or predicted initial state does that mapping have the power to falsify the QCD equation of state.

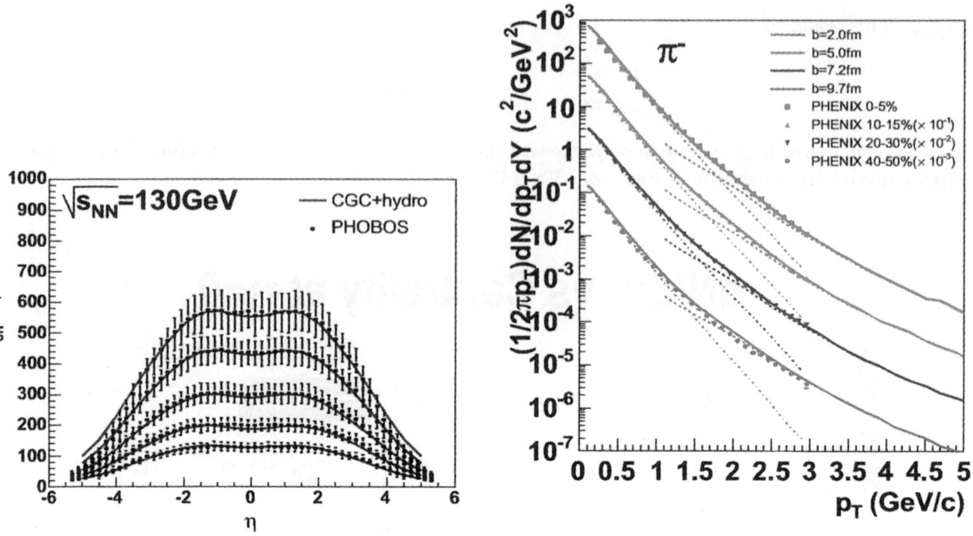

Figure 17: Results of hydrodynamic calculations[74] with a CGC initial condition. Left is the centrality dependence of the rapidity density from PHOBOS[143]. Right shows the transverse momentum distribution at zero rapidity for different centralities from PHENIX[89]. For $p_T > 2$ GeV jet quenching[105] through the evolving QGP is also taken into account.

In Figs 15 and 16, one can see that several phenomenological models, such as HIJING[108], could also account qualitatively for some of the global multiplicity observables. However, the surprising very weak centrality and beam energy dependence observed[76, 81, 143, 144] is most satisfactory explained and predicted by the CGC as arising from the slow $1/\alpha_S \sim \log Q^2_{sat}(N_{part}, \sqrt{s})$ in eqs.(9,10). This is one of the strongest lines of empirical evidence from RHIC that the CGC initial state (with its predicted N_{part} and $\sqrt{s}$ dependence) is formed and that it is the seed of the QGP that evolves from it.

6.4 High p_T Particles

The early results from RHIC on gold-gold collisions revealed that the high p_T production cross sections were almost an order of magnitude below that expected for jet production arising from incoherent parton-parton scattering.[145] This could be either due to initial state shadowing of the gluon distribution inside the nuclei,[146] or to final state jet quenching.[147] For centrally produced jets, the x of the parton which produces a 5-10 GeV particle is of order 10^{-1}, and this is outside the region where on the basis of the HERA data one expects the effects of the CGC to be important. Nevertheless, nuclei might be different than protons, so it is not a priori impossible.

The crucial test of these two different mechanisms is the comparison of dA scattering to pp. If there is suppression of jet in dA collisions, then it is an initial state effect. The experiments were performed, and all there is little initial state effect for centrally produced jets.[148] The suppression of centrally produced jets in AA collisions at RHIC is indeed due to final state interactions, that is jet quenching.

This is not in contradiction with the existence of a CGC. The particles which control the multiplicity distribution in the central region are relatively soft, and arise from $x \sim 10^{-2}$. To probe such small x degrees of freedom at high transverse momentum at RHIC requires that one go to the forward region.[149]-[150]

If one uses naive Glauber theory to compute the effects of shadowing by multiple scattering, one expects that if one goes into the forward region of the deuteron, the probe propagates through more matter in the nucleus. This is because we probe all of the gluons with x greater than the minimum x of the nucleus which can be seen by the deuteron. Going more forward makes this minimum x smaller. Now multiple scattering will produce more particles at some intermediate value of p_T. (At very high p_T, the effects of multiple scattering will disappear.) This is the source of the Cronin peak and it is expected to occur at p_T of $2-4\ GeV$. Clearly the height of this peak should increase as one goes more forward on the side of the deuteron, and should increase with the centrality of the collision.[151] A result of such a computation is shown in Fig. 18.

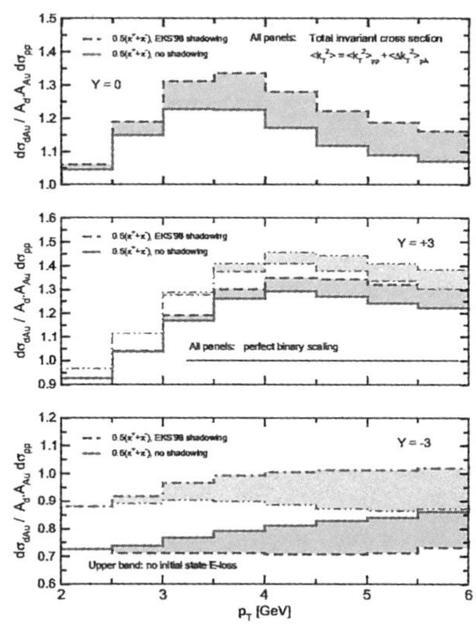

Figure 18: The expectations of classical multiple scattering for the rapidity and p_T dependence (with and without shadowing) for the p_T distribution in $D+Au$ collisions from [116].

Classical rescattering effects are included in the computation of the properties of the CGC.

There is another effect however and that is quantum evolution generated by the renormalization group equations. It was a surprise that when one computed the evolution of the gluon distribution function including both effects, the quantum evolution dominated. This means that the height of the Cronin peak, and the overall magnitude of the gluon distribution decreased as one went from backwards to forward angles.[152]- [154] The results of one such computation are shown in Fig. 19

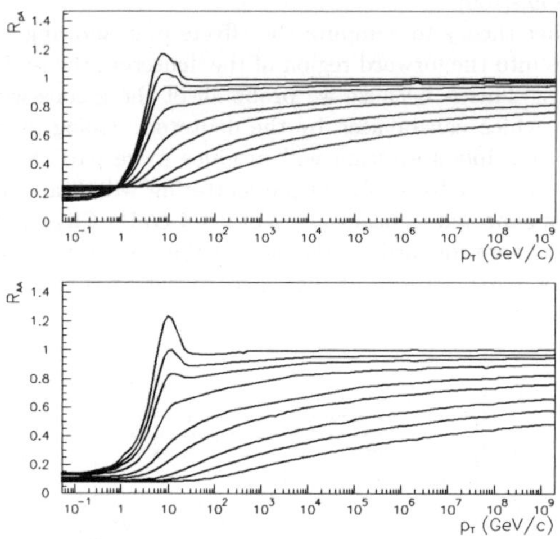

Figure 19: The gluon intrinsic gluon distribution function as a function of p_T for different pseudo-rapidities[154].

It was also a surprise how rapid the effect set in.

The Brahms experiment at RHIC recently presented data[155] on the ratio of central to peripheral transverse momentum distributions.The ratio R_{CP} is defined in such a way that if the processes were due to incoherent production of jets, then $R_{CP} = 1$. A value less than one indicates suppression, and a value larger than one indicates a Cronin type enhancement. The results for a variety of forward angles for R_{CP} as a function of p_T is shown in Fig. 20 a. There is clearly a decrease in R_{CP} as one goes to forward angles, in distinction from the predictions of classical multiple scattering. The effect is very rapid in rapidity, as was expected from computations of the CGC. In Fig. 20, the ratio R_{CP} is shown as a function of p_T for the forward pseudorapidity $\eta \sim 3$ for less central and more central events. The ratio decreases for more central collisions, against the expectation of classical multiple scattering and consistent with the CGC hypothesis.

Preliminary data[156]-[158] from all four experiments on the rapidity dependence of the transverse distributions in $D + Au$ suggest striking effects consistent with CGC. This is a very active area of research both theoretically and experimentally at this time.

This data suggest a Cronin enhancement on the gold side and the depletion on the deuteron

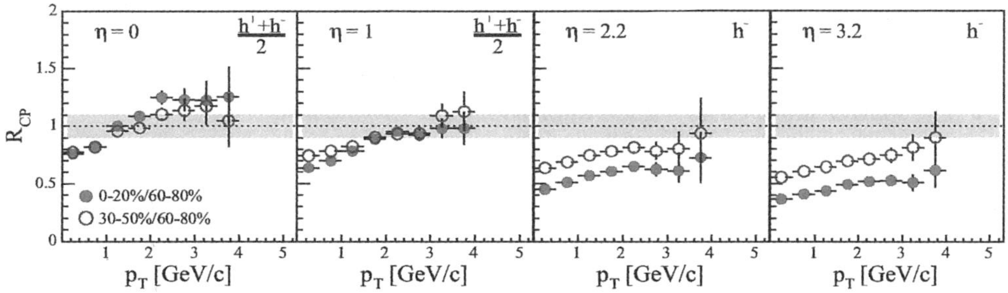

Figure 20: The central to peripheral charged and negative hadron ratios R_{CP} as a function of p_T for various forward pseudorapidities in 200 AGeV $D + Au$ from BRAHMS[155].

side, and as well a definite dependence on centrality. When these $D + Au$ data become finalized, they could prove that classical multiple scattering dominates on the gold side (large $x > 0.01$), but quantum evolution - i.e., deep gluon shadowing- on the deuteron side (small $x < 0.01$).

By a "happy coincidence", these effects nearly cancel in the mid-rapidity region, making RHIC well suited for studying QGP effects at midrapidity for hard probes.

6.5 The Developing Case for the CGC

In addition to the results described above, the Color Glass Condensate will be the subject of further experimental study at RHIC, LHC and eRHIC. At RHIC, one can study forward backward correlations in the forward direction in analogy with what was done for centrally produced jets. At LHC, in the forward region one measures relatively large p_T jets at $x \sim 10^{-6}$. This provides a direct measurement of the very small x gluon distribution function. Eventually eRHIC would be required to provide precision measurements of quark and gluon distribution functions at small x in a variety of nuclei.

The Color Glass Condensate hypothesis describes remarkably well generic features of *ep* measurements of properties of protons at small x. It also successfully predicted the the previously unexpected slow growth of multiplicity of produced particles with $\sqrt{s}$ and centrality at RHIC. The data from Brahms on the forward particle production appear to be qualitatively in accord with prediction of the CGC, and the preliminary data from Phobos, Star and RHIC on this subject await submission for publication. While the CGC hypothesis successfully describes the data from disparate experimental measurements, it can be further tested in a variety of new environments.

7 Conclusions

Our criteria for the discovery of the Quark Gluon Plasma at RHIC are:

- **Matter at energy densities so large that the simple degrees of freedom are quarks and gluons. This energy density is that predicted by lattice gauge theory for the existence of a QGP in thermal systems, and is about 2 GeV/fm^3**
- **The matter must be to a good approximation thermalized.**
- **The properties of the matter associated with the matter while it is hot and dense must follow from QCD computations based on hydrodynamics, lattice gauge theory results, and perturbative QCD for hard processes such as jets**

All of the above are satisfied from the published data at RHIC. A surprise is the degree to which the computations based on ideal fluid hydrodynamics agree so well with elliptic flow data. This leads us to conclude that the matter produced at RHIC is a strongly coupled Quark Gluon Plasma (sQGP) contrary to original expectations that were based on weakly coupled plasma estimates.

The case for the Color Glass Condensate is rapidly evolving into a compelling case. Much of the exciting new data from RHIC presented at QM2004 has yet to be published. Nevertheless, the data from HERA taken together with the data on particle multiplicities, and the data submitted for publication by Brahms make a strong case, which may become compelling with further reinforcement from results from the other experiments at RHIC, and future experimental tests at LHC and eRHIC. This area continues to evolve rapidly both experimentally and theoretically.

Although in our opinion, the case for the sQGP at RHIC is now overwhelming, there are of course many important scientific issues not yet addressed in the first three years of data. The experiments have demonstrated that a new form of matter, the sQGP, exists. The harder long term task of mapping out more of its novel properties can now confidently proceed at RHIC and soon LHC.

8 Acknowledgments

This work was supported by the Director, Office of Energy Research, Office of High Energy and Nuclear Physics, Division of Nuclear Physics, of the U.S. Department of Energy under Contracts DE-FG-02-93ER-40764 and DE-AC02-98H10886.

References

[1] Riken-BNL-Research-Center Worshop on " New Discoveries at RHIC the current case for the strongly interactive QGP", http://www.bnl.gov/riken/May14-152004workshop.htm May 14-15, 2004.

[2] M. Gyulassy and L. McLerran, "New forms of QCD matter discovered at RHIC," Nucl. Phys. A **750**, 30 (2005) [arXiv:nucl-th/0405013].

[3] T.D. Lee, "The Strongly Interacting Quark-Gluon Plasma and Future Physics," Nucl. Phys. A **750**, 1 (2005)

[4] T. Ludlam, "Experimental Results from RHIC," Nucl. Phys. A **750**, 1 (2005)

[5] H. Stoecker, "Collective Flow signals the Quark Gluon Plasma," Nucl. Phys. A **750**, 121 (2005) [arXiv:nucl-th/0406018].

[6] E. V. Shuryak, "What RHIC experiments and theory tell us about properties of quark-gluon plasma?," arXiv:hep-ph/0405066.

[7] X. N. Wang, "Discovery of jet quenching and beyond," Nucl. Phys. A **750**, 98 (2005) [arXiv:nucl-th/0405017].

[8] J. P. Blaizot and F. Gelis, "Searching evidence for the color glass condensate at RHIC," Nucl. Phys. A **750**, 148 (2005) [arXiv:hep-ph/0405305].

[9] B. Muller, "Hadronic signals of deconfinement at RHIC," Nucl. Phys. A **750**, 84 (2005), arXiv:nucl-th/0404015.

[10] R. C. . Hwa and X. N. . Wang, "Quark-gluon plasma. Vol. 3," (World Scientific, Singapore,2004), p 1-777.

[11] K. Adcox *et al.* [PHENIX Collaboration], arXiv:nucl-ex/0410003.

[12] B. B. Back *et al.*, arXiv:nucl-ex/0410022.

[13] I. Arsene *et al.* [BRAHMS Collaboration], arXiv:nucl-ex/0410020.

[14] J. Adams *et al.* [STAR Collaboration], arXiv:nucl-ex/0501009.

[15] *Report of the workshop on BeV/nucleon collisions of heavy ions – how and why*, Bear Mountain, New York, Nov. 29 – Dec. 1, 1974 (BNL-AUI, 1975); G. Baym, Nucl. Phys. A **698**, XXIII (2002) [arXiv:hep-ph/0104138].

[16] T. D. Lee and G. C. Wick, Phys. Rev. D **9**, 2291 (1974).

[17] J. Hofmann, H. Stocker, W. Scheid and W. Greiner, Bear Mountain Workshop, New York, Dec 1974; H. G. Baumgardt *et al.*, Z. Phys. A **273**, 359 (1975).

[18] J. C. Collins and M. J. Perry, Phys. Rev. Lett. **34**, 1353 (1975); G. Baym and S. A. Chin, Phys. Lett. B **62** (1976) 241; B. A. Freedman and L. D. McLerran, Phys. Rev. D **16**, 1169 (1977); G. Chapline and M. Nauenberg, Phys. Rev. D **16**, 450 (1977).

[19] E. V. Shuryak, Sov.Phys.JETP 47:212-219,1978, Zh.Eksp.Teor.Fiz.74:408-420,1978; Phys. Lett. B **78**, 150 (1978); Phys.Rept.61:71-158,1980; O. K. Kalashnikov and V. V. Klimov, Phys. Lett. B **88**, 328 (1979); J. I. Kapusta, Nucl. Phys. B **148** (1979) 461.

[20] J. D. Bjorken, FERMILAB-CONF-83-070-T; Phys. Rev. D **27**, 140 (1983).

[21] RHIC: http://www.bnl.gov/rhic/default.htm

[22] For a complete list of RHIC publications see QPIRES: http://www.slac.stanford.edu/spires/hep/ using "find cn brahms or cn phobos or cn star or cn phenix and ps published"

[23] BRAHMS exp. homepage: http://www4.rcf.bnl.gov/brahms/WWW/brahms.html

[24] PHENIX exp. homepage: http://www.phenix.bnl.gov/

[25] PHOBOS exp. homepage: http://www.phobos.bnl.gov/

[26] STAR exp. homepage: http://www.star.bnl.gov/

[27] CERN symposium, Feb. 10, 2000. See, e.g, CERN Courier **40** (May 2000) 13; B. Schwarzschild, Phys. Today **53** (May 2000) 20.

[28] M. Gyulassy, Proc. NATO/ASI: Structure and Dynamics of Elementary Matter (Kemer, Turkey, 2003, ed. W. Greiner), arXiv:nucl-th/0403032.

[29] L. McLerran, Proc. NATO/ASI: Structure and Dynamics of Elementary Matter (Kemer, Turkey, 2003, ed. W. Greiner), arXiv:hep-ph/0402137.

[30] S. A. Bass, M. Gyulassy, H. Stocker and W. Greiner, J. Phys. G **25**, R1 (1999) [arXiv:hep-ph/9810281].

[31] C. R. Allton et al, Phys. Rev. D **68**, 014507 (2003) [arXiv:hep-lat/0305007]. F. Karsch, E. Laermann and A. Peikert, Phys. Lett. B **478**, 447 (2000) [arXiv:hep-lat/0002003]. F. Karsch, Lect. Notes Phys. **583**, 209 (2002) [arXiv:hep-lat/0106019].

[32] C. W. Bernard *et al.* [MILC Collaboration], Phys. Rev. D **55**, 6861 (1997) [arXiv:hep-lat/9612025].

[33] S. Gupta, Pramana **61**, 877 (2003) [arXiv:hep-ph/0303072].

[34] Z. Fodor and S. D. Katz, arXiv:hep-lat/0402006. F. Csikor et al arXiv:hep-lat/0401022.

[35] F. Karsch, K. Redlich and A. Tawfik, Phys. Lett. B **571**, 67 (2003) [arXiv:hep-ph/0306208].

[36] M. A. Halasz et al, Phys. Rev. D **58**, 096007 (1998) [arXiv:hep-ph/9804290]. M. A. Stephanov, K. Rajagopal and E. V. Shuryak, Phys. Rev. Lett. **81**, 4816 (1998) [arXiv:hep-ph/9806219].

[37] D. H. Rischke, Prog. Part. Nucl. Phys. **52**, 197 (2004) [arXiv:nucl-th/0305030].

[38] J. Breitweg et. al. *Eur. Phys. J.* **67**, 609 (1999).

[39] L. V. Gribov, E. M. Levin and M. G. Ryskin, *Phys. Rept.* **100**, 1 (1983).

[40] A. H. Mueller and Jian-wei Qiu, *Nucl. Phys.* **B268**, 427 (1986); J.-P. Blaizot and A. H. Mueller, *Nucl. Phys.* **B289**, 847 (1987).

[41] L.N. Lipatov, *Sov. J. Nucl. Phys.* **23** (1976), 338;
E.A. Kuraev, L.N. Lipatov and V.S. Fadin, *Sov. Phys. JETP* **45** (1977), 199;
Ya.Ya. Balitsky and L.N. Lipatov, *Sov. J. Nucl. Phys.* **28** (1978), 822.

[42] L. D. McLerran and R. Venugopalan, *Phys. Rev.* **D49**, 2233(1994); 3352 (1994); **D50**, 2225 (1994).

[43] E. Iancu, A. Leonidov and L. D. McLerran, *Nucl. Phys.* **A692**, 583 (2001); E. Ferreiro E. Iancu, A. Leonidov and L. D. McLerran, *Nucl. Phys.* bf A710,373 (2002).

[44] E. Iancu and L. McLerran, *Phys.Lett.* **B510**, 145 (2001).

[45] A. Kovner, L. D. McLerran and H. Weigert, *Phys. Rev.* **D52**, 6231 (1995); 3809 (1995); A. Krasnitz and R. Venugopalan, *Phys. Rev. Lett* **84**, 4309 (2000); *Nucl. Phys.* **B557**, 237 (1999); A. Krasnitz, Y. Nara and R. Venugopalan, *Phys. Rev. Lett.* **87** , 192302 (2001).

[46] B. B. Back *et al.* [PHOBOS Collaboration], Phys. Rev. Lett. **88**, 022302 (2002) [arXiv:nucl-ex/0108009].

[47] B. B.Back et. al. [PHOBOS], *Phys. Rev. Lett.* **85**, 3100 (2000);

[48] H. Stocker, J. A. Maruhn and W. Greiner, Z. Phys. A **293**, 173 (1979); Phys. Rev. Lett. **44**, 725 (1980); L. Csernai and H. Stocker, Phys. Rev. C **25**, 3208 (1981).

[49] H. Stocker *et al.*, Phys. Rev. C **25**, 1873 (1982).

[50] H. Stocker and W. Greiner, Phys. Rept. **137**, 277 (1986).

[51] W. Reisdorf and H. G. Ritter, Ann. Rev. Nucl. Part. Sci. **47**, 663 (1997).

[52] S. Voloshin and Y. Zhang, Z. Phys. C **70**, 665 (1996) [arXiv:hep-ph/9407282].

[53] J. Y. Ollitrault, Phys. Rev. D **46**, 229 (1992).

[54] S. A. Voloshin and A. M. Poskanzer, Phys. Lett. B **474**, 27 (2000) [arXiv:nucl-th/9906075].

[55] C. Alt *et al.* [NA49 Collaboration], Phys. Rev. C **68**, 034903 (2003) [arXiv:nucl-ex/0303001].

[56] G. Stoicea *et al.*, arXiv:nucl-ex/0401041.

[57] J. Adams *et al.* [STAR Collaboration], arXiv:nucl-ex/0310029, Phys. Rev. Lett. 92 (2004) 062301.

[58] P. R. Sorensen, hadronization of the bulk partonic matter created in Au + Au collisions at arXiv:nucl-ex/0309003. Ph.D. thesis.

[59] J. Adams *et al.* [STAR Collaboration], arXiv:nucl-ex/0306007, Phys. Rev. Lett. 92 (2004) 052302

[60] C. Adler *et al.* [STAR Collaboration], Phys. Rev. C **66**, 034904 (2002) [arXiv:nucl-ex/0206001].

[61] S. S. Adler *et al.* [PHENIX Collaboration], Phys. Rev. Lett. **91**, 182301 (2003) [arXiv:nucl-ex/0305013].

[62] P. F. Kolb, P. Huovinen, U. W. Heinz and H. Heiselberg, Phys. Lett. B **500**, 232 (2001).

[63] P. Huovinen, P. F. Kolb, U. W. Heinz, P. V. Ruuskanen and S. A. Voloshin, Phys. Lett. B **503**, 58 (2001).

[64] P. F. Kolb, U. W. Heinz, P. Huovinen, K. J. Eskola and K. Tuominen, Nucl. Phys. A **696**, 197 (2001).

[65] P. Huovinen, Published in Quark Gluon Plasma 3, editors: R.C. Hwa and X.N. Wang, (World Scientific, Singapore,2004) p.600; arXiv:nucl-th/0305064.

[66] P. F. Kolb and U. Heinz, Published in Quark Gluon Plasma 3, editors: R.C. Hwa and X.N. Wang, (World Scientific, Singapore,2004) p.634; arXiv:nucl-th/0305084.

[67] D. Teaney, J. Lauret and E. V. Shuryak, nucl-th/0104041.

[68] D. Teaney, J. Lauret and E. V. Shuryak, arXiv:nucl-th/0110037.

[69] D. Teaney, Phys. Rev. C **68**, 034913 (2003). D. Teaney, arXiv:nucl-th/0301099.

[70] T. Hirano and Y. Nara, Phys. Rev. Lett. **91**, 082301 (2003) [arXiv:nucl-th/0301042].

[71] T. Hirano and Y. Nara, Phys. Rev. C **68**, 064902 (2003) [arXiv:nucl-th/0307087].

[72] T. Hirano and Y. Nara, arXiv:nucl-th/0307015.

[73] T. Hirano, arXiv:nucl-th/0403042.

[74] T. Hirano and Y. Nara, arXiv:nucl-th/0403029.

[75] E. Iancu and R. Venugopalan, arXiv:hep-ph/0303204. L. D. McLerran and R. Venugopalan, Phys. Rev. D **49**, 2233 (1994) [arXiv:hep-ph/9309289]. D. Kharzeev and E. Levin, Phys. Lett. B **523**, 79 (2001) [arXiv:nucl-th/0108006].

[76] B. B. Back *et al.* [PHOBOS collaboration], Nucl. Phys. A **715**, 65 (2003) [arXiv:nucl-ex/0212009].

[77] G. Agakichiev *et al.* [CERES/NA45 Collaboration], arXiv:nucl-ex/0303014.

[78] D. Molnar and M. Gyulassy, Nucl. Phys. A **697**, 495 (2002) [Erratum-ibid. A **703**, 893 (2002)] [arXiv:nucl-th/0104073]. B. Zhang, M. Gyulassy and C. M. Ko, Phys. Lett. B **455**, 45 (1999) [arXiv:nucl-th/9902016].

[79] Y. Cheng, F. Liu, Z. Liu, K. Schweda and N. Xu, Phys. Rev. C **68**, 034910 (2003). N. Xu *et al.* [NA44 Collaboration], Nucl. Phys. A **610**, 175C (1996).

[80] S. A. Bass and A. Dumitru, Phys. Rev. C **61**, 064909 (2000) [arXiv:nucl-th/0001033].

[81] I. G. Bearden *et al.* [BRAHMS Collaboration], Phys. Rev. Lett. **88**, 202301 (2002) [arXiv:nucl-ex/0112001].

[82] P. Danielewicz, R. Lacey and W. G. Lynch, Science **298**, 1592 (2002) [arXiv:nucl-th/0208016].

[83] G. Brown et al, arXiv:hep-ph/0402068

[84] P. Danielewicz and M. Gyulassy, Phys. Rev. D **31** (1985) 53.

[85] M. Gyulassy, I. Vitev, X. N. Wang and B. W. Zhang, Published in Quark Gluon Plasma 3, editors: R.C. Hwa and X.N. Wang, (World Scientific, Singapore,2004) p.123; arXiv:nucl-th/0302077. M. Gyulassy, Lect. Notes Phys. **583**, 37 (2002) [arXiv:nucl-th/0106072].

[86] R. Baier, D. Schiff and B. G. Zakharov, Ann. Rev. Nucl. Part. Sci. **50**, 37 (2000) [arXiv:hep-ph/0002198].

[87] D. Kharzeev, Phys. Lett. B **378**, 238 (1996) [arXiv:nucl-th/9602027]. S. E. Vance et al, Phys. Lett. B **443**, 45 (1998) [arXiv:nucl-th/9806008]. I. Vitev and M. Gyulassy, Phys. Rev. C **65**, 041902 (2002) [arXiv:nucl-th/0104066].

[88] P. Csizmadia, et al J. Phys. G **25**, 321 (1999) [arXiv:hep-ph/9809456]. R. J. Fries, B. Muller, C. Nonaka and S. A. Bass, Phys. Rev. Lett. **90**, 202303 (2003) [arXiv:nucl-th/0301087]. D. Molnar and S. A. Voloshin, Phys. Rev. Lett. **91**, 092301 (2003) [arXiv:nucl-th/0302014]. V. Greco, C. M. Ko and P. Levai, Phys. Rev. C **68**, 034904 (2003) [arXiv:nucl-th/0305024]. Z. W. Lin and D. Molnar, Phys. Rev. C **68**, 044901 (2003) [arXiv:nucl-th/0304045].

[89] K. Adcox *et al.*, Phys. Rev. Lett. **88**, 022301 (2002); P. Levai *et al.*, Nucl. Phys. A **698**, 631 (2002).

[90] K. Adcox *et al.* [PHENIX Collaboration], Phys. Lett. B **561**, 82 (2003) [arXiv:nucl-ex/0207009].

[91] S. S. Adler *et al.* [PHENIX Collaboration], Phys. Rev. Lett. **91**, 072301 (2003) [arXiv:nucl-ex/0304022].

[92] J. Adams *et al.* [STAR Collaboration], Phys. Rev. Lett. **91**, 172302 (2003) [arXiv:nucl-ex/0305015].

[93] C. Adler *et al.*, [STAR Collaboration] Phys. Rev. Lett. **89**, 202301 (2002) [arXiv:nucl-ex/0206011].

[94] P. Jacobs and J. Klay [STAR Collaboration], arXiv:nucl-ex/0308023.

[95] C. Adler *et al.* [STAR Collaboration], Phys. Rev. Lett. **90**, 082302 (2003) [arXiv:nucl-ex/0210033].

[96] D. Hardtke [The STAR Collaboration], Nucl. Phys. A **715**, 272 (2003) [arXiv:nucl-ex/0212004].

[97] C. Adler *et al.* [STAR Collaboration], Phys. Rev. Lett. **90**, 032301 (2003).

[98] D. d'Enterria, arXiv:nucl-ex/0403055.

[99] A. Accardi, "Cronin effect in pA : A survey of theoretical models," arXiv:hep-ph/0212148.

[100] M. Gyulassy and M. Plumer, Nucl. Phys. A **527**, 641 (1991). M. Gyulassy, M. Plumer, M. Thoma and X. N. Wang, Nucl. Phys. A **538**, 37C (1992)

[101] X. Wang and M. Gyulassy, Phys. Rev. Lett. **68**, 1480 (1992).

[102] J. D. Bjorken, FERMILAB-PUB-82-059-THY and erratum (unpublished); M. H. Thoma and M. Gyulassy, Nucl. Phys. B **351**, 491 (1991); E. Braaten and M. H. Thoma, Phys. Rev. D **44**, 2625 (1991); M. H. Thoma, J. Phys. G **26**, 1507 (2000) [arXiv:hep-ph/0003016].

[103] M. Gyulassy, P. Levai, and I. Vitev, Phys. Lett. B **538**, 282 (2002): E. Wang and X.-N. Wang, Phys. Rev. Lett. **89**, 162301 (2002); C. A. Salgado and U. A. Wiedemann, Phys. Rev. Lett. **89**, 092303 (2002);

[104] I. Vitev and M. Gyulassy, Phys. Rev. Lett. **89**, 252301 (2002) [arXiv:hep-ph/0209161].

[105] M. Gyulassy, P. Levai and I. Vitev, Nucl. Phys. B **594**, 371 (2001) [arXiv:nucl-th/0006010]. Phys. Rev. Lett. **85**, 5535 (2000) [arXiv:nucl-th/0005032]; Phys. Rev. D **66**, 014005 (2002) [arXiv:nucl-th/0201078]; Nucl. Phys. B **571**, 197 (2000) [arXiv:hep-ph/9907461].

[106] D. d'Enterria [PHENIX Collaboration], arXiv:nucl-ex/0401001.

[107] K. J. Eskola, K. Kajantie, P. V. Ruuskanen and K. Tuominen, Phys. Lett. B **543**, 208 (2002) [arXiv:hep-ph/0204034]; Phys. Lett. B **532**, 222 (2002) [arXiv:hep-ph/0201256]; *Nucl. Phys.* **B570**, 379 (2000); *Phys. Lett.* **B497**, 39 (2001).

[108] V. Topor Pop *et al.*, Phys. Rev. C **68**, 054902 (2003) [arXiv:nucl-th/0209089]. X. N. Wang and M. Gyulassy, Phys. Rev. D **44**, 3501 (1991).

[109] S. S. Adler *et al.* [PHENIX Collaboration], Phys. Rev. Lett. **91**, 072303 (2003) [arXiv:nucl-ex/0306021].

[110] J. Adams *et al.* [STAR Collaboration], Phys. Rev. Lett. **91**, 072304 (2003) [arXiv:nucl-ex/0306024].

[111] I. Arsene *et al.* [BRAHMS Collaboration], suppression," Phys. Rev. Lett. **91**, 072305 (2003) [arXiv:nucl-ex/0307003].

[112] B. B. Back *et al.* [PHOBOS Collaboration], Phys. Rev. Lett. **91**, 072302 (2003) [arXiv:nucl-ex/0306025].

[113] Transverse Dynamics at RHIC, BNL March 6-8, 2003, "http://www.phenix.bnl.gov/phenix/WWW/publish/rak/workshop/int/program_TD.htm"

[114] X. N. Wang, Phys. Rev. C **61**, 064910 (2000) [arXiv:nucl-th/9812021].

[115] X. N. Wang, Phys. Rept. **280**, 287 (1997) [arXiv:hep-ph/9605214].

[116] I. Vitev, Phys. Lett. B **562**, 36 (2003) [arXiv:nucl-th/0302002]; X. N. Wang, Phys. Lett. B **565**, 116 (2003) [arXiv:nucl-th/0303004]; A. Accardi and M. Gyulassy, arXiv:nucl-th/0308029; P. Levai, G. Papp, G. G. Barnafoldi and G. I. Fai, arXiv:nucl-th/0306019.

[117] J. w. Qiu and I. Vitev, arXiv:hep-ph/0309094; arXiv:hep-ph/0401062.

[118] D. Kharzeev, E. Levin and L. McLerran, Phys. Lett. B **561**, 93 (2003) [arXiv:hep-ph/0210332].

[119] X. N. Wang, arXiv:nucl-th/0305010.

[120] X. N. Wang, Phys. Lett. B **579**, 299 (2004) [arXiv:nucl-th/0307036].

[121] D. Schildknecht and Bernd Surrow, *Phys. Lett.* **B499**, 116 (2001)

[122] A. M. Stasto, K. Golec-Biernat and J. Kwiecinski, *Phys. Rev. Lett.*, **86**, 596 (2001).

[123] E. Levin and K. Tuchin, *Nucl. Phys.* **A691**, 779 (2001)

[124] E. Iancu, K. Itakura and L. McLerran, Nucl. Phys. A **708**, 327 (2002).

[125] A. H. Mueller and V. N. Triantafyllopoulos, *Nucl.Phys.* **B640**, 331 (2002). D. N. Triantafyllopoulos, Nucl. Phys. B **648**, 293 (2003).

[126] E. Iancu, K. Itakura, and S. Munier, to be published in *Phys. Lett. B*, hep-ph/0310338

[127] L. Frankfurt, V. Guzey, M. McDermott, *Phys.Rev.Lett.* **87**, 192301 (2001) L. Frankfurt, M. Strikman and M. Zhalov, *Phys.Lett.* **B537**, 51 (2002); T. Rogers, V. Guzey, M. Strikman and Z. Xu, hep-ph/0309099.

[128] K. Golec-Biernat and M. Wüsthoff, *Phys. Rev.* **D59** (1999), 014017; *ibid.* **D60** (1999), 114023; *Eur. Phys. J.* **C20** (2001) 313.

[129] W. Buchmuller and A. Hebecker, *Nucl.Phys.* **B476**, 203 (1996). W. Buchmuller, T. Gehrmann and A. Hebecker, *Nucl.Phys.* **B537**, 477 (1999)

[130] Y. Kovchegov and L. McLerran, *Phys.Rev.* **D60**, 054025 (1999).

[131] J. Bartels, K. Golec-Biernat and H. Kowalski, *Phys.Rev.* **D66**, 014001 (2002);

[132] A. H. Mueller, S. Munier and A. Stasto *Nucl.Phys.* **B603**, 427 (2001)

[133] H. Kowalski and D. Teaney, *Phys. Rev.* **D68**,114005 (2003).

[134] A. Kovner, L. McLerran and H. Weigert, Phys. Rev **D52** 3809 (1995); **D52** 6231 (1995).

[135] A. Krasnitz and R. Venugopalan, Nucl. Phys. **B557** 237 (1999); *Phys. Rev. Lett.* **84** (2000), 4309; Phys. Rev. Lett. **86** (2001).

[136] T. Lappi, *Phys. Rev.* **C67**, 054903 (2003).

[137] A. Krasnitz, Y. Nara, and R. Venugopalan, *Phys. Rev. Lett.* **87** (2001) 192302.

[138] K. J. Eskola, Nucl. Phys. A **698**, 78 (2002) [arXiv:hep-ph/0104058].

[139] K. Eskola, K. Kajantie and K. Tuominen, *Phys. Lett.* **B497**, 39 (2001).

[140] D. Kharzeev and M. Nardi, *Phys. Lett.* **B507**, 121 (2001); D. Kharzeev, E. Levin, *Phys. Lett.* **B523**, 79 (2001).

[141] B.B. Back et al [PHOBOS] *Phys. Rev.* **C65**, 31901R (2002).

[142] K. Adcox et. al. *Phys. Rev. Lett.* **86**, 3500 (2001)

[143] B. B. Back *et al.* [PHOBOS Collaboration], Phys. Rev. Lett. **85**, 3100 (2000) [arXiv:hep-ex/0007036].

[144] K. Adcox *et al.* [PHENIX Collaboration], Phys. Rev. Lett. **88**, 242301 (2002) [arXiv:nucl-ex/0112006]; Phys. Rev. Lett. **86**, 3500 (2001) [arXiv:nucl-ex/0012008]; Phys. Rev. Lett. **87**, 052301 (2001) [arXiv:nucl-ex/0104015].

[145] B. Back. et. al. *Phys. Lett.* **B578**, 297 (2004); C. Adler et. al. *Phys. Rev. Lett.* **89** 202301, (2002).

[146] D. Kharzeev, E. Levin and L. McLerran, *Phys. Lett.* **B561**, 93 (2003)

[147] J. D. Bjorken, Fermilab-Pub-82-059-THY; D. Appell, *Phys. Rev.* **D33**, 717 (1986);M. Gyulassy and M. Plumer, *Phys. Lett.* **B243**, 432 (1990); M. Gyulassy, M. Plumer, M. Thoma and X. N. Wang, *Nucl. Phys.* **A538**, 37C (1992);

[148] I. Arsene et. al. *Phys. Rev. Lett.* **91**, 072305 (2003); B. Back et. al. *Phys. Rev. Lett.* **91**, 072302 (2003); S.S. Adler et al., *Phys. Rev. Lett.* **91** , 072303 (2003); J. Adams, *Phys. Rev. Lett.* **91**, 072304 (2003); M. Gyulassy, I. Vitev and X. N. Wang, *Phys. Rev. Lett.* **86**, 2537 (2001)

[149] A. Dumitru and J. Jalilian-Marian, *Phys.Lett.* **B547**, 15 (2002); *Phys.Rev.Lett.***89**,022301(2002)

[150] F. Gelis and J. Jalilian-Marian, *Phys.Rev.* **D67**, 074019 (2003).

[151] I. Vitev and M. Gyulassy, *Phys. Rev. Lett.* **89**, 252 (2002); M. Gyulassy, P. Levai and I. Vitev, *Phys. Rev.* **D66**, 014005 (2002); *Nucl. Phys.* **B594**, 371 (2001); *Nucl. Phys.* **B571**, 197 (2000) I. Vitev, *Phys. Lett.* **B562**, 36 (2003).

[152] D. Kharzeev, Y. Kovchegov and K. Tuchin, *Phys.Rev.* **D68**, 094013 (2003).

[153] R. Baier, A. Kovner and U. Wiedemann, *Phys.Rev.* **D68**, 054009 (2003).

[154] J. Albacete, N. Armesto, A. Kovner, C. Salgado and U. Wiedemann, hep-ph/0307179

[155] I. Arsene *et al.* [BRAHMS Collaboration], arXiv:nucl-ex/0403005.

[156] Zhangbu Xu for the Star Collaboration at Quark Matter 2004.

[157] Rachid Noucier for the Phobos Collaboration;

[158] Talks presented at Quark Matter 2004: R. G de Cassagnac for the Phenix Collaboration at Quark Matter 2004, M. X. Liu for the Phenix Collaboration;

CHAIRMAN: M. GYULASSY

Scientific Secretaries: C. A. Aidala, C. Klein-Bösing

DISCUSSION I

- *Loktik:*

At the beginning of the talk you mentioned that there is a group of experimentalists who are composing some sort of paper that argues that everything you say in the next few days is wrong. Is it a fight between two camps?

- *Gyulassy:*

Actually a PHENIX paper will soon appear that is very positive. Other experimental groups remain less optimistic. It started out more negative when we were initiating discussions in May. I attribute that mainly to the fact that experimentalists were too busy measuring things to stop to review the physics of their results. The May workshop at RBRC arose because at Quark Matter 2004 in January experimentalists presented data without conclusions.

This motivated us to organize a workshop at RIKEN/RBRC (encouraged by T.D. Lee) to get the RHIC community to evaluate the physics implications of the data. At that time there was quite a bit of a controversy about what I discuss here, but I feel that there is more convergence now. This was a very interesting sociology of how 1000 experimentalists and scientific leadership interact to converge to a few clear ideas and conclusions. It's a sociology I've never seen before.

- *Loktik:*

But that's what science is about, no? Most people agree.

- *Gyulassy:*

This contrasts with table-top experiments. A good example of this contrast is at Duke University [Science 298 (2002) 2180], where a small group has been studying the flow of 10^5 Li-atoms at micro-Kelvin temperatures and observed very similar elliptic flow phenomena as observed at RHIC at 10^{12} Kelvin. One professor with a few graduate students were able to converge quickly on the new physics because he does not need the consensus of a thousand people.

- *Salur:*

So we have really a bunch of signatures for weakly coupled QGP. Now you say it's actually strongly coupled plasma. So what essentially needs to be reconsidered, which of these signatures?

- *Gyulassy:*

I haven't really finished that part of the elliptic flow, which is really the spectacular part. Just seeing the elliptic flow like for example in the crab nebula is nice, but so what? That just happens to be an asymmetric collective flow of gas from a supernova. The interest at RHIC comes in the next pile of transparencies where one discovers that that degree of collectivity is way higher than one would expect based on any microscopic transport theory. The existence of this essentially dissipative-free or minimally dissipative system, which is many orders of magnitude more perfect than water or any other liquid is a very surprising result. I think we haven't reached the end of where that will lead. The number of publications that may be relevant from string theory in this connection is increasing rapidly. This is an unexpected and very strange thing to me. Ed Shuryak has been also noting it in many of his talks. In the strong-coupling limit with t Hooft's g^2N going to infinity, string duality calculations seem to reproduce some of the the anomalous features seen in the data. How this influences other important pieces of data yet to come remains to be seen.

One of the students at this conference is doing a thesis on direct gamma production at RHIC. Photons measure EM current correlators whereas viscosity measures the stress $T_{\mu\nu}$ correlator. $T_{\mu\nu}$ is the source of the gravitational waves, which the fancy GW detectors discussed here will try to measure. Therefore supergravity is the natural animal to compute $T_{\mu\nu}$ correlators and therefore viscosity is the first thing that plops in their lap. However, other gauge-invariant correlators like electromagnetic correlators may also be calculable.

Prof. Veneziano can do one of these calculations and predict what the direct photon signature should be in this extreme point of view? This is radically different, 180 degrees different, than how people have thought about it for the last 25 years, where we thought we could just simply compute ordinary plasma like radiative processes. In the old view, basically free quarks bounce around and thermal current fluctuations radiate photons. Hence, people are now trying to measure thermal photons. This is analogous to the study of the Cosmic Microwave Background. But it may turn.out that past thermal estimates based on the weak coupling QCD, the simple dielectric view of QGP, are completely wrong as with the viscosity by an order of magnitude. Maybe this could perhaps test another strong coupling limit of the superworld down at our low-energy scale of the standard model.

So the answer to your question is still unclear, we just know that we have hit literally a brick wall. Something very strange [hydro saturated elliptic flow] has now been seen rather than something simpler. It may have profound consequences on all the other signatures that we've been dreaming about for so long.

-.*Ferrara:*
I know the calculation of glue-ball masses using adS conformal field theory, but now what are you are you referring to?

- *Gyulassy:*
This is the viscosity, the shear viscosity, which via the Kubo formula is the static long wavelength correlator of T_{xy}. It is compared to the entropy density that in turn is observable via the hadronic multiplicity produced. This viscosity/entropy density ratio from Policastro, Starinets, Son [PRL87 (2001) 081601] and other people is predicted to be this universal number $1/4\pi$. It is about one tenth as predicted by weak coupling estimates. That lower bound is as much viscosity as is tolerable in hydrodynamics so as not to spoil the agreement with the observed elliptic deformation at RHIC. This success has motivated several colleagues down in Brian Greene's corner of Columbia to start looking at other predictions from the adS/CFT conjecture.

We don't live in this conformal supersymmetric world. How could this have anything to do with anything? But now there are calculations that show that the $1/g^2N^{3/2}$ correction is small as well. It somehow survives having finite values instead of infinite values of g^2. Similarly one can start looking at flavor breaking, so there are people trying to compute N_f deviations. I personally have ZERO understanding at present of how those calculations are done, but the string people should realize that it may be possible to test experimentally their calculations via the QGP produced in AA. I think that this is going to stimulate a whole industry. I'm trying to learn from string experts.They're trying to teach a nuclear physicist string theory relevant to nuclear physics. So far it has been a very slow convergence.

- *Simon:*
You have shown the phase diagram of strongly interacting matter in the presentation of temperature vs. chemical potential. Within nuclear collisions we try to chart out this phase diagram by following trajectories. RHIC and LHC essentially go into the regime of high temperature and low chemical potential and this is a regime which has very much resemblance to the early universe. But here are other regimes: Low temperature and high chemical potential. And it's an interesting question what to do, to reach this regime: How should one design a heavy ion

reaction to have low temperature and high chemical potential? GSI is trying this route.

- *Gyulassy:*
I know that project. It evolves from 30 years in the history, starting With T.D.'s suggestion to start perturbing the vacuum. Initially it was 0.2-1.0 GeV per nucleon first done at the BEVALAC. Then SIS/GSI was able to continue that with much more refined detector systems. In that energy range one basically learned (a very useful lesson for us nuclear physicists) that the momentum- and density- dependent nuclear forces used in nuclear structure calculations also worked well to account for modest increases in the nuclear density. Calculations can reproduce the collectivities that are observed up to two or three times nuclear density. There were no surprises indicating that that standard [nuclear theory] model fails. However, we also learned that the nuclear system is extremely dissipative and that was expected, just like the QGP was expected to be dissipative. But in the case of the nuclear matter it turned out to be right. There are many transport theories that have been developed over the last decades that provide a quantitative understanding of data in the 1-GeV range.

There is one anomaly as the energy was increased by a factor of 20 at the SPS: the K/π horn. This is directly relevant to your question about what could happen as you go to higher baryon densities. Wilczek and Rajagopal and others have speculated about a possible critical point at some finite density and moderate temperatures. If one could produce reactions that passed by that point, there could be anomalous fluctuations in some quantities. The data at the SPS initially at 200 GeV overshot the critical point, essentially getting closer to RHIC. It wasn't that far away from RHIC in a sense of generating high energy densities but with low baryon densities.

However, at lower energies near 20 GeV in the lab, then some interesting signals seem to be coming out from the CERN experiments. In particular there is the famous K/π ratio that seems to have a peak somewhere around that energy range. This is where the new GSI-type of facility could help. At this energy, it has always been thought that AA produces the maximum baryon density regime. At higher energies the valence baryons pass through each other leaving a rubble of more pure vacuum excitation. This is what RHIC is doing now. Energies on the order of 20-30 GeV are where the maximum baryon densities that can be reached with heavy ion reactions should be achieved. I think this is a promising open area of research.

- *Korthals-Altes:*
This morning you mentioned this lower bound on the viscosity in units of the entropy density, and you showed examples. One example was liquid helium. Now

my question is, was this superfluid helium, where you would say, well, the bound will be probably saturated?

- *Gyulassy:*

It was superfluid, and it has viscosity because it's not a perfect ideal superfluid, it has some small finite viscosity. However, it is still an order of magnitude higher than the adS/CFT lower bound. So this bound is really very small. It is very hard to reach. You have to have a highly coupled system, in order to drive the system down to essentially the uncertainty principle limit. So that's what's remarkable about it. So far it is only expected from the point of view of the $g^2N \rightarrow \infty$ limit and stringy duality. Something rather weird has to happen in that limit, and apparently the viscosity gets as small as possible. That's my interpretation based on just that kinetic theory argument I presented. Why it's $1/4\pi$ vs some other number is some property of the gravitational calculation, perhaps the area of a sphere.

- *Loktik:*

You just flashed a slide with the Bjorken estimate of something, and there was this term of "quark-gluon matter". So I just wondered, what is it? Is it just high energy density, or is a new sort of state? I just never heard about it.

- *Gyulassy:*

This was Larry McLerran's cartoon. There are other versions of this cartoon. The only point that's really important is the issue of the time scale of the initial conditions. The incident gluonic fields decohere through multiple scattering processes, either via two-to-one or two-to-two processes. Then there's a period between that production phase of "matter" and the freely streaming out of the partons produced in the interactions. Typically, kinetic calculations indicate that it takes about three collisions to reach equilibrium. This intermediary, transient phase before equilibration is what was labeled as "quark-gluon matter". It denotes the non-equilibrium phase before hydrodynamics.

The time corresponding to equilibration is on the order of 0.6-1 fermi. In fact, one of the tests of thermalization is to see happens when the themalization time is increased. What Heinz and Kolb showed is that it cannot be bigger than about a fermi/c. Otherwise, hydro cannot push matter up to the observed elliptic deformations. That time scale is reasonable based on the uncertainty principle -- the thermal wavelength 1/3T multiplied by three to reach equilibrium.

- *Salur:*

I think I understood how important it is in terms of flow as being represented as hydrodynamic models, but the fact that SPS does not see this thing really makes SPS not say anything about their observation of QGP? Does this really by default cancel their observation?

- *Gyulassy:*

No. I think in fact one can sort of reverse engineer backward from RHIC to SPS now. The theory which accounts for the SPS deficit, the fact that it differs by a factor of two from pure hydrodynamics is successful because it takes into account the highly dissipative hadronic corona. That corona physics is encoded in a combined hydrodynamic-kinetic description. So the fact that this calculation works at SPS supports in restrospect the claim that SPS reached the boundary of the phase transition at about 3 GeV/fm^3. At least part of the action at SPS is due to QGP formation I believe. Unfortunately the QGP signal is camouflaged and distorted by the hadronic kinetic background. What the RHIC data show is that one moves into a deeper domain above the deconfinement at collider energies and reaches early local equilibrium over a larger volume. Elliptic flow becomes a cleaner barometer of the pressure in the QGP phase, less polluted by hadronic corona effects. Most people feel that the SPS really went over the boundary, but a convincing proof was lacking. By reverse engineering from what we now know from RHIC and trying to improve the theory and the interplay between hadronic kinetic and QGP hydrodynamics, one can try to pinpoint better exactly what form of matter was produced at the SPS energies.

- *Salur:*

But the fact that the QGP is then hidden under this hadron gas, doesn't that mean that at the SPS they should never observe the QGP because they are losing the signatures?

- *Gyulassy:*

No, you get to see part of that -- you do see elliptic flow, it's just not as high as it would be if the system reached local equilibrium. So it's a quantitative issue of exactly how much of the hydrodynamic phase is influencing that yield, and how much is being distorted by the hadronic phase. That will require a lot more analysis, because it's a lot harder calculation than at RHIC. It's very much like the theoretical studies of supernovas. People have been trying to blow up supernovas on a computer for decades. This is also a coupled hydrodynamic-radiative transport problem. Woosley has organized a SciDac activity, similar to lattice QCD SciDac initiative, to assemble experts from a variety of specialties to try to make supernova explode on a computer. Most of the time calculations still fail. It is very hard to solve non-linear radiative transport and hydrodynamic equations correctly. The same is true for heavy ion reactions. When you have to couple signals from both the non-equilibrium part, which is the hadronic gas in our case, to the equilibrium part, the QGP, then the details matter a lot. This is especially true below RHIC energies. At the moment one has only a qualitative picture in terms of this hybrid, HYDRO+RQMD model, but the details are still to be worked out, and that will require much more work, to evaluate what the SPS data really imply.

\- *Ferrara:*

It seems to me that N = 4 super Yang-Mills theory hasn't anything to do with ordinary QCD, like N = 1. What is the logic of using N = 4 Yang-Mills theory to compute these numbers? This author must have used some argument why these calculations are reliable for ordinary QCD.

\- *Gyulassy:*

That's what I asked string experts. That is what I hope to learn at this school, from you and other experts -- why in hell the Maldacena conjecture could have any relation to our real-world QCD physics. I would never have paid attention to these calculations, except for the fact that these are the only ones at the moment that come anywhere near the experimental findings at RHIC. I would love to learn, from you or anybody here, how one can better justify this conformal N = 4 supersymmetry theory conjectured duality. It could be that all this is basically understandable from just the uncertainty principle argument I presented (and details of the CFT do not matter). If the coupling becomes huge all rates have to be h-bar over the typical energy scale, and no matter what the theory is, you're constrained to get to this type of number like one tenth. So it could be accidental, but there are a number of other accidents, such as the entropy density magical 3/4 number, which pop out of the calculation and seem to agree with the lattice QCD results. They again suggest, well, maybe there's something more to it after all. I really don't understand how one can justify a priori that this duality could be used, but it seems to be getting more experimental push to make it worth thinking about.

It could be that in fact the most ideal situation is this $T = T_c$ arena, not for calculating glue-balls and real-world stuff, but in the critical region where there's some fundamental change in the nature of the vacuum. The adS/CFT technology may capture some critical universal feature of extremely strongly coupled matter. I was hoping that Professor 't Hooft would help enlighten me on this. He suggested to try $1/g^2N$ analytic calculations at large 't Hooft coupling to see whether you can reproduce in some form some set of coefficients to get this $1/g^2N^{3/2}$ in QCD itself. It's a resummation that may be doable, like a plasmon loop in weak coupling, where the perturbation theory is not $1 + g^2 + g^4$, but it contains g^3 etc. This well-known phenomenon in QED comes from the dielectric screening, and the plasmon loop, the collectivity. Now a similar resummation in the $1/(g^2N)^n$ could lead to some useful insight.

- *Ferrara:*

You know, supersymmetry can be used and give you information also in many cases for non-supersymmetric theories. I think Gabriele Veneziano will talk about it. But for instance, in a famous example, supersymmetry allows you to prove the positive-energy theorem in gravity, which was done by Witten, and others, and that uses supersymmetry in a non-supersymmetric theory.

- *Gyulassy:*

Well, Witten himself is the one who calls the QGP a black hole and is very comfortable with that statement. I would love to understand what he means. Except for somehow a belief that this kind of calculation with the adS near a background black hole is reproducing lattice QCD, non-perturbative correlators. At least some of them. So one of the questions we discussed at lunch is what would be the analog for current correlators. Somehow $T_{\mu\nu}$ seems like an ideal thing for gravity, but where is the electromagnetic current hiding -- that's something you can do on a lattice because you can figure out what the operator is -- but what is the conjectured analog? What do I have to do in this adS theory to get current fluctuations?

- *Ferrara:*

The answer is very simple. In the case of the stress tensor, the homographic dual is the graviton. The stress tensor corresponds to the graviton. In the case of electromagnetic current, it would be a vector field in the bulk. So it would be the correlation function of some vector, graviphoton, in the bulk. That would be electromagnetic current.

- *Gyulassy:*

So is that a doable task?

- *Ferrara:*

Yes, it's computable.

- *Gyulassy:*

Now, my question to you again: What experimentalists can do is to measure high-momentum, short-wavelength, perturbative photons, which are current correlators in a special light cone limit. These are not the large-wavelength, static type of limits like the Kubo formula. But looking really on the light cone, where the weak coupling and the running coupling matter presumably. Is there a domain in applying this duality that prevents you from calculating via adS/CFT above some scale? You can calculate very large-wavelength phenomena, but when the coupling starts to vary a lot, as at short wavelengths, then this duality may break down. What is the parameter of the adS type of calculation that must break down if I want to push myself along the light cone in Minkowski space rather than being stuck in the static limit in Euclidean space? Does anybody have any insight into these?

- *Veneziano:*

Did you say you have to go to the small-coupling region?

- *Gyulassy:*

Well, in order to pull out signatures which are not products of pion decay, pi0's and etas and all sorts of rubble -- there's a huge rubble of hadronic debris -- I have to get out far in p_T to be able to get the signal out of the noise. One of the nice things is that the jet quenching is able to remove a lot of the hard photons coming from fragments of jets that happen to have pi0's, which make photons, and that helps. It's a question of being able to see the signal.

- *Veneziano:*

What I wanted to say is that the rule of thumb is that the smaller the coupling, the smaller this adS radius, and if you go to small coupling, you enter the high-curvature regime of adS, where you don't know how to solve string theory. This is for N = 4, where the coupling is constant, because there is no running. You can tune the coupling to whichever value you like, and you can describe the theory of different couplings. If you go to QCD, the coupling runs, and therefore, you see different regimes in QCD seem to correspond to adS radii.

- *Gyulassy:*

On the other hand, g^2N is always about 10. So it's not small. It never really gets below 1 or even near 1. We are never in the very weak coupling in that sense. It's in this region, g^2N on the order of ten, where all these remarkable string results with 3/4 and only 10% corrections are coming out. What we think of naively in perturbative QCD as weak coupling is really strong coupling from the g^2N-expansion, and so maybe it still is possible. There's a very nice curve which I didn't show, from Shuryak, where you put together the strong coupling expansion and the weak coupling, and there really is a funny divergence around $g^2N = 1$. That weird region is just not explorable in any experiment.

- *Ferrara:*

If I understand youı logic, you are saying that QGP corresponds to strongly-coupled QCD.

- *Gyulassy:*

That's right.

- *Ferrara:*

And strongly-coupled QCD can be studied in adS conformal theory?

- *Gyulassy:*

That is the advertised result, which I don't understand. If it's true, it would be spectacular.

- *Ferrara:*

Does that mean that all the people who have not used strongly-coupled QCD have made nonsense predictions for QGP?

- *Gyulassy:*

That's possible. That's what we've been trying for quite a long time, using alpha of 0.3, thinking that's small enough to be able to do an expansion. One is encouraged by the fact that pQCD works at 4-5 GeV scales remarkably well. Next-to-leading-order calculations are semiquantitative. It's not that we are off by some giant factor. So at least for non-thermal dynamical short wavelength phenomena, alpha = 0.3 is a reasonably small number. But thermal matter properties of the QGP above T_c are not perturbatively computable with alpha = 0.3.

The difference is in matter, where the screening scale is $\mu_E \approx g_T$. Unfortunately $g \approx 2$. So the expansion in g's of matter properties seems to be wrong near T_c. I think that's what we learned from the very long and hard process of trying to get resummations in the hard thermal loop program that Pisarski and Bratten invented and elaborated with Rebhan and others. That approach, which is a resummation in powers of g, for thermodynamic properties, screening masses and things, doesn't work. The same alpha expansion doesn't work for QGP matter properties in contrast to short-wavelength dynamical properties.

- *Clark:*

You showed quite a lot of lattice results over the past couple days. When you have a QGP, you have chiral symmetry. All the lattice results you showed were from the MILC collaboration, which use a fermion formulation, which doesn't have true chiral symmetry. If you did the same calculations with a more chirally symmetric lattice formulation would you expect the results to change much?

- *Gyulassy:*

The thermodynamic quantities haven't changed much with improved actions as far as I know. This is one of the problems. Many lattice groups have turned to ε/ε and other interesting things. The convergence to the thermodynamic limit is essentially carried by Karsch and his group in Bielefeld. Those are the nice curves I showed. The old MILC calculations, which I got from Tom Blum almost ten years ago now, illustrate well the basic features of the transition. They haven't changed much, but the curves themselves have just gotten smaller error bars.

Another remark in answer to your question is the absence of numerical lattice data on hadronic matter below T_c. Part of the problem is that the pion is still too heavy in all lattice formulations so far. I think it's 300 MeV in the best of the calculations. The lattice matter below T_c is not our known nuclear world. The matter above T_c is matter that we'd like to see, but we don't know yet if it's true. So in a sense that is a dilemma. We believe that the lattice calculations are better above T_c, where the coupling should be smaller. One is in a regime where one can test the scaling laws, and that works fine. This does not work in the hadronic sector, which is the strongly coupled, confined, chirally-broken world we live in.

- *Arcelli:*

I was just wondering whether with the current data and the status of the theoretical predictions it is possible to link the long-wavelength fingerprints for QGP to the hard probes like jet quenching, so that you can make sort of quantitative estimates and say that the signals come from the same state.

- *Gyulassy:*

This is the subject of the next lecture. Essentially, the other limit that is computable is the perturbative high-p_T limit, or the heavy-quark mass limit. These are a complementary set of short-wavelength probes. There is a consistency between the long- and short- wavelength probes of QGP at RHIC that helps to support our claim that RHIC experiments have produced a deconfined QCD plasma. That is the strength of the argument at the moment. There are of course lots of tests that still have to be done. For example, the charm quark measurements still have large errors but this will be vastly improved in the upcoming year.

CHAIRMAN: M. GYULASSY

Scientific Secretaries: E. Bruna, S. Salur

DISCUSSION II

-Achim Geiser:

How exactly is the jet algorithm defined for the yields of the famous "mono-jet" events? Energy and momentum conservation should ensure that the momentum flows also into the opposite hemisphere. Are you only looking at the leading particles which eliminate all the other balancing opposite side particles?

-Miklos Gyulassy:

The primary jet is triggered by at least one charged particle with p_T>5 GeV. The corresponding associated backward going particles have to carry at least 2 GeV transverse momentum. In the p+p collisions, when jet fragments, the away side jet is visible. The disappearance of the away side jet in central Au+Au collision systems can be studied with different momentum selections. This analysis is in progress and it is very suggestive and very exciting since as you lower the momentum selection criteria of the associated particles, there is an increase in the number of particles over a broader azimuthal angle in the away side jet distribution. The energy is lost via showering of particles like in cosmic rays and this measurement can be related to the broadening of the jet. This will be addressed in the new data this year and in the future with much higher statistics.

-Luisa Cifarelli:

Could you please comment on quarkonia suppression and open flavour production from SPS to LHC?

-Miklos Gyulassy:

The problem with the SPS NA50 results on J/Ψ suppression is the lack of adequate control experiment, in particular centrally triggered p+A collisions. In Prof. Zichichi's "universality" domain, the T=0 vacuum, the production of the charm should look similar in e+e- and p+p. However, in charm production with a nuclear target, a gluon from the incoming proton turns into a virtual $c\bar{c}$ pair that must interact prior to hadronization with the local color fluctuations in the nuclear medium. These interactions reduce the probability that the $c\bar{c}$ pair will emerge as a bound state since the probability that the pair fits inside a J/Ψ orbital is reduced. In the p+A interactions, the J/Ψ suppression is indeed observed to increase as the size of the nucleus and the penetration depth. The problem is that those studies have not been extended to central triggered p+A where the $c\bar{c}$ pair has to traverse the full diameter of the nucleus. In A+A collisions some parts of the reaction always probe the maximum penetration depth L~10 fm. This is roughly twice the value measured thus far in p+A interactions.

The suppression of J/Ψ varies exponentially with L. In QCD calculations energy loss in a cold nuclear medium is predicted to depend quadratically on L. J. Qiu and J. Vary estimated that this nonlinear effect should cause an additional suppression comparable to observed in A+A. Centrally triggered p+A at SPS would be a wonderful test of nonlinear medium induced QCD effects to resolve this question. The linear extrapolation of the normalized J/Ψ suppression from small L cannot be true since QCD is nonlinear. At RHIC interactions in the cold-nucleus causes the diffusion of charm and bottom pairs due to both elastic and inelastic processes. This can only be falsified or proved with dedicated experiments in p+A collisions.

Open charm is an external probe with a kinematical mass effect in terms of radiation pattern which leads to a dead cone. Dokshitzer and Kharzeev predict a radiation hole produced by heavier masses such that heavier the particle, bigger is the hole.
My student Magdalena Djordjevic has quantified this effect for both RHIC and LHC. The bottom and charm propagation can be used as a lever arm for LHC because the very high densities completely all light quarks and gluons and even charm will be strongly suppressed. At LHC the charm will be close to the limit while the bottom will be a wonderful probe since it still has some leveraging room in its quenching pattern.

-Silvia Arcelli:

Which range of energy losses in the plasma could be expected at LHC energies and are the current parametric descriptions available in heavy ion Monte Carlo simulations adequate to describe this lost?

-Miklos Gyulassy:

The energy loss at RHIC simulated by the Hjing Monte Carlo 2 GeV/fm is based on an overly simplified algorithm. In these simulations, jet's path length is subdivided into L/mean free path segments with a constant dE/dx used in each. In reality the medium is expanding both in longitudinal and transverse directions. The dynamical line integral through this evolving medium needs to be calculated the GLV jet tomography. Ivan Vitev and I attempted to solve these integrals analytically. Instead of the time dependent dE/dx, it is more useful to express the results in terms of the initial dN/dy of the produced gluons that form the QGP. This means that jet tomography determines the initial entropy indirectly through its effect on the radiative energy loss. We found that the results are not sensitive to the details of the transverse expansion, but depend on the longitudinal Bjorken expansion.

Unfortunately there is as yet no extension of HIJING or other full event simulators that incorporates the correct energy loss physics. Only analytic calculations for the inclusive cross sections exist up to now. Parenthetically, in fact no asymptotic analytic formula reproduces accurately the GLV terms that we evaluate numerically. Ultimately, I hope that an event generator will be constructed to generalize our results to complete event simulations. HIJING can still be used for quick estimates at RHIC but for LHC analytic calculations are more reliable.

.

-Luisa Cifarelli:

Can one figure out an enhancement of quarkonia production at LHC according to a hard energy loss process where a generated high energy gluon produces a $c\bar{c}$ pair?

-Miklos Gyulassy:

That is an interesting question that I have not thought about it. This is actually a very good idea since the showering of the high momentum jet can occasionally produce extra charm in the QGP. A $c\bar{c}$ pair excitation can be produced easily by the induced several ~ 10 GeV gluons radiated from a high momentum (100 GeV) jet. Similarly another good observable is to look for the photons from an induced electromagnetic shower in the wake of the jet to search for an enhancement of induced reactions of $q\bar{q}$ pairs from these showers.
An observation of an enhancement of charm content in a 100 GeV jet at LHC with respect to $e^{+}+e^{-}$ and p+p collisions can give a nice piece of information about the change of the T=0 "Universality" discussed by Prof Zichichi in QGP conditions with $T>Tc^{0}$.
Since the charm content of jet fragmentation is small, maybe the induced $c\bar{c}$ pair production in the jet showers could lead to a dramatic enhancement relative to fragmentation in the vacuum. This would be a good test also of the different approximations used by the GLV, WW, BDMS and SW groups working on jet quenching phenomenology.

Lattice QCD

Richard Kenway
School of Physics, The University of Edinburgh,
Edinburgh EH9 3JZ, Scotland

Abstract

This article is aimed at non-experts and tries to explain without any technicalities how lattice QCD is helping us to understand the Standard Model and to search for physics beyond it.

1. Beyond Small Parameters

Up to the highest energy scales available to experimental particle physics today, our world is described by relativistic quantum field theories of point-like particles, together with the classical theory of General Relativity. Specifically, the strong, electromagnetic and weak interactions are described by the so-called Standard Model. There must be physics beyond this to tie gravity into a consistent quantum framework, but also, hopefully, to explain the 25 or so otherwise fundamental parameters of the Standard Model. Theories of the physics beyond these energies postulate exotic new fundamental entities like strings, or *p*-dimensional "branes", which may exist in more than four space-time dimensions and offer the unification of all four forces.

However, at the relatively low energies available today, these entities manifest themselves as particles, i.e., quanta of a relativistic field theory, and the Standard Model plus General Relativity seem likely to remain the most appropriate description of their physics that we have. As such, the Standard Model is an effective theory, only valid below a certain energy scale. It contains just those degrees of freedom that are relevant below that scale. The more fundamental physics at higher energies is encapsulated in the masses, strengths and particular forms of the interactions between the Standard Model particles.

Lattice QCD attempts to provide a computational framework for the strong interaction between coloured quarks and gluons, described by Quantum Chromodynamics (QCD). Taken with perturbation theory for the electroweak interaction, it should enable arbitrarily precise calculations of Standard Model processes to be carried out, given sufficient computer power. The key point is that these calculations can be performed for all values of the input parameters of QCD. These parameters are the masses of the quarks and the gauge coupling, which is the strength of the interaction between quarks and gluons. They originate in the higher-energy, or 'beyond the Standard Model', theory. At our current level of understanding, they might just as well come 'from God', and the best we can do is to measure them experimentally and look for clues for something more fundamental within them.

The main point of the lattice approach to quantum field theories is that there is no need to assume that any parameter is small, when, in the real world, it may not be. This is in contrast to perturbation theory, which assumes that some parameter is small enough that physical quantities may be computed reliably by the first few terms of a series expansion in powers of that parameter.

The computational framework of lattice QCD rests on a rigorous formulation of the quantum field theory that began with Ken Wilson's 1974 discovery of how to maintain local gauge symmetry on a discrete space-time lattice. It was only completed during the last decade with the understanding of how also to preserve chiral symmetry on the lattice. The local non-abelian gauge symmetry of QCD is responsible for quark confinement, i.e., the fact that no isolated single quark (or any other coloured object for that matter) has ever been observed experimentally. Chiral symmetry is responsible for pions being very much lighter than any other hadrons (actually, this is due to the mechanism whereby chiral symmetry breaks spontaneously). The computational framework itself comprises a set of numerical algorithms and approximations that we control and that we can, in principle, make arbitrarily good. How good is just a matter of how much computer power can be applied.

Computer simulation has become the third methodology of science, alongside experiment and theory. It has grown rapidly over the past 50 years, driven by exponential growth in computer power and the need in many areas of science to understand complexity. The idea is to embody our knowledge of a particular system into a computer code that simulates its behaviour and to confront the resulting virtual reality with experimental measurements. That may invalidate the simulation and hence the underlying theory, exposing shortcomings in our knowledge. Hopefully, such an outcome would also point towards new physics, eventually leading to a better understanding. This is one of the main objectives of lattice QCD.

Alternatively, the simulation may accurately reproduce experiment. Then the simulation becomes a virtual laboratory in which we can explore aspects of the system inaccessible to experiment (e.g. the interior of the Sun), or forecast its behaviour in the future (e.g. our climate). In QCD, this might enable us to predict the behaviour of strongly-interacting matter at extremely high baryon densities, such as might exist in gravitationally collapsed objects like neutron stars.

2. The Search for New Physics

One Standard Model particle, the Higgs boson, is yet to be discovered. Its properties have been so well constrained by high precision experiments that if it is not found by CERN's Large Hadron Collider (LHC), which should begin operation in 2007, then the Standard Model must itself be seen to break down in some unexpected way. Thus, it seems that the LHC must discover something. The Higgs boson could take two or three years of painstaking measurements and analysis to expose, but there are plenty of candidates for other new particles that might appear immediately LHC opens up the new multi-TeV energy frontier.

Today, we have to invoke new fields, or particles, to explain the early phase of rapid inflation that produced a flat and isotropic Universe, the dark matter that is responsible for gravitational lensing, and the dark energy that is responsible for the current accelerated expansion of the Universe. Supersymmetry could provide candidates for many of these, if Nature chooses to use it, while helping to explain why the Higgs boson is light on the scale of the Planck mass and how the strong and electroweak forces can be unified at energies around 10^{16} GeV. The tiny neutrino masses recently discovered also point to new particles with masses at these sorts of energies. We are poised to enter a period of discovery that could radically change our picture of particle physics. So where does lattice QCD fit in?

Putting aside the use of lattice field theory to study supersymmetric theories directly, the answer is our ability to constrain the possible new physics at high energies through a combination of precision calculations and experimental measurements at low energies. These might reveal a breakdown of the Standard Model itself, e.g., through an inconsistency between two different experimental determinations of one of its fundamental parameters, or restrict processes which receive (small) quantum effects from as yet undiscovered particles at energies that are experimentally accessible today, thereby constraining the properties of those particles, such as their masses.

Lattice QCD is needed to do this because of quark confinement. The strong interaction causes all quarks and gluons, the fundamental particles which carry colour charges, to appear in Nature as colourless bound states, called hadrons (except at extremes of temperature or baryon density that are only beginning to come within our experimental reach). Thus, we experiment on baryons, comprising three valence quarks, and mesons, comprising a valence quark-antiquark pair. (There are many other types of colourless bound states that may be allowed by QCD, but whose experimental verification is not yet certain, such as pure gluon bound states called glueballs and pentaquarks states with four valence quarks and one valence antiquark.) The binding energy involved is huge, for instance, providing 99% of the proton's mass. In order to relate experimental measurements on hadrons to processes involving the (fundamental) quarks and gluons, we need reliable and accurate calculations of these binding effects, and that is what lattice QCD enables us to do.

Since quarks participate in all four interactions of Nature, they play an inevitable part in particle physics experiments and yet we must infer their interactions through those of hadrons. We can do this by guessing the values of the quark parameters to use as input to our simulation, computing the resulting hadron properties, comparing them with experiment, and then adjusting the input quark parameters until the simulation and experimental results match. When they do, we can infer that the values of the fundamental parameters entering our simulation really are those in Nature. Hence, we have translated experimental measurements on hadrons into measurements on quarks.

With the limited computer power available today, we have to surround the simulations by an extensive theoretical scaffold. This uses perturbation theory for all but the low-energy strong interactions of quarks and gluons, simulates these over a restricted range of parameter values, and then uses various effective theories to extrapolate the results (if necessary) to parameter values beyond the simulation range. As computer power grows, the range of validity of our simulations should eventually extend sufficiently far into the domain of validity of these effective theories that the entire parameter range can be covered with high precision.

One of the biggest puzzles in particle physics today is why there are three generations of quarks and leptons, each with identical sets of charges and differing only in their masses. The electroweak interaction mixes the three generations of quarks, allowing one flavour of quark to decay into another, e.g., via a generalisation of β decay. However, if there really are just three generations, then the strengths of these decays can be related to one another – they are not all independent. Experiments which measure quark decays in several different ways can test the consistency of this picture and, hence, the consistency of the Standard Model. This is the goal of B Factories that are studying decays of B mesons (those containing a single valence b antiquark) with unprecedented precision. Their results are challenging lattice QCD calculations to match their precision, so that the decays of b quarks can be known just as accurately as those of B mesons.

Within the Standard Model, three generations permit the symmetry between matter and antimatter to be broken by an amount controlled by a single parameter. Breaking of this CP symmetry is necessary for the Universe to be matter dominated. The electroweak decays of one quark flavour into another allow us to measure this parameter, and this forms part of the consistency test of the three-generation picture. Again, to relate meson decays to quark decays, lattice QCD is essential.

Despite the dramatic improvements in precision achieved by the B Factories, all experimental measurements still support the three-generation Standard Model. In several key cases, the largest uncertainty in the analysis comes from the lattice QCD input. This has been one of the strongest motivations behind the development of lattice QCD, and it will continue to be so until we understand the puzzle of the three generations. It is a pity that Nature has hidden its secrets so well.

Even if the level of precision needed to expose an inconsistency in the Standard Model continues to elude us, the physics beyond the Standard Model may affect experiments at currently available energies in measurable ways. Rare decays of one quark flavour into another are those that are forbidden in the Standard Model at leading order in perturbation theory, but are possible at higher orders. As such, these decays are purely quantum effects and may receive contributions from virtual particles that are not contained in the Standard Model. If we can compute the Standard Model decay rate and measure the actual decay rate precisely enough, any difference between them places a bound on new physics.

Also, most theories of physics beyond the Standard Model, particularly Grand Unified Theories (GUTs) that attempt to unify the strong and electroweak interactions, contain processes that are completely forbidden within the Standard Model alone. An example is proton decay. Experiments that place lower bounds on the proton lifetime already rule out some GUTs. A significant uncertainty in the bound comes from QCD effects because the process involves quarks decaying within the proton bound state. These bound-state effects can be separated off from the beyond Standard Model physics and computed using lattice QCD. A precision calculation would tighten significantly the experimental constraints on models of new physics. Given that this physics involves particles with masses of order 10^{16} GeV, the use of these indirect methods, employing high-precision low-energy calculations, is the only option.

3. Objectives of Lattice QCD

Lattice QCD aims to provide a means of computing to some desired level of accuracy many, perhaps most, of the properties of hadrons, such as their masses, structures, decays and scattering, from first principles, i.e., directly from the QCD Lagrangian. The essential challenge that must be overcome is to achieve control over all sources of uncertainty in these calculations. During its 30-year history, much progress has been made through theoretical breakthroughs in understanding quantum field theories outside perturbation theory, new numerical algorithms and faster computers. We are still short of our goal, but, for a range of quantities of real significance to experiment, we have reached the point where it is achievable given only more computer power. Furthermore, the power required should be technologically possible over the next few years.

There are a variety of formulations of lattice QCD, each with its own pros and cons. The choice of which to use depends on the physics objectives. We learnt a lot during the earlier years of the subject by using an approximation called quenched QCD, which omits the effects of virtual quark-antiquark pairs on the vacuum (except for that part of the effect which can be modelled by adjusting the quark-gluon coupling and quark masses), simply because our computers were not powerful enough to do anything else. Quenched QCD turned out to be a good model for a surprisingly wide range of phenomenologically interesting quantities at the 10-20% level. As we include quark pair creation and annihilation effects in the simulation, the computational cost grows very rapidly with decreasing quark mass. Today's computers can simulate s quarks directly and are pushing the u and d quark masses down towards a tenth of the s quark mass (within a factor of two-to-three of their physical values), but not with full control of all the sources of uncertainty.

The most ambitious simulations today use a formulation called improved staggered quarks that has not been fully justified theoretically. However, it has been used to compute a range of experimentally well known quantities and gives agreement with the experimental values at the few-percent level. This is an important milestone, because it demonstrates empirically the consistency of QCD within the Standard Model, i.e. unique values of the quark masses and the quark-gluon coupling can be found for which at least a subset of hadron properties are consistent with experiment to high precision.

The next step will be to perform rigorously justified simulations also to the few-percent level of precision that can, in principle, be used to falsify the Standard Model. At the moment, the prospects for really finding a disagreement with experiment seem slim. But the ability to perform such simulations will provide a firm foundation for constraining new physics and for predicting new phases of strongly-interacting matter.

4. Methods of Lattice QCD

The path integral formulation of a quantum field theory expresses expectation values of products of the elementary fields in terms of a functional integral over all possible field configurations in space-time. We work in Euclidean space-time, where time is taken to be purely imaginary, because then the path integral has a probabilistic interpretation. Formally, it looks like equilibrium statistical mechanics in four dimensions. This brings useful insights, for instance, from critical phenomena, but most importantly it allows the path integral to be evaluated by Monte Carlo methods. The physical quantities we are interested in computing, like hadron masses and matrix elements of operators in the effective electroweak Lagrangian, may be obtained directly from the expectation values in the Euclidean formulation without analytical continuation back to Minkowski space, so there is no embarrassment in using imaginary time.

The next step is to replace four-dimensional Euclidean space-time by a regular four-dimensional hypercubic lattice of points. This renders the number of degrees of freedom countable and the theory finite, because the lattice spacing, a, is the shortest distance and this is equivalent to a momentum cut-off of order π/a. Only a finite range of momentum is allowed on the lattice, so there is nowhere that divergences can enter. The choice of hypercubic geometry is arbitrary – other lattice geometries also work, but are more complicated.

Ken Wilson discovered not only how to maintain local gauge invariance at each site of the space-time lattice, but also that it was possible to prove that quarks are confined in the lattice theory if the quark-gluon coupling is large. In this limit, the potential energy of a static quark-antiquark pair grows linearly with their separation. Thus, a fundamental property of QCD, i.e. that its spectrum consists of hadrons, emerges naturally on the lattice, suggesting that it is the right starting point for trying to understand the low-energy properties of the theory. Subsequently, Monte Carlo simulations indicated that nothing happens to change this confinement picture as the coupling is made small and the quarks are treated dynamically – a quark-antiquark pair is bound by a string of flux whose energy grows with its length. When the energy stored becomes big enough to pop a quark-antiquark pair out of the vacuum, the string breaks and the new quark and antiquark combine with the original antiquark and quark, each joined by flux strings, to form two mesons.

The use of a lattice, or grid, for numerical simulations of continuous systems is standard, e.g. in engineering. It is intuitively obvious that if the lattice spacing is much smaller than any of the interesting physical length scales, then the discretisation effects should be negligible. In lattice QCD, we aim to perform simulations at several different lattice spacings and extrapolate the results to the continuum limit, i.e. zero lattice spacing. As we approach this limit, physical lengths diverge in units of the lattice spacing, which is the signal that the lattice theory is approaching a critical point, where the four-dimensional statistical mechanical system undergoes a continuous phase transition.

Taking the continuum limit, then, is equivalent to finding values of the parameters of the lattice theory for which it has a critical point. In QCD, we know from perturbation theory that this occurs where the quark-gluon coupling vanishes, due to asymptotic freedom. This property of QCD, which means that the effective interaction between quarks and gluons gets weaker at short distances (i.e. at high energies, and in the opposite extreme to that in which confinement sets in), has, as a side effect, that zero coupling is a critical point of the lattice theory. Since perturbation theory works increasingly well in this limit, we have very good analytical control of the continuum limit of lattice QCD.

Lastly, to put lattice QCD onto a computer requires that we limit the space-time lattice to a finite volume, because, of course, the computer can only store a finite number of variables. Provided we keep the volume big enough to contain the hadrons of interest, any distortion of the physics due to finite volume effects tends to decrease exponentially with the linear size of the lattice and is relatively easy to control.

Thus, a particular simulation involves first choosing input values for the quark-gluon coupling and quark masses, and for the number of lattice sites. Next the Monte Carlo method is used to generate an ensemble of gluon field configurations that are distributed according to the probability distribution defined by the path integral. This enables us to approximate expectation values by averages taken over this ensemble, but incurs a statistical error which falls inversely with the square root of the number of configurations. The computational cost therefore increases by roughly a factor of four if we want to halve the statistical uncertainty in our results. From the behaviour of these expectation values in Euclidean time we can extract the spectrum and various matrix elements. To do this reliably, our lattice must be sufficiently long in the time direction.

At this point, we have a set of physical quantities evaluated with a certain level of statistical uncertainty at fixed values of the quark masses, in a finite volume and with fixed lattice spacing. How close these results are to the real world depends on how close the quark masses are to their physical values, whether the lattice spacing is small enough and whether the volume is large enough.

5. Continuum Physics

The definition of the QCD path integral on a Euclidean space-time lattice involves replacing the QCD action with a lattice version. There is considerable freedom of choice here. The lattice action must become the continuum QCD action in the limit that the lattice spacing is taken to zero, so that we recover the correct classical theory. In order to get the correct quantum theory, the lattice theory should, as far as possible, respect the symmetries of the continuum theory.

Having the correct classical limit means that, if we expand the lattice action in powers of the lattice spacing about the continuum action, then the higher-order terms, being multiplied by positive powers of the lattice spacing, all vanish in the continuum limit. Nevertheless, these terms affect results at non-zero lattice spacing. They can be chosen freely so that, for instance, the discretisation effects in physical quantities are minimised. This procedure is called the improvement programme for lattice QCD. It is possible to choose the dimension 5 terms in the lattice action (which are multiplied by one power of a) so that all the $O(a)$ contributions to physical quantities are cancelled and discretisation effects enter first at $O(a^2)$. Using an improved lattice action of this sort should enable continuum physics to be approximated better at a given lattice spacing, or with a given accuracy at larger lattice spacings, than with an unimproved action for which discretisation effects are typically $O(a)$.

Fixing up the lattice action so that it reproduces the classical continuum action in the zero-lattice-spacing limit is necessary, but does not guarantee that the lattice path integral has the correct quantum continuum limit. For this, we need to show that all the symmetries of the continuum theory are recovered.

The most obvious concern is whether we recover the full Lorentz (or Poincaré) space-time symmetries from a path integral defined on a hypercubic lattice. As it happens, Lorentz symmetry is what is called an 'accidental' symmetry of the lattice theory. It is not possible to construct a combination of the lattice fields that breaks Lorentz symmetry and survives in the continuum limit – any interaction terms in the effective theory that describes the lattice theory close to the continuum limit which break Lorentz symmetry are multiplied by a positive power of a and so get driven to zero in the limit. Simulations show that, at lattice spacings which are reasonably small on the scale of hadrons, rotational symmetry holds to a good approximation for quantities, like the inter-quark potential, at distances bigger than a few lattice spacings.

The next issue is quark confinement. Wilson showed that quarks are confined at strong coupling (equivalent to large lattice spacings). We achieve the continuum limit by tuning the quark-gluon coupling to zero, but there is no guarantee that quarks remain confined. In fact, simulations show that there is no sign of a phase transition occurring as the quark-gluon coupling is reduced. This suggests that the theory remains in the same, i.e. confining, phase for all values of the coupling and, hence, is confining also in the continuum.

A more subtle symmetry is chiral symmetry. This is a property of a theory with massless fermions and, in QCD it is the reason why pions are so much lighter than other hadrons. QCD has an approximate chiral symmetry, because the u and d quarks are almost massless. We know that chiral symmetry is spontaneously broken by the QCD vacuum, because, otherwise, hadrons would have nearly degenerate partners and, in the real world, they do not. If chiral symmetry was exact (the u and d quarks truly massless) then the pion would be the Goldstone boson associated with the spontaneous breaking of chiral symmetry and it would be massless. To the extent that the u and d quark masses are small but non-zero, the pion acquires a small mass.

It is important that this symmetry breaking mechanism is correctly reproduced by lattice QCD. Until relatively recently, we did not have a lattice action with the full chiral symmetry and, at the same time, the correct quark flavours. Consequently, most of the simulations to date have used lattice actions for which full chiral symmetry is only recovered in the continuum limit. This has made it difficult to disentangle discretisation effects from those due to incomplete chiral symmetry. Now we have a set of lattice actions which satisfy a condition called the Ginsparg-Wilson relation and which can be shown to have full chiral symmetry (and the correct quark flavours) at non-zero lattice spacing. These actions are only local (a necessary condition for a well-defined continuum limit) at small lattice spacings (or with sufficiently improved actions that the gluon fields are smooth enough) and they are very costly to simulate with our current algorithms. However, they are the last piece in the jigsaw which means that we can, in principle, formulate a lattice theory whose quantum continuum limit is QCD (in fact, they amount to a rigorous construction of QCD).

Finally, our lattice action must satisfy the Osterwalder-Schrader conditions of constructive field theory, which ensure that the continuum limit satisfies all the axioms to be a local relativistic quantum field theory. While proofs that these conditions are satisfied exist for some simple lattice formulations, they are tricky to carry through for the improved actions which are the basis of most simulations today. This is not thought to be anything more than a technical problem though.

6. Renormalisation

To summarise where we have got to, we know how to formulate and simulate (within some statistical accuracy that is under our control) QCD for any chosen set of input values for the quark masses and the quark-gluon coupling on a finite four-dimensional space-time lattice. Next, we need to fix the quark masses to correspond to their actual physical values, fix the normalisation of the lattice fields to match those in the continuum, and vary the coupling so that we can extrapolate our results to the continuum limit.

For each flavour, the quark mass is fixed by choosing a hadron that has that flavour and matching its computed mass to the experimental value. The simulation is carried out in terms of dimensionless variables and so the computed hadron mass is also dimensionless. Strictly speaking, the simulation only predicts values for dimensionless ratios and it is these that are matched to experiment. We can introduce a scale by selecting one extra dimensionful experimental quantity and matching its value to the computed value, with a suitable number of powers of the physical lattice spacing to balance dimensions. This gives a value for the lattice spacing in some physical units, and so gives us an idea of how close to the continuum limit we are on the scale of hadronic physics. Then we can adjust the coupling so that the physical lattice spacing decreases.

When we are close enough to the continuum limit that hadronic physics becomes insensitive to the lattice, we should see dimensionless ratios becoming independent of the coupling. This is called scaling. The precise way in which individual quantities approach the continuum limit is given by the renormalisation group. It can be computed for QCD using weak-coupling perturbation theory, provided we are close enough to the continuum limit, since the continuum limit for QCD occurs at zero coupling.

The final part of this renormalisation process is to define the normalisations of the lattice fields, and of composite products of the lattice fields, so that we can relate the values of matrix elements of composite operators computed in lattice QCD to the same matrix elements defined in a given continuum renormalisation scheme. This matching is needed because the analysis of experimental data uses matrix elements defined according a continuum perturbative scheme. In order to input the lattice results, we need to compute the relationship between the perturbative scheme and that used to fix the normalisation of the lattice operators.

Renormalisation introduces an unphysical momentum scale that, through a process called dimensional transmutation, eventually sets the scale for dimensionful quantities like hadron masses. The normalisation of the lattice fields must be done by introducing a momentum scale that fits on the lattice and so cannot exceed the momentum cut-off imposed by the inverse lattice spacing. Continuum renormalisation schemes rely on perturbation theory that is only good at relatively high momentum scales. One way to match the lattice normalisations to the continuum normalisations is to compute both in perturbation theory. However, the two results are needed at the same momentum scale and there may not be a momentum range in which both the lattice and continuum perturbation theory calculations are sufficiently under control. For this to be the case, requires simulations at small lattice spacings. A more sophisticated approach carries out a sequence of lattice renormalisations and matches them to each other at momentum scales differing by a factor of two. Iterating this matching enough times, gives renormalisation factors at arbitrarily high momentum scales that can then be matched accurately to continuum schemes using perturbation theory.

7. Computational Cost and Matching to Effective Theories

It is obvious that the computational cost is proportional to the number of lattice sites. In trying to approach the continuum limit, we are interested in changing the lattice spacing while holding the volume fixed – just halving the lattice spacing increases the number of lattice sites by a factor of 16, because we are in four dimensions, so even on its own this modest step towards the continuum involves a big increase in cost.

Decreasing the quark mass also increases the cost substantially, because our algorithms involve repeatedly inverting the Dirac matrix (the lattice version of the Dirac operator) in the evolving gluon field and this becomes more ill-conditioned as we approach zero quark mass. Furthermore, lighter quarks form hadronic bound states that are physically larger and are therefore more spread out across the lattice. So, if we are to avoid increasing the finite-volume distortions, we must increase the physical volume as we decrease the quark mass. If we are not going to make the discretisation errors bigger, this means increasing the number of lattice sites.

The finite computing power available to us at any time therefore places constraints on the range of quark masses and coupling that can be simulated on lattices that are big enough to avoid large finite-volume effects. Unfortunately, even ignoring the top quark, the masses of the other five quark flavours, and their associated low-lying hadron states, span a huge range compared with the range available for simulation. The b quark mass in Nature is well above the lattice cut-offs possible today, and the u and d quark masses drive the computational cost beyond our current reach.

The solution to this problem is to perform simulations within limited quark mass ranges, where the other systematic effects (discretisation and finite volume) are under control, and to match the results onto effective theories of QCD that become valid when the quark mass is sufficiently large (heavy-quark effective theory), or sufficiently small (chiral perturbation theory). If we can show, by varying the quark masses in the simulation, that the simulation results accurately reproduce the corresponding effective theory in some range, then we are justified in using the effective theory to extrapolate the results beyond the simulation range. In this way, lattice QCD enables us, in principle, to achieve the goal we set at the outset of performing reliable calculations for any values of QCD's input parameters.

In practice, we do not yet have enough computer power to match onto effective theories convincingly. Fortunately, two aspects of computer technology development mean that the computer power we can apply to QCD is growing exponentially. Moore's Law, which says that microprocessor speeds double every 18-24 months, has held for the past 50 years. Even though Moore's Law must eventually run out of steam, lattice QCD is translationally invariant and largely a local calculation, so it can be decomposed very efficiently onto a parallel computer, with each processor computing the fields in a localised sub-lattice. Our increasing ability to integrate many processors onto a single chip, and to couple many chips into a massively parallel system, should deliver the computing power we need for lattice QCD within the next decade.

8. Status of Simulation Results Today

The first lattice QCD calculation of the nucleon mass, by Hamber and Parisi in 1982, achieved agreement with experiment to within a claimed error of about 10%. It employed the quenched approximation, in which the effects of virtual quark-antiquark pairs are ignored, so it could not, in principle, have produced the correct result if the simulation had been carried out with high enough precision. In fact, it took a further 17 years before a Japanese group, called CP-PACS, with vastly greater computer power and improved algorithms, showed that quenched QCD has the wrong light hadron spectrum, although the deviation from experiment is only around 5-10%. This was an important result, because it demonstrated how computer simulation could falsify a theory (even though we knew the theory had to be wrong).

It also showed that quenched QCD can be quite a good model of strong interaction physics, if we are satisfied with something that works at the 10-20% level. While some quantities that are particularly sensitive to virtual quark-antiquark pairs are completely wrong in quenched QCD, it has provided a valuable testing ground for lattice QCD methods and has produced phenomenologically useful results for many quantities that are not sensitive to quenching, or were essentially unknown previously.

Today, the biggest computational challenge is to control the chiral extrapolation to the physical values of the *u* and *d* quark masses in full QCD simulations. This often represents the largest source of uncertainty in the results. Our algorithms become so costly, as we decrease the quark masses, that we have not yet been able to penetrate far enough into the chiral regime to see the behaviour predicted by chiral perturbation theory, except using the relatively 'cheap' formulations that do not have full chiral symmetry and somewhat cloud the interpretation of the results.

The lattice formulation that has reached the furthest towards zero quark mass is called 'improved staggered quarks'. At lattice spacings around 0.1 fm and in boxes around 3 fm across, the US MILC Collaboration has been able to simulate QCD in this way, with an *s* quark mass close to its experimental value and *u* and *d* quark masses around 10 MeV, i.e. only 2-3 times bigger than experiment. Their results for a range of well-known quantities, like the omega minus baryon mass, the pion and kaon decay constants, and mass-splittings in heavy quark bound states, show impressive agreement with experiment to within a few percent. This is compelling evidence that lattice QCD reproduces experiment.

However, there are open theoretical questions about the improved staggered lattice formulation that could call into question its use for quantities which are not well-known experimentally and where there is the possibility that the results could be used to falsify the Standard Model. If we're not sure about our theoretical foundation, we can't be sure that any discrepancy with experiment is a signal of new physics. So, the next step is either to put the improved staggered formulation onto a more solid footing, or to use one of the new Ginsparg-Wilson formulations with full chiral symmetry.

9. Results for Light Hadrons

Important Standard Model parameters that have to be determined at the outset of any lattice QCD calculation are the quark masses. The masses of the light quarks, *u*, *d* and *s*, are so non-linearly encoded in the light hadron spectrum that, before lattice QCD, their values were uncertain at the level of a factor of two or more. Yet, they have major implications for the way our Universe works. Lattice QCD is the only route to their precise values.

Currently, since electromagnetic effects which split the *u* and *d* quark masses are not included, the measurement which comes from fitting the pion mass computed in lattice QCD to experiment is for the average of their masses. A value around 3 MeV with an uncertainty of about 10% is obtained. The *s* quark mass, from fitting the kaon mass to experiment, turns out to be around 80 MeV, also to within 10%, very much at the lower end of what was expected. A light *s* quark has important phenomenological implications, e.g. for the size of *CP* violation (matter-antimatter asymmetry) in kaon decays. Also, whether or not there is a phase transition from the confined phase of QCD into a quark-gluon plasma at high temperatures is very sensitive to the value of the *s* quark mass.

Ultimately, of course, although they are fundamental parameters of the Standard Model, the quark masses should be derivable from some more complete theory. When we have some idea what this is, their precision determination through lattice QCD will be an important test of that theory.

The electroweak decays of pions and kaons, along with the mixing of neutral kaons with their antiparticles, being well-determined experimentally, pose important tests for lattice

QCD. Since these calculations involve simulating the light quarks and correctly incorporating flavour breaking (due to the *u*, *d* and *s* quarks having different masses), they are at the limits of our computational ability today. This is where efforts to control chiral extrapolations are focussing. There are encouraging quantitative results for relatively simple quantities like the leptonic decay constants.

The calculations for neutral kaon mixing are technically very difficult, involving delicate cancellations between QCD matrix elements, each of which must be computed accurately if the result is not to be completely unstable. The key quantity is ε'/ε, which measures the relative size of direct and indirect *CP* violation effects in the decay of a kaon into two pions. The current state of the art for this 'holy grail' of lattice QCD is that sensible results have been obtained for the individual matrix elements in quenched QCD (along with some initial results using dynamical quarks), after many years of effort, but, when combined, although the valued obtained for ε'/ε has the right magnitude, it has the wrong sign compared with experiment. This suggests that the calculation is possible, but we're not yet controlling the cancellations well enough.

Nucleon structure, which is probed in deep inelastic scattering experiments, can be measured accurately with the help of QCD perturbation theory. Lattice results are not needed by phenomenology, but the ability to compute nucleon structure functions from first principles is obviously an important test of QCD. Here it is the moments of structure functions that can be related directly to matrix elements computable using lattice QCD. The first moment gives the average momentum carried by partons in the nucleon. The result from quenched QCD is larger than experiment, presumably because the valence quarks carry more momentum than in full QCD. This picture doesn't change when dynamical quarks with fairly large masses are included in the simulation. It is essential to extend these simulations to lighter quarks, if we are to make contact with the conventional parton distributions extracted from experiment and put this aspect of lattice QCD to the test.

One of the most exciting predictive capabilities of lattice QCD is the ability to compute masses of as yet undiscovered hadrons, particularly pure gluon bound states ('glueballs') and those with exotic quantum numbers that are not allowed in the quark model. It may also be possible to resolve the interpretation of exotic resonances observed experimentally, like the recent claims to have seen a pentaquark state. This is a narrow resonance, with unknown spin and parity, seen by some (but not all) experiments in a nucleon-kaon channel at 1540 MeV. The challenge is to distinguish between a new type of five-quark hadron, which is not forbidden in QCD, and a scattering state of a nucleon and a kaon. There are many ways to construct pentaquark states in QCD, but, while most of the initial lattice calculations see a spin ½ negative parity state at about the right mass, it seems to be a scattering state. The high level of interest in the existence of pentaquarks is motivating a sustained effort by the lattice community and it is possible that lattice QCD could resolve the question before experiment.

10. Heavy Quark Physics

The *c* and *b* quarks present separate and special challenges to lattice QCD. Their masses can be obtained, with few percent uncertainties, from mesons containing one of them as a valence quark, in the usual way. With a mass around 1.3 GeV, the *c* quark is below, but quite close to the cut-off in most of today's simulations. Thus, it is particularly sensitive to discretisation effects and these must be carefully controlled via the improvement programme.

The *b* quark mass is 4.2 GeV, well above the lattice cut-off, so it cannot be simulated directly on present-day lattices. Instead, we must either treat the *b* quark non-relativistically, or extrapolate results for lighter relativistic quarks (with masses around charm) up to the *b* mass using heavy-quark effective theory. Again, we must be careful, because these heavy quarks can be badly affected by the discretisation. In neither case are simulations precise enough to see the effects of virtual *c* or *b* quark-antiquark pairs, so both heavy-quark flavours are implemented in the quenched approximation.

Charm physics has also recently experienced the experimental discovery of puzzling narrow resonances. These have the quantum numbers of *p*-wave D_s mesons. The puzzling aspect is that the quark model predicts the latter states to be above the threshold for them to fall apart into a *D* meson and a kaon, and therefore they should be broad. Perhaps, the narrow resonances are something new? This speculation spawned many papers. But QCD should be our guide, not the quark model, and, although the simulation data are rather noisy, they are consistent with the conventional *p*-wave D_s mesons being below threshold and hence narrow resonances. If so, exotic interpretations are not needed.

The up-coming CLEO-c experiment aims to measure *D* and D_s leptonic and semileptonic decays to an unprecedented 2% accuracy. This presents lattice QCD with an opportunity to make high-precision predictions ahead of the experimental measurements. It will also allow us to test our lattice methods for computing semileptonic decay form factors, the accuracy of which, particularly for *B* mesons, is crucial to checking the consistency of the three-generation Standard Model. Semileptonic decays are those in which the electroweak interaction changes one flavour of quark into another – there is just one hadron and a pair of leptons (from the decay of the emitted *W* boson) in the final state. The strengths of these flavour-changing decays are fundamental parameters of the Standard Model.

The calculation of decay form factors is affected by large discretisation errors if the momentum transfer involved gets close to the lattice cut-off, i.e. large enough to resolve the lattice spacing. When a *D* meson decays semileptonically into a pion or a kaon, the range of momentum of the pion or kaon is around 1 GeV and this can be accommodated on existing lattices. Thus, although discretisation effects are potentially large and must be controlled, the whole decay is within the reach of the current generation of simulations. The results so far are encouraging and we can look forward to charm physics becoming the proving ground where lattice QCD justifies its claim to be a model-independent high-precision tool for phenomenology.

Lattice QCD should have a big payoff when it comes to the mixing, leptonic and semileptonic decays of *B* mesons, coupled with the high-precision measurements being carried out at *B* Factories, since the corresponding Standard Model parameters are amongst the least well known and could be hiding signals of new physics.

However, some of the most important processes for checking the consistency of the Standard Model, such as the semileptonic decay of a *B* meson into a pion, present the biggest challenges of all to our ability to control the systematic errors in lattice QCD. The *b* quark is too heavy and the *u* and *d* quarks (in the pion) are too light for direct simulation. So we must control the matching with (different) effective theories at both ends of the quark mass range. Also, the recoil momentum of the pion can exceed the maximum momentum allowed on the lattice, so we cannot directly simulate the full kinematic range of the decay. This may not be an issue if sufficiently precise experimental data are obtained for the differential decay rate, which would allow the strength of the decay of a *b* quark to a *u* quark to be extracted only from data at recoil momenta where the lattice results can be trusted.

Experimental measurement of the mixing of a neutral *B* meson with its antiparticle, when the light valence quark they contain is either a *d* or an *s* (the latter has not yet been observed), taken with the lattice QCD calculation of the mixing matrix element, permits the determination of the strength of the decay of a *t* quark into a *d* or an *s* quark respectively. This is because the mixing process involves an intermediate virtual *t* quark at leading order in the electroweak interaction. In particular, the ratio of the *d* and *s* matrix elements should be a number that can be computed very reliably, once we have control of the heavy-quark and chiral limits, because systematic uncertainties tend to cancel in such ratios.

11. Beyond the Standard Model

The QCD matrix elements in rare decays, such as the decay of a *B* meson into a *K** meson and a photon, corresponding to a *b* quark decaying into an *s* quark, are no more difficult to compute than *B* semileptonic decays. They have received less attention because the experimental data is relatively poor. However, any difference between the Standard Model and experimental values due to new physics could be large, because both effects occur at leading order in perturbation theory, via the exchange of a virtual particle. In the Standard Model, the particle exchanged is a *W* boson, but it could just as well be a new particle like a squark (the supersymmetric partner of a quark). Today, the observation of a discrepancy with the Standard Model prediction would signal new, but unknown, physics. If the resulting squarks are produced at the LHC, then pinning down their couplings to Standard Model particles could help us start to understand it.

While, in a similar vein, the observation of proton decay would signal new physics, its non-observation is beginning to rule out candidate Grand Unified Theories already. The experiments, like SuperKamiokande, involve watching a large tank of water for one of the protons to decay. The current lower bound on the lifetime of around 10^{33} years comes about because a kiloton of water contains roughly 10^{33} protons. Watching nothing happen for several years generates the bound. Since we are looking for the quarks inside a proton to decay, the lifetime is proportional to (the square of) a QCD matrix element. The uncertainty in the proton lifetime coming from the lattice QCD calculation is around a factor of 10. Reducing this uncertainty could tighten the constraint on new physics by an amount equivalent to several more years of patient observation, and is well within reach of the current simulations.

So, what if the LHC discovers supersymmetry? Can lattice simulations play a role when attention shifts to understanding this new physics? Assuming there are important non-perturbative effects, we would need to formulate supersymmetric quantum field theories on a space-time lattice. The obstacle is that the anticommutator of two supercharges is the generator of translations and the lattice theory breaks (continuous) translational invariance. So it seems inevitable that the lattice formulation breaks supersymmetry.

Perhaps supersymmetry could be an accidental symmetry of the lattice theory, just as Lorentz symmetry is an accidental symmetry of lattice QCD? Supersymmetric theories with scalar fields would be a problem, because it is easy to generate scalar mass terms which survive in the continuum limit and break supersymmetry. There needs to be enough of a remnant of supersymmetry on the lattice to forbid scalar masses. Kaplan has given a general construction of supersymmetric lattice field theories. These have exotic lattice geometries compared with QCD and contain exactly massless fermions. So they will be very costly to simulate. Some do not have positive definite path integration measures, so Monte Carlo methods would fail. Yet, we are starting to understand how we might apply lattice simulations if Nature has chosen supersymmetry.

12. Conclusions

After 30 years, we have a rigorously correct implementation of QCD on a space-time lattice that enables us to simulate it over a wide range of values of the quark masses and the quark-gluon coupling. We know in principle how to extend this range to all parameter values by matching the simulation results onto effective theories that can be treated analytically and whose domain of applicability can be validated by the simulations themselves. The simulations are very expensive, but computer power has reached the point where they are feasible and they are now underway using a new generation of computers. We are also beginning to understand how to formulate chiral and supersymmetric quantum field theories on a lattice, although their simulation will be even more computationally demanding than QCD.

Using more approximate lattice formulations of QCD, a huge body of results for a wide range of phenomenologically important quantities has been built up. Although for many of these we have not yet achieved control of all sources of uncertainty, some have been determined to few-percent accuracy. The totality gives us confidence that correctly formulated lattice QCD can fulfil its goal of producing high-precision results that, in conjunction with precision low-energy experiments, can be used to test, and possibly even falsify, the Standard Model and to constrain the physics beyond it.

CHAIRMAN: R.D. KENWAY

Scientific secretaries: G. Imponente and J. Kamenik

DISCUSSION I

- *Bourjaily*:

Surely the limit a → 0 is not the 'continuum limit' because an infinity of points is approached which is countable (i.e. homomorphism to Q^4). How robust is the connection between the continuum description of a gauge field theory – i.e. Euclidean path integral over R^4 of a space-time manifold – and that of the a → 0 limit of gauge theory defined on the lattice? In particular, are the non-perturbative effects of gauge theory defined correctly by the lattice theory?

- *Kenway:*

First of all, there are two limits we need to understand. You described quite well the first one, which is the classical continuum limit, at the level of the action, showing that as you take the lattice spacing to zero you recover the continuum action. The continuum limit of the quantum theory is a much more subtle process. I would go as far as to say that does define the theory non-perturbatively. In other words, the whole idea of the lattice is that it gives you a non-perturbative way of regularizing QCD, so that at non-zero lattice spacing you have a perfectly finite definition of QCD. The process of taking the continuum limit in the quantum theory means that you have to tune the parameters of the theory to get to a critical point where the physical length scales, Compton wavelengths of the observable hadrons etc, diverge on the scale of the lattice spacing, so that as the lattice spacing is going to zero you can hold physical quantities fixed. This is the way of defining the theory that is valid for all values of the gauge coupling, quark masses, etc. At the moment, we do not have an alternative prescription. If we did, then you could ask yourself the question "does this alternative prescription actually give you the same physics as you get from the limit in which the lattice spacing goes to zero?".

- Loktik:

First a comment on the previous question: We do not know if space-time is a continuum manifold. It might well have some finite scale built in, like the Planck scale. And the question is: Can you give some intuitive explanation of the Ginsparg-Wilson relationship?

- Kenway:

In the context of the lattice theory that I have been describing, I am not making any statement about the nature of space-time at small scales. The lattice is purely a mathematical artifact as far as anything that I have described is concerned. And of course, the reason is that we know quite a lot about QCD as we approach the continuum limit. In the limit in which g^2 goes to zero we have control of QCD via perturbation theory. We are approaching a critical point, so that we can use the framework of the renormalization group for second order phase transitions to control what we mean by approaching the continuum limit.

Regarding your question, I am going to discuss that in tomorrow's lecture. All I was trying to point out this morning was that we do now have a prescription which enables us to formulate lattice theories with exact chiral symmetry and that solved a long standing problem in the field.

- Salur:

What is the definition of the vacuum and how do you describe the gluon-gluon interactions?

-Kenway:

The definition is exactly the same as in standard quantum field theory, because all I have used is the relationship which gets you from the path integral to vacuum expectation values of time-ordered products of operators. That is a statement which is true in the continuum theory that I have used to define the matrix elements, so it is not really necessary to talk about the vacuum in the lattice theory.

Concerning gluon-gluon interactions: the quarks live on the sites of the lattice. I also introduced link variables which, in continuum language, are the parallel transporters that tell you how to rotate the color fields so as to maintain gauge invariance when you move a color field from one point in space to another. These play the role of the gauge fields on the lattice because they are directly related to the A^μ s. What I have just described defines an element of the gauge group on the link, that is, the line between one lattice point and its neighbour. The gauge interactions come about through the so called plaquette term in which you have the product of these link variables around an elementary square on the lattice. It takes quite a bit of algebra, but if you work out what the trace of the product of these link variables is, and expand it in powers of the lattice spacing, you will find that that is equal to ¼ $F^{\mu\nu} F_{\mu\nu}$ plus higher order terms in the lattice spacing. So the gluon-gluon interaction is there at leading order as you do a Taylor expansion in the lattice spacing; there are higher order (in the lattice spacing) effects but, since they all vanish as we go to the continuum limit, they are not a problem. The quark-gluon coupling is in the fermionic part of the action, which I haven't as yet said a great deal about; I just wrote it down as (i Dslash + m) but the Dslash, very much like the continuum Dslash, has a gauge field in it which gives you the quark-gluon coupling, if you expand it out in powers of the lattice spacing.

- *Wagner:*

What is the way to estimate the error in your calculation?

- *Kenway:*

I described three sources of uncertainty or error in the calculation. One comes from the fact that you are working at non-zero lattice spacing. Let us deal with this one first: there is a fairly simple, practical way of estimating how big that error is in a particular quantity. You calculate the quantity with some fixed lattice spacing and then you re-do the calculation with a smaller lattice spacing and see how much that physical quantity changes. So that can give you a feel for the lattice spacing dependence. The other systematic uncertainty comes from the fact that you are working inside a box with fixed volume and there again the same sort of approach will work; if you do the calculation in one volume and then in a bigger volume or in a smaller volume you can get a quantitative feel for the dependence on the volume. The third source of error is the statistical error coming about from the fact that we approximate the path integral by a sum over a set of sample field configurations. There are fairly sophisticated statistical analysis techniques which we use to directly derive the standard deviation coming from that average over the sample. The reason that we need to use quite sophisticated techniques is that when you use simple minded statistics you assume that your samples are all independent. Typically, because we use a Monte Carlo algorithm to generate the field configurations, our field configurations are not statistically independent. So we have to look carefully at correlations from one gauge configuration to the next and make sure that we have factored in the so called auto-correlations that measure the extent to which our algorithm has failed to produce independent samples. But if you forget about that technicality, then the statistical errors are obtained using the same formulae as you normally use for standard deviation and errors of the mean, etc.

- *Zichichi:*

The best answer would probably be to show the graph where the predictions of lattice QCD are plotted versus the decreasing lattice spacing.

- *Kenway:*

The systematic error due to the lattice spacing is something that we work very hard to have good control of. So typically what happens in most calculations of important physical quantities is that we do the calculations at three different lattice spacings or more and do an extrapolation to lattice spacing equals zero. That is typically quite well controlled. The other source of error, the box size, is less of a problem. The reason for that is simply that we tend to work with box sizes that are fairly big relative to the scale of physics we are dealing with, so the uncertainty coming from the finite box is not particularly large. Once you have got a box that is big enough to contain your physics (which has a particular length scale) then the edges of the box do not really affect things very much.

- *Schleifenbaun:*

In lattice calculations the string tension approaches a non-zero constant in the continuum limit, which accounts for confinement. For high temperature, the string tension should drop to zero to satisfy deconfinement. If it does not, then does a phase exist that can be really called "deconfined"? Furthermore, how can one relate this to the findings of the Bielefeld group where it is shown that the equation of state does not quite reach the Stefan-Boltzmann limit?

- *Kenway:*

I did not give you the complete story about the string tension. That linear potential, first of all, comes about from a calculation of the expectation value of Wilson loops, which are products of the link variables around large rectangles in the lattice. The expectation value of those quantities in the confining phase follows what is called an area law, that is, the expectation value is proportional to the area of the loop, and it is that that gives you the linear potential. In a deconfined phase, the same expectation values do not follow an area law. They follow a perimeter law. And that does not give you a string tension. So the expectation values of these quantities can be regarded as an order parameter which signals which phase – confined or deconfined – you are in. The other thing that is subtle about the linear potential is that, as you separate the quark and anti-quark, you have to pump more and more energy into the flux tube. At some point, when you get enough energy in the flux tube to allow a quark – antiquark pair to pop out of the vacuum, the flux tube will break and the potential flattens out. That has proved to be quite difficult to demonstrate in the numerical simulations, but is now pretty adequately demonstrated. So the picture of confinement is actually quite a bit more complicated than simply a linear potential, due to these string breaking effects. Concerning the discrepancy with the Stefan-Boltzmann law, I don't think anybody understands this.

- *Gyulassy:*

Can you try to test predictions of AdS/CFT correspondence conjectures for strong coupling limits of CFT via some suitable lattice implementation of not QCD but the conformal supersymmetric continuum CFT? For example, can the mysterious entropy at g^2N=infinity being only 3/4 of the perturbative entropy at $g^2=0$ be directly numerically verified by a suitable lattice calculation?

- *Kenway:*

There are a number of things that can now be done, in principle, with supersymmetric theories on the lattice. In practice, in order to maintain supersymmetry on the lattice you need to have massless fermions, which is ok, but very expensive. Now that we understand how to formulate massless fermions, it's becoming more feasible to try to test some of these conjectures. However, the simulations that have been done so far are rather crude and exploratory.

- *Gyulassy:*

What is the problem of being able to control the continuum limit? In QCD you rely on the fact that, when the coupling gets weak, then you have control of the functional behaviour. Is there a technical problem of knowing how bad the lattice approximation is to a conformal theory in the continuum?

- *Kenway:*

You need to come up with the prescription for defining the continuum limit that works for that particular theory. The techniques that we have rely on identifying critical points in the phase diagram of the system, the second order phase transitions, where you can take a limit in which the physical length scale diverges relative to the lattice spacing. That is particularly easy in QCD, as you pointed out, because that critical point occurs in the perturbative regime. It doesn't have to occur in the perturbative regime if you take some other theory, and then the way in which you approach that critical point is going to be much more difficult to control. What you get from that limit may not be the theory that you want. It is something which you have to establish on a case by case basis.

- *Kamenik:*

When you renormalize quark masses on the lattice using hadron mass matching conditions, do you derive any consistency conditions, since for N_f quarks you can have more than N_f hadrons which can be used?

- *Kenway:*

If QCD describes the real world and we have implemented QCD correctly on the lattice, it shouldn't matter in principle which hadron masses you use to fix the quark masses. It's a good idea to use a hadron that has the quark whose mass you are trying to fix as one of its valence quarks so that the hadron mass depends quite strongly on that mass, because you are trying to tune the quark's mass so that the hadron mass matches experiment. But having said that, if QCD is not quite the theory of the real world or, as a more realistic example, if the lattice formulation is not quite right, then this is precisely where you suspect to see something going wrong. In quenched QCD, if you try to fix the strange quark mass with the phi meson or the K meson mass, you will get different answers for the strange quark's mass and that is a signal that your theory is inconsistent. Similarly, if you try to estimate the lattice spacing from two different physical quantities, you will get different estimates, another signal that something is wrong. Hopefully QCD is the right description, at least to the level that we can do our calculations. Then, once N_f masses are fixed, the entirety of the hadron spectrum has to be in agreement with the experiments.

- *Bourjaily:*

The practice of tuning the quark masses on the lattice so that hadron masses agree implicitly assumes that all the binding energy of the hadrons comes from the Standard Model. Are these masses robustly determined, i.e. how stable are these masses when supersymmetry, QED, technicolor, etc, are added?

- *Kenway:*

All the results I will give are in an approximation to the SM where there is just QCD and the rest of the SM is treated in perturbation theory. So when I calculate the proton mass, for example, what we do is compute it in QCD in isolation, so we are missing the electromagnetic effects and so forth. But these are small on the scale of the QCD contribution to the proton mass. The other interactions are not difficult in principle to factor into the calculations perturbatively, it's just that they produce small effects so that, within the uncertainties that we are able to compute at the moment, it's not worth trying to include them.

Eventually, maybe, we will be able to "push" these lattice calculations down to fractions of a percent accuracy where the other interactions matter. I think it's much more likely that lattice QCD will reach a level of maturity when we have control at the few percent level for a lot of these quantities. This will satisfy many people. You will have to work very hard to get much more accuracy than that. I think that it's questionable whether it will be worthwhile.

- *Ferrara:*

In N=1 pure Yang-Mills theory many perturbative and non-perturbative properties are known thanks to supersymmetry. For instance, chiral symmetry is broken by gaugino condensate VEV's. Do you think that some of these properties can be checked in lattice gauge theory?

- *Kenway:*

Yes, some people are actually trying to do this. At the moment, the problem is that the simulations have not been pushed near the level of sophistication that we have in QCD. The lattices are small and the lattice spacings are big, and we haven't varied enough parameters to understand the systematic effects. So what we typically get is a gaugino spectrum, say, that is a snapshot for a particular set of parameters on a particular small and rather coarse lattice.

It is quite difficult to conclude very much from that, other than that these simulations are becoming possible and, if we have enough energy and there are enough people around interested in doing it, then the sort of technology that is now being employed in QCD could be applied in that area. Right now, not many people are working on this topic,

because the big computers are being funded to do QCD rather then supersymmetric theories.
You have to have massless fermions and right now in QCD we can't go to the limit of massless quarks because of the computational costs. So it's very difficult to even get a starting point for a supersymmetric theory that is feasible for today's computers. It's going to be four or five years before we will have powerful enough and cheap enough computers. If supersymmetric particles were discovered tomorrow, the funding agencies might well give us large sums of money to devote computer time to doing this sort of calculation. At the moment they are not doing that.

-Alwall:

How is a renormalization scale introduced in LQCD (as in the comparison of matrix elements with the MS-bar scheme)? Is not the LQCD calculation done in "all orders of perturbative theory"? (i.e. non perturbatively)

-Kenway:

I introduced the momentum scale to define renormalization conditions for vertex functions in the lattice theory; thus we do need to introduce some sort of physical scale to do the renormalization.
But there is a second stage in the renormalization process. Having got quantities defined relative to a lattice scheme, you need to match that scheme to a scheme that is useful for experimentalists such as MS-bar. And that ultimately is a perturbative calculation to relate the renormalization constants in the lattice scheme to the MS-bar scheme. This requires that the renormalisation scale in the lattice scheme is sufficiently large to justify perturbative calculations, but also not too large that it's close to the lattice cut off.
An approach that avoids having to have a small enough lattice spacing to satisfy both of these conditions, takes the renormalisation scale to be the linear size of the box. We can then apply a sort of finite size scaling, where we go to ever smaller and smaller boxes through a sequence of renormalizations, which corresponds to higher and higher energy scales. In this way the lattice quantities are eventually defined at a momentum scale

which is sufficiently high to guarantee that the perturbative matching with the standard MS-bar scheme has essentially no error in it.

- *Salur:*

What is the error in the results if we don't know precisely the valence bare quark masses?

- *Kenway:*

The point is that many quantities are very sensitive to the number of light quarks flavors and their particular masses (I think this is what is happening with those plots of the equation of state with different numbers of flavors of quarks that we saw this morning). Of course, what we are doing is reverse engineering the quark masses by trying to match a finite set of hadronic properties to experiment. Because those are chosen to depend strongly on the quark masses, this gives you a rather sharp definition of the values of the quark masses. The numbers that are coming out are really quite precise and we have numbers, for instance, for the up quark mass that is around 3 MeV with 10% errors. It's clear, for instance, that it is not zero. When we finally have bigger computers, I think quark masses will be known to very high precision.

CHAIRMAN: R. D. KENWAY

Scientific Secretaries: M. Clark, O. Loktik

DISCUSSION II

- Kamenik:

When connecting the lattice QCD operators to the continuum via lattice spacing expansion, how general have these attempts been in terms of utilising all possible correction terms which vanish in the continuum limit?

-Kenway:

This is called the improvement programme in lattice QCD, the idea being that you add higher dimensional operators to cancel the leading lattice spacing dependencies. An example would be where you have a leading error of O(a), you add a dimension 5 operator and adjust the coefficients to cancel this leading term. This would leave you with a leading $O(a^2)$ error.

This programme was laid out by Symanzik in the early days of lattice QCD. He systematically listed the operators you would need to add in order by order to cancel the lattice spacing effects.

The difficulty in going beyond the leading order terms, is that you need to have a method of calculating the coefficients of the terms you add in. The improvement programme has to give a method of calculating these coefficients which are dependent on parameters in the action. This can be done in perturbation theory, but this is typically only good to cancel $O(\alpha_s a)$ or $O(\alpha_s a^2)$ effects. Ideally you would want to compute this non-perturbatively, but because of the number of operators involved, this becomes impractical beyond leading order in a.

- 't Hooft:

How does the Ginsparg Wilson operator behave in the presence of instantons? You mentioned the anomaly problem in your introduction, but it is not clear to me what happens to these anomalies when using this operator.

- Kenway:

Instantons are not very well defined objects on the lattice because they are not topologically stable. The Ginsparg Wilson operator obeys the Atiyah-Singer index theorem, and so it can be used as a definition of topological charge on the lattice.

- *'t Hooft:*

But that theorem gives you integers, which should jump discontinuously as the gauge field configurations vary continuously. How do such theta jumps come about?

- *Kenway:*

The Ginsparg Wilson operator will give integer values for the topological charge, but I believe it has not yet been compared with other lattice definitions. This is something that will be done soon.

- *'t Hooft:*

Also related to instantons, you said that the assumption that $m_u=0$ has been ruled out. Now, due to instantons, the value of m_u would get renormalised by terms proportional to m_d, so that it could sum to zero in the far ultra-violet, yet be non-zero in the near ultra-violet. Has that also been ruled out?

- *Kenway:*

I'm not aware of any results for that. Whether $m_u=0$ has only been looked at in the context of computing the low energy constants of the chiral effective Lagrangian.

- *Alwall:*

How are the valence quarks treated in e.g. the calculation of a hadron mass? Do you use an ensemble of field configurations as for the gauge fields?

- *Kenway:*

Once we have integrated out the quark fields, we have two terms in the path integral that involve the massive Dirac operator: the determinant piece and the inverse of the fermion matrix, which occurs in the source term. In fact the sources are the valance quarks, and the determinant is the contribution to the vacuum coming from the sea quarks. When we do the Monte Carlo, that is replace the integral over the gauge fields by a sum over a finite set of representative gauge fields, we build in this determinant factor. That's what the words "including dynamical quarks" mean. They mean that we include the sea quark effects in the gauge fields. The valence quarks enter when we try to calculate a particular expectation value via the choice of fields, whose expectation values we are calculating. So if we choose to calculate a correlation function that involves, say, a pion, we'll choose the fields to have a non-zero overlap with the pion, i.e. we choose a combination of the fields to have pion-like quantum numbers. And then, once we've integrated out the quarks, differentiating with respect to the valence quark sources brings down inverses of this fermion matrix. This is the valence quark propagator. So this is how valence quarks come in to the calculation.

- Bourjaily:

The hadron mass spectrum is used to demonstrate the superiority of staggered quarks over quenched QCD for example. However, if beyond the standard model effects have significant effects on binding energies, we wouldn't expect lattice calculated spectra to agree with experiment. At what point do spectra discrepancies become falsifiable tests of QCD?

- Kenway:

Fundamentally, if you want to be in the position to do a falsification test, you need a rigorous implementation of QCD on the lattice. The current approach, using staggered quarks, seems to work at few percent accuracy, but is not good enough for a falsification test unless you can convince yourself that there is nothing wrong with that formulation which could give you a small deviation in, say, a hadron mass or matrix element. If there were any doubts that your formulation was not really QCD, then you would not be justified if you saw a small deviation from experiment to say that that is physics beyond the standard model.

Ginsparg-Wilson formulations, as far as we understand them, enable us to control all of the systematic uncertainties in putting QCD on the lattice. To the extent to which they are an approximation, we know which knobs to turn to make the approximation arbitrarily good. This provides a framework in which we can do the sort of test you are suggesting. Whether in practice that will ever be done is of course an open question. It comes back to a comment Professor Zichichi made in my first lecture about the CKM matrix. Before the B factories switched on, many people in the lattice community hoped that the unitarity triangle would have big discrepancies in it. When experiments got better we would find that the big uncertainties actually were pointing towards big discrepancies, i.e. that the triangle doesn't close. Had that been the case, once we had the results from the B factories, the dominant uncertainty in the unitarity triangle would be coming from theory, that is QCD matrix elements. There would have been a strong motivation to make the lattice results more precise, in order to ascertain whether the remaining discrepancies we were seeing were just due to theory or to new physics. Unfortunately, nature has been unkind. The main result so far, which is the angle sin 2β, appears to put the vertex right in the middle of the allowed region. So already it looks like any deviations from the SM that will be exposed within the CKM scene are likely to be very small. Although there are still places within CKM physics where significant discrepancies might show up. The tremendous vindication of the SM, which Belle and BaBar achieved, was also in a sense a disappointment for those who hoped that they would provide strong and urgent motivation to improve the precision of lattice QCD.

That may be the end of the story, as far as discovering new physics with regards to CKM and B physics is concerned, but theories beyond the SM could be constrained by low energy calculations which are sufficiently precise. An example is the amplitude for proton decay, which is involves a QCD matrix element that is unknown within a factor of ten. Over the next five to ten years, as Super Kamiokande gathers more data, there should be an improvement on that bound by an order of magnitude. However, since there is already a factor of ten uncertainty in the lifetime from QCD calculations, we could achieve a similar improvement in the bound by improving these calculations, although it might take almost as long!

- Salur:

Can you describe the evolution of quark gluon plasma such as the hadronisation with lattice QCD?

- Kenway:

The question is whether we can get a handle on the process when the QGP cools to produce hadrons. That's difficult because you are going from a regime where the QGP is in equilibrium to a regime where the QGP is out of equilibrium and described by a non-zero chemical potential and low temperature. All of the lattice calculations are assuming that we have thermodynamic equilibrium. Working in Euclidean geometry you don't have the time dependent Green function technology to enable you to look at non-equilibrium effects.

- Korthals-Altes:

In principle you could sit at T_c, and then cool to get hadronisation, but I don't know how to see that on the lattice.

-Kenway:

The lattice does not do everything. The numerics of it require that we work in Euclidean space. The technology is really equilibrium thermodynamics. If you are out of that limit then you could use other techniques such as quantum Monte Carlo. There's perhaps a lot of new physics you could do there, but I'm not aware that anyone is doing it.

- Simon:

The success of lattice calculations is intimately linked to computing power. We had exponential growth in the past and expect exponential growth to continue in the future. But there will be fundamental limits for classical computers. How about quantum computers?

-Kenway:

Referring to the classical computer, we're currently at 10 Teraflops. For full Ginsparg-Wilson dynamical results, good to the few percent level, Petaflops is probably what's required. Given the technology available today, then it is possible to build a Petaflops scale computer within the next three to five years, and it will happen subject to funding. At that point, most likely we will be able complete the job with regards to lattice QCD. Many matrix elements required by experimentalists will be known to the three percent level, as well as the hadron spectrum. If you accept these errors we'll be able to say "this is QCD".

I don't think we need quantum computers at this level. Feynman, when he put forward the idea of quantum computers had in mind that simulating quantum systems on a classical computer was very difficult. This difficulty arises because the determinant which appears in the path integral need not be positive, and leads to an inability to use Monte Carlo algorithms because you would have negative probability. This is known as the "sign problem" and it occurs also in many quantum many body problems. The solution is to put this fluctuating sign into the observables being measured, but this increases the noise in results requiring that you have to calculate for much longer. It was Feynman's idea, that if you had quantum computers which could implement the quantum mechanics that is giving the sign problem, then the problem which is very expensive on classical computers would scale as the size of the system on a quantum computer.

I have no idea when we will have quantum computers, but I'm inclined to believe that we'll still have quantum field theory problems that will have a sign problem and will need them. A specific example is SUSY on the lattice. There are two major challenges: firstly, you require massless fermions, which is very expensive; and secondly some SUSY models on the lattice exhibit the sign problem.

- Simon:

Schemes have been designed for quantum computers to factorise large numbers or to search huge databases. These schemes outperform any algorithm on classical computers. Could the formalism of lattice calculations be adapted to quantum computers?

- Kenway:

The people that invent these algorithms for quantum computers do write papers about how to simulate quantum systems, e.g. the Hubbard model of interacting electrons. So this is being done, we just don't have a quantum computer to run these algorithms.

- *'t Hooft:*

At the new talent session today we had an argument concerning the gluon propagator. Whether the gluon propagator will go to zero at low momentum or go to infinity. The claim was that the lattice theory supported the gluon propagator going to zero. What do you think the lattice says?

- *Kenway:*

This is not my field. I am skeptical about such non-gauge invariant calculations. I know there are people doing this, but I just don't know what the results are.

- *'t Hooft:*

If you use different arguments, you get totally different results. The claim was that the lattice backs up their results.

- *Kenway:*

You have to show that the lattice is exhibiting robust results, for instance, when you fix the gauge in different ways. Even if that can be done, it may be difficult to interpret the results.

- *Fazid:*

It has been done by Zakharov of Moscow, there are publications about the behaviour in infrared Landau gauge and this shows suppression of the gluon propagator and enhancement of the ghost propagator. In 1978 Gribov predicted this kind of behaviour.

- *Gyulassy:*

One of the old questions for lattice is when will the pion mass become realistic with dynamical quarks? You showed a plot where current results with domain wall and staggered quark implementations were still too heavy to allow ρ decay. What is the status of the ρ on the lattice now? It cannot be the good old $\pi\pi$ resonance of experiment. How much more computational effort will be necessary to allow the π and ρ to come out more physical?

- *Kenway:*

That plot of computational cost is for simulations of dynamical quarks. There have been calculations using Ginsparg Wilson fermions with quenched configurations, achieving much lighter quarks. It's only been in the last year or so that anyone has attempted to perform dynamical Ginsparg Wilson simulations, with relatively heavy quarks. The data in the plot is for the domain wall approximation, which in it's current implementation, still has significant chiral symmetry breaking. We don't know what the computational cost for realistically light quarks is going to be, though we believe it to be somewhere between one hundred and a thousand more than the simulations we can do today.

It is expected that as we move to the chiral limit, the other lattice QCD formulations will explode in computational cost and they will become more expensive than the Ginsparg Wilson approach.

A test for lattice QCD is to see ρ to $\pi\pi$ decays happening on the lattice. In almost all of the simulations, except some of the recent 2+1 flavour staggered results, the pion masses are just too high for the ρ to decay. Furthermore the pions in the final state are in a p-wave, so you have a unit of momentum in there, which makes it even harder for the ρ to decay. The results for the ρ meson mass are open to question for that reason. We don't actually know what the signal for ρ decay will be on the lattice. The formulism we have at the moment doesn't work well when you have two hadrons in the final state. Strong decays, where you have more than one final state hadron, is a technology that's yet to be developed.

\- *Schleifenbaum:*

I must relate to 't Hooft's question, and come back to the infra-red suppression of the gluon propagator. The lattice results that exist do not reach the infra-red momentum regime of the gluon propagator, but only cover the intermediate and ultra-violet region. What is the reason for that?

\- *Kenway:*

There are two effects which can affect the range of momentum accessible on the lattice. One is the lattice spacing, which gives you an ultraviolet cutoff at high momentum. The other is the box size which gives an infrared cutoff at low momentum. I assume that these results will have been in a small box, since the people doing these simulations don't have access to particularly large computers. It's probably the case that this is what is cutting off the data in the infrared. I'm not really convinced that the lattice is being helpful here. If someone could make a good statement about what could be computed on the lattice with respect to the gluon propagator, and how to interpret the results, then we could devote more computing power to this problem. This would produce bigger lattices, giving a much wider and finer grained momentum range. It's just a question of how much precision you can justify, given the cost.

CHAIRMAN: R. D. KENWAY

Scientific Secretaries: R. J. Tweedie, D. de Gruttola

DISCUSSION III

- *Salur:*

Is it possible to check if already known bound states, such as Λ*(1520) or Ξ*(1530), are bound states on the lattice in comparison to the Θ^+ being a scattering state?

- *Kenway:*

The technique that was used, to see whether that particular pentaquark state was bound or not, is new. The preprint did include some checks to show that, by varying the boundary conditions, if the state is unbound you can lift the energy by a unit of momentum, whereas if it is a bound state it is unaffected. I'm not aware of any checks using the same technique on other systems. Generally the results for pentaquarks are rather confusing on the lattice. 4 or 5 groups have produced results during the past six months. They have used somewhat different operators on the lattice to create the pentaquark states. Using different operators can give you a better or worse overlap onto the physical state. If you choose an operator that is not very good you get rather a poor signal in the lattice calculations, and that, of course, can make the interpretation of the results more difficult. There is a bit of witchcraft going on as regards what is the best operator to use and certainly there has not been an exhaustive analysis of all the possible pentaquark states that you can construct on the lattice. Hence, you have to regard the lattice results as exploratory which, of course, parallels the uncertainty in the experimental situation. All but one agree that the parity of the lightest pentaquark state is negative.

-*Korthals-Altes:*

Which, I think, is the opposite of the original calculation by Diakonov. The lattice gets negative parity but I think the original calculation gives positive parity.

-*Kenway:*

Even the negative parity is not unambiguous within the lattice results. It is certainly proving difficult on the lattice to get a signal for a bound state. The situation is quite similar to the situation experimentally. There is a lot of confusion. The good thing is that it is encouraging the lattice to develop techniques to answer this question of whether a resonance is tightly bound or not.

-Kamenik:

While presenting the results of B $\rightarrow \pi$ form factor calculations on the lattice, you mentioned that there exists no parallel, model independent determination of the form factors. Couldn't one match the lattice results to heavy meson chiral perturbation theory at high q^2?

-Kenway:

Yes, but it cannot give you the form factor over the whole range of q^2 and its range of validity is not known a priori. The main point about using the lattice was in contrast to the B $\rightarrow$ D decays, which do seem to be well described by HQET. There HQET gives you the form factor normalization at zero recoil. That is almost all you need to extract $|V_{cb}|$ from experiment. The lattice is only adding the small percentage correction due to the fact that the b and c quarks are not infinitely massive. This is important for precision tests, but it is not really a qualitatively important contribution from the lattice. On the other hand, for B $\rightarrow \pi$, because you have a heavy quark decaying into a light quark, there is no symmetry to give you even the normalisation of the form factors in some limit.

-Kamenik:

What kind of model did you use for extrapolating the lattice results of the B $\rightarrow \pi$ form factors to low q^2?

-Kenway:

There are 3 or 4 different parametrisations on the market – they all basically make an assumption that the form factor has a particular pole structure. The details are not important. They are needed to extend the lattice calculations down to small q^2, where currently the simulations don't work. The model dependence you get by assuming a particular functional form is not significant considering the uncertainties that exist within the lattice calculation. The advertised objective for the lattice is to do something model independent – to free up phenomenology from having to use models of QCD. In this particular case, it is difficult to remove that model dependence if you want the total width. Fortunately, the experiments are starting to produce differential rates, so we can focus on the experimental results within the range of q^2 that you can simulate directly on the lattice. That is good enough, for example, for determining $|V_{ub}|$ as long as the experimental errors are reasonably small.

-Bourjaily:

Can spin-physics be determined on the lattice? Specifically can you compute the spin content of the proton or the decay asymmetry parameters of polarized hyperons?

-Kenway:

The structure function calculations that have been done are just the lowest moments. The spin content connects to the anomaly and so you need to understand topology on the lattice which is not technically straightforward. How to formulate those sorts of calculations and do some preliminary numerical work is currently being studied.

Two things work against you in hyperon decays. One is that classifying spin states on the lattice can be difficult because we don't have the full space-time symmetries and you can get mixing of higher spin states. Secondly, for hyperons, you are getting into the regime where the spectroscopy on the lattice is tricky and the signals are not so clean as for light hadrons.

-Alwall:

In your first slide you made a comparison to experimental data for the CKM matrix element $|V_{us}|^2$. Isn't lattice data necessary to extract this, i.e. by calculating f_K^2 and f_π^2 and using the ratio of decay widths of K and π?

-Kenway:

The CKM matrix elements are given by God, at least for those of us doing standard model physics. They have to be determined by experiment. I have tried to explain the comparison of lattice results with experiment by extracting the corresponding CKM matrix element using the lattice data and then comparing with the PDG's best fit, which is an average of all sorts of fits to experiment mostly using non-lattice techniques. You can measure various quantities to get a handle on the CKM matrix elements, provided you make appropriate theoretical input. We are really doing a set of consistency checks on the determination of $|V_{us}|$ and, particularly, the less well known CKM matrix elements.

-Ferrara:

Spontaneously broken gauge theories as well as QCD can exhibit abelian or non-abelian "monopoles". Can you see these configurations in lattice QCD?

-Kenway:

You can certainly see configurations of the gauge fields that look like monopoles. The difficulty usually is, because you don't have your fields defined in a space-time continuum, you don't have topological stability for these sorts of objects. That usually means there are several different ways you can define topological charges. At some particular value of the gauge coupling, typically you have a lot of noise in the gauge fields, i.e. the monopoles tend to be obscured by quantum fluctuations. The technique used to expose the monopoles is to cool down the configurations. That means getting rid of the short distance fluctuations in the gauge configurations without destroying the larger objects. Then you can try to estimate the extent to which monopoles saturate some physical quantity, for instance whether configurations with monopoles give a linearly rising potential and explain confinement.

-Korthals-Altes:

Sergio mentioned monopoles in SUSY theories, and asked whether those may be visible on the lattice. The answer was: they are individually visible once you take out the noise due to short distance fluctuations. For monopoles in confining theories like QCD at non-zero temperature this "cooling" is, alas, not an option. In that case you have to look at collective effects of the monopoles. One rather obvious way of realising this idea is to look at the flux that the monopoles shine through a Wilson loop. Especially at high temperature such a lattice simulation is quite feasible since the periodic temperature direction is then quite small so the system is three dimensional, and Michael Teper's group has done an exhaustive set of simulations of Wilson loop averages. These are consistent within 1-2%, with a dilute monopole gas with the monopoles in the adjoint representation of the SU(N) magnetic group, as suggested some time ago by my student Giovannangeli and me.

-Cifarelli:

Could you tell us more about the approach used to compute Supersymmetry on the lattice and could you better specify the meaning of "accidental" symmetry?

-Kenway:

Let us look at how Poincaré invariance is recovered in the continuum limit of the lattice theory. I am going to argue that it is the same mechanism that can work in the case of supersymmetry. When we put any field theory on a space-time lattice we break translational invariance, but we break it down to a finite group of translations and rotations. The words "accidental symmetry" mean that the Poincaré symmetry emerges naturally without fine tuning in the continuum limit. There are no relevant operators on the lattice that break Poincaré symmetry. The only operators that you can write down that break Poincaré symmetry are irrelevant, that is higher than dimension 4 and so they come in multiplied by positive powers of the lattice spacing. As you take the lattice spacing to zero they disappear.

A simple example is the energy-momentum dispersion relation:

$$E^2 = p^2 + m^2 = \sum_i p_i^2 + m^2$$

Since we only have the hypercubic translation and rotation group to determine the types of terms that can come in, we can imagine two possibilities: firstly corrections like

$$E^2 = \sum_i p_i^2 + m^2 + a^2 c \sum_i p_i^4$$

This is allowed by the lattice symmetry group. The factor of a^2 is for dimensional reasons. This is an example of an irrelevant operator in the sense that, as the lattice spacing goes to zero, the a^2 factor drives this term to zero, and it doesn't appear in the continuum limit (the momentum components are in physical units and are held constant in this limit).

The other thing that could happen is the addition of another term, for example:

$$E^2 = \sum_i p_i^2 + m^2 + a^2 c \sum_i p_i^4 + \frac{c'}{a} \sum_i p_i$$

For dimensional reasons this gets multiplied by 1/a. If these terms were allowed on the lattice (and they are not due to rotational symmetry) you would have a fine tuning problem, because, as you try to take the lattice spacing to zero, this coefficient would blow up unless you keep tuning c' to zero. Because we have remnant space-time symmetries on the lattice there are no terms that come in with inverse powers of the lattice spacing – the only corrections come in with positive powers of the lattice spacing. So accidental symmetry means that there is enough symmetry in the lattice theory itself to recover Poincaré symmetry in full, accidentally. This is an approach that is being promoted as a way of putting some SUSY models on the lattice. Can we come up with lattice versions of SUSY theories that have enough remnant SUSY to forbid any relevant operators that in the continuum limit break SUSY? Some explicit constructions have been carried out in 1+1 and 2+1 dimensions. In 3+1 dimensions, it has not yet been shown that fine tuning can be avoided, except for N=1 SYM, which comprises just gluons and gluinos (i.e. no scalar fields), and the GW formulation of massless fermions permits enough symmetry to be retained.

-Bourjaily:

The current techniques of lattice QCD make critical use of Wick rotation to a Euclidean action – this allows the powerful techniques of statistical mechanics. Is there any reason why one cannot in principle use the Lorentzian signature to study dynamical systems?

-Kenway:

The only way you could proceed with a Minkowski type formulation is to go to a Hamiltonian formulation where you keep time continuous and replace space by a three dimensional lattice. What you are doing then is looking at the quantum problem defined by that Hamiltonian. There are various techniques such as strong-coupling expansions and variational methods that you can use to study that system, but they haven't been applied successfully to QCD.

-Gyulassy:

Several years ago (~2001) two lattice groups, Takahashi et al. and Forcrand et al., studied the baryon flux configurations to differentiate between baryon junction Y configurations, proposed long ago by Veneziano, versus "delta" flux configurations. They reported opposite conclusions. What is the current status of this question? This is of interest to RHIC because several baryon observables are anomalous and we have speculated that the physics of baryon junctions may play a role. These observables pertain to baryon number transport over many units of rapidity as well as high $p_\perp$.

-Kenway:

The type of flux tube model you describe only makes sense for baryons made out of very massive quarks. It is a picture you would naturally have with static colour sources, although I am not exactly sure whether this was the case for the results you mention. I am doubtful that it tells you anything at all about baryons composed of light quarks.

-Gyulassy:

Well actually it is relevant. The way of making the light quarks heavy is to give them an enormous energy, e.g. give the three valence quarks 100GeV each then the valence quarks end up as mesons and where the baryon number finally migrates to in this triangle of three fast quarks is very suggestive in at least the baryon junction idea.

-Kenway:

It will be very tough on the lattice to do anything which is really close to the real world situation you just described, because giving the valence quarks large amounts of momentum on the lattice is not going to happen for a long time. It is not obvious that three static quarks is giving you the right intuition, but it may be the best we can do at the moment.

The J/Ψ as a probe of Quark-Gluon Plasma

Luciano MAIANI
Università di Roma "La Sapienza" and
INFN. Roma. Italy

Introduction

Confinement means that the heavy quarks in a c-cbar pair are bound by an asymptotically constant attractive force (i.e. a linearly rising potential). This is what happens in normal vacuum.

Several theoretical arguments suggest that by increasing the temperature, normal vacuum gives rise to a new phase where quarks and gluons are not confined in hadron bags. In the deconfined phase, the attractive force between c and c-bar is screened by the Quark-Gluon Plasma (QGP). Charmonia bound states "melt" as temperature rises, starting from the less bound higher resonances down to the more deeply bound lower states. Thus, te onset of J/Ψ suppression in relativistic heavy ion collisions would signal the formation of QGP, a suggestion originally made by T. Matsui and H. Satz[i].

The method can work, however, only if we are able to control all other sources of J/Ψ absorption in heavy ion collisions, both nuclear and hadronic [ii]. To this aim, several calculations of the J/Ψ dissociation cross-sections have been performed, e.g. for the process:

$$h + J/\Psi \rightarrow D + \overline{D}$$
$$(h = \pi, \rho, ...)$$

J/ Ψ dissociation by hadrons was originally believed to be very small, on the basis of perturbative QCD calculations, but more recent studies have shown its importance.

I will report on the results of a recent calculation by our group[iii] and apply them to the NA50 data obtained at the SPS.

The present Lectures give a "bottom up" presentation[iv], going (slowly) from low to high temperature. The 1st lecture contains an elementary introduction to the basic concepts and in the 2nd lecture I shall present the results of our calculations and their application to the data taken at the CERN SPS by the NA50 Collaboration[v].

The main issue I will address is: ***did quark-gluon plasma show up at the SPS?***

On the basis of our results, I conclude: ***yes, most likely !!***

But we need to know better...and to study QGP more, at RHIC and LHC.

Before closing the Introduction, I would like to express my gratitude to U. Heinz, U. Wiedemann and F. Becattini for interesting discussions. The help of F. Becattini in constructing the 2004 hadron level spectrum is gratefully acknowledged. Finally, I want to thank my younger collaborators F. Piccinini, A. Polosa and V. Riquer, who introduced me to the problem of J/Ψ dissociation and shared the enthusiasm of exploring the new (for us) world of heavy ion collisions.

LECTURE 1

Summary

1. *A simple view of the collisions*
2. *Does the fireball thermalise?*
3. *Hadron gas*
4. *Hagedorn gas, the phase transition from below*
5. *Deconfined Quarks and Gluons*
6. *Debye screening*
7. *Summing up*

1. A simple view of the collisions

Snapshots of the high-energy collision of two (equal) heavy ions, taken in the c.o.m. frame before and after the collision, would look like the cartoons in Fig.1. The center of mass energies of the projectiles are 8.5 GeV/A at the SPS, 100 GeV/A at RHIC and will be 2.5 TeV/A at the LHC.
Nuclei are compressed in the direction of flight by the relativistic contraction, the closest distance of approach (impact parameter) is indicated by *b*.
The value of *b* in each collision can be measured by observing the energy carried by the fragment of nucleus which has gone on unperturbed, represented with a solid shape in Fig. 1. In the SPS, fixed target, experiments the unperturbed fragment of the projectile nucleus goes forward and its energy is measured in NA50 by a Zero Degree Calorimeter. The fragment has the same Energy/A ratio as the original projectile, thus we can get A from the energy and deduce its size, i.e. *b*, from nuclear models (see below).

The overlap region of the two nuclei after collision is represented with a fuzzy shape: nucleons are now unbound and mixed with the forward and backward fragments of individual collisions.

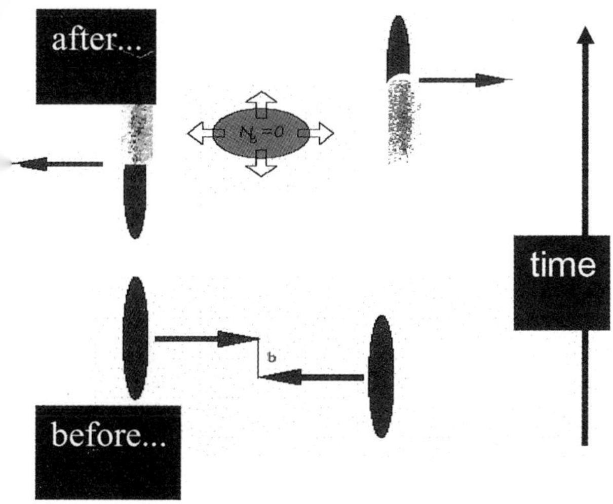

Fig. 1.1. Snapshot of a relativistic heavy ion collision, center of mass frame.

Hadrons produced in the central region of the rapidity plateau form a fireball which expands rapidly under the pressure exerted by the momentum of the inner particles.. The central plateau is separated from the fragmentation region by a rapidity interval which increases with the c.o.m. energy. Therefore, the central region is more and more baryon number free. The fireball starts as a state of dense hadronic matter with a transverse size which we denote by *l* (*l* =2*R*-*b* where *R* is the radius of each nucleus).

Shortly after the collision, at a time τ_0 = 0.1-1 fm/c, the energy density of the fireball for central collsions, *b*=0, can be estimated in terms of the particles produced in elementary nucleon-nucleon collision[vi]:

$$\varepsilon_{central} = \frac{A}{S}(\frac{dE}{dy})\frac{1}{c\tau_0} \tag{1.1}$$

here A/S is the average baryon number surface density of the incoming nuclei while the energy per unit rapidity (per nucleon-nucleon collision), can be estimated at the SPS according to:

$$\frac{dE}{dy} = \frac{dN_{ch}}{dy}(1+\frac{N_{neutrals}}{N_{charged}}) < E > \cong 3 \cdot 1.5 \cdot 400 MeV = 1.8 GeV \tag{1.2}$$

For central Pb-Pb collision, one has:

$$\frac{A}{S} = \frac{A}{\pi R^2} = \frac{A^{1/3}}{\pi r_0^2} = 1.5 fm^{-2} (for : Pb)$$

For collisions with a generic impact parameter, the baryon density per unit surface is reduced by a geometrical factor:

$$\frac{A(b)}{S(b)} = \frac{A(0)}{S(0)} g(b)$$

where g(*b*) is the nucleon number density per unit area, averaged over the region of overlap of the colliding nuclei:

$$g(b) = \frac{\pi}{2}[\frac{(1-b/2R)^2(1+b/4)}{\arccos(b/2R) - b/2R\sqrt{1-b^2/4R^2}}] \tag{1.3}$$

For Pb-Pb collisions, *R*= 6 fm and g(*b*) decreases from 1 to about 0.6 when *b* goes from 0 (central collisions) *to b*=8 fm. The latter value corresponds to a transverse diameter of the fireball: *l*=2*R*-*b*=4 fm where the interestingly central collisions begin, as we shall see in Lecture 2. From the estimates above, we obtain

$$\begin{aligned} \varepsilon &= 1.6 GeV fm^{-3} (\frac{1 fm}{\tau_0}) \cdots (l = 4 fm) \\ \varepsilon &= 2.7 GeV fm^{-3} (\frac{1 fm}{\tau_0}) \cdots (l = 12 fm) \end{aligned} \tag{1.4}$$

2. **Does the fireball thermalise?**

Particles produced in the primary collisions are mainly pions, with an average energy of 400 MeV each. If the density of the early fireball is sufficiently high, pions will scatter many times off each other and relax to a state with some definite temperature, T. For $\tau_0 < 1$fm/c and $l = 4$fm, we can estimate an average initial density: $\rho > 4\ fm^{-3}$. Taking a typical strong interaction cross section: $\sigma \cong 40mb = 4fm^{-2}$, we get the mean free path:

$$\lambda^{-1} = \rho\sigma \geq 40mb \cdot 4\,fm^{-3} = 16\,fm^{-1};$$
$$\lambda \leq 0.06\,fm$$

λ is much smaller than the size of the fireball produced (a few Fermi). Thermal equilibrium is a reasonable guess.

Some words of caution are appropriate here. Pions at zero momentum have a vanishing interaction since they are the quasi-Goldstone-bosons of chiral symmetry. Calculations based on chiral perturbation theory provide cross sections that could be much smaller than what estimated above, and predict no thermalization at all[vii]. However, if the original particles have an average energy E= 400 MeV, with random distributed directions the average center of mass energy in the pion-pion system is about 560MeV ($<s> = 2E^2$) and are rather far from the soft pion limit for Chiral perturbation Theory to be applicable. As an alternative, we may compute the P-wave pion-pion cross section using a Breit-Wigner formula for the $\rho(770)$, to find $\sigma_\rho(560\text{MeV}) \sim 33$ mb. If we add an S-wave resonance, $\sigma(480)$ with $\Gamma \sim 300$ MeV, as suggested by recent data[viii], we obtain a total cross section of about 50 mb, in line with the previous estimate.

What about experiment? The hadrons that are observed originate at a later time, the so-called freeze-out time, when the expansion of the fireball has reduced the density such that the hadrons do not interact anymore among themselves. Coherently with the thermalization assumption, hadrons at freeze-out exibit a thermal distribution[ix], with T=170-180 MeV.

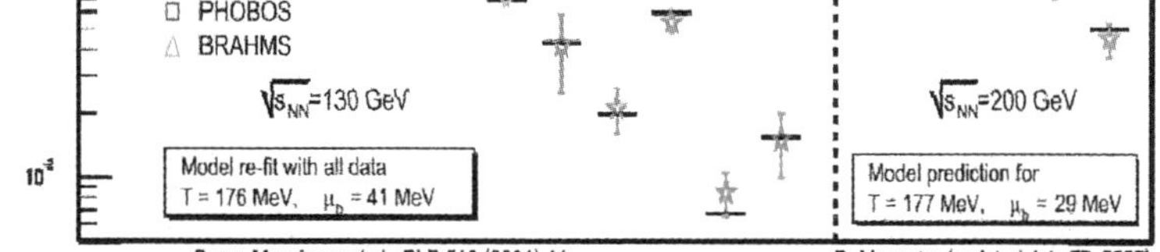

Fig. 2.1 Hadron abundances and thermdynamical fits at RHIC.

3. Hadron gas

At thermal equilibrium and low energies, we describe the fireball as a gas made by hadron resonances. Each particle species is treated as a free particle. Interactions are rather introduced in the form of resonances which appear at increasing energies: we have pions only at low energy, then kaons, etas, (i.e. the strange quark flavour) then ρ, ω, ϕ, K*, etc. (see e.g. Ref.[x]).

The thermodynamical partition function, Z is easily obtained if we know the hadronic gas composition, i.e. the density of hadronic level as a function of the mass:

$$\ln Z = \int \rho(m)dm[V \int \frac{d^3p}{(2\pi)^3} \ln(z_i)];$$
$$\ln(z_i) = \pm\ln(1 \pm e^{+\mu_i} e^{-\beta E_i(p)}); (+ = fermions, - = bosons) \tag{3.1}$$

$$\rho(m) = \sum_i (2J+1) N_{ch} \delta(m - m_i) \tag{3.2}$$

(β=1/T, μ=chemical potential, k=Boltzmann's constant=1) and N_{ch} represents the charge multiplicity (3 for pions). For simplicity, in what follows we assume vanishing chemical potential. This is appropriate for the fireball produced in the central region of very high energy collisions. It is a rough approximation at SPS energies, getting better and better at RICH and LHC.

In the Boltzmann limit all exponents are <<1, so that: $\ln z \approx e^{+\mu} e^{-\beta E}$. The energy density of the hadron gas and the average species abundances are obtained from the partition function as usual:

$$\frac{U}{V} = \varepsilon(T,\mu) = -\frac{\partial \ln Z}{\partial \beta};$$
$$n_i = \frac{\partial \ln Z}{\partial \mu_i}$$

It is convenient to introduce the ratio of the energy density to that of a gas made by one specie of spin zero, massless bosons. We call N_{eff} the energy density thus normalized:

$$N_{eff}(T) = \frac{\varepsilon(T)}{\varepsilon_o(T)};$$
$$\varepsilon_o(T) \equiv \frac{T^4 \pi^2}{30} \approx 0.33 T^4 \tag{3.3}$$

Justifying its name, N_{eff} gives the number of degrees of freedom that are "active" at temperature T. At low temperatures, of the order of 150 MeV, one would expect only pions, namely $N_{eff} \sim 3$. However, in spite of higher mass, higher resonances contribute significantly to the energy density even at these temperatures, because of their increasing spin and charge multiplicities, as we discuss in the next Section.

4. Hagedorn gas: the phase transition from below

The central ingredient of the fireball thermodynamics is the density of the hadron levels, ρ(m). Of course, we are unable to compute it from first principles, at present.

Repeating the exercise found in Ref.[10], I report in Fig. 4.1, black solid curve, a

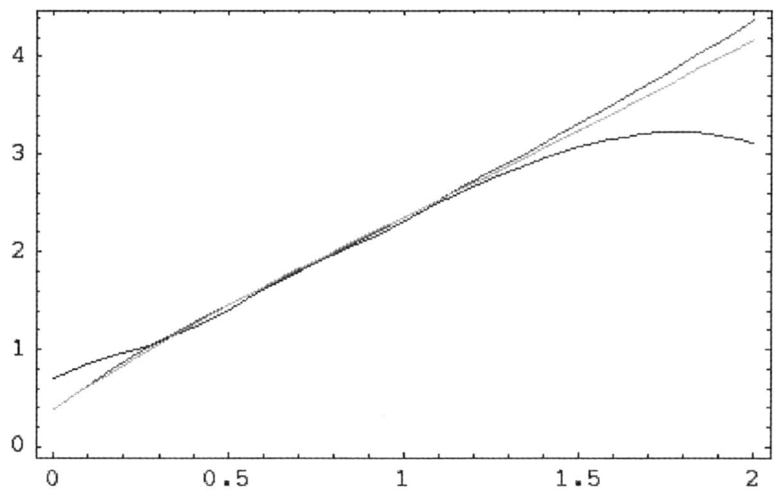

Fig. 4.1. Hadron level density functions (see text). In ordinates, the values of $\log_{10}[\rho(m)]$ are reported, with m in GeV and ρ in GeV^{-1}. For the meaning of the different curves, see text.

determination of the "experimental" level density, $\rho_{2004}(m)$ corresponding to the hadron levels listed in the Particle Data Book of 2004. Pions are not included. For the other particles, each delta-function in eq.(3.2) is smeared out in a gaussian to get a regular behaviour:

$$\delta(m - m_{res}) \rightarrow \frac{1}{\sigma\sqrt{2\pi}} e^{-\frac{(m-m_{res})^2}{2\sigma^2}}$$

with 200 MeV resolution. The level density ρ_{2004} drops for masses above approximately 1.5 GeV. This is probably because we are unable to identify highly excited resonances with increasing angular momentum and increasingly large widhts. However, in the region where we can reasonably assume that we know almost all the hadronic levels, the trend supports an exponentially increasing density, as predicted by Hagedorn, in the prehistory of modern hadron theory, on the basis of his bootstrap principle. The exponential rise of ρ(m) is supported by hadron models based on quarks (e.g. the bag model).

The blue line in Fug. 4.1 was obtained in ref.[10] by a fit of the smeared data from the 1996 Particle Data Group with a theoretical level density of the form:

$$\rho(m) = C(m_0^2 + m^2)^{-3/2} e^{+m/T_H} \tag{4.1}$$

The power -3/2 of the pre-exponential factor is predicted by Hagedorn's "bootstrap principle" and it fits well the data. The resulting parameters are:

$$\begin{aligned} C &= 0.7GeV^2; m_0 = 0.66GeV; \\ T_H &= 158MeV \end{aligned} \tag{4.2}$$

The value of T_H thus obtained is in agreement with the original determination by Hagedorn, which was based on the p_T cutoff observed in the distribution of the pions produced in hadronic collisions. However, T_H is too low to comply with the temperatures obtained from hadronic distribution at freeze-out, which we have seen to be in the range 170-180 MeV, and also with the lattics QCD results (see below) which indicate a transition temperature around 170 MeV.

However, T_H is not so well determined by the mass distribution of the observed resonances, due to the falling of the black curve. We can obtain a good fit with a larger temperature by readjusting the coefficients of the pre-exponential terms. The purple curve shows the result[3] for:

$$\begin{aligned} C &= 2.55GeV^2; m_0 = 1.01GeV; \\ T_H &= 180MeV \end{aligned} \tag{4.3}$$

The exponentially rising spectrum leads to a limiting temperature for the hadronic matter, equal to T_H.

Intuitively, this can be understood as follows. An increase of the energy of the gas gives rise to two competing effects: increase of the average energy of the particles present in the gas, thereby increasing the temperature, increase of the number of species present in the gas, by creating heavier particles. With a spectrum limited in mass, the first mechanisms takes over at large temperatures and leads to the T^4 behaviour of the energy of a gas made by a fixed number of species of relativistic particles. If the level density increases exponentially, however, the second mechanisms dominates more and more and at some point it will be impossible to increase the temperature further, since any newly provided energy is spent in creating new particles.

The argument is well exemplified by Fig. 4.2. The lowest red curve gives the contributions of pions alone, flattening at the value 3. The next higher, red, curve gives N_{eff} for a hadron gas made by the pseudoscalar and vector nonets. Here, ρ(770) and ω(770) make a sizeable contributon, nothwitstanding that the values of their masses are quite larger than T. Still higher is N_{eff} computed with ρ_{2004} (green curve), while, finally the blue curve represents N_{eff} for the Hagedorn gas with the parameters given in eq.(4.3), diverging at T_H=180 MeV.

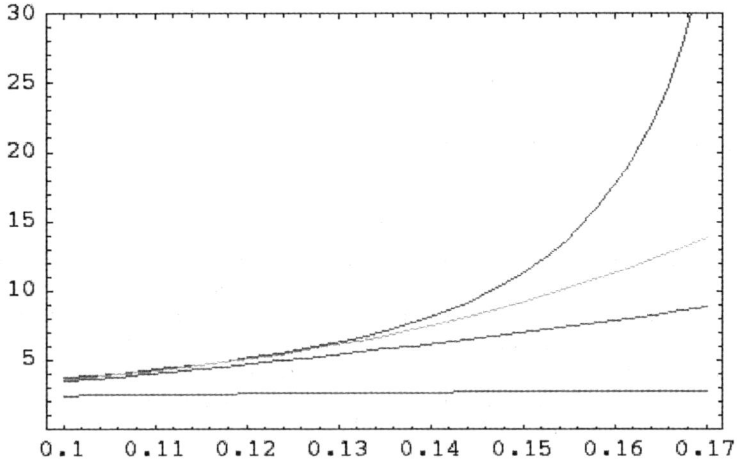

Fig. 4.2. The behavior of N_{eff}, Eq. (3.3), vs. temperature (GeV) and different hadron gases. Red lowest curve: pions; red higher curve: pseudoscalar and vector mesons; green curve: all particles in Particle Data Group 2004, with the smeared level density $\rho_{2004}(m)$; blue curve: Hagedorn gas with T_H=180 MeV.

The existence of a limiting temperature can be seen at once from the fact that the partition function:

$$Z = \sum \rho(m) e^{-\frac{m}{T}} \propto \sum m^{-k} e^{+m(\frac{1}{T_H} - \frac{1}{T})} \tag{4.4}$$

does not converge for $1/T<1/T_H$. In more recent times, looking at the same formula *from below*, the Hagedorn temperature has been interpreted as the transition to a new phase of matter[xi].

This is easily seen by computing the behaviour of the thermodynamical functions while approaching T_H from below. We use the non-relativistic Boltzmann approximation, since the critical behaviour is determined by the high masss part of the spectrum, m>>T, to find (β=1/T, β_C=1/T_H):

$$\ln Z = \frac{V}{(2\pi\beta)^{3/2}} \int dm \frac{C}{(m_0^2 + m^2)^{3/2}} m^{3/2} e^{-m(\beta-\beta_c)} \approx V[A(\beta-\beta_c)^{1/2} \int_{E_0(\beta-\beta_c)}^{+\infty} dx x^{-3/2} e^{-x} + reg.]$$

where A is some constant and "reg." represents terms analytic at β_c. The integral is evaluated as follows:

$$\int_{E_0(\beta-\beta_c)}^{+\infty} dx x^{-3/2} e^{-x} = -2 \int_{E_0(\beta-\beta_c)}^{+\infty} dx x^{-1/2} e^{-x} + \frac{2}{E_0(\beta-\beta_c)^{1/2}} \approx -2 \int_0^{+\infty} dx x^{-1/2} e^{-x} + \frac{2}{E_0(\beta-\beta_c)^{1/2}}$$

so that we find, in conclusion, for pressure and for energy density :

$$P = \frac{\ln Z}{V} = reg. - 2A(\beta - \beta_C)^{1/2};$$
$$\varepsilon = -\frac{\partial}{\partial\beta}\ln Z \propto (\beta - \beta_C)^{-1/2}; \tag{4.5}$$

Near the critical point, the hadron gas is very soft: the velocity of sound vanishes at β_C:

$$v^2 = \frac{dP}{d\varepsilon} \propto (\beta - \beta_C) \tag{4.6}$$

The behavior of ε is quite reminiscent of the critical behaviour near a second order phase transition, which Cabibbo and Parisi interpreted as quark and gluon de-confinement. However, there are several differences with respect to a second order phase transition. Until now we have considered poinlike particles only. If one introduces a finite radius of the hadron levels, the behaviour becomes less critical (see again ref.[10]). In fact, we could rather be in presence of a smooth cross-over from the hadron gas state to the liberated quark and gluon plasma.

5. Order parameters for the phase transition

The order parameter for the deconfinement transition, in the pure gauge system without fermions, is the Wilson- Polyakov loop:

$$L = < \exp(-F_{Q\bar{Q}}/T) > \tag{5.1}$$

F is the free energy of a pair of static point sources (heavy quarks) at distance r. F and L have the limiting behaviours for r → ∞:

$T < T_c$: $F \propto r \to \infty$, **$L \to 0$** confinement

$T > T_c$: $F \to 0$, **$L \to Const. \neq 0$** deconfinement

Fig. 5.1 shows indeed a rapid rise of L at the critical temperature[xii].

According to many indications, chiral symmetry is restored in the deconfined phase, indeed calculations support a drop to zero of the chiral symmetry order parameter, $<\bar{\psi}\psi>$, correlated to the raising of L , as shown in Fig. 5.2.

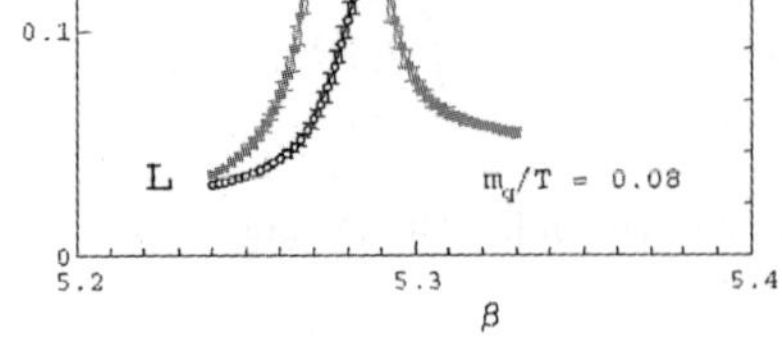

Fig. 5.1. Lattice QCD calculation of the behaviour of the Polyakov loop, Eq. (5.1), as function of Temperature (blue curve). Ref.[12].

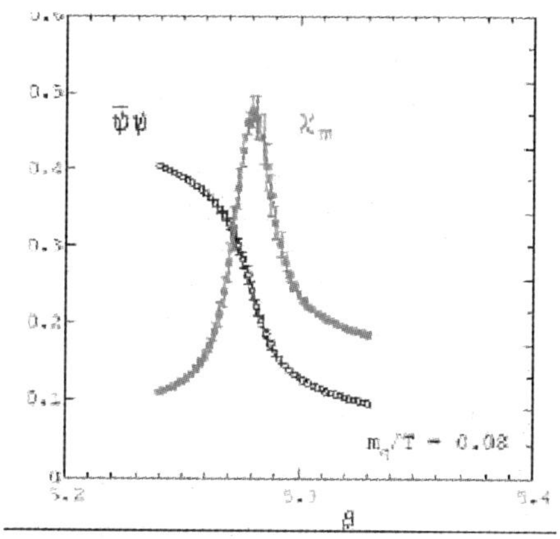

Fig. 5.2. Lattice QCD calculation of the behavior of the chiral order parameter as function of temperature (blue curve). Ref.[12].

Lattice QCD calculations indicate a first order phase transition for the pure gauge system. In the presence of quarks, it is not yet clear if the transition stays, or if it attenuates in a simple crossover, a rapid but continuous transition between a gas of hadrons and a gas of quarks and gluons. According to recent speculations, the transition would exist for non vanishing chemical potential, to stop at a tricritical point, where all phases coexist, see point E in Fig. 5.3.

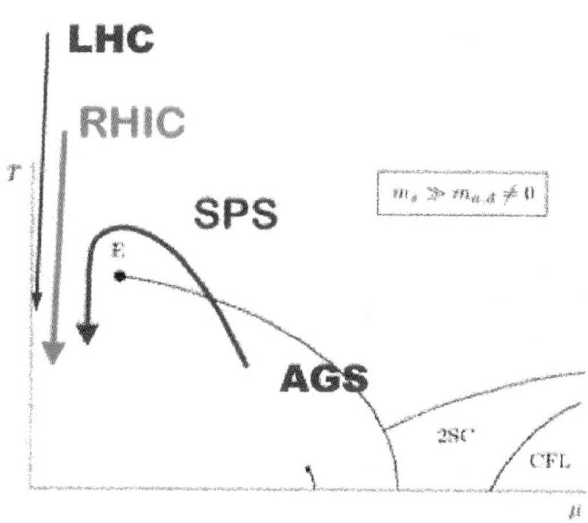

Fig. 5.3. Schematic view of predicted and/or hypothesized phase regions of hadronic matter.

A rapid increase in the energy density is shown by lattice QCD calculations with unquenched quarks (see again F. Karsch, quoted under Ref. [12]), until the ratio $\varepsilon(T)/T^4$ stabilizes to a constant, as it would be the case for a gas of non interacting massless particles. In the latter case, the value of the constant is predicted to be:

$$\varepsilon = N_{eff}\frac{\pi^2}{30}T^4 \approx \frac{N_{eff}}{3}T^4$$

$$N_{eff.QGP}(n_f) \cong 8\cdot 2 + (\frac{7}{8})(4\cdot 3)n_f \ldots i.e.: \frac{N_{eff.QGP}(3)}{3} \cong 16 \qquad (5.2)$$

Lattice QCD calculations give a value around 12, about 80% of the predicted value. The deviation could be due to (expected) corrections to the relativistic perfect gas, which die only logarithmically with the increasing temperature, due to asymptotic freedom. It is unclear at the moment if the difference can be accounted by perturbative QCD calculations, or if it indicates more radical, non-perturbative departures from the perfect gas (see the Lectures by M. Gyulassy at this School[xiii]).

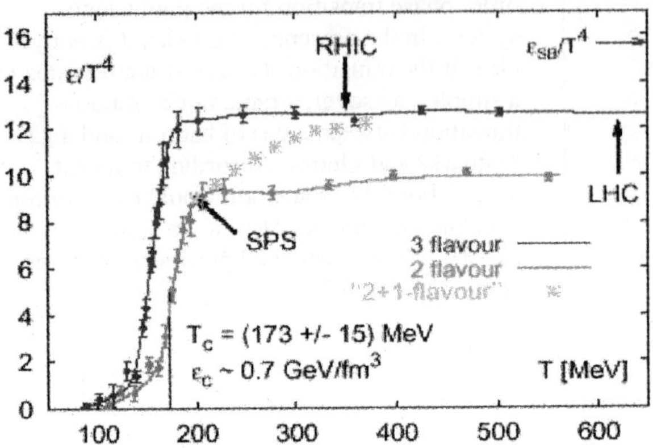

Fig. 5.4. Lattice QCD calculations of the energy density of hadronic matter vs. T (Ref. [12]).

6. Debye screening of charmonia

The spectrum of charmonium states is well reproduced by a charm-anticharm potential, which is the superposition of a Coulomb-like term (i.e. a massless gluon) and a rising, confining, term:

$$V(r) = \sigma r - \frac{\alpha_c}{r} \tag{6.1}$$

In the de-confined phase, a phenomenon analogous to Debye screening is supposed to take place, with the gluon taking a non-vanishing mass (like the photon in a Coulomb plasma). One can describe the new situation by a modified potential:

$$V(r) = \frac{\sigma}{\mu_D(T)}(1 - e^{-\mu_D(T)r}) - \frac{\alpha_c}{r}e^{-\mu_D(T)r} \tag{6.2}$$

the lattice QCD calculations described in the last Section suggest precisely this to happen. The exponential of the free-energy diverges for large r in the confined phase, corresponding to the rising term in Eq.(6.1), while it goes to a constant in the de-confined phase, corresponding to the finite range potential in (6.2).

The function $\mu_D(T)$ determines the interpolation between the two cases. μ_D vanishes below the critical temperature, and Eq. (6.2) goes back to (6.1), while it rises with T above T_c. For large enough μ_D, the screening can prevent charmonium formation. We give in the Table, see Ref. [2], the values of μ_D for which the lowest lying charmonium states become unbound and the corresponding values of the temperature, T_D.

	J/Ψ	Ψ'	$\chi_c(1P)$
M(GeV)	3.096	3.686	3.415
r(fm)	0.89	1.5	2.0
μ_D	0.699	0.357	0.342
$T_D(n_f=3)$(MeV)	406	189	178

The J/Ψ disappears at quite higher temperatures than the critical temperature. However, a good fraction of the J/Ψ observed in absence of nuclear effects originate from the decay of higher charmonium states, about 40% from Ψ' and χ_c states which melt right above the critical temperature. Thus the estimates presented here are consistent with a sizeable reduction of J/Ψ as soon as the critical temperature is reached.

7. Summing up

The fireball produced in collisions with low energy density is essentially a pion gas thermalized at some temperature,T, below the Hagedorn temperature;

Increasing the energy density ε, by increasing the c.o.m. energy and/or centrality, several phenomena take place in succession:

- higher resonances appear in the fireball;
- due to the increasing multiplicities of higher states, it becomes more and more difficult to increase the temperature;
- we get in the Hagedorn gas limit:

$$dT/d\varepsilon \sim (\beta-\beta_c)^{3/2}$$

- hadron bags get in contact, bags fuse and quarks and gluons start being liberated.

A cartoon representing all this could be as in Fig. 7.1.

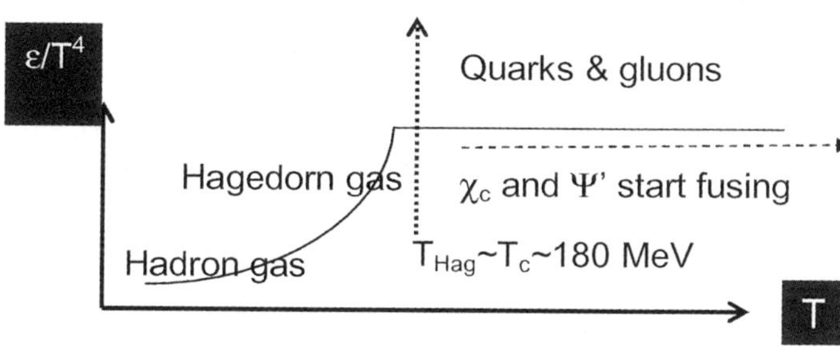

Fig. 7.1.

LECTURE 2

Summary

1. *J/Ψ dissociation cross-sections in the Constituent Quark Model*
2. *Attenuation factors*
3. *Results for the hadron and Hagedorn gases*
4. *A bold speculation*
5. *J/Ψ as a probe of QGP: some conclusions*

1. J/Ψ dissociation cross-sections in the Constituent Quark Model

I consider in this Lecture the interaction of the J/Ψ with the hadrons that can be excited from vacuum at the huge energy densities reached already at the SPS in heavy ion collisions. Each of these hadrons is able to interact with the J/Ψ, transferring its c-cbar content to open charm final states:

$$h + J/\Psi \to D^{(*)}_{(s)} + \overline{D^{(*)}_{(s)}}$$
$$(h = \pi, \eta, K, \rho, \omega, \phi)$$

Even restricting to tree-level diagrams, the interactions involve a large number of multi-meson couplings which cannot be derived from first principles, at present. Some couplings we can control from experimental data, for example the D*Dπ coupling, but for most of them we have no experimental input . One has to resort to models. Calculations available in the literature can be classified in four classes:

- perturbative QCD (pQCD) based calculations;
- quark interchange models;
- QCD sum rules calculations;
- meson exchange models.

For a nice review of these topics see Ref.[xiv].

The approach I shall describe is closely related to the meson exchange method. The total amplitude is written in terms of tree-diagrams involving effective trilinear and quadrilinear couplings such as:

$$g^{(3)}{}_{MD\overline{D}} = (\pi, \eta, K, \rho, \ldots) D\overline{D}; g^{(3)}{}_{J} = JD\overline{D};$$
$$g^{(4)}{}_{JMD\overline{D}} = J(\pi, \eta, K, \rho, \ldots) D\overline{D}$$

In turn, multi-meson couplings are evaluated by the so-called Constituent Quark Model, a model originally devised to compute exclusive heavy to light meson decays and tested on a quite large number of such processes[xv].

The essence of the model is shown in Fig.1.1 , for the case of the $J\pi D\overline{D}$ coupling. One starts by computing the quark loop on the left-hand side, c and q being the charm and light quarks, respectively, and then equates the result with the right-hand side diagram, computed with the vector dominance ansatz:

$$j_\mu^{e.m.} = \frac{M_{J/\Psi}^2}{f_{J/\Psi}} \Psi_\mu$$

Calling $\zeta(q^2)$ the result of the Feynman diagram on the left-hand side, we derive $g_{J\pi D\bar{D}}$ from the equation:

$$\zeta(q^2) = \frac{M_{J/\Psi}^2}{f_{J/\Psi}} \frac{1}{q^2 - M_{J/\Psi}^2} g_{J\pi D\bar{D}} \tag{1.1}$$

with both sides evaluated at $q^2=0$. I give a few details on the calculation, mainly to illustrate the parameters involved and the related uncertainties.

D meson-charm quark vertices are represented by a quark-meson coupling of the form: $\sqrt{(M_D Z_D)}\,\gamma_5$, pions appear with a pseudovector coupling, $(f_\pi)^{-1} k^\mu \gamma_\mu \gamma_5$, as appropriate to their nature of quasi Goldstone bosons. Quark propagators depend upon internal momenta and an additional parameter, Δ,

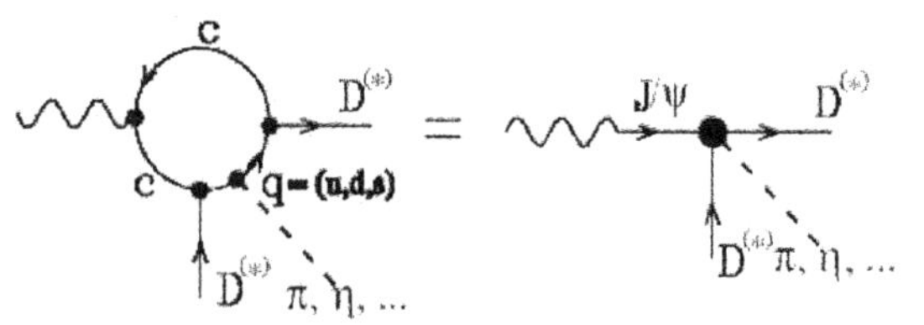

Fig. 1.1. Diagrammatic basis for the computation of $g_{J\pi DD}$.

representing the difference bewteen quark mass and the corresponding meson mass. Δ is the main free parameter of the model. It varies in the range Δ =0.3-0.5GeV for u,d quarks and 0.5, 0.6, 0.7 GeV for strange quarks[xvi]. The variation of results with Δ allows to estimate the theoretical error.

Integration over the internal momentum in the loop requires both an infrared and an ultraviolet cut-off. The ultraviolet cut-off , Λ, has been set to the chiral expansion scale, $\Lambda = 4\pi f_\pi$. The infrared cut-off, μ, prevents momenta to access the confinement energy region, $\mu \approx \Lambda_{QCD}$. Z_D is, essentially, the meson wave-function renormalization constant and is determined from self-energy diagrams in terms of the quark mass and the ultraviolet and infrared cutoffs[xvii].

The extension of pion couplings to the pseudoscalar octet is done by flavour SU(3) symmetry, ρ couplings are computed by insertion of a vector current in the quark loop and vector dominace, the extension to the other light vector mesons is done by SU(3) nonet symmetry.

In the computation of cross sections, thresholds are expressed exactly, in terms of the physical masses. Thresholds determine the energies at which the particles in the hadron gas become effective to dissociate the J/Ψ and are very relevant since they are of the same order as, sometime larger than, the thermal energies involved.

Pion cross-sections have relatively high thresholds, which disfavours them with respect to ρ s. Similarly, P-wave processes such as: π+ J/Ψ $\rightarrow$D+ Dbar are disfavoured with respect to S-wave processes such as: π+ J/$\Psi$$\rightarrow$ D* Dbar, D D*bar.

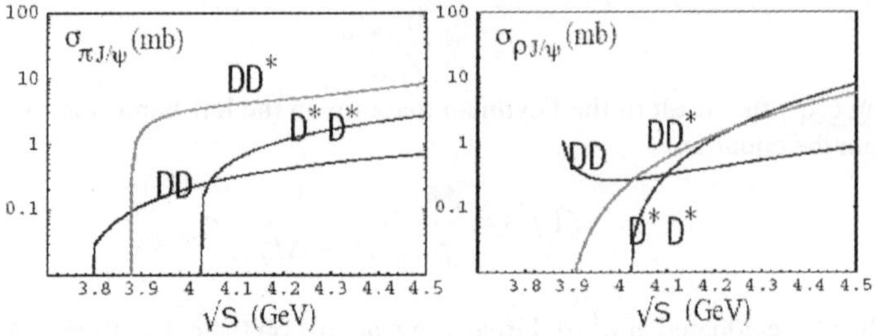

Fig. 1.2. J/Ψ dissociation cross-sections obtained in the CQM vs. c.o.m. energy

Fig. 1.2 shows the J/Ψ dissociation cross-sections obtained in the CQM as functions of c.o.m. energy[xviii]. Fig. 1.3 gives a compilation of the cross sections obtained with other methods, see Ref. [14].

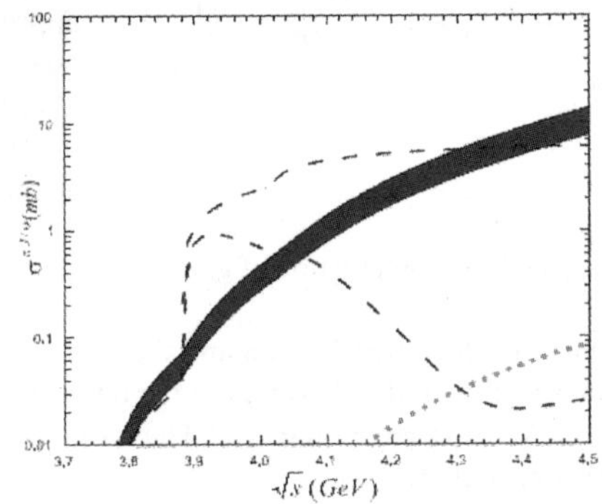

Fig. 1.3. J/Ψ dissociation cross-sections obtained with different approaches, from Ref.[14]: QCD sum rules (band), short-distance QCD (dotted line), meson-exchange models (dot-dashed lines), non-relativistic constituent quark model (dashed line).

It is reassuring to see that different approaches, all affected by their model dependencies, with the exception of the perturbative QCD calculation, give consistently non negligible dissociation cross sections:

$$\sigma_{hadr.abs.} = (1-10)\dots mb$$

This is true, in particular for initial π's and ρ's, the most studied cases. Hadron dissociation cross sections are quite comparable to the nuclear absorption cross-section, experimentally determined[xix] from p-A production of J/Ψ:

$$\sigma_{nucl.abs.} = (4.3 \pm 0.6)\dots mb$$

In conclusion, there is a clear indication that a reliable calculation of J/ Ψ absorption needs to take into account both the effects of nuclear matter and of the hadron fireball.

2. Attenuation factors

J/Ψ produced by nucleon-nucleon collisions are detected by their decay into muon pairs. One usually gives the production rate normalized to the Drell-Yan cross section:

$$R_{J/\Psi} = \frac{\sigma(A+A \to J/\Psi + ...)}{\sigma(A+A \to \mu^+ + \mu^- + ...)}$$

Prompt muon pairs are not affected by propagation in the nuclear and hadronic medium, so that this ratio is sensitive to any mechanism which suppress the J/Ψ in the medium, that is Quark-Gluon Plasma formation as well as nuclear and hadronic absorption.

The absorption lenght in nuclear matter has been estimated by NA50 from the data on J/Ψ production in p+A collisions. The effect is parametrized by NA50 according to:

$$(R_{J/\Psi})_{nucl} = \exp(-\rho_0 \sigma_{nucl.abs.} L)$$

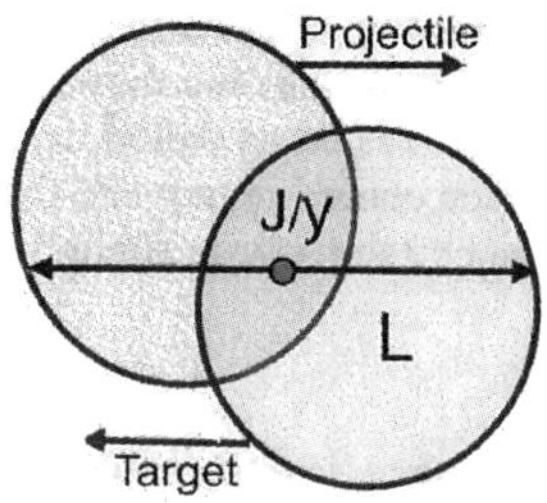

Fig. 2.1. The longitudinal lenght L, which regulates the nuclear J/Ψ absorption. Lorentz contraction is not reported (see text).

where:

$$\rho_0 = 0.17 fm^{-3}$$

and L is the (average) longitudinal lenght[1] to be traversed by the J/Ψ, Fig. 2.1. Using nuclear models, NA50 gives L as function of *b*.

After traversing the nuclear matter, the J/Ψ finds itslef in the hadron fireball, which is just starting to expand. The lenght that J/ Ψ has to traverse to get out of the fireball depends, among other factors, from the direction of the J/ Ψ and the shape of the fireball. Assuming for simplicity a spherical fireball and isotropic J/Ψ, the average leght that a J/ Ψ produced anywhere inside the sphere has to traverse before escaping is:

$$l_{av.} = \frac{3}{8} l$$

where l is the transverse dimension of the fireball[2], defined in Lecture 1, see Fig. 1.1. Correspondingly, we get the attenuation factor:

[1] Note that the decrease of L due to the Lorentz contraction in the c.o.m. system, $1/\gamma$, is compensated by a factor γ in the nucleon density.

[2] for a flat-disk fireball of diameter *l*, we would get a slightly different numerical factor, $l_{av.} = 4/(3\pi)l$.

$$A_{hadrons} \propto \exp[-\sum_i < \rho_i \sigma_i >_T \frac{3}{8} l]$$

where ρ_i and σ_i are densities and dissociation cross-sections of the different hadron species. The product is averaged in the thermal bath represented by the hadron gas at temperature T.

Using nuclear models, NA50 gives L as function of *b*, as shown in Fig. 2.2. Note that L flattens out around $l=2R$ (corresponding to maximum centrality, $b = 0$). If there were nuclear effects only, the ratio of J/Ψ production to Drell-Yan pairs should flatten for $l \to 2R$ (L → maximum value) while data continue to decrease at large l, as we show later. The variable l is better suited than L, to describe the central region and the absorption by the hadron fireball.

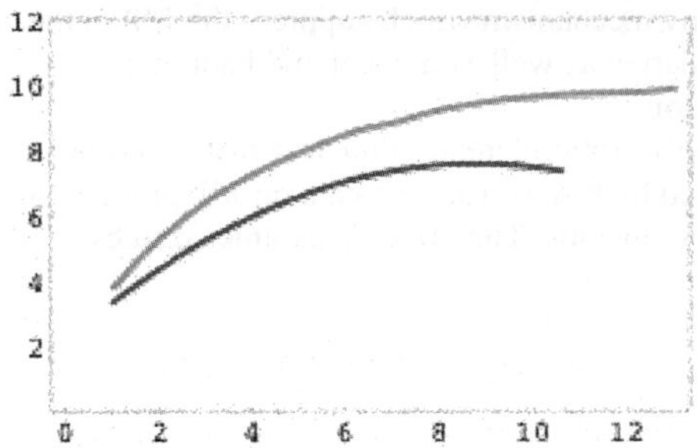

Fig. 2.2. The average longitudinal lenght L vs the centrality variable, l, for Pb (red) and for In. Note the flattening of L around the maximum value of l, correponding to central collisions.

Using further the geometrical relation: $l=2R-b$, we express L in terms of l, thus obtaining the total attenuation factor as function of the fireball transverse size l, in the form:

$$A_{Tot} = N \cdot \exp[-\rho_0 \sigma_{nucl.abs.} L(l)] \cdot \exp[-\sum_i < \rho_i \sigma_i >_T \frac{3}{8} l] \tag{2.1}$$

The absorption lenght in the fireball is quite comparable to $\lambda_{nucl.abs.} \sim 14$ fm and the absorption increases quite strongly with temperature. We find that the contributions of vector mesons are very important, due to the low threshold of the cross-sections and to their large multiplicity. Including some other resonance like e.g. $a_1(1230)$ should not make much of a difference, since they do not present further advantages and are unfavoured by the higher mass.

3. Results for the hadron and Hagedorn gases

We fit the NA50, S-U and Pb-Pb data for l <5 fm with the previous formula, including fireball absorption as function of temperature, T, as well as nuclear absorption. We find good fit for the range: 165 MeV< T<185 MeV, see Fig. 3.1.

Fig. 3.1. J/Y production normalized to Drell-Yan muon pairs vs the centrality variable *l*.NA 50 data on Pb-Pb (boxes) and S-U (stars) collisions. Superimposed the nuclear absorption curve (red) and nuclear+ fireball absorption (blue), without geometrical, centrality dependent, effects (see text).

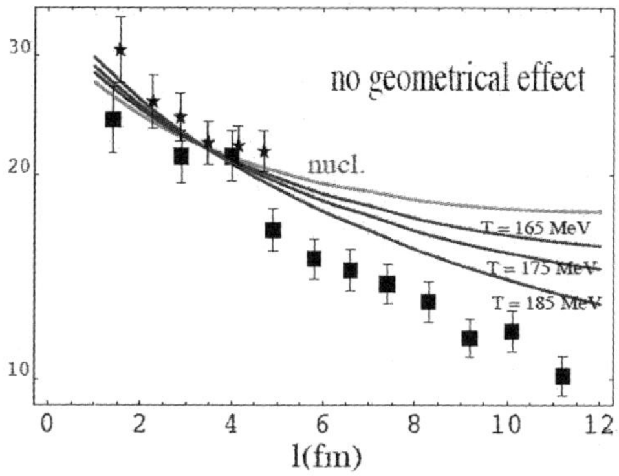

It is quite encouraging that a microscopic calculation produces a temperature which is quite consistent with everything we know about the not-so-central collisions. The fit becomes worse and worse with increasing centrality.

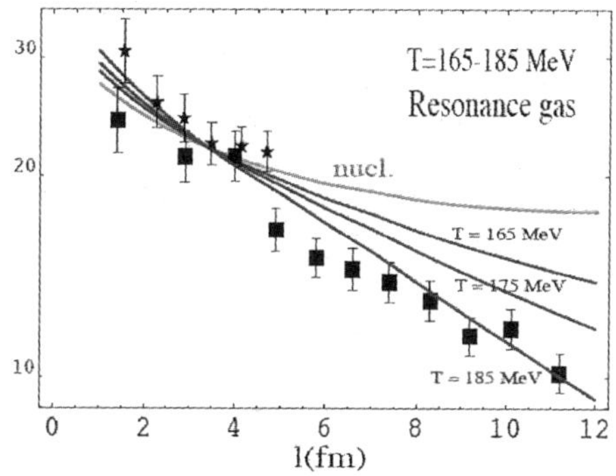

Fig. 3.2. Same as in Fig. 3. with geometrical effects taken into account.

However, we should take into account that the energy density deposited in the fireball increases with centrality. Thus we may expect that T increases with centrality and, correspondingly, the opacity becomes larger. To estimate this effect, we:

- allow an increase in energy density from l-5 fm onwards, by scaling with the factor g(*b*)=g(2*R*-*l*) introduced in Lecture 1, Sect. 1;
- use the energy density-temperature relation of the hadron gas to derive the temperature appropriate to *l* to compute the opacity.

The procedure introduces in Eq. (2.1) a further dependence from *l*, via the temperature of the thermal bath. We find curves much closer to the data, but still a marginal fit not reproducing, in particular, the discontinuity that data show around *l* =5. Even worse, the temperatures that are required at small centrality are quite large: if we start with T = 185 at *l* = 5 fm, we need to go as high as T = 205 MeV at *l* = 12 fm. It is unlikely that a hadron gas can reach these temperatures.

To analyse the situation further, we introduce the concept of a limiting temperature. Instead of using a pseudoscalar- vector Meson gas we fix the temperature for *l* < 5fm at T = 175 MeV and extrapolate to higher centrality with the above procedure applied ***to a full Hagedorn gas with T= 180 MeV***. The result is shown in Fig. 3.3. Corrections due

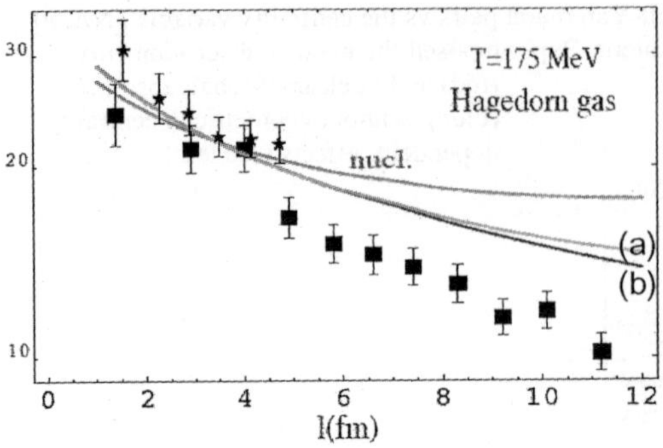

to increasing energy density with centrality are not accounted in (a) and introduced in (b).

Fig. 3.3. Same as in previous figures, assuming the fireball to be a Hagedorn gas with T_H=180 MeV. Initial temperature is fixed at T=175 MeV to fit the data for *l*<5fm.

The sharp rise of degrees of freedom near the Hagedorn temperature makes so that the temperature does not rise appreciably with centrality, the dissociation curve cannot become harder and the predicted absorption falls short from explaining the drop observed by NA50.

To appreciate better the significance of the effect, we plot in Fig. 3.4 the ratio Observed/Expected vs. *l* for the cases: (a) T(l < 5fm) =175 MeV; (b) T(l < 5fm) =185 MeV; (c) T(l < 5fm) =175 MeV, Hagedorn gas with T_H = 180 MeV.

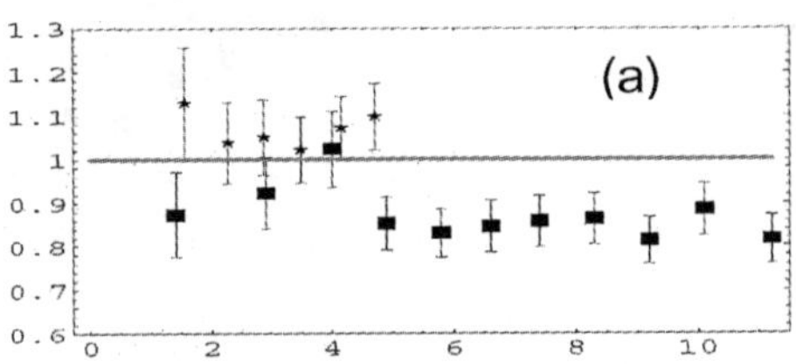

T(*l*<5fm) = 175 MeV
T(*l*=12fm)>190 MeV

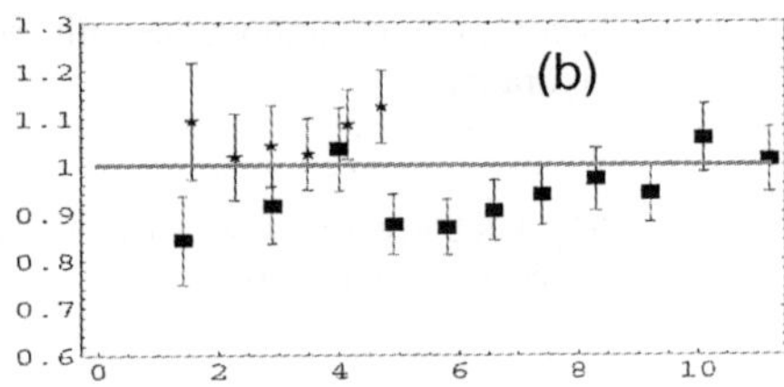

T(*l*<5fm) = 185 MeV
T(*l*=12fm)>200 MeV

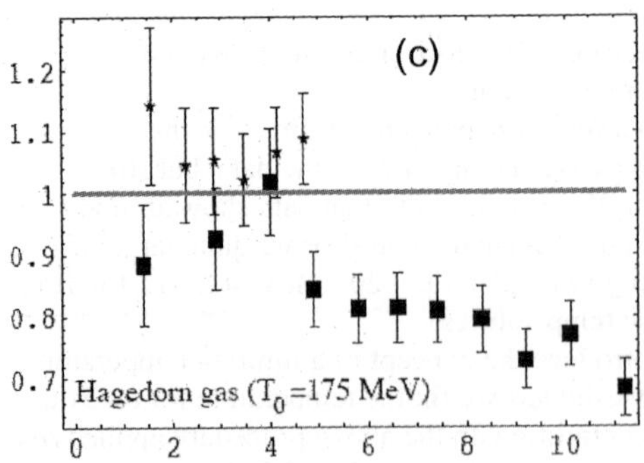

T(*l*<5fm) = 175 MeV
Hagedorn gas, T_H=180 MeV

Fig. 3.4. Ratio: Observed/Expected for J/Ψ production normalized to Drell-Yan muon pairs vs the centrality variable *l*. NA 50 data on Pb-Pb (boxes) and S-U (stars) collisions. Figs. (a) and (b): fireball is a pseudoscalar + vector meson gas, with initial temperatures (for *l*<5fm) as indicated; also indicated the temperatures reached at *l*=12fm; (c) Hagedorn gas with T_H=180 MeV, initial temperature T=175 MeV.

Some comments. The curve shown in Fig. 3.4 (c) embodies the limiting absorption from a hadron gas, anything harder could be due to the dissociation of the J/Ψ in the quark-gluon plasma phase. We must stress, however, that J/Ψ dissociation due to higher resonances than vector mesons has been neglected. The decreasing couplings of the higher resonances could eventually resum up to a significant effect, which would change the picture. Something similar happens e.g. in deep inelastic scattering, where the cross section due to any individual resonance falls down because of its form factor, but the total cross section is mantained by the new resonances appearing at larger values of the energy. However, the underlying reason for this is that we are entering a regime in which the phenomenon is described by a new picture, namely the scattering off elementary partons. In our case, this would mean going over to a description where charmonium dissociation is due to the interaction with quarks and gluons, which is precisely the signal of the deconfined plasma, above the Hagedorn temperature.

4. A bold speculation

Going back to Fig. 5.4 in Lecture 1, it is tempting to interpret the rise of ε/T^4 as due to the excitation of more and more hadron resonances, as it happens in the Hagedorn gas. The transition starts when the number of effective degrees of freedom in the hadron gas equals approximately the degrees of freedom of a gas of quarks and gluons. For a Hagedorn gas with T_H=180 MeV, the transition temperature and energy density are numerically found to be:

$$\varepsilon(T)/T^4 \approx 12 \Leftrightarrow T = T_{trans} = 168\,MeV;\ldots\varepsilon(T_{trans}) = 2\,GeV/fm^3 \qquad (4.1)$$

The value of the energy density agrees with what can be estimated from the Bjorken formula, Lecture 1. Encouraged by this, we may assume, as a bold hypothesis, that Pb-Pb collisions at l = 5 fm produce a fireball with exactly ε = 2 GeV/fm^3. For l > 5fm we may trasform the lenght scale in Fig. 3.4 (c) into an energy density scale, by scalig the energy density with the geometrical factor g(b).

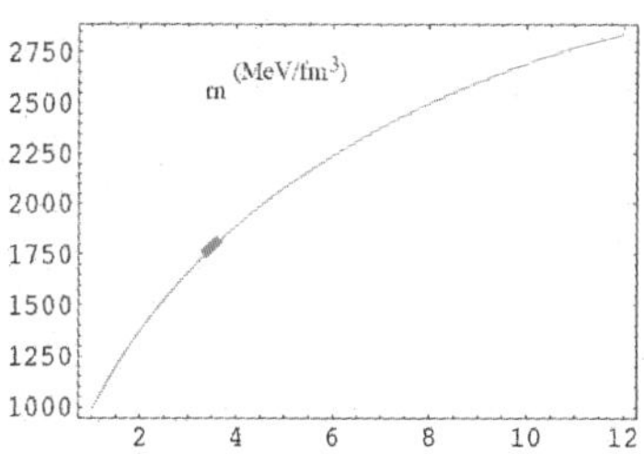

Fig. 4.1. Energy density, according to the Bjorken formula, vs. the centrality variable l, as obtained from the geometrical function g(b). Normalization is given by the assumption that ε=2 GeV/fm^3 for l = 5 fm.

Recall that g(b) embodies the increase of the nucleon surface density with increasing centrality. Fig. 4.1 gives the l-ε relation implied by g(b) and Fig. 4.2 gives the result of the recalibration: the ratio Observed/Expected for J/Ψ given as a function of the energy density. For l>5 fm we may also transform energy densities into temperatures, using the relation: $T \approx (\varepsilon/12)^{1/4}$, approximately valid above the transition point. We have indicated in Fig. 4.2 the temperatures at which Ψ' and χ states melt, according to the results presented in Sect. 6 of Lecture 1. The reduction of 20-30% in the expected J/Ψ yield is quite compatible with the disappearence of these states.

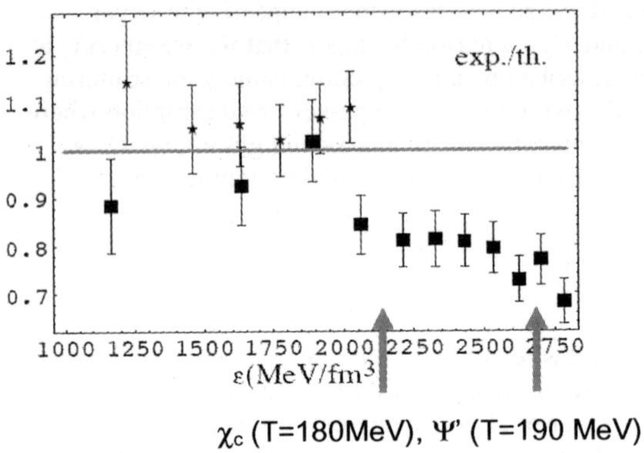

Fig. 4.2. The ratio: Observed/Expected for J/Ψ production normalized to Drell-Yan muon pairs vs the energy density ε. NA 50 data on Pb-Pb (boxes) and S-U (stars) collisions.
The fireball is assumed to be a Hagedorn gas with initial temperature T=175MeV and T_H=180 MeV. Arrows indicate the temperatures where χ_c and Ψ' melt, according to the Table in Lecture 1, sect. 6.

5. J/Ψ as a probe of QGP: some conclusions

When the idea was proposed, it was believed that J/Ψ would suffer little absorption from nuclear matter and from the "comoving particles" ($\sigma<1$ mb), hence that there would be very little background to the QGP signal;

Nuclear absorption measured from p-A cross sections (but uncertainties still remain!) is instead ~ 4-5 mb, with a corresponding attenuation lenght ~ 14 fm and therefore a signal-to-noise ratio ~ 1. Absorption cross sections by comoving particles point also to the few mb range, as has been estimated with several methods involving rather different approximations.

In the work I have described in these Lectures, we have made a new calculation of the J/Ψ dissociation cross-sections by pseudoscalar and vector mesons in a reliable model tested in other processes (the Quark Constituent Model) and we have applied the results to a hadron gas made of these particles.

The effects of comovers are (i) non negligible and (ii) strongly temperature dependent.

If we allow temperatures in excess of 200 MeV in this hadron gas, we can fit the NA50 results, albeit marginally.

On the other hand, if we accept that there is a limiting temperature to the hadronic phase, estimated to be 170-180 MeV from independent sources (hadron spectrum, lattice QCD calculations) comovers ***cannot explain*** the drop in J/Ψ production seen at large centralities by NA50.

The picture that QGP sets in at centrality $l \sim 5$ fm is consistent with known temperature and energy density ranges. The observed drop in J/Ψ would be due first to χ_c and, later, to Ψ' melting.

The limiting temperature being a quite reasonable assumption, we conclude that the SPS has most likely seen the QGP. It would be interesting to correlate the J/Ψ signal to the other hint of new phase seen at the SPS, namely the increase in strangeness production. This has not been done, thus far, but work is in progress.

RHIC data on J/Ψ would be extremely valuable, to check the signal against other signatures and so would the beauty signal. LHC data are eagerly wanted !

The study of charmonia in QGP is not concluded with the demonstration that QGP exists. The level spectrum vs. T could give interesting information on the dynamics of quark and gluons, on the non-perturbative corrections to the picture of a free quark gluon gas and it would probe deeply the new phase of matter.

REFERENCES

[i] T. Matsui and H. Satz, Phys. Lett. **B178**, 416 (1986).

[ii] For a review see e.g. R. Vogt, Phys. Rep. **C 310** (1999) 197.

[iii] L. Maiani, F. Piccinini, A.D. Polosa, V. Riquer, Nucl. Phys. **A 741** (2004) 273; **A 748** (2005) 209.

[iv] For a recent and extended review see: U. Heinz, Concepts of Heavy-Ion Physics, lecture Notes, hep-ph/0407360.

[v] M.C. Abreu et al., Phys. Lett. **B450**, 456 (1999); M.C. Abreu et al., Phys. Lett. **B477**, 28 (2000). Latest analysis: http://na50.web.cern.ch/NA50.

[vi] J. D. Bjorken, Phys. Rev. **D27** (1983) 140.

[vii] See e.g. J. L. Goity and H. Leutwyler, Phys.Lett.B228 (1989) 517.

[viii] E791 Collaboration (E.M. Aitala et al.) Phys. Rev. Lett. **89** (2002)121801.

[ix] SPS data have been analysed recently in F. Becattini, A. Keranen, L. Ferroni, T. Gabbriellini, nucl-th/0507039; for RHIC data see P. Braun-Munzinger, D. Magestro, K. Redlich , J. Stachel Phys. Lett. **B518** (2001) 41.

[x] J. Letessier, J. Rafelski, Hadrons and Quark–Gluon Plasma, Cambridge Monogr. Part. Phys. Nucl. Phys. Cosmol.,vol.18, 2002.

[xi] N. Cabibbo, G. Parisi, Phys. Lett. **B59** (1975) 67.

[xii] F. Karsch, Lattice QCD at High Temperature and Density, hep-lat/0106019.

[xiii] M. Gyulassy, in these Proceedings.

[xiv] T.Barnes, Charmonium Cross Sections and the QGP, nucl-th/0306031.

[xv] A. Deandrea, N. Di Bartolomeo, R. Gatto, G. Nardulli and A.D. Polosa, Phys. Rev. **D58** (1998) 034004 (1998); See also A.D. Polosa,``The CQM Model", Riv. Nuovo Cim. Vol. 23, N. 11 (2000).

[xvi] A. Deandrea, R. Gatto, G. Nardulli, N.A. Tornqvist, Phys. Lett. **B502** (2001) 79.

[xvii] D. Ebert, T. Feldmann, R. Fredrich, H. Reinhardt, Nucl. Phys B434 (1995) 619

[xviii] see the second paper quoted in [3].

[xix] B. Alessandro et al., NA50 Collaboration, Nucl. Phys. **A715** (2003) 679c.

CHAIRMAN: Luciano Maiani

Scientific Secretaries: Verónica Riquer and Diego Casadei

DISCUSSION I

— *Sevil Salur*:
The statistical models do not describe resonances and most of the stable particles that you observe are actually coming from resonances. How can you be sure that the fireball is really thermalized?

— *Luciano Maiani*:
There are actually two different questions. The first is whether you can describe the fireball as something in thermal equilibrium, and this depends on the density and time scale—particles need to have several interactions to reach the equilibrium—and on the many degrees of freedom of the fireball. The second is about the nature of the fireball: we consider it a gas of free pions and we add resonances to represent interactions. In the last stages of the evolution, when the particles that you see are produced, there may be significant distortions to the equilibrium distribution, resonances that decay are no more reformed, etc.

— *Miklos Gyulassy*:
The measurements of resonances like K*, Δ, etc.—that Sevil says in disagreement with thermodynamic models—are actually a different thing from the resonances that are included in the fireball to represent interactions: we cannot see them directly. What we measure is instead the result of the hadronization, at later stages.

—*Gerardus 't Hooft*:
I think that this phase transition of the QGP has not to be a sharp transition, but it could be gradual.

— *Luciano Maiani*:
This is a real possibility. It could not even be a phase transition in a classical sense. In pure QCD, there is clear evidence of a 1st order transition, but the situation is much less clear when one introduces quarks.

— *Richard D. Kenway*:
The current issue in lattice QCD is whether the transition is really a second order transition, a first order process or a cross-over. The deconfinement transition does actually depend on the strange quark mass, and currently we are still not able to carry on simulations at the physical strange quark mass. Also, it is not clear what the order parameter for a transition would be, whether it is the Polyakov line or chiral condensate.

— *Christian P. Korthals-Altes*:
Strictly speaking, an order parameter is only defined if there is a symmetry that is going to be spontaneously broken. The Polyakov loop is the order parameter only in the infinite mass limit for quarks. On the other hand, you can have zero masses and chiral symmetry, and in this case $\langle \bar{q}q \rangle$ is an order parameter. But unfortunately we are in the middle, so that strictly speaking we have no order parameter.

— *Luciano Maiani*:
On the lattice, it appears that $\langle \bar{q}q \rangle$ is really an order parameter, even for $m_q > 0$.

— *Christian Klein-Bösing*:

Thermal models are used also to describe e^+-e^- collisions, in addition to p-p or A-A collisions. Do you think they are "universal"?

—*Luciano Maiani*:

It's very difficult to say *a priori* if the system will go to thermal equilibrium or not, but it's a fact that you can describe the data in this way.

— *Miklos Gyulassy*:

I'd like to remark that in e^+-e^- and p-p collisions there is a fugacity parameter that can be adjusted to fit the data, whereas in ion-ion collisions at SPS there is no need for this parameter. It is one of the most important observations at SPS that the strangeness degrees of freedom are fully saturated.

—*Antonino Zichichi*:

The QGP is a very exciting new state of matter. Do you think that the thermodynamic description is a good description, even though this state is very different from the systems that are usually described by thermodynamic laws?

—*Luciano Maiani*:

Well, this question belongs to the discussion we had on Monday, that the upper level (thermodynamics) depends very little on the detailed aspects of the lower level. Any system with many degrees of freedom and whose constituents can exchange energy can be described by statistical thermodynamics. Actually, in the case of the QGP, things should be even simpler because of asymptotic freedom, which assures that the QGP at very high energies behaves like a free particle gas.

CHAIRMAN: Luciano Maiani

Scientific Secretaries: Verónica Riquer and Diego Casadei

DISCUSSION II

— *Diego Casadei:*

We just learned that there is evidence in SPS data of the interaction between the J/Ψ with the fireball above a critical temperature or—equivalently—a critical density. In addition, data from RHIC are coming. What are the properties of the QGP that you expect that will be understood first?

— *Luciano Maiani:*

We know very well the potential of a charm or beauty pair in the vacuum and we can compute the levels with great precision. The knowledge of level spectrum variation with temperature would give important clues on the dynamics of the medium. We have hints of variation of particle properties because of the medium, like the effects on the rho or Delta masses, but level melting would be the first really clear case.

— *Sevil Salur:*

$T = 200$ MeV is needed to describe the attenuation of the J/Ψ data. Can we say that it is the critical temperature?

— *Luciano Maiani:*

200 MeV is the temperature needed to fit the opacity at large centralities with a simple hadron gas. However, It is unlikely that hadrons can reach this temperature. The fact that the fitted temperature from J/Ψ production looks higher than what it is "expected" implies that the suppression of the J/Ψ really comes from another mechanism, that is most likely the QGP.

— *Christian Klein-Bösing:*

How can you compute the fireball energy density?

— *Luciano Maiani:*

One way is to use the Bjorken formula for the different centralities. The alternative way is to fit T from the opacity, then get ε from T via thermodynamics, assuming a hadron gas, or a Hagedorn gas. This should work at low centralities. From ℓ=5fm onwards, we scale ε with the baryon surface density. The fact that this second method gives values consistent with the first is quite encouraging.

— *Tiehui Ted Liu:*

[With respect to a plot shown at the end of the 2nd lecture.] How can Ψ' melt after χ_c? One would naïvely expect that Ψ' will melt first, then χ_c and finally the J/Ψ.

—*Luciano Maiani:*

I am not an expert in these calculations and I was also surprised to see this counterintuitive ordering. It may have to do with small effects related to spin-spin interaction. You should consult the original papers.

— *Tiehui Ted Liu:*

The Ψ"(3.77 GeV) normally decays into open charm (D D-bar), having a mass slightly above the threshold. As temperature increases, its mass would start decreasing, possibly reaching the threshold (i.e. it would decrease faster than the D D-bar threshold). In this case, its decay would not be possible, thus increasing the J/Ψ yield.

—*Luciano Maiani*:

This is a good question but I am not expert enough to reply quantitatively. However, soon after going below threshold, the Ψ" becomes unbound so the effect, if any, should disappear very soon. Again, you should consult the specific literature.

—*Elena Bruna*:

What do you mean by "absorption factor by comoving particles"?

—*Luciano Maiani*:

I mean absorption by the particles in the fireball at zero rapidity in the center of mass frame. In the laboratory system, these particles move with the same speed as the projectile nucleus, hence the name of "commoving particles".

—*Luisa Cifarelli*:

Together with heavy-flavoured quarkonia, don't you think that open states will bring a lot of information as far as quarkonia suppression is concerned?

—*Luciano Maiani*:

Of course, open charm production can give a lot of information. However, to realte it to quarkonium suppression is a difficult matter. In principle, a measurement of the production of hidden and open charm should show that their sum doesn't change. However, the production of charmonium is much less than open charm, so as to make it unlikely that a variation of the former reflects in a measurable variation of the latter.

—*Miklos Gyulassy*:

How well is the cold nucleus "normal" absorption curve known? It seems that the curve shown in your plots overestimates somewhat the low-energy data, and there is no indication of the error size. It is very important to make "reference" measurements of the nuclear absorption curve, for example using proton-nucleus collisions. This would be extremely useful for the interpretation of both SPS and RHIC data.

—*Luciano Maiani*:

I fully agree with you. Measurements of nuclear absorption could be done at a relatively low cost and we should urge the SPS and RHIC committees to consider recommending it.

THE BLACK HOLE INFORMATION PROBLEM

Gerard 't Hooft

Institute for Theoretical Physics
Utrecht University
and

Spinoza Institute
Postbox 80.195
3508 TD Utrecht, the Netherlands

e-mail: g.thooft@phys.uu.nl
internet: http://www.phys.uu.nl/~thooft/

Astract

A resumé is presented on the issue of information and information loss during the process of black hole formation and evaporation. An important question is whether fundamental interactions at the Planck scale can be described by an exactly unitary scattering matrix or not. According to Superstring Theory, the so-called holographic principle holds for black holes, and this supports the view that unitarity is preserved. In a recent conference, S. Hawking announced that he modified his position on this issue so that he, too, adheres to a unitary theory. This author agrees with the conclusion, but not with the arguments.

1. Introduction: The black hole horizon

The Planck scale is defined by the units

$$\begin{aligned} L^{\text{Planck}} &= \sqrt{\hbar\, G_N/c^3} &= 1.616 \times 10^{-33}\ \text{cm} \\ T^{\text{Planck}} &= \sqrt{\hbar\, G_N/c^5} &= 5.39 \times 10^{-44}\ \text{sec} \\ M^{\text{Planck}} &= \sqrt{\hbar\, c/G_N} &= 21.8\ \mu\text{g}\ . \end{aligned} \tag{1.1}$$

No single theory exists, as yet, to explain the local nature of the Laws of Physics at the Planck scale. This does not stop theoreticians from speculating about such a theory. What will the basic ingredients of matter be? Slightly below the Planck scale, one often assumes, as Superstring Theory does, that the basic ingredients are strings, to be complemented by branes of higher dimensionality, whose dynamics is controlled by that of the surrounding strings. But, since the gravitational force is fundamentally unstable against collapse, one also expects black holes, or black hole like objects, which must be almost pointlike (since the dimensions of their horizons will be not much larger than the Planck size, L^{Planck}). These black holes will be much heavier than the lower energy modes of the strings, and, in the limit where their masses are much greater than the Planck mass, M^{Planck}, their behaviour should be well-described by the conventional geometrical expressions from General Relativity. Will they form a discrete spectrum, like all conventional elementary particles do, or perhaps a continuum, as seems to be implied by General Relativity? It is worth-while to investigate the restrictions imposed by purely logical reasoning.

The first part of the lecture presented at the Erice School of 2004 was a resumé of the discussion presented at Erice in 2003, and we refer to the proceedings of that School[1] and Ref. [2] for further details. In our summary, of the four coordinates $(r,\ \theta,\ \varphi,\ t)$ for a black hole space-time, we suppress the angular coordinates θ and φ. Then, we perform a conformal transformation in the $(r,\ t)$ space, such that radial light follows trajectories in the $\pm 45^\circ$ direction. This leads us to the *Penrose diagram*, Fig. 1, a conformal two-dimensional representation of space-time.

The black hole horizon is indicated in the Figure. Matter trajectories are indicated by arrows. We see that part of the initial matter disappears behind the horizon, to be absorbed by the black hole singularity, and parts of it may escape to infinity. The part that does escape to infinity turns into an indefinite amount of radiation called Hawking radiation. In an approximation where

- the back reaction of the surrounding space time metric is neglected, and
- field equations are all assumed to be linear, so that interactions between in- and out going radiation may be ignored,

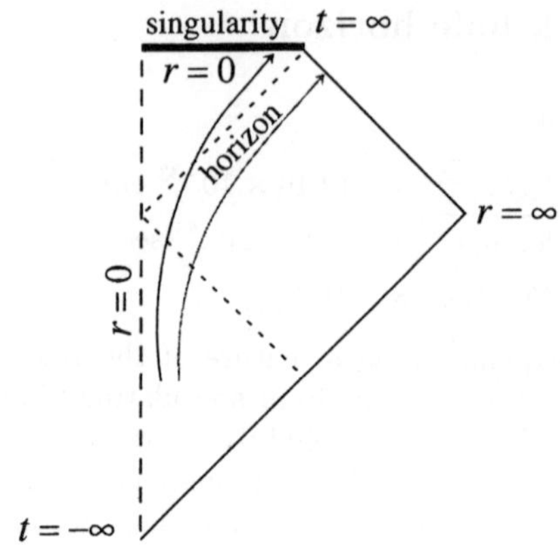

Figure 1: Penrose diagram for the classical black hole. The angles θ and φ are suppressed.

one finds that in-going and out-going states are not related by a scattering matrix.

In last year's lecture, however, it was explained that these assumptions cannot be valid, even approximately, and the existence of a unitary scattering matrix was considered to be inevitable. Its consequences were studied in detail. We found that, first of all, there is no fundamental disagreement with the starting points of General Relativity, *provided* that one takes gravitational interactions into account between in-going and out-going matter near the black hole horizon. One also has to assume that the out-going radiation is not exactly thermal; this is surprising since a background field calculation appears to generate a perfectly thermal density matrix, without any room for non-thermal corrections[1].

Indeed, experts in General Relativity often stress the fact that, geometrically, black holes act as genuine sinks of data for wave functions, and in doing so, they should turn pure quantum mechanical wave functions into mixed states[3]. If you want to describe entire black holes, you should work with equations for the density matrix

$$\varrho = \sum_i W_i |\psi_i\rangle\langle\psi_i| \ , \tag{1.2}$$

where W_i are probabilities, not with Schrödinger equations for pure wave functions $|\psi\rangle$. With a little effort, one can write down simple models for density matrices where in-

[1]Non-thermal corrections have often been suggested, and there will definitely be thermal *isolation* due to barriers such as the centrifugal barrier, but once it is put inside a heat bath, General Relativity would predict that a black hole reaches perfect thermal equilibrium, at exactly the Hawking temperature.

deed mixing (in practice: decoherence) occurs, while essential conservation laws such as conservation of energy, are maintained.

We claim, however, that there is a severe difficulty with such a view. Even though energy is conserved in simple models, it cannot be conserved in field theories where decoherence takes place, if this decoherence, or mixing between the wave functions $|\psi_i\rangle$ inside the density matrix, occurs via *localized* evolution equations. We see this as follows. The time evolution of a density matrix in a theory without decoherence is exactly as in a product Hilbert space of two universes, one spanned by the kets $|\psi\rangle$ and one by the bras $\langle\psi|$. The total Hamiltonian H is then given by

$$H = H_I - H_{II} \ . \tag{1.3}$$

Decoherence would then be described by an interaction Hamiltonian $H_{\rm int}$ between these two universes. Due to this interaction, energy would be exchanged, but this would lead to a rise of the energy H_I as well as H_{II}, since only their *difference* is protected by conservation of the total Hamiltonian (1.3), due to time translation invariance. We now insert a statistical argument: phase space of the higher energy states $|E_i\rangle\langle E_i|$ is very large, while we have only a single vacuum state $|\Omega\rangle\langle\Omega|$. Therefore transitions towards higher energy states would occur frequently while returns to the vacuum would be suppressed. Such a universe would therefore heat up spontaneously; even the vacuum state would be unstable.[2] This is why quantum field theories based on evolution equations for the density matrix (1.2) that are more generic than (1.3) could never be constructed in practice.

2. The Brick Wall and the S-matrix

The assumption that black holes, just like any other forms of matter, are described by a scattering matrix acting on pure states, has far-reaching consequences. The first consequence is that the thermodynamic *entropy*, to be derived directly from the energy-temperature relation, reflects the density of quantum states. These states are evenly and universally smeared across the black hole horizon, as if the horizon possesses a dynamical system of its own. The statistical features of this system are neatly reproduced if we introduce a *cut-off* at the horizon, usually referred to as a "brick wall"[4]: close to the horizon, at a distance of the order of the Planck length, incoming waves may be assumed to be reflected rather than transmitted. Upon closer examination, this reflection process must violate all global conservation laws that arise from continuous symmetries, such as

[2]A complication in this argument is that, in the density matrix (1.2), conservation of probability implies conservation of $\sum_i W_i$, not that of $\sum_i W_i^2$, so that $H_{\rm int}$ is not hermitean; but even so, we are deprived of a definition of an inert vacuum state.

baryon number conservation or isospin. This is surprising, not only because General Relativity does not seem to allow for the existence of any such wall, but also because General Relativity is usually not assumed to interfere with symmetries among particles.

One can do — a bit — better. Of all interactions between ingoing and outgoing beams of particles, it is the gravitational one that has the most drastic effects. It diverges on the horizon! The calculation has been done in detail in [2], and can be summarized as follows. If, in the coordinate frame of the Penrose diagram of Fig. 1, an ingoing particles has a momentum distribution $p^\mu(\tilde{x})$, where $\tilde{x}$ stands for the transverse, angular, coordinates, then one of the effects on the outgoing particles is a tiny displacement $\delta y^\mu(\tilde{x})$, related to p^μ by an equation of the form[3]

$$\tilde{\partial}^2 \delta y^\mu(\tilde{x}) = -8\pi G p^\mu(\tilde{x}) \,, \tag{2.1}$$

where G is Newton's constant. We picture this situation in Fig. 2.

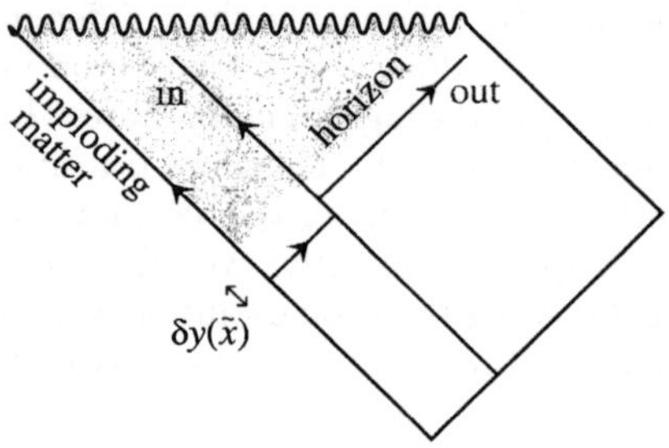

Figure 2: Effect of ingoing matter on an outgoing test particle.

Now let us *assume* that a scattering matrix S exists that relates the complete set of out-states to the complete set of possible in-states. Consider one element $\langle\text{out}|\text{in}\rangle$ of this S-matrix. Perform a very slight modification of the state $|\text{in}\rangle$ by adding an ingoing particle with momentum δp^{in}. Then all outgoing states are shifted by the same amount δy^{out}. This is effected by the shift operator

$$\exp\left(i\int \mathrm{d}^2\tilde{x}\, P^{\text{out}}_\mu(\tilde{x})\delta y^\mu(\tilde{x})\right) . \tag{2.2}$$

Therefore, the new out-state is

$$\exp\left(i\int \mathrm{d}^2\tilde{x}\int \mathrm{d}^2\tilde{x}'\, P_\mu(\tilde{x})G(\tilde{x}-\tilde{x}')\delta p^\mu(\tilde{x}')\right)|\text{out}\rangle \,, \tag{2.3}$$

[3]This equation was slightly simplified, valid as such for flat transverse coordinates $\tilde{x}$. For angular coordinates there is an additional curvature term.

where $G(\tilde{x}-\tilde{x}')$ is the Green function for Eq. (2.1). We can suppress the Lorentz index μ here, because only the components in one light cone direction contribute. Repeating this for many new particle insertions, we find that the total scattering matrix will be

$$\begin{aligned}&\langle\text{out}|\text{in}\rangle = \\ &\mathcal{N}\exp\Big(i\int \mathrm{d}^2\tilde{x}\int \mathrm{d}^2\tilde{x}'\, P^{\text{out}}(\tilde{x})G(\tilde{x}-\tilde{x}')P^{\text{in}}(\tilde{x}')\Big) \;=\; \int \mathcal{D}X^{\text{out}}(\tilde{x})\int \mathcal{D}X^{\text{in}}(\tilde{x}) \\ &\exp\Big(\int \mathrm{d}^2\tilde{x}\big[-i\tilde{\partial}X^{\text{out}}\tilde{\partial}X^{\text{in}} + iX^{\text{in}}P^{\text{in}} + iX^{\text{out}}P^{\text{out}}\big]\Big)\,, \end{aligned} \tag{2.4}$$

where $\mathcal{N}$ is a total normalization factor. We find the surprizing solution that this S-matrix is indeed unitary provided that we choose the normalization $\mathcal{N}$ correctly, and provided that the in- and out-states are *specified entirely* by their longitudinal momentum distributions $p(\tilde{x})$. In fact, the functional integral expression in Eq. (2.4) looks suspiciously much like a string theory amplitude, *even though we never assumed string theory!*.

3. Holography

The functional integral expression (2.4) appears to describe a dynamical system on the horizon, yet it entirely includes the interactions outside the black hole. It seems as if there is a layer of dynamical degrees of freedom with the thickness of a few Planck lengths, stretching all over the horizon. This was the starting point of the idea of 'holography', the fact that a dynamical system in two spacelike dimensions can account for the dynamics in a three dimensional world.

We could repeat the calculation for the case that in- and outgoing particles have electric charge in addition to their gravitational interactions. One then finds that ingoing particles not only shifts the outgoing ones, but also gauge rotates them. The gauge rotation is exactly what one would expect if electro-magnetism is regarded as gravity in a fifth, Kaluza-Klein dimension. Thus, besides the variables $X(\tilde{x})$, one gets variables $\Lambda(\tilde{x})$. This prompted us to formulate a more general conjecture:

> Every 'gauge' field ($g_{\mu\nu}(\vec{x},t)$ or $A_\mu(\vec{x},t)$) in the bulk theory, generates a field on the horizon with one Lorentz index less, corresponding to the generators of local the gauge transformations ($X^\mu(\tilde{x})$ or $\Lambda(\tilde{x})$ in the above examples). Thus, a non-Abelian gauge theory in the bulk, with gauge generators in some group $U(\vec{x},t)$, is expected to generate degrees of freedom on the horizon of the form $U(\tilde{x})$. Gravitinos in the bulk are expected to generate spin $\frac{1}{2}$ fermions on the horizon.

What effect does a scalar field theory in the bulk have on the horizon? This we could figure out by noticing that the Higgs mechanism in the bulk generates massive Λ fields on the horizon. This mass breaks explicitly local gauge invariance on the horizon. We concluded that, therefore, scalar fields in the bulk generate non-dynamical symmetry breaking interactions on the horizon. These were called *schizons* in the old days of model building. So, scalar fields in the bulk generate schizons on the horizon. Indeed, since scalar fields are Lorentz-invariant, they have fixed values on the horizon, not dependent on the Rindler or Schwarzschild time parameter. So indeed they break symmetries there.

This mapping between 3+1 dimensional theories onto two-dimensional systems on a horizon is called 'holography', and somewhat modified versions of it are being used in recent approaches involving String Theory. There is an apparent paradox associated with holography. If we move together with a freely falling observer, we see a world of particles that should be mapped onto the two-dimensional surface that plays the role of a black hole horizon. However, this surface can be anywhere, because General Relativity does not allow us to single out a "preferred" coordinate frame that singles out any particular two-dimensional surface as being more special than others. On the other hand, we expect that, even on a black hole horizon, there should be some version of *locality* in its quantum structure. But how can there be locality on all these different possible surfaces? We don't we experience our world all together as two-dimensional, and how can locality in 3+1 dimensions be interpreted in terms of the variables on the horizon?

These questions, though perhaps not so well-put, cannot easily be dismissed as being rhetoric, or 'irrelevant, as long as the theory works'. In a world where the concept of locality is not well-defined, we cannot expect any useful form of physical law, and I am convinced that restoration of locality is a must in our journey towards a better understanding of Nature. This alley of reasoning led this author to conjecture that we should be able to identify a dynamical theory underlying Quantum Mechanics that is based on more conventional logic[5]. In that case, the apparent conflict disappears that arises when we try to reconcile counting of quantum states with strict locality in three plus one dimensions. Thus, issues concerning the very foundations of Quantum Mechanics itself may well be linked to questions concerning Quantum Gravity and Black Holes.

4. Black holes and the topology of space-time

Quantum mechanical amplitudes can be defined in terms of functional integrals, and it often has turned out to be useful to perform the so-called Wick rotation in these amplitudes. The Wick rotation is associated to the analytic extrapolation of time-dependent amplitudes to imaginary time. Thus, instead of matrix elements of the form

$$\langle U(t)\rangle = \langle\psi_1|\, e^{-iHt}|\psi_2\rangle \ , \tag{4.1}$$

one may consider

$$\langle U(-i\beta)\rangle = \langle\psi_1|\, e^{-\beta H}|\psi_2\rangle \ . \tag{4.2}$$

In ordinary quantum field theories, these expressions are better-behaved than the real-time amplitudes, because the higher-energy contributions are strongly suppressed. This is reflected in the functional integral expressions. If a real-time amplitude is written as

$$\langle U(t)\rangle = \int \mathcal{D}\phi(\vec{x},t)\, e^{i\int \mathcal{L}(\vec{x},t)\mathrm{d}^3\vec{x}\mathrm{d}t} \ , \tag{4.3}$$

one finds in quantum field theories,

$$\langle U(-i\beta)\rangle = \int \mathcal{D}\phi(\vec{x},x^4)\, e^S \ ; \qquad S = \int \mathrm{d}^4x\, \mathcal{L}(x) \ , \tag{4.4}$$

where $\mathcal{L}(x)$ is defined for Euclidean values for x and it is bounded from above, such as

$$\mathcal{L} = -\tfrac{1}{2}(D_\mu\varphi)^2 - V(\varphi) \ , \qquad \text{and} \qquad \mathcal{L}(x) = -\tfrac{1}{4}F_{\mu\nu}(x)F_{\mu\nu}(x) \ , \tag{4.5}$$

for scalar fields $\varphi(x)$ and vector fields $A_\mu(x)$, respectively. The procedure is even more useful when applied to systems at finite temperature (see for instance Refs. [6][7][8]). Then, one may substitute $\beta = 1/kT$, where T is the temperature, and Eq. (4.2) gives us directly the Boltzmann weights, so that $\langle U(-i\beta)\rangle$ is an expression for the free energy.

In Quantum Gravity, one often tries to do the same thing [9]. A fundamental problem is the encountered: the Einstein-Hilbert action for the gravitational interactions,

$$S = \frac{1}{16\pi G}\int \sqrt{g}R \ , \tag{4.6}$$

is not bounded from below. Since this leads to a serious divergence already at the leading order in perturbative Quantum Gravity, one can look at the perturbative theory to find both the cause and the remedy of this problem[10]. The cause of this difficulty lies in the conformal component of the metric, but it cannot be cured simply by imposing a gauge condition that keeps the conformal term equal to one. The cure is actually very straightforward: *One may chose the contours of the functional integral in complex extensions of Minkowski space, but only in such a way that the action stays bounded from above.* This works perfectly in the perturbative theory, and is obviously a correct prescription in any 'non-perturbative' formulation of the theory, in as far as that means anything.

Thus, the correct procedure for perturbative gravity requires contours that continue to be complex even in 'Euclidean' space-time. However, one does find that the leading contributions to the functional integrals may lie in the Euclidean section. One then writes

$$g_{\mu\nu}(x) = g_{\mu\nu}^{\mathrm{backgr}}(x) + \delta g_{\mu\nu}(x) \ , \tag{4.7}$$

where $g_{\mu\nu}^{\text{backgr}}(x)$ is in Euclidean space-time, but $\delta g_{\mu\nu}(x)$, which is to be functionally integrated over, is still complex. The background metric $g_{\mu\nu}^{\text{backgr}}(x)$ is a solution of the classical Einstein equations. In the case of a black hole, it becomes the Eclidean extrapolation of the Schwarzschild metric, which is surprisingly simple. As the Schwarzschild time near the horizon normally corresponds to the generator of Lorentz boosts there, the imaginary component of time generates simple rotations. This means that the Euclidean time coordinate for the external observer is periodic. Indeed, the period is β, the inverse of Hawking's temperature times Boltzmann's constant, $kT_{\text{H}} = 1/8\pi GM_{\text{BH}}$.

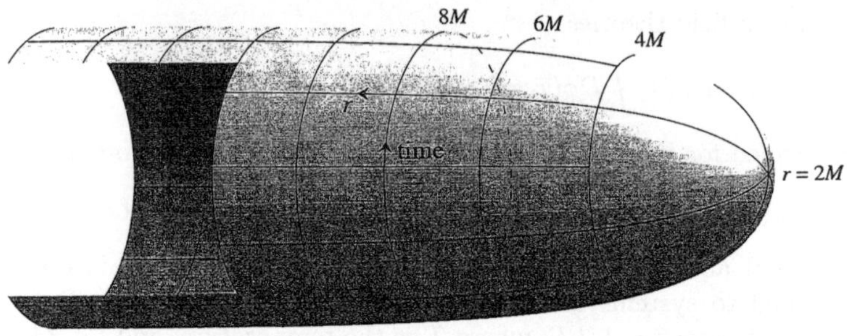

Figure 3: The black hole metric in Euclidean space-time.
This figure is geometrically accurate, that is, apart from tiny rounding errors it is a precise rendition of what the longitudinal part of the metric would look like when embedded in a three dimensional flat space.

In Fig. 3 the Euclidean black hole metric is shown. The tip of the cigar is the horizon. Here, we see explicitly that requiring this to be singularity-free implies that the Euclidean time variable is to be periodic everywhere, so that the temperature everywhere must be the inverse of $k\beta$.

Let us now consider the functional integral at finite temperature. The free energy is given by $\text{Tr}\, U(-i\beta)$, where $U(-i\beta)$ is given by Eq. (4.2). Taking the trace means that we consider all states at $t = 0$ and choose then to be equal to the states at $t = -i\beta$. This implies that we have *periodic* boundary conditions, with period $(-)i\beta$ in time. In Quantum Gravity, it is important to include the fluctuations of the space-time metric. This means that we cannot keep a fixed length of the period at all points $\vec{x}$ in three-space, but we can impose a period $(-)i\beta$ at $|\vec{x}| \to \infty$, as a boundary condition. Thus, we consider the metric on a very large sphere S_2 corresponding to $|\vec{x}| \to \infty$, and have time form a circle there, called S_1. In short, the free energy of the vacuum at temperature T is given by a space of the form $S_2 \otimes S_2$, where the period of S_2 is given by the inverse of kT.

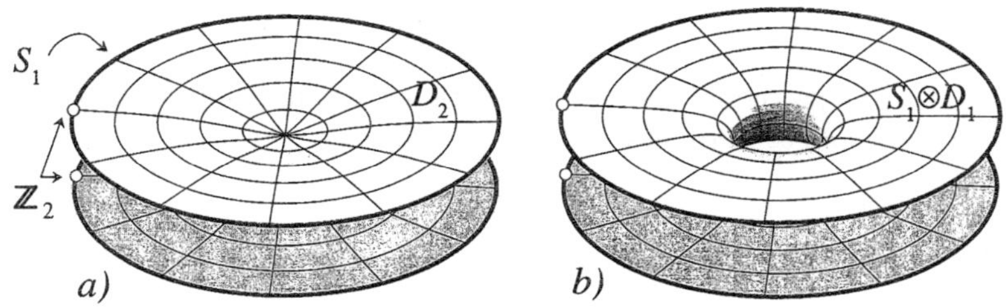

Figure 4: Two different topologies for a given boundary.

$S_2 \otimes S_1$ is a 3-dimensional boundary of 4-dimensional Euclidean space-time. There are topologically distinct ways one can now fill up the region within this boundary[11]. Two of these are standing out: either

(i) One fills up the sphere S_2 with a (3-dimensional) domain D_3, and leaves S_1 as it is, which gives $D_3 \otimes S_1$, or

(ii) One fills up S_1 with a (2-dimensional) domain D_2, leaving S_2 as it is, finding $S_2 \otimes D_2$.

in the first case, time takes the form S_1 everywhere, so that it can easily be extended to a different periodicity; making the temperature arbitrary. This is the functional integral for a space-time at any temperature, but without a black hole. In the second case, we get a space-time with, somewhere, a black hole in it. To see why this is so, it helps to visualize the situation. We found that it is convenient to remove one dimension from S_2 and one from S_1, so that 4-dimensional space-time becomes two dimensional, which is much easier to imagine. Thus, instead of filling S_2 with a domain D_3, we consider the boundary of a lower dimensional space of the form S_1, to be filled by D_2. The original S_1 is replaced by $\mathbb{Z}_2$, the set of two points. It is the boundary of a domain D_1, that is, a line segment. Fig. 4a) shows what happens in case (i), or $S_1 \otimes \mathbb{Z}_2 \to D_2 \otimes \mathbb{Z}_2$. One simply obtains two disks. Fig. 4b) shows the other case (ii), where $S_1 \otimes \mathbb{Z}_2 \to S_1 \otimes D_1$. Observe that the difference is a 'wormhole', connecting the two disks. Indeed, this wormhole is a black hole.

We see that only at finite temperature, when the boundary contains an S_1, or $\mathbb{Z}_2$ in Fig. 4, black holes can contribute. There appears to be no contribution for functional integrals that describe a black hole being formed out of ordinary matter or a black hole evaporating completely, at external temperature zero. Can one conclude from this that 'pure black holes' cannot be formed at all? Of course not. The classical black hole formation process does not go away for these formal reasons, and the information problem, combined with the locality problem, is not resolved this way. We still need to know how to count states. The procedure described in Section 2 is still required, and it demands for a revision of our views concerning Quantum Mechanics.

References

[1] G. 't Hooft, *Horizons*, Proceedings of the School of Subnuclear Physics, Eice, Italy, 41st Course: *From Quarks to Black Holes: Progress in Understanding the Logic of Nature*, September 2003, gr-qc/0401027.

[2] G. 't Hooft, *J. Mod. Phys.* **A 11** (1996) 4623; gr-qc/9607022.

[3] W.G. Unruh, private communication.

[4] G. 't Hooft, *Nucl. Phys.* **B256** (1985) 727.

[5] G. 't Hooft, *Determinism beneath Quantum Mechanics*, presented at *Quo Vadis Quantum Mechanics*, Temple University, Philadelphia, September 25, 2002; quant-ph/0212095, ITP-UU-02/69, SPIN-02/45.

[6] L. Dolan and R. Jackiw, *Phys. Rev. D* **9** (1974) 3320.

[7] S. Weinberg, *Phys. Rev. D* **9** (1974) 3357.

[8] B. Simon, *Euclidean (Quantum) Field Theory*, Princeton University Press, 1974, ISBN 0691081441.

[9] G. W. Gibbons, *Euclidean Quantum Gravity*, World Scientific Publishing Company, 1993, ISBN 9810205163.

[10] G. 't Hooft, *Perturbative Quantum Gravity*, in *From Quarks and Gluons to Quantum Gravity*, Proceedings of the International School of Subnuclear Physics, Erice 2002, Subnuclear Series Vol. 40, ed. A. Zichichi, World Scientific, pp 249 - 269.

[11] S.W. Hawking, Speech given at the 17th International Conference on General Relativity and Gravitation, Dublin, July 2004.

The Hidden SUSY face of QCD

G. Veneziano
CERN

Only the discussions are available for these lectures

CHAIRMAN: G. VENEZIANO

Scientific secretaries: S. Actis, A. R. Fazio

DISCUSSION I

- *C. Klein-Bősing:*

Has this generalization of fundamental *QCD* to orientifold *QCD* any implications for usual *QCD*? Can you imagine any observables to measure?

- *G. Veneziano:*

We do not propose to modify the *QCD* that describes our world, which contains three colors. We just extrapolate ordinary *QCD* with three colors to *QCD* with an arbitrary number *N* of colors and study the large-*N* limit. We thus replace, as usual, the gauge group *SU(3)* with *SU(N)* but we assign quarks to the antisymmetric-tensor representation (rather than to the fundamental); we claim that in this case the large-*N* limit is more manageable. This replacement is irrelevant for N=3, hence there is no change for *QCD* with three colors. Instead, at large *N* colors orientifold QCD gets related to a supersymmetric theory, super Yang-Mills, in we can compute, for instance, the gaugino condensate obtaining the result mentioned by S. Ferrara in his lecture. Finally, by interpolating between this result at $N=\infty$ and the known (vanishing) value at $N=2$, we obtain the fermion condensate for *QCD at N=3*. The details will be given in my second lecture.

- *G. 't Hooft:*

You mentioned "common wisdom" that at $N \to \infty$ the *U(1)* problem is solved by terms *O(1/N)*, so the η mass is *O(1/N)*. At large N_f, it is $O(N_f / N)$. But in perturbation expansion, also at large *N*, it is zero at all orders. Instanton effects go as $\exp(-N_f / g^2 N)$. Everybody cites Witten here. Can you give the argument in your own words?

- *G. Veneziano:*

The equation of the anomaly is:

$$\partial_\mu j^5_\mu \propto g^2 N_f F_{\mu\nu} \widetilde{F}_{\mu\nu} \, .$$

If $g^2 N$ is fixed, then $g^2 \sim O(1/N)$, and the anomaly formally disappears as N → ∞ if N_f is kept fixed.

The *U(1)* problem (the problem of the splitting of the π, the η and the η') can be discussed by looking at correlation functions of various pseudoscalar operators. There are two kinds of diagrams. Those without quark-antiquark annihilation

contribute to all three channels while those with quark-antiquark annihilation into gluons only contribute to the flavour singlet and thus, a priori, only to the the η and the η'.

These latter diagrams get a contribution from the anomaly and go like $\propto N_f / N$, so the mass splitting between η and η' comes out to be $m^2{}_\eta = cN_f / N$. At finite N_f / N_c the *U(1)* problem will be solved if and only if *c≠0.* The question is to find out whether this is the case. Fortunately, because of their non trivial topology, these diagrams can be related, at leading order in 1/N, to the purely gluonic correlator $< F_{\mu\nu} \widetilde{F}_{\mu\nu} F_{\mu\nu} \widetilde{F}_{\mu\nu} >$, evalutated in the absence of quarks. The lattice calculation of this correlator gives an estimate for the mass splitting of η and η' in agreement with the experimental results.

- *G. 't Hooft:*

But *c=0* in all orders of perturbation theory.

- *G. Veneziano:*

It has been found in lattice calculations that $<F_{\mu\nu}\widetilde{F}_{\mu\nu}(x)F_{\mu\nu}\widetilde{F}_{\mu\nu}(0)>$ is not zero; it is proportional to the expectation value of the topological charge $<\nu^2>$. On the lattice, this expectation value appears not to depend on N. Calculations have been performed from *N=3* to *N=9*, I believe.

- *C. P. Korthals-Altes:*

And for *N=6*.

- *J. L. Bourjaily:*

What is the status of the conjectured *O(1)* equivalence between QCD_{orb} and QCD_{adj} in the large N limit?

- *G. Veneziano:*

We have a perturbative argument and a non-perturbative one in favour of it; it would be important to prove rigorously this equivalence. A first possibility, suggested to us by a mathematician and by Alvarez-Gaume', would consist in write the fermionic determinants of both theories as products of eigenvalues. One then should show that the eigenvalues agree at large *N*. We could also approach this problem numerically i.e. by lattice calculations. We do have numerical codes to compute fermionic determinants and should be able to show that those two become closer and closer as *N* is increasing.

- *A.Zichichi:*

When you say "agreement with the experiments for η' ", do you mean the mixing between η and η' ?

- *G. Veneziano:*

Yes.

- *A. Zichichi:*

Do you get the right number?

- *G. Veneziano:*

Yes.

- *A. Zichichi:*

I can't believe.

- *G. Veneziano:*

Let's consider the full M^2_{PS} pseudoscalar mass matrix:

$$M^2_{PS} = \begin{matrix} u \\ d \\ s \end{matrix} \begin{pmatrix} a + \Lambda m_u^2 & a & a \\ a & a + \Lambda m_d^2 & a \\ a & a & a + \Lambda m_s^2 \end{pmatrix}$$

written in the quark basis, u, d and s. Only the neutral scalars π^0, η, η', can mix. In the limit $N_f / N_c \to 0$ we get:

$$a \sim \int d^4x < F\tilde{F}(x) F\tilde{F}(0) >_{YM}$$

which is to be computed in pure Yang-Mills theory without quarks. If you know the parameter a you can determine M^2_{PS} completely and in particular the η and the η' masses and their mixing angles, since the entries $m_I \Lambda$ can be expressed in terms of the charged or strange mesons (pions and kaons). If a is large enough the diagonalization of M^2_{PS} gives almost pure *SU(3)* states. It seems therefore that the origin of the small pseduscalar mixing angle (about 10 degrees) is due to the large value of a, due to the anomaly and instantons.

This kind of reasoning can be generalized, in principle, to isospin.

If $a \gg m_d - m_u \Lambda$ then the M^{2}_{PS} is diagonalized in the isospin basis; if instead $a \ll m_s \Lambda$, an "ideal mixing" prevales. I mean that it is possible to have mesons which are (almost) pure "u" or pure "d" states . However, this doesn't happen for ordinary mesons.

- A. Zichichi:

Why not for ordinary mesons?

- G. Veneziano:

This is because $a \gg m_d - m_u \Lambda$ is very small and thus one is usually in the first case I mentioned. It is very difficult to make a so small that the second case is realized, because, for ordinary mesons these annihilation diagrams are not so suppressed. However, if the system under consideration is very heavy, asymptotic freedom suppresses annihilation into gluons, gives a small "a" and can allow an "ideal mixing squared" , I mean ideal mixing both at the SU(3) and at the SU(2) level.

- L. Maiani:

A phenomenological comment. The scalar mesons f_0 (960) (I=0) and a_0 (960) (I=1) have very closed masses. The difference is compatible with zero with an error of 10 MeV. The u-d quark mass difference is ~ 5 MeV, so f_0 and a_0 could very well be more aligned to $\bar{u}u$ and $\bar{d}d$ rather than to the isotopic eigenstates. This shows that already at about 1 GeV annihilation diagrams are rather small indeed.

- S. Ferrara:

Why the pseudoscalar mass matrix has two vanishing eigenvalues for large *a*?

- G. Veneziano:

Because the contribution proportional to *a* gives a rank 1 mass matrix i.e. has only one non-vanishing eigenvalue. This is a consequence of the Ward identities.

- *M. Gyulassy:*

What is known about the temperature dependence of the a parameter? This could have observable consequences on the particle ratios measured in heavy ion reactions that L.Maiani spoke about.

- *G. Veneziano:*

People have investigated this dependence on lattice. If at some finite temperature a drops to zero, but the quark condensate remains non-vanishing, then in an interval of temperature there will be nine rather than eight Goldstone bosons. This could give one more interesting signal for the quark-gluon plasma. However, the results from lattice calculations are not definitive.

- *S. Salur:*

What is the quark content of η ?

- *G. Veneziano:*

It can be written as follows:

$$\eta = \cos\vartheta[\frac{1}{\sqrt{6}}(u\bar{u} + d\bar{d} - 2s\bar{s})] + sin\ \vartheta[\frac{1}{\sqrt{3}}(u\bar{u} + d\bar{d} + s\bar{s})]$$

where $\theta \sim 11°$.

In the limit of zero quark mass, $\sin\theta \to 0$, and η becomes a pure octet, while η' becomes a pure singlet

$$\eta' = \frac{1}{\sqrt{3}}(u\bar{u} + d\bar{d} + s\bar{s})$$

In reality the physical η and of η' are slightly rotated with respect to the pure octet and the pure singlet due to the strange quark mass. This is important in the computation of the decay of η and η' into two gammas.

- *M. Gyulassy:*

Is there an analogous AdS/CFT calculation for the connection between the $F\tilde{F}$ correlator and the T^{xy} correlator?

- *G. Veneziano:*

I am not sure I understood your question. But let me take the opportunity to point out that, in N = 1 supersymmetry, the previously given diagrammatic argument remains valid. Since supersymmetry is not broken, scalars and pseudoscalars must have the same mass. Therefore in the supersymmetric case annihilation diagrams should push the η' mass up to the scalar mass. In my second lecture I will present some lattice results going in this direction.

- *S. Ferrara:*

By , $F\widetilde{F}$ do you mean trace of $F\widetilde{F}$?

- *G. Veneziano:*

I mean the color trace of $F\widetilde{F}$. Its correlator at zero momentum, computed in lattice, is not zero for pure Yang Mills but must becomes zero in presence of massless quarks. My guess therefore is that, in N=4 supersymmetry, this correlator is zero as well.

CHAIRMAN: G. VENEZIANO

Scientific Secretaries: G. De Risi, A. Polosa

DISCUSSION II

C. Klein-Bosing:

You had a K parameter which is 1 when N goes to infinity and 0 when N=0...

G. Veneziano:

N goes to 2...

C. Klein-Bosing:

Ok...then you said that when N goes to 3 you assume it to be 1 ± 30%. Can it be larger than one?

G. Veneziano:

I plot $K_f=(1-2/N)k(1/N)$ as a function of 1/N. If the interpolation between 1 and 1/2, obtained respectively for $N\to\infty$ and N=2, where fermion decouples, is linear, that's it. But it has not to be linear. With k=1±30 % at N=3 one has the factor 1/3 in front which makes the agreement to data better...but something crazy could happen and the linear approximation be not linear.

S. Ferrara:

I would like to know if you can make a similar analysis for theories with N=1 matter and what is the use of them.

G. Veneziano:

This is one of the generalizations that I did not discuss. You can explore several extensions, but we did not consider none of them to the same level of the detail. The one you mention is probably closer to this one. We can try to play this large N game to relate not only QCD like theories with with N =1 Supersymmetry but also to relate an N=1 and N=2 supersymmetric theories. In this framework we thought of the following trick. We take the Supesymmetric theory I have been talking today, just gluons and gluinos (supersymmetric Yang-Mills theory). N =1 SUSY allows to add matter fields. In particular there is SUSY QCD in which I can add quarks and squarks, in the fundamental representation. Now we play again this trick: for $N_c=3$ being in the fundamental is the same as being in the antisymmetric. Let's generalize this SUSY QCD to arbitary N by keeping the matter superfields in the antisymmetric, and then the field content becomes closer and closer, as we increase N_c, to the one of an N =2 theory which is the Seiberg-Witten theory with two gluinos and with a scalar in the adjoint.
This case is more difficult of the case I have presented in detail because the Seiberg-Witten theory has a bigger space of vacua. Moreover it seems that the massless limit and the large N limit do not commute. We think that we can do some work in this direction.

S. Ferrara:

Have you found the moduli space in this other theory? The space of the Seiberg-Witten theory is rather complicated, even if its known…

G. Veneziano:

We looked into that. In the orientifold theory the moduli space non-perturbatively is much more trivial so that's why we have to put a mass. If you put the mass you go to some point in the moduli space and you can compare the N =1 theory with a small mass to the S-W in a particular point of their moduli space.

S. Salur:

What is the effect of mixing of states in multi-quark states to the decay width of these particles?

G. Veneziano:

If the following diagram where important then the 5-quark states could mix not only with another multiquark sate, but also it could mix with an ordinary three quark state, the intermediate state in the diagram. But then it is hard to understand how it can be so narrow. The narrowness of these states leads us to believe that these diagrams are suppressed. Then m_u-m_d and $O(\alpha)$ effects push the eigenstates in the pure quark flavor direction. The fact that they are mixed doesn't change the narrowness of these states but suppresses or enhances some decay channels.

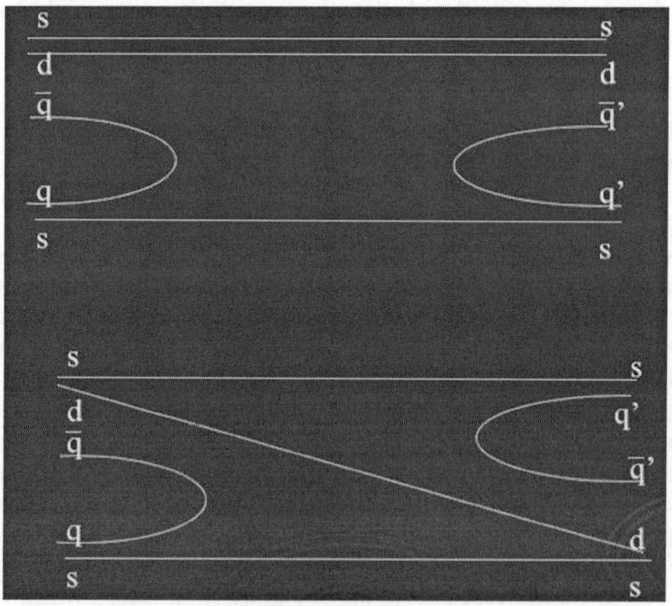

fig. 1

C. Aidala:

You said 1-flavor QCD should be confining and have a mass gap. But if there's just one flavor, what is the mass gap between?

G. Veneziano:

The mass gap is the mass gap in the true spectrum of the theory, the one you obtain after you take into account confinement chiral symmetry breaking and so on, not in the quark content.

J. Bourjaily:

You outlined an argument to prove confinement in N =1 SUSY with $N_f=1$.
Is this a proof that exists?

G. Veneziano:

The best argument we have for confinement in SYM theories comes from SW theory. They showed that if you add a small mass to one of the gluinos and to the scalar, the theory becomes N =1 supersymmetric, in a way that is under control, and the magnetic monopoles condese. This realizes the mechanism that 't Hooft and Mandelstam have proposed as a criterium for confinement. To get to the theory I'm interested in, you have to take this mass parameter to be much bigger than the scale of the theory. It is hard to think that, thing will get worse if you increase the mass, because the theory becomes even "more" asymptotically free. In other words, if the theory is confined at low mass, it would be confined even at infinite mass, which is the theory I describe...

S. Ferrara:

I think you mean that the scalar is frozen...

G. Veneziano:

The scalar is frozen, but, if I can remember correctly, there is also an order parameter that is related to monopole condensation...

S. Ferrara:

It depends on the mass you have...

G. Veneziano:

Yes, it depends on the mass you have. I think it is proportional to the mass, so there are no reason that it will disappear if you increase the mass and reach a N =1 theory. Now this theory is connected to QCD with one flavor, so you are led to think that even this theory is confining.

A. Zichichi:

But the only way to have masses in our real world is by spontaneous symmetry breaking. You put them in "ad hoc" instead...

G. Veneziano:

There is no problem in having masses as far as pure QCD is concerned. Problem arise when you consider weak interaction and the *SU(2)* gauge group. Then the fermions become chiral and it is not possible to put masses by hand.

G. 't Hooft:

You have made a quantitative comparison with experimental data, such as the mass of the pseudoscalar. In your theory there is only one pseudoscalar and its mass should increase because of quark masses. I agree with you as far as the *u* and *d* quarks are concerned, but the *s* quark should actually lower its mass, so the agreement with the η mass doesn't seem so great, unless you get a big contribution from the *u* and *d* quarks.

G. Veneziano:

I was looking at this quite differently: I tried to discuss what would happen in an ideal world with one flavor (of course you could do the calculation on the lattice). To compare the mass of the pseudoscalar with the experimental mass of the lower scalar, which is this broad π–π resonance…

G. 't Hooft:

This is the isosinglet pseudoscalar…I thought you were going to compare this with the mass of the η…

G. Veneziano:

I will not compare this mass with any of the physical pseudoscalar, neither the η nor the η'. What I want to do is to compare the mass of the would-be Goldstone boson in our theory with only one flavor, to a scalar in that same theory because SUSY would tell you that they should be degenerate. Since I don't know how to compute the mass of the scalar, for it I used data.

The σ presumably has no s quark content because it does not match with *K*, decaying mainly to $\pi\pi$. But in a world with one flavor there would be no pions and the sigma would be perfectly stable. Some lattice arguments seem to support the same idea. The point here is that as one turns on this annihilation diagrams:

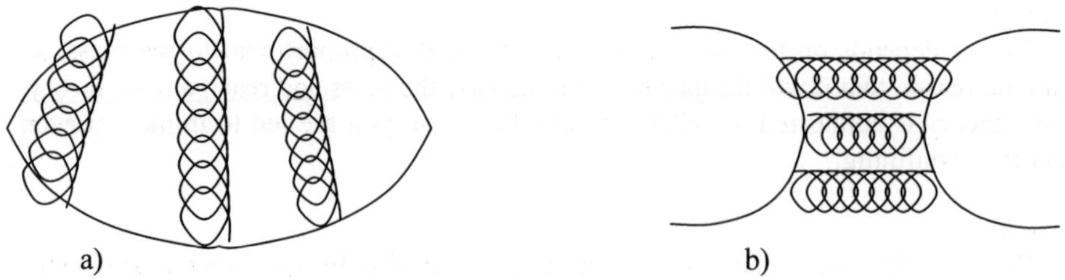

fig. 2

which split the singlet from the non-singlet, one gets an opposite effect on the scalars and pseudoscalars. If you neglect those diagrams, the psudo-scalar is massless, and is a Goldstone boson, and the scalar is very massive. But this effect seems to bring them together.

The best thing would be to do a real, very hard, unquenched calculation with full dynamical fermions. What one usually does, like for the η', is to compute this Zweig rule violating diagram but in the quenched approximation, without the fermion determinant. My understanding is that one can easily compute these two topologies on the lattice. Now the diagram in fig. 2(a) would give $m_P=0$, $m_S\neq 0$, the scalar would be massive. With this diagram there is no way to have degeneracy between scalar and pseudoscalar. If you add the diagram in fig. 2(b) it should push up the pseudoscalar mass, as it is expected from the resolution of the $U_A(1)$ problem, and will also renormalize the scalar mass. In summary the scalar mass is pushed down and the pseudoscalar one is brought up so that eventually they will be almost degenerate.

G. `t Hooft:

Can you remind me please what kind of symmetry requires these masses to be degenerate? It could not be SUSY because they both are bosons.

G. Veneziano:

A supersymmetric supermultiplet has necessarily both a scalar and a psudoscalar; It is a complex scalar, which can be decomposed in scalar and psudoscalar; of course there is also a fermion.

Our connection between the supersymmetric and non-supersymmetric theory only works in the bosonic sector; the fermion is not degenerate in the orientedfold theory or in QCD but it will be of course degenerate in the supersymmetric case. All this supersymmetry should be visible in the lattice calculations. The price to pay is that you have to be unquenched. You have to add fermion loops in order not to broke supersymmetry.

C. Korthals Altes:

In your first lecture you showed that in the large N_c limit the orientifold and the adjoint representation converge. Now the fermions F in QCD the antisymmetric representation have N-ality k = 2. If I consider a meson state $\overline{F}F$ in this theory, it is a meson state with a $k = 2$ string in between $\overline{F}$ and F in the limit of large N_c the $k = 2$ string becomes two $k = 1$ strings. On the other hand QCD_{adj} has for any finite N_c a $k = 0$ string (i.e. no strings at all) between a $\psi_i^{\bar{j}}$ and $\psi^{\bar{i}}{}_j$ pair. So how can it become the large N_c version of QCD_{OR}.

S. Ferrara:

Maybe it has something to do with the fact that the two representation behave differently under the center of the group *SU(N)*.

G. Veneziano:

The adjoint theory has Wilson loops that for large N_c start to factorize:

$$\frac{1}{N^2}\left\langle \mathrm{Tr}W\mathrm{Tr}W^+\right\rangle = \left|\frac{1}{N}\left\langle \mathrm{Tr}W\right\rangle\right|^2 + O\left(\frac{1}{N^2}\right)$$

The l.h.s. for any finite N_c has no area law, whereas the first term on the r.h.s. has an area law with twice the string tension. So it is this term that large N_c QCD shows up, at least formally.

COMPLEXITY EXISTS AT THE FUNDAMENTAL LEVEL

Antonino Zichichi
INFN and University of Bologna, Italy
CERN, Geneva, Switzerland
World Federation of Scientists, Beijing, Geneva, Moscow, New York

CONTENTS

COMPLEXITY EXISTS AT THE FUNDAMENTAL LEVEL

1 PURPOSE

Purpose of this Lecture is to show that Complexity exists at the fundamental level.

To prove this it is necessary:

i) to identify the experimentally observable effects which call for the existence of Complexity;

ii) to analyse how we have discovered the most advanced frontier of Science;

iii) to construct the platonic version of this frontier: i.e. what would be the ideal platonic Simplicity.

We start with a note on the Logic of Nature and Complexity (chapter 2). In chapter 3 we present seven definitions of Complexity and on chapter 4 a view on the whole of our knowledge.

In chapter 5 we discuss the experimentally observably quantities needed for the existence of Complexity, and compare History and Science on the basis of 'What if?'.

In chapter 6 we review the scale which bring us from the basic quantities to Life and History.

Chapter 7 is devoted to examples of AFB phenomena going from Beethoven to nuclear physics, subnuclear physics and to the superworld.

In chapter 8 there is a set of UEEC events strictly related to the fundamental steps which have brought us to the discovery of the most advanced frontier of Science. For completeness we start from the origin of Science (Galileo Galilei) and go to the present frontier of our knowledge.

We review in chapter 9 the existence of Complexity at all scales and in chapter 10 how progress is made in advanced research.

In order to see how far from Simplicity are the foundations of scientific knowledge, we present in chapter 11 the platonic Grand

Unification, in chapter 12 the Platonic Supersymmetry and in chapter 13 the Deviations from Simplicity needed in order to describe the basic Logic of Nature which has in chapter 14 its exact formulation in terms of the Grand Unification. In chapter 15 there is a summary of the main points and in chapter 16 the Conclusion.

We will see that the two experimental basic elements which motivated the birth of Complexity exists at all levels, including the fundamental one. Therefore, if we want to understand Complexity, the only way is to continue with the Reductionistic analysis of the phenomena, no matter the Mass-Energy and Space-Time scales involved. In fact, it is thanks to Reductionism that we can arrive at the result that Complexity exists in all fields of our existence including Science.

2 PRELIMINARY NOTES ON THE LOGIC OF NATURE AND COMPLEXITY

The logic of nature allows the existence of a large variety of structures with their regularities and laws which appear to be independent from the basic constituents of nature and the fundamental laws which govern their interactions. But, without these laws it would be impossible to have the real world which is in front of us and of which we are part of.

Let me say a few words on what I have done since the year 2000 in order to follow the interesting developments concerning Complexity in the various fields going from my own field of activity, subnuclear physics, to history.

My field of activity is considered to be the asymptotic limit of Simplicity where the reductionist method is dominating.

History is considered the asymptotic limit of Complexity and here the holistic approach is unavoidable.

During the last six years I have presented, on different occasions, my views and the results of my analysis concerning Complexity [1–4] to a large variety of research workers engaged, not only in my own field, but also in other disciplines. The purpose of this activity was to investigate how colleagues working in different fields react to an exact definition of Complexity.

The fields where I have been working in terms of a direct study in order to learn more about Complexity are listed in the 'Complexity–Table' below. In this way I have attempted to see how Complexity was (and is) defined in the various research fields.

The results are in the next chapter 2 where seven definitions of Complexity, coming from seven different research fields, are reported.

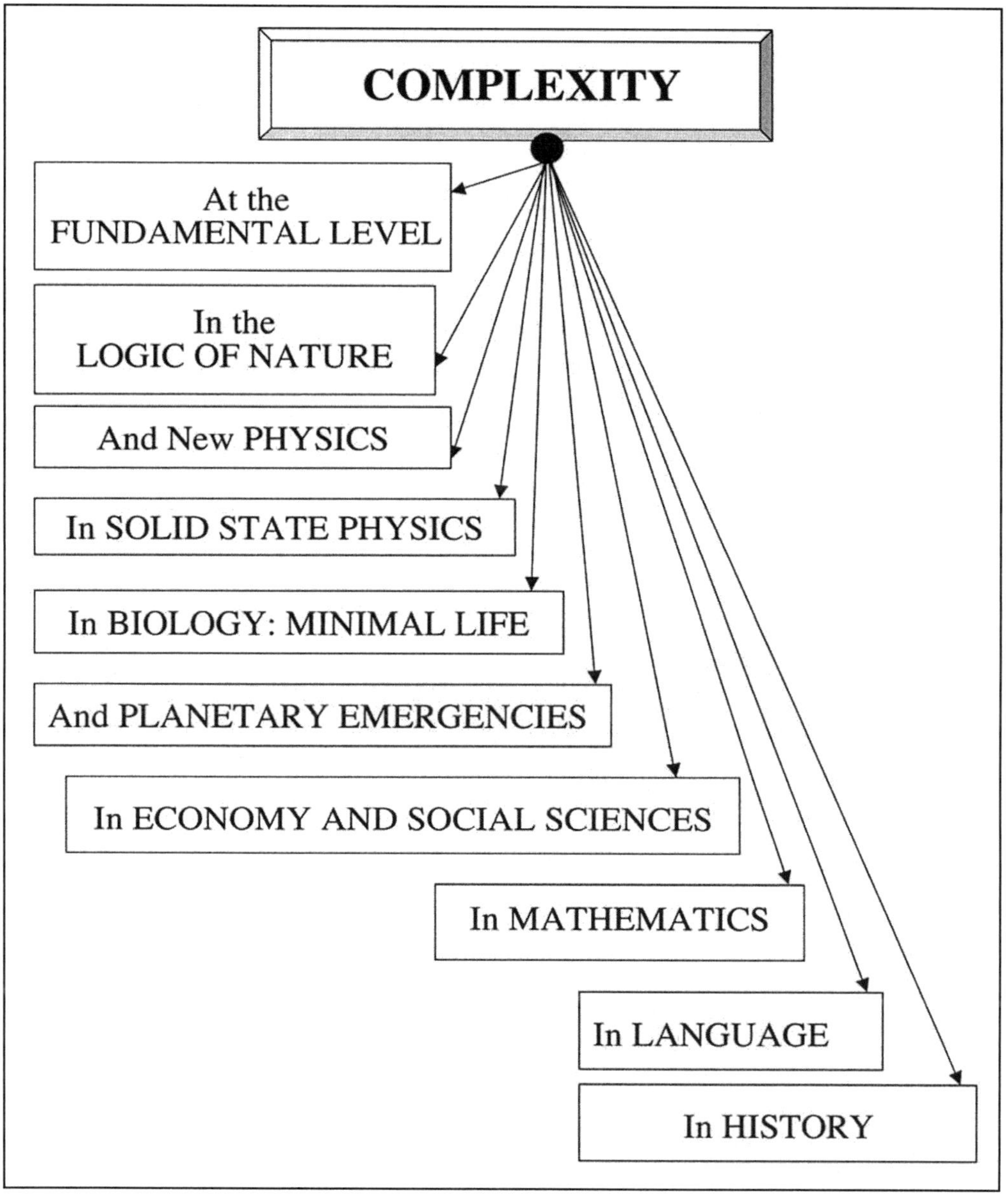

Complexity–Table

3 SEVEN DEFINITIONS OF COMPLEXITY

DEFINITION NUMBER 1

Complexity is a property of systems that are somewhere in between a completely random and a completely regular state, often described by a highly non linear set of equations but sometimes not describable by equations at all.

DEFINITION NUMBER 2

Bad ones:

1) Chaos
2) The need for lengthy calculations
3) The need for many distinct variables.

Better ones:

4) Unexpected difficulty when attempting to describe something in a precisely formulated theory.
5) What is left over after all systematic approaches failed.

But it could also be that: Complexity is an excuse for sloppy thinking.

DEFINITION NUMBER 3

The Complexity of a theory (problem) is the minimum amount of computer time and storage required to simulate (solve) it to a specified level of precision.

DEFINITION NUMBER 4

If we admit that biological or linguistic evolution, or financial dynamics are complex phenomena, then their typical dynamics is somehow between strong chaos (i.e. positive Lyapunov exponents) and simple orbits (i.e. negative Lyapunov exponents).

In other words, Complexity (or at least some form of it) is deeply related to the edge of chaos (i.e. vanishing maximal Lyapunov exponent).

Since the edge of chaos appears to be related paradigmatically to an entropy index 'q' different from unity, there must be some deep connection between Complexity and generalized entropies such as 'S_q'.

DEFINITION NUMBER 5

From the mathematical point of view:

- A problem can be polinomial, which means that it is not to hard to predict surprises.

- A problem can be NP or NP-complete, which represent different degrees of difficulty in predicting surprises.
- •• Surprises means: UEEC event (see later).
- •• That degree of difficulty can be associated with the level of Complexity.

<u>*DEFINITION NUMBER 6*</u>

A system is 'complex' when it is no longer useful to describe it in terms of its fundamental constituents.

<u>*DEFINITION NUMBER 7*</u>

The simplest definition of Complexity: '*Complexity is the opposite of Simplicity*'. This is why we have studied the platonic Standard Model and its extension to the platonic superworld.

These seven definitions of Complexity must be compared with the whole of our knowledge in order to focus our attention on the key features needed to study our real world.

The whole of our knowledge is synthetically exposed in the next chapter 4.

4 THE WHOLE OF OUR KNOWLEDGE

The fundamental quantities and their properties, needed to build up the Universe, together with the basic conceptual structure of all our intellectual activities are shown in figure 1. Here are the basic elements.

The Universe could have existed and no Life; the fundamental quantities and the structures needed to build the Universe would have been exactly the same; Universe and Life could have existed but no Conscience; the next step is Creativity which could have not been there. In the same figure 1, it is indicated a place where in the study of the structures without Life, the cosmological structures, Complexity seems to show up.

Finally Reason, which has produced the three greatest achievements of the human intellect: Language, Logic and Science [5].

The time-sequence of Language, Logic and Science is shown in figure 2. It is thanks to Language (i.e. to its written expressions) that Permanent Collective Memory (PCM) exists. It has recently been discovered that what we call 'Language' consists of two levels [5].

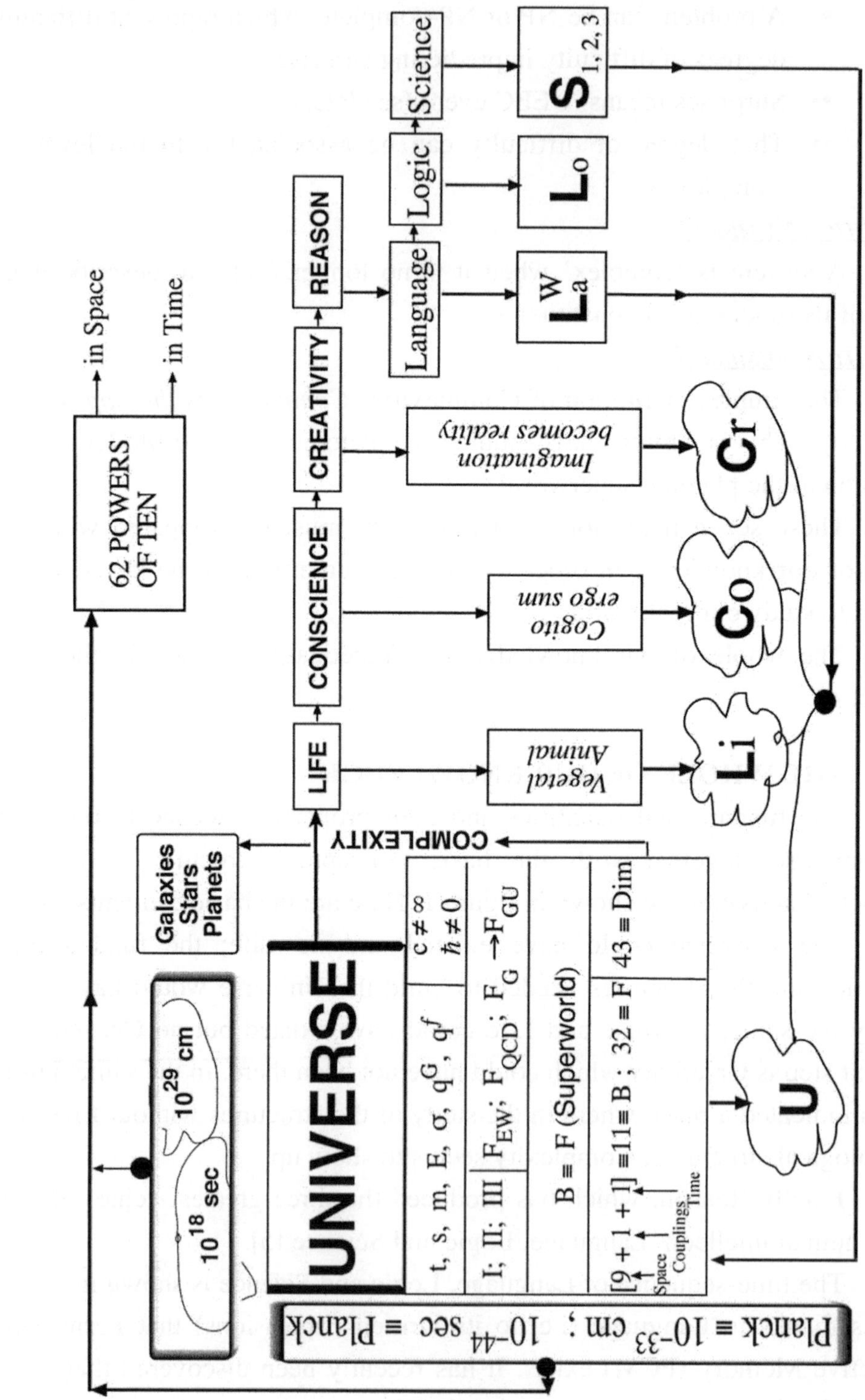

62 POWERS OF TEN
in Space
in Time
10^{29} cm
10^{18} sec
Galaxies Stars Planets
COMPLEXITY
UNIVERSE
LIFE
CONSCIENCE
CREATIVITY
REASON
Language
Logic
Science
Vegetal Animal
Cogito ergo sum
Imagination becomes reality
Li
Co
Cr
L_a^W
L_o
$S_{1,2,3}$
U
t, s, m, E, σ, q^G, q^f
$c \neq \infty$
$\hbar \neq 0$
I; II; III
F_{EW}; F_{QCD}; $F_G \rightarrow F_{GU}$
B ≡ F (Superworld)
[9 + 1 + 1] ≡ 11 ≡ B ; 32 ≡ F
43 ≡ Dim
Space
Couplings
Time
Planck ≡ 10^{-33} cm ; 10^{-44} sec ≡ Planck

Figure 1

The lowest one is the one needed in order to understand a 'message' (i.e. a group of words constructed on the basis of appropriate rules).

We can call this level 'Language-understanding'. The other level is at a much higher degree of intellectual ability. It is the one which is needed in order to elaborate a 'message'. Our species is the only one which is able to elaborate 'messages'.

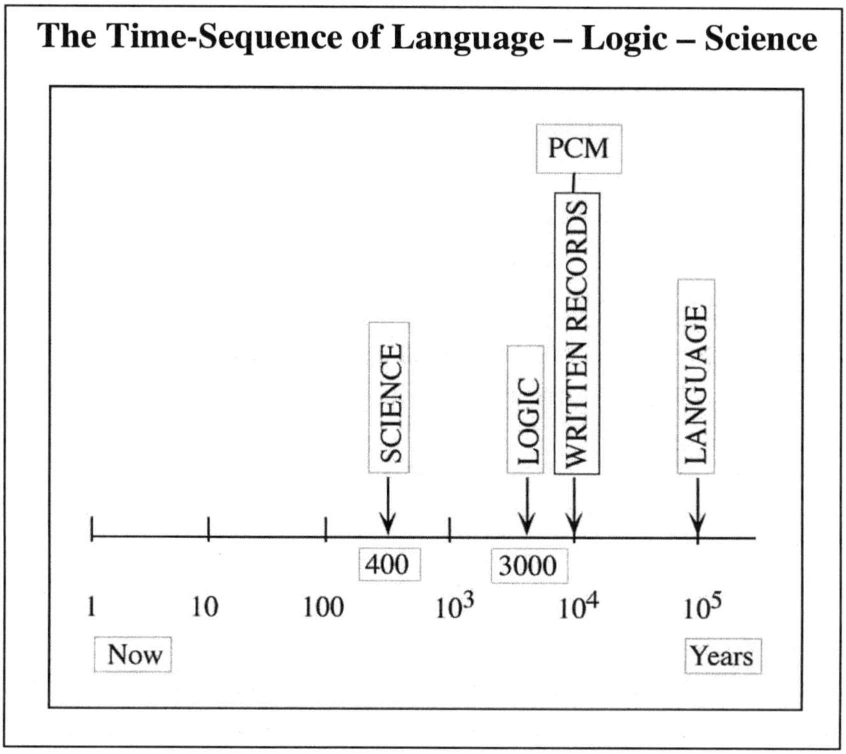

Figure 2

In figure 3 we refer to Language at its highest level where its various components become effective. The intellectual achievement, called Language, is due to the fact that our species is endowed with Reason, as illustrated in figure 1.

The best definition of the activities which build 'Language' can be obtained by realizing that all these activities would exist even if neither Rigorous Logic nor Science had been discovered.

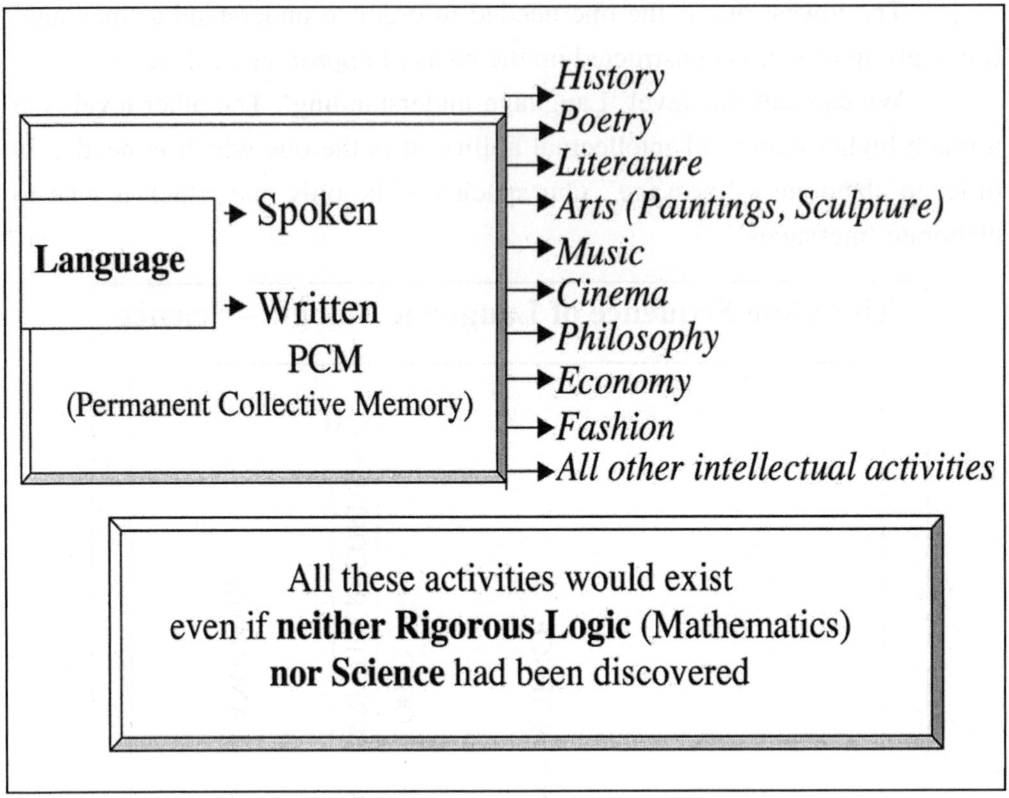

Figure 3

In figure 4 the main achievements of Rigorous Logic are reported. All these achievements would exist even if Science had never been discovered.

In the following figures (5, 6, 7) the point to notice is the vital condition which allows the three achievements to exist; i.e. 'to be fascinating' for Language, the 'non-contradiction' for Logic and the 'reproducibility' for Science.

In figure 5 there is an attempt to express, the language in terms of a mathematical formalism.

The symbols refer to sum '$\sum$' and product '$\prod$' of the various functions 'f' describing the large number of constituents of a linguistic structure, as indicated by the symbols R, Cr, Co, Li and U, whose meaning is reported.

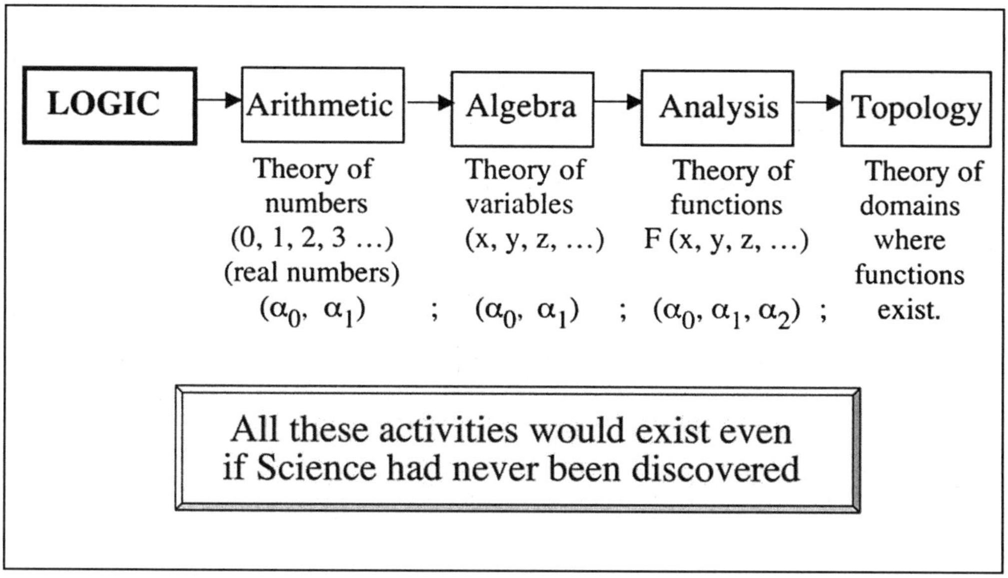

Figure 4

$$La \equiv F_{La}(R, Cr, Co, Li, U) \equiv \sum_{j,k,l,m,n} \left[f^{j}_{La}(R) \oplus f^{k}_{La}(Cr) \oplus f^{l}_{La}(Co) \oplus f^{m}_{La}(Li) \oplus f^{n}_{La}(U) \right] \oplus$$

$$\oplus \prod_{j,\ k,\ l,\ m,\ n} \left[f^{j}_{La}(R) \otimes f^{k}_{La}(Cr) \otimes f^{l}_{La}(Co) \otimes f^{m}_{La}(Li) \otimes f^{n}_{La}(U) \right] \Rightarrow$$

$\Rightarrow$ **Be Fascinating**

Reason	$\equiv$ R	$\Rightarrow$	$j =$	$1, 2, 3, \dots\dots\dots n_j$
Creativity	$\equiv$ Cr	$\Rightarrow$	$k =$	$1, 2, 3, \dots\dots\dots n_k$
Self Conscience	$\equiv$ Co	$\Rightarrow$	$l =$	$1, 2, 3, \dots\dots\dots n_l$
Life	$\equiv$ Li	$\Rightarrow$	$m =$	$1, 2, 3, \dots\dots\dots n_m$
Universe	$\equiv$ U	$\Rightarrow$	$n =$	$1, 2, 3, \dots\dots\dots n_n$

LANGUAGE

Figure 5

$$\mathbf{Lo} \equiv \mathbf{F}_{\mathbf{Lo}}\,(\mathbf{Arithmetic};\ \mathbf{Algebra};\ \mathbf{Analysis};\ \mathbf{Topology}) \Rightarrow$$

$\Rightarrow$ **Non-Contradiction**

LOGIC

Figure 6

$$\mathbf{S}_{1,2,3} \equiv \mathbf{F}_{\mathbf{S}_{1,2,3}}\,(\mathbf{Inventions};\ \mathbf{Discoveries};\ \mathbf{Measurements}) \Rightarrow$$

$\Rightarrow$ **Reproducibility**

SCIENCE

Figure 7

One should realize that there are three levels of scientific credibility. They are illustrated in figure 8 and quoted in figure 1 as $S_{1,2,3}$.

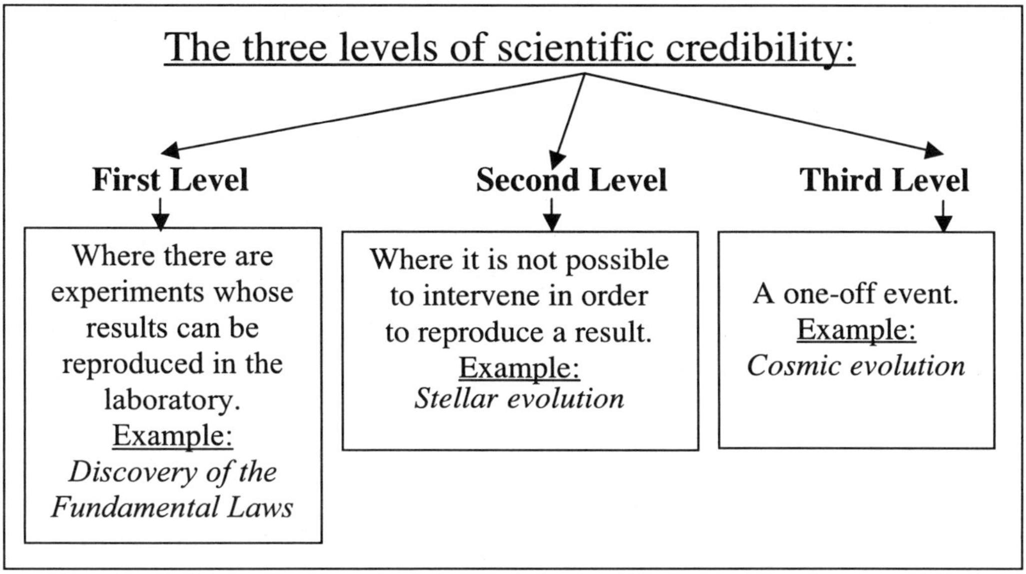

Figure 8

Let me say a few words on Creativity in basic Science. Contrary to the other two cases, here Creativity implies the need to overcome the principle of the reproducible experimental proof.

And this means to check our imagination with the being who created the three basic structures (three families of elementary particles) and the three fundamental forces of the Universe (the electroweak, the strong subnuclear and the gravitational). As Isidor I. Rabi used to say, '*He is smarter than all of us, as correctly pointed out by Galileo Galilei. Never forget that this is how Science started 400 years ago*'.

Thus Creativity in basic Science is the most difficult one, when compared to basic Logic and basic Language, since basic Science means the Logic of Nature and this Logic has been chosen by a being who is 'smarter than all of us'. This brings us to the crucial problem we want to investigate: if Complexity exists at the fundamental level.

5 COMPLEX SYSTEMS: THE EXPERIMENTALLY OBSERVABLE QUANTITIES AND A COMPARISON BETWEEN HISTORY AND SCIENCE

The first point we have to investigate concerns the experimental evidences which are at the origin of our concern on the existence of Complexity. In fact the existence of Complexity in the real world – according to my own view – emerges from two experimentally well-established basic elements:

1) the Anderson-Feynman-Beethoven-type phenomena (AFB) i.e. phenomena whose laws and regularities ignore the existence of the Fundamental Laws of Nature from which they originate;
2) the Sarajevo-type effects, i.e. Unexpected Events of quasi irrelevant magnitude which produce Enormous Consequences (UEEC).

But '**Complexity**' – as we have seen in chapter 3 – is ill-defined; nevertheless people speak of 'Complexity' as a source of new insights in physics, biology, geology, cosmology, social sciences and in all intellectual activities which look at the world through the lens of a standard analysis in terms of either Simplicity or Complexity.

The real world seems in fact characterized by two basic features, which are one on the opposite side of the other: *Simplicity* and *Complexity*.

It is generally accepted that *Simplicity* is the outcome of *Reductionism*, while *Complexity* is the result of *Holism*.

The most celebrated example of *Simplicity* is *Science* while the most celebrated example of *Complexity* is *History*.

Talking about asymptotic limits, we can consider *History* as the asymptotic limit of *Holism* and of *Complexity*, but *Science* as the asymptotic limit of *Reductionism* and of *Simplicity*.

Four centuries of Galilean research work based on Reductionism, i.e. on the identification of the simplest elements in the study of Nature, has allowed us to get the greatest achievement of Science, i.e. the so called **Standard Model** and its extension.

This extension predicts GUT (the Grand Unification Theory), the existence of the superworld and the resolution of the quantum-gravity problem via the powerful theoretical structure of RQST (Relativistic Quantum String Theory).

Following the Majorana approach to understand the Laws of Nature

(I am referring to the Majorana paper no. 10[1], totally unknown to physicists but known to social scientists) it is necessary to review the present status of all we know of the Reductionistic achievements, together with our present understanding of the rigorous attempts towards the basic features which allow Complexity to exist.

In fact, the Logic of Nature allows the existence of Science (the asymptotic limit of Simplicity) and of History (the asymptotic limit of Complexity), which share a property, common to both of them.

It is therefore interesting to define Science and History in terms of this property, probably the only one, which they share; i.e. Evolution.

- Science is the Evolution of our Basic Understanding of the laws governing the world in its Structure ≡ EBUS.
- History is the Evolution of the World in its Real Life ≡ EWRL.

Let us compare these two supposedly asymptotic limits — History and Science — on the basis of 'What if?'; a condition elaborated by the specialists in what is now known as 'virtual history'. On the basis of 'What if?' these specialists conclude that the world would not be as it is, if one, or few, or any number of 'What if?' had not been as History tells us. This is not the case for Science. The world would have exactly the same laws and regularities, whether Galileo Galilei or somebody else had discovered $F = mg$, and so on for all the other scientific discoveries.

It is in the consequences of 'What if?' that the two asymptotic limits of Simplicity and Complexity seem to diverge, despite the fact that the sequence of 'What if?' in Science belongs to the 'totally unexpected events' (UEEC) exactly like the others listed in the column of History.

The table of 'What if?' shows that UEEC events exist in Science, since its origin.

[1] Ettore Majorana, Il valore delle leggi statistiche nella fisica e nelle scienze sociali, *Scientia*, Quarta serie, Febbraio-Marzo 1942 pp. 58. English translation [6]. Majorana had a knowledge of the physics of his time so complete that few others could compete with him [7]. This is why Enrico Fermi considered him a genius like Galilei ad Newton. His knowledge of the reductionistic approach brought Majorana to study what relation could exist between physics and social sciences. Translated into the language of present day this means reductionistic versus holistic approach. The holistic approach needs to be taken into account once we have a comprehensive view of the reductionistic achievements. Thus the Majorana case is very illuminating.

'WHAT IF ?'

	In History ▪ EWRL		In Science ▪ EBUS
I	What if Julius Caesar had been assassinated many years before?	*I*	What if Galileo Galilei had not discovered that $F = mg$?
II	What if Napoleon had not been born?	*II*	What if Newton had not discovered that $F = G\frac{m_1 \cdot m_2}{R_{12}^2}$?
III	What if America had been discovered few centuries later?	*III*	What if Maxwell had not discovered the unification of electricity, magnetism and optical phenomena, which allowed him to conclude that light is a vibration of the EM field?
IV	What if the 1908 Tunguska Comet had fallen somewhere in Europe instead of Tunguska in Siberia?	*IV*	What if Planck had not discovered that $h \neq 0$?
V	What if the killer of the Austrian Archduke Francisco Ferdinand had been arrested the day before the Sarajevo event?	*V*	What if Lorentz had not discovered that space and time cannot be both real?
VI	What if Lenin had been killed during his travelling through Germany?	*VI*	What if Einstein had not discovered the existence of time-like and space-like real worlds? Only in the time-like world, simultaneity does not change, with changing observer.
VII	What if the first nuclear weapon had been built either by Japan before Pearl Arbour (1941) or by Hitler in 1942 or by Stalin in 1943?	*VII*	What if Rutherford had not discovered the nucleus?
VIII	What if Nazi Germany had defeated the Soviet Union?	*VIII*	What if Dirac had not discovered his equation, which implies the existence of the antiworld?
IX	What if Karol Wojtyla had not been elected Pope, thus becoming John Paul II?	*IX*	What if Fermi had not discovered the weak forces?
X	What if USSR had not collapsed?	*X*	What if the 'strange particles' had not been discovered in the Blackett Lab?

Table

In addition to Science (the asymptotic limit of Simplicity) and History (the asymptotic limit of Complexity), Nature allows the existence of many other structures whose Complexity seems to be in between the two extreme asymptotic limits. Let me show (figure 9) a sample of systems, which, according to the present way of looking at the real world, are considered as being complex.

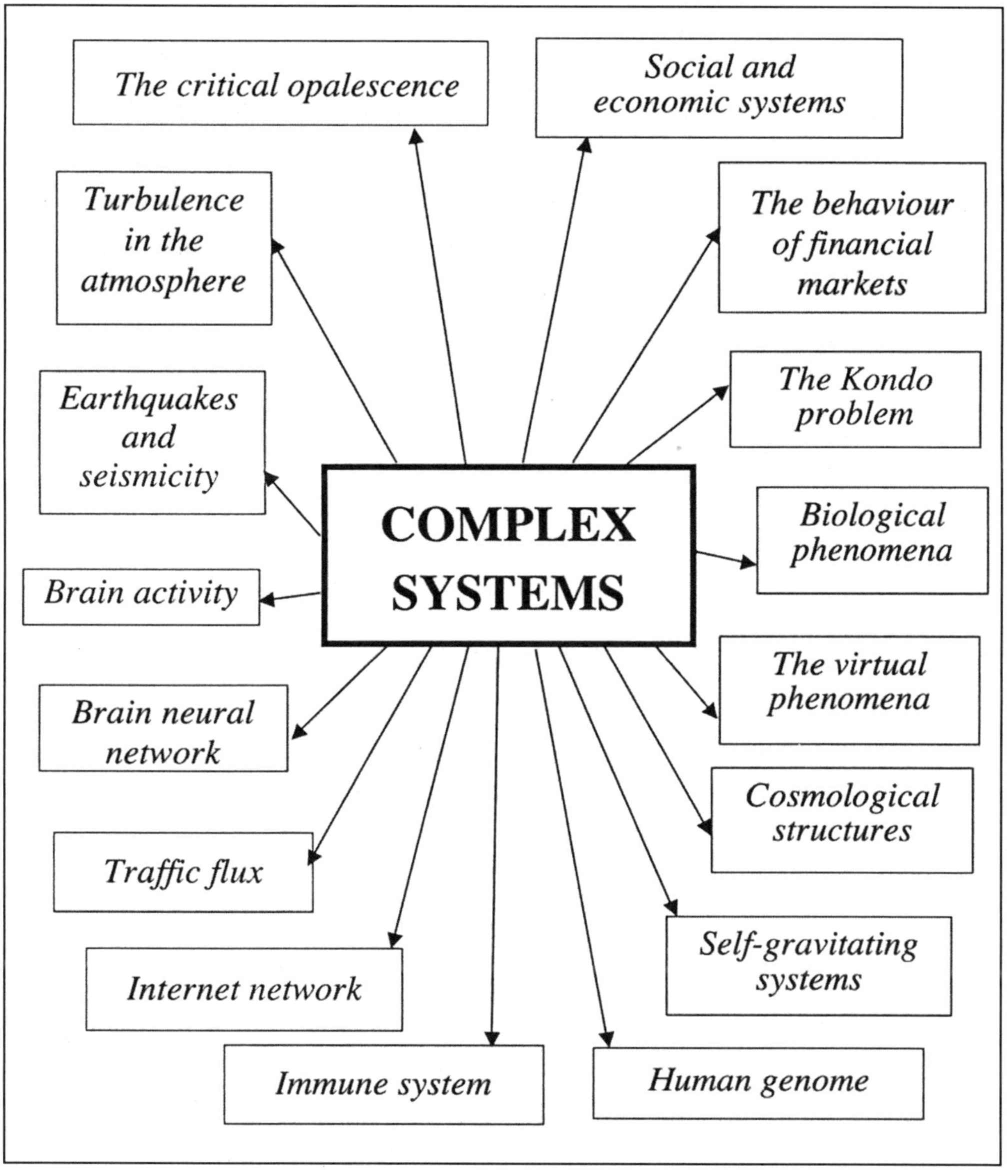

Figure 9

As you can see, we go from traffic flux, to the internet network, to earthquakes and seismicity, to social and economic systems, to the behaviour of financial markets, to the study of cosmological structures, and so on.

During the last years, we have had at the Ettore Majorana Centre (Erice) an intense activity, on many of these fields, with seminars fully devoted to the specific topic of Complexity with world leaders participating. I have been also personally engaged in the study of Complexity, as already mentioned in chapter 2 [1–4].

6 FROM THE BASIC QUANTITIES TO LIFE AND HISTORY

As you know my field of scientific activity is in the frontier of Simplicity: subnuclear physics. It is thanks to this field of Science that it has been possible to identify the Basic Quantities needed to build the world where we live, as shown in figure 10.

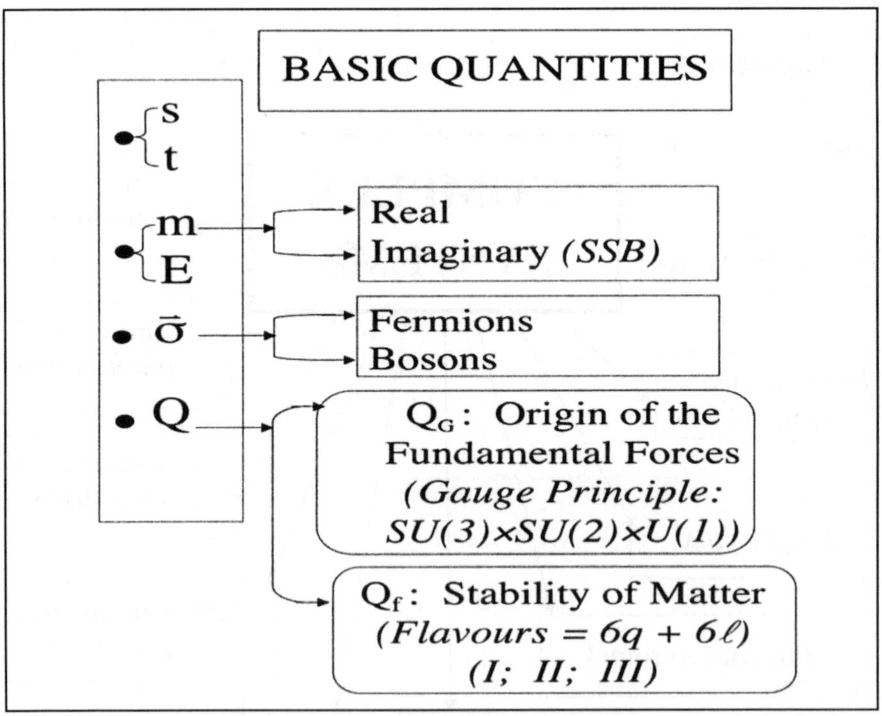

Figure 10

The logic structure of these '**Basic Quantities**' is in the modern synthesis of scientific knowledge, called **SM&B** (Standard Model and

Beyond) (figure 11). This is how it has been possible to go from the world to the superworld (figure 12). We will return to this topic in chapter 8 where we discuss the totally unexpected events (UEEC) from Galilei to the **SM&B**. All we need for the time being, is to point out that the **'Basic Quantities'** in figure 10 are the origin of all we can think off, including Life and History, as illustrated in figure 13.

SM&B

THE STANDARD MODEL AND BEYOND

① RGEs (α_i $(i \equiv 1, 2, 3)$; m_j $(j \equiv q, l, G, H)$) : $f(k^2)$.

- GUT ($\alpha_{GUT} \cong 1/24$) & GAP ($10^{16} - 10^{18}$) GeV.
- SUSY (to stabilize $m_F/m_P \cong 10^{-17}$).
- RQST (to quantize Gravity).

② Gauge Principle (hidden and expanded dimensions).

— How a Fundamental Force is generated: SU(3); SU(2); U(1) and Gravity.

③ The Physics of Imaginary Masses: SSB.

— The Imaginary Mass in SU(2)×U(1) produces masses ($m_{W^\pm}$; m_{Z^0}; m_q; m_l), including $m_\gamma = 0$.

— The Imaginary Mass in SU(5)⇒SU(3)×SU(2)×U(1) or in any higher Symmetry Group (not containing U(1)) ⇒ SU(3)×SU(2)×U(1) produces Monopoles.

— The Imaginary Mass in $SU(3)_c$ generates Confinement.

④ Flavour Mixings & CP ≠ , T ≠ .

— No need for it but it is there.

⑤ Anomalies & Instantons.

— Basic Features of all Non-Abelian Forces.

Note:

q ≡	quark and squark;	m_F ≡	Fermi mass scale;
l ≡	lepton and slepton;	m_P ≡	Planck mass scale;
G ≡	Gauge boson and Gaugino;	k ≡	quadrimomentum;
H ≡	Higgs and Shiggs;	C ≡	Charge Conjugation;
RGEs ≡	Renormalization Group Equations;	P ≡	Parity;
GUT ≡	Grand Unified Theory;	T ≡	Time Reversal;
SUSY ≡	Supersymmetry;	≠ ≡	Breakdown of Symmetry Operators.
RQST ≡	Relativistic Quantum String Theory;		
SSB ≡	Spontaneous Symmetry Breaking.		

The five basic steps in our understanding of nature. ① The renormalization group equations (RGEs) imply that the gauge couplings (α_i) and the masses (m_j) all run with k^2. It is this running which allows GUT, suggests SUSY and produces the need for a non point-like description (RQST) of physics processes, thus opening the way to quantize gravity. ② All forces originate in the same way: the gauge principle. ③ Imaginary masses play a central role in describing nature. ④ The mass-eigenstates are mixed when the Fermi forces come in. ⑤ The Abelian force QED has lost its role of being the guide for all fundamental forces. The non-Abelian gauge forces dominate and have features which are not present in QED.

Figure 11

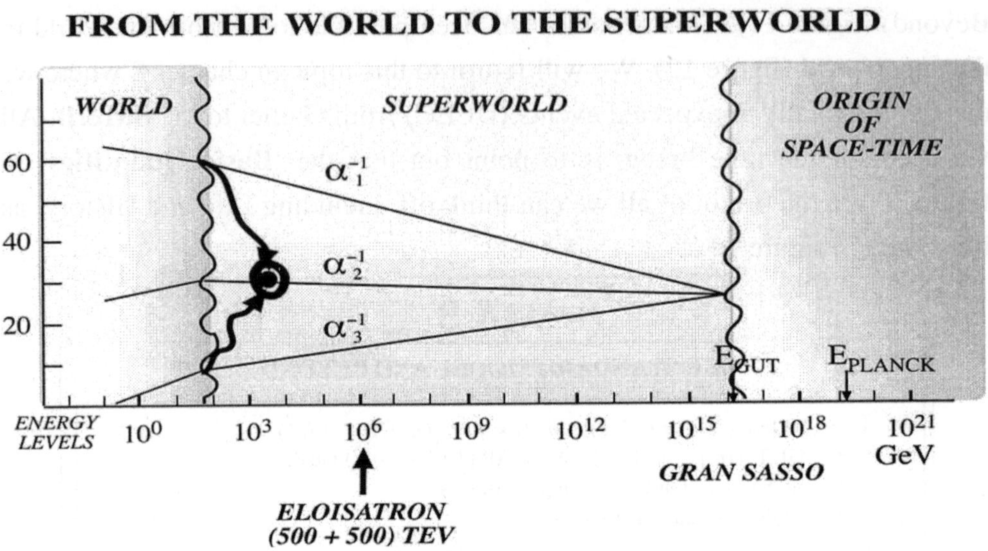

Figure 12

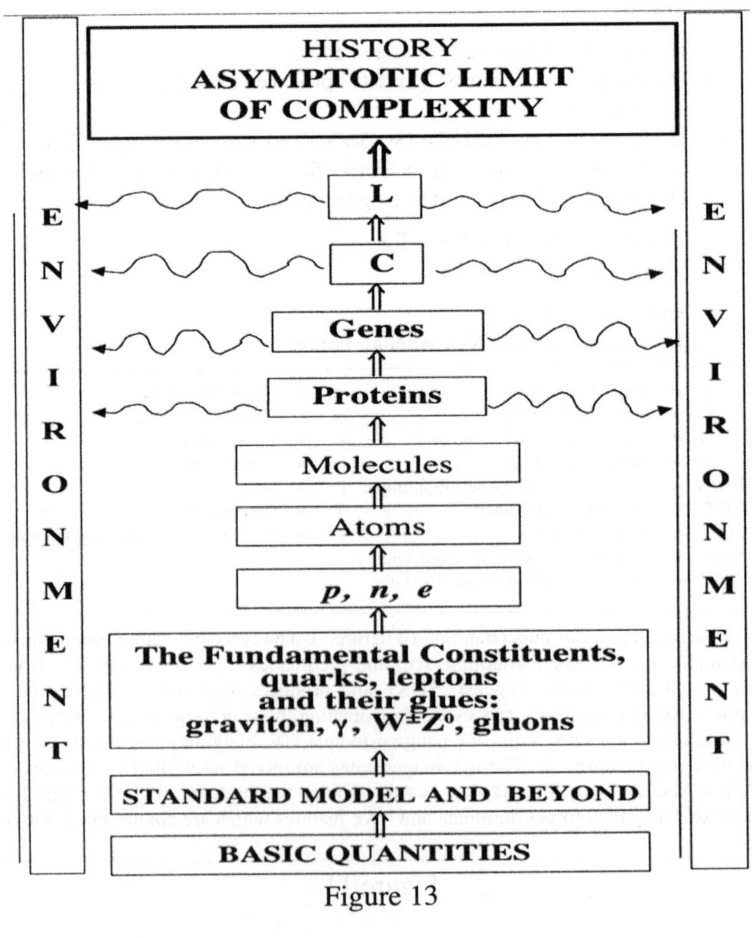

Figure 13

7 A FEW EXAMPLES OF AFB PHENOMENA: FROM BEETHOVEN TO THE SUPERWORLD

Let me now mention a few examples of AFB phenomena in Science.

Beethoven and the laws of acoustics.

Beethoven could compose superb masterpieces of music without any knowledge of the laws governing acoustic phenomena. But these masterpieces could not exist if the laws of acoustics were not there.

The living cell and QED.

To study the mechanisms governing a living cell, we do not need to know the laws of electromagnetic phenomena whose advanced formulation is called Quantum ElectroDynamic, QED. All mechanisms needed for life are examples of purely electromagnetic processes. If QED was not there Life could not exist.

Nuclear physics and QCD.

Proton and neutron interactions appear as if a fundamental force of nature is at work: the nuclear force, with its rules and its regularities. These interactions ignore that protons and neutrons are made with quarks and gluons. Nuclear physics does not appear to care about the existence of QCD, although all phenomena occurring in nuclear physics have their roots in the interactions of quarks and gluons. In other words, protons and neutrons behave like Beethoven: they interact and build up nuclear physics without 'knowing' the laws governing QCD.

The most recent example of Anderson-Feynman-Beethoven-type phenomenon. the world could not care less about the existence of the superworld.

A few words on how is the world made. A flower, the sea, the air we breathe, the Moon, the Stars, the Sun itself, whatever we call World is made of Fermions and Bosons.

- Fermions: (with 'half-integer' spin like the electron). These particles – called quarks and leptons – are the 'bricks', (shown in figure 14), more precisely, the tiny 'tops', of our material existence.
- Bosons: (with 'integer' spin like the photon). Particles having integer 'spin' value are the 'glues' of the fundamental forces (shown in figure 15), acting between 'bricks'. The 'glues' are also

tiny ‘tops’. The ‘electroweak’ and the subnuclear forces have the minimum value of integer spin: one in Planck unit, the ‘gravitational’ glue has twice this integer value of the spin.

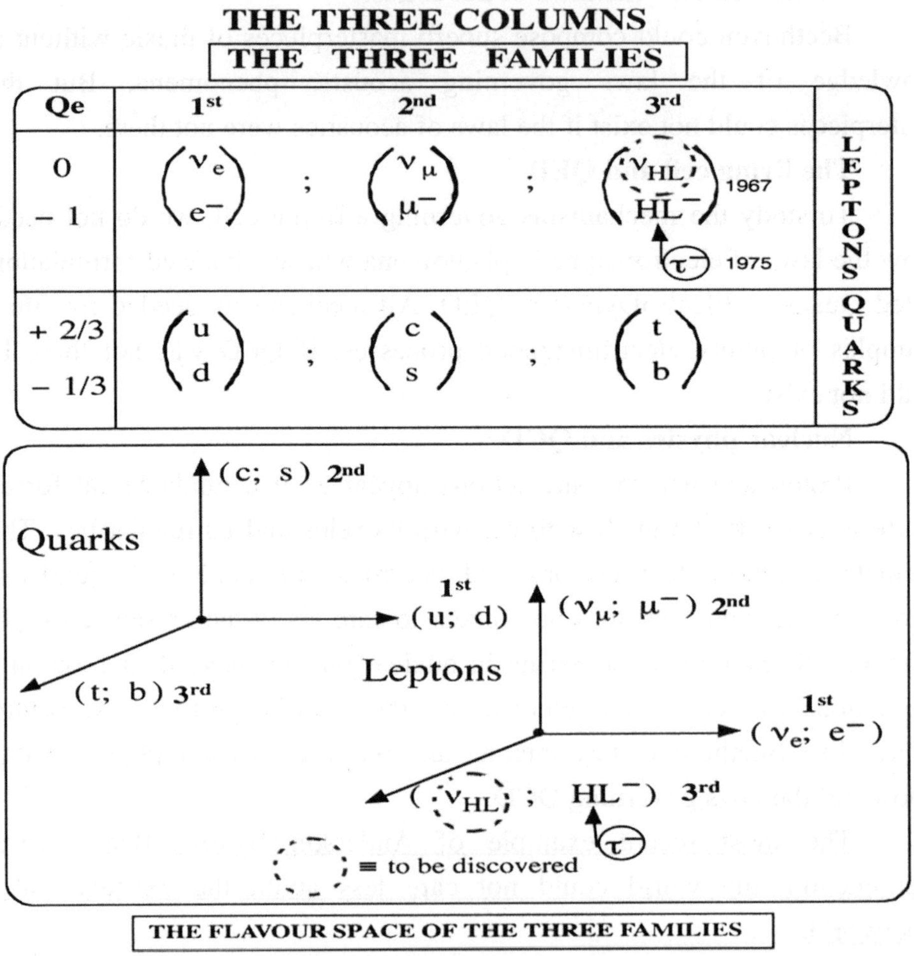

Figure 14

The upper panel of figure 14 shows the three families of quarks and leptons, together with the respective values of their electric charge, Q_e.

The lower panel shows the spaces of the so-called ‘subnuclear flavour’ charges (with additively conserved quantum numbers).

Notice that the third family lepton τ^- was originally (1967) named HL^- by the BCF (Bologna-CERN-Frascati) Group [8–11]. Its partner ν_{HL}, the existence of which was postulated in 1967 in the experimental proposal by the BCF group to search for heavy leptons at Frascati, would have been directly observed only recently.

The upper panel of figure 15 shows a standard Lorentz Space-Time to illustrate how gravitational Forces originate from the fact that, in any point $(i\vec{x}, t)$ of such a Space, a change of reference system is possible without altering any physical result. In the lower panel the origin of the other forces, the [SU(2) × U(1)] electroweak one and the [SU(3)] strong one, is sketched.

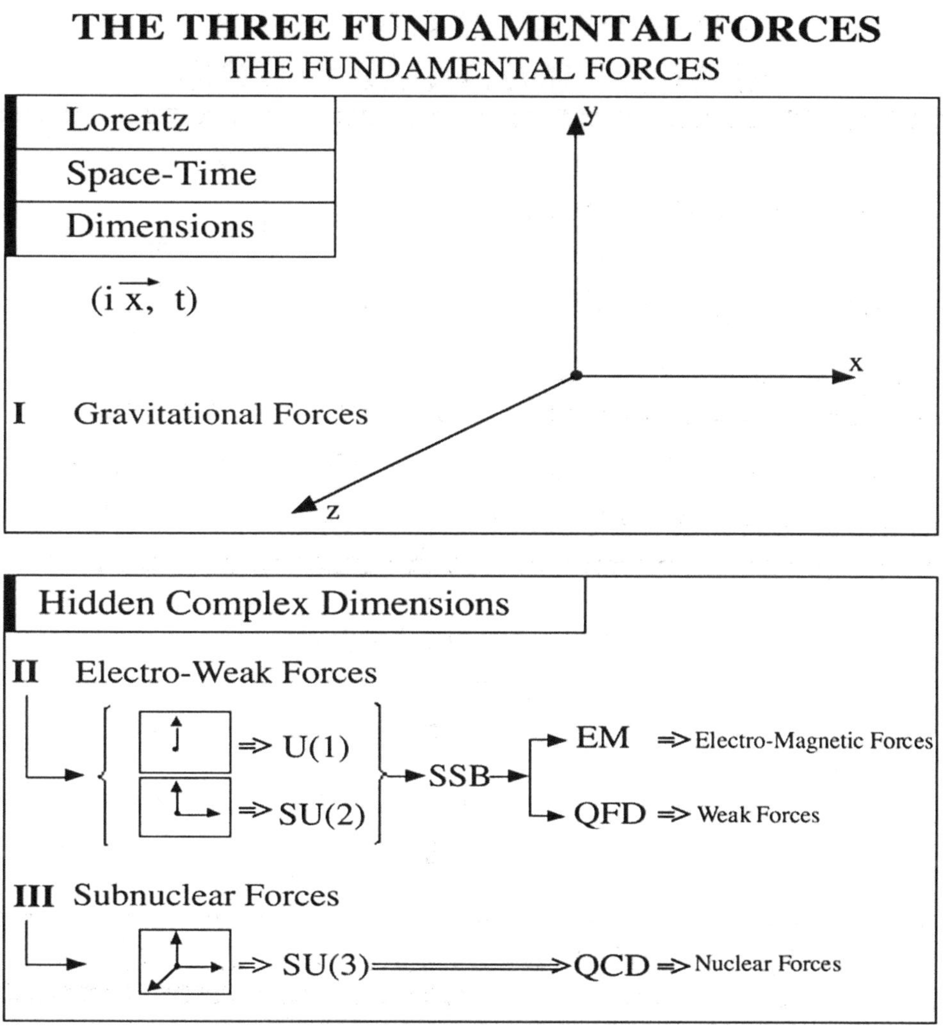

Figure 15

These forces originate from the possibility to operate in fictitious Spaces with **one** [U(1)], **two** [SU(2)] and **three** [SU(3)] **complex dimensions**, under the condition of local invariance of the physics results. This is the reason why the glues have integer spin.

Why do the columns have half-integer spin we do not know: it is an experimental unexpected discovery. Can columns with integer spin and forces with half-integer spin exist? The answer is Yes if we introduce supersymmetry whose foundation is superspace (figure 16).

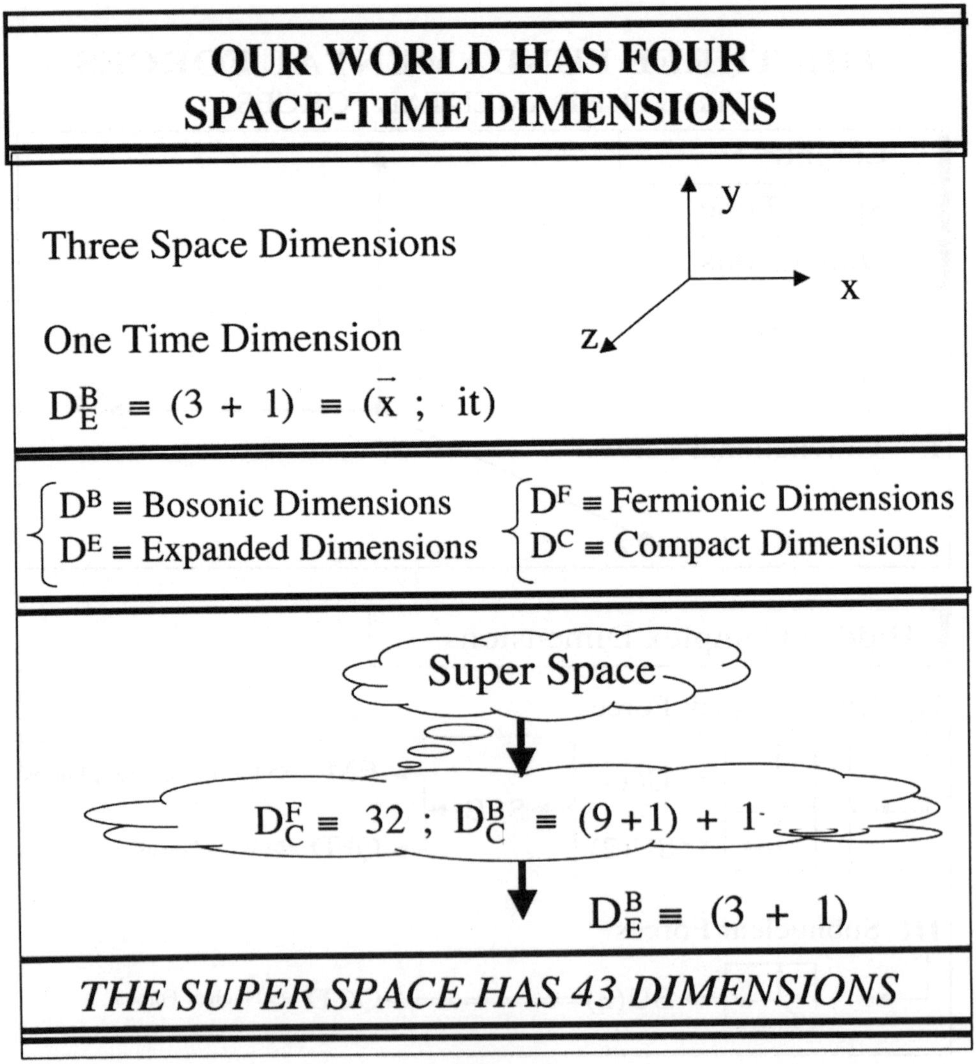

Figure 16

Our space-time has (3+1) bosonic dimensions; superspace has [(9+1) + 1] bosonic and 32 fermionic dimensions.

The (9+1) dimensions are 9 for Euclidian space and 1 for time, while the additional + 1 is the compactified dimension which produces the gravitational gauge coupling.

Where could be the proof that we come from the superworld?

The ashes of the superworld (the so-called neutralinos) could explain the compactness of our Galaxy (as in the drawing of figure 17).

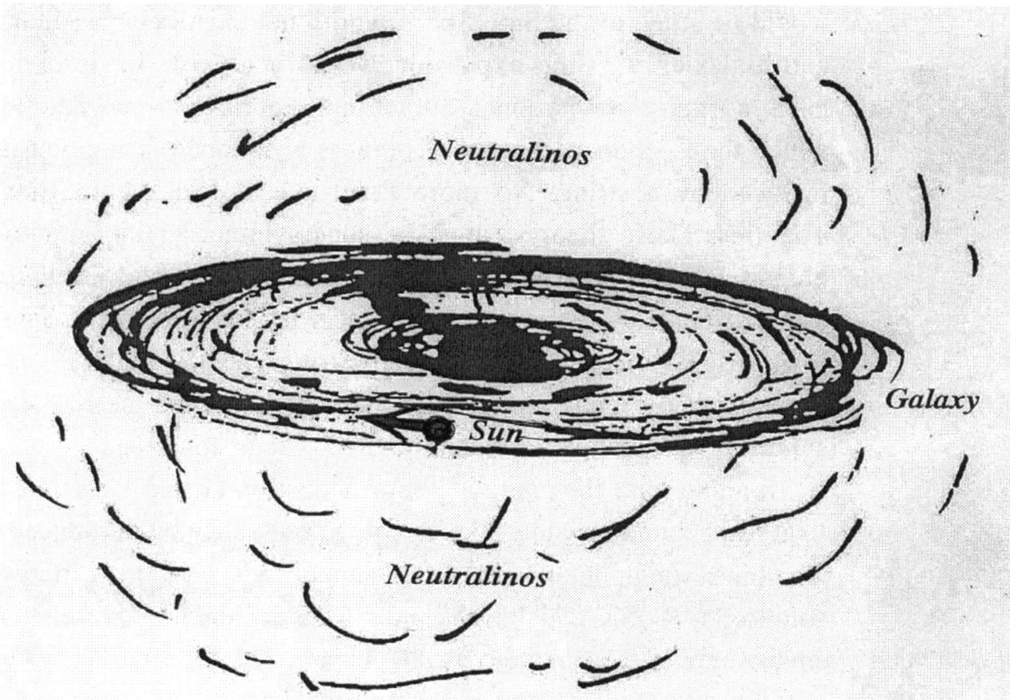

Figure 17

Neutralinos cannot aggregate into Stars since, being neutral, they lose little energy. This would allow neutralinos to remain in a sphere concentric with our Galactic centre.

Even though they aggregated into Stars, neutralinos could not emit light, like ordinary Stars do.

Fire needs the plasma of protons and electrons. This is why super Stars cannot exist.

WHY DO WE NEED THE SUPERWORLD

There are fundamental reasons making the superworld a need.

1) The two energy scales must be kept separate: 10^{19} GeV (Planck) and 10^{2} GeV (Fermi).
2) The gravitational attraction of light must be prevented from being infinite. Otherwise we could see neither the light from Stars nor our light. The 'gravitino' (supergravity) allows the gravitational attraction of light to be finite.
3) Gravitational attraction is powerful but it cannot be infinite. We would be stuck to the Sun. Space would not exist between Stars and Galaxies. Cosmic expansion would not exist. In order to have a finite gravitational attraction, theories are needed in which the Euclidean concept of point is abandoned. The point is replaced by a string. No more Point-like Theories but String Theories. These theories must be supersymmetric: the already quoted supersymmetry law ($F \equiv B$) must be valid in their mathematical structure. Otherwise 'tachions' would appear. This is the origin of Relativistic Quantum String Theory (RQST).
4) If we want the unification of all fundamental phenomena – the synthesis of which is provided by three 'gauge couplings', α_1 α_2 α_3, running with the energy (Renormalization Group Equations) – the supersymmetry law ($F \equiv B$) must necessarily be introduced.
5) An interesting detail: no scale-supergravity is an infrared solution of RQST. This might allow to understand the extremely small value of the Cosmological Constant.
6) Finally: why Three Columns and Three Forces? The answer to this question should come once we will be able from the compactification of the 43-dimensional superspace to go to our present world with (3+1) Space-Time dimensions.
7) **Note**: **A big problem**. Supersymmetry does not show up at our energy scale. Hence the problem arises to compute the energy above which the ($F \equiv B$) Law starts to act. Thanks to the EGM effect, this energy level is 700 times more accessible than thought so far.

These six reasons are the evidence why we need the superspace with 43 dimensions. But the space we live has only 4 dimensions and all the properties of the real world described in chapter 4 could not care less about the fundamental structure needed for us to exist as we do.

8 TOTALLY UNEXPECTED DISCOVERIES: UEEC EVENTS FROM GALILEO GALILEI TO THE STANDARD MODEL AND BEYOND

In this chapter we briefly recall the sequence of UEEC events from Galileo Galilei to the SM&B. In figure 18 these is the group of UEEC events from Galilei to 1947, already used to compare History and Science on the basis of 'What if?'. In fact, all 'What if?' are UEEC events, as already pointed out in chapter 5.

'UEEC'
TOTALLY UNEXPECTED DISCOVERIES
FROM GALILEI TO FERMI-DIRAC AND THE 'STRANGE' PARTICLES

I	Galileo Galilei discovery of $F = mg$.
II	Newton discovery of $$F = G\frac{m_1 \cdot m_2}{R_{12}^2}$$
III	Maxwell discovers the unification of electricity, magnetism and optical phenomena, which allows him to conclude that light is a vibration of the EM field.
IV	Planck discovery of $h \neq 0$.
V	Lorentz discovers that space and time cannot be both real.
VI	Einstein discovers the existence of time-like and space-like worlds. Only in the time-like world, simultaneity does not change, with changing observer.
VII	Rutherford discovers the nucleus.
VIII	Dirac discovers his equation, which implies the existence of the antiworld.
IX	Fermi discovers the weak forces.
X	The 'strange particles' are discovered in the Blackett Lab.

Figure 18

In figure 19 we show the other impressive sequence of UEEC events which has brought us to the construction of the Standard Model. A description of the content of figure 19 would take us out of the main purpose of this work. The interested reader can find it in the volume 'Subnuclear Physics' quoted in chapter 10 [12].

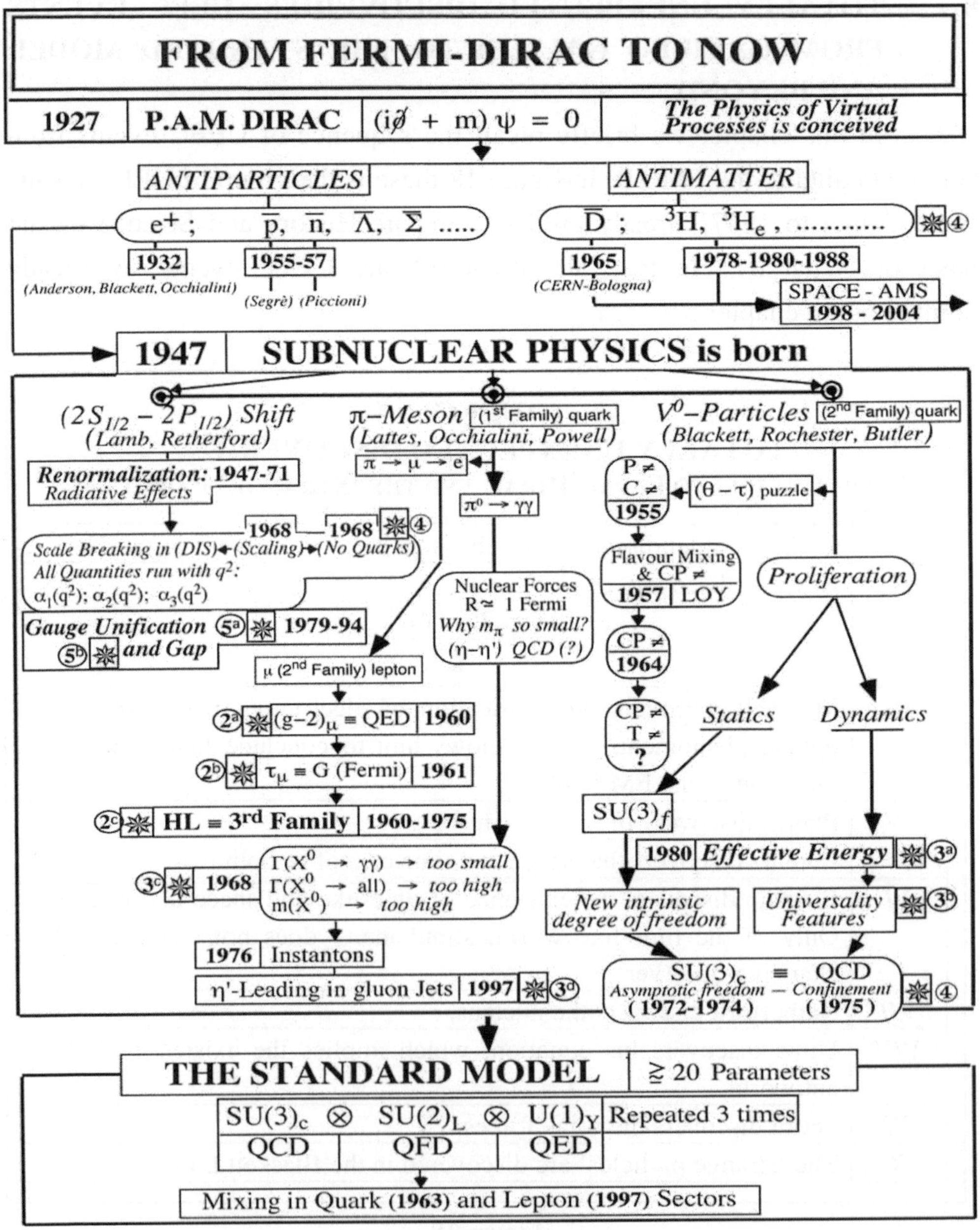

Figure 19

In figures 20 and 21 there are some further relevant details on QFD, QED and QCD which are the 3 boxes on the bottom of figure 19. Finally in figure 11 (from chapter 6), we show the sequence of UEEC events which have produced the new frontier of our knowledge, i.e. **SM&B**.

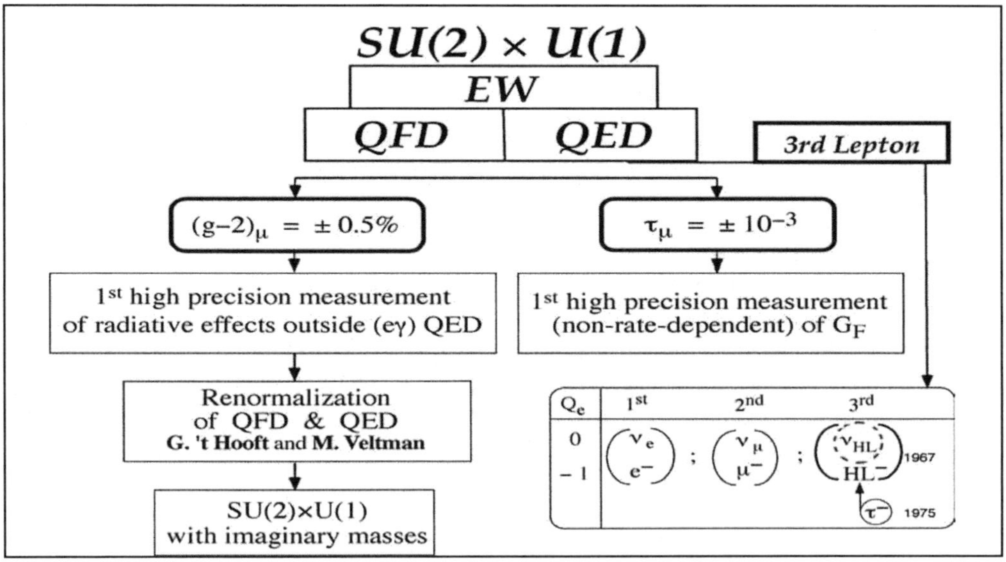

Figure 20

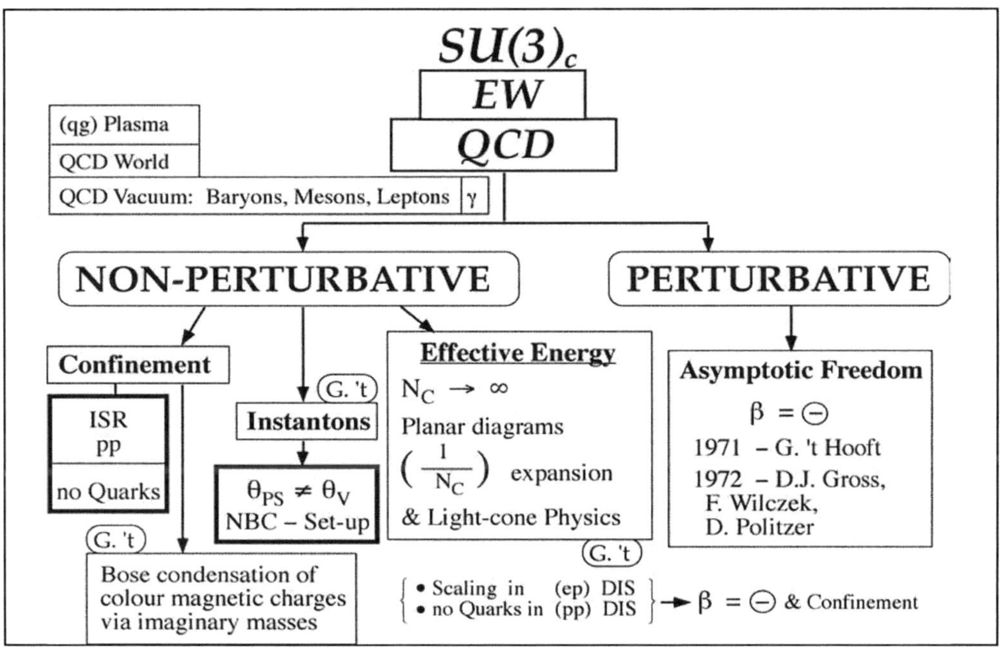

Figure 21

SM&B

THE STANDARD MODEL AND BEYOND

① RGEs (α_i $(i \equiv 1, 2, 3)$; m_j $(j \equiv q, l, G, H)$) : $f(k^2)$.

- GUT ($\alpha_{GUT} \cong 1/24$) & GAP ($10^{16} - 10^{18}$) GeV.
- SUSY (to stabilize $m_F/m_P \cong 10^{-17}$).
- RQST (to quantize Gravity).

② Gauge Principle (hidden and expanded dimensions).

— How a Fundamental Force is generated: SU(3); SU(2); U(1) and Gravity.

③ The Physics of Imaginary Masses: SSB.

— The Imaginary Mass in SU(2)×U(1) produces masses ($m_{W^\pm}$; m_{Z^0}; m_q; m_l), including $m_\gamma = 0$.

— The Imaginary Mass in SU(5)⇒SU(3)×SU(2)×U(1) or in any higher Symmetry Group (not containing U(1)) ⇒ SU(3)×SU(2)×U(1) produces Monopoles.

— The Imaginary Mass in $SU(3)_c$ generates Confinement.

④ Flavour Mixings & CP ≠ , T ≠ .

— No need for it but it is there.

⑤ Anomalies & Instantons.

— Basic Features of all Non-Abelian Forces.

Note:

q	≡	quark and squark;	m_F	≡	Fermi mass scale;
l	≡	lepton and slepton;	m_P	≡	Planck mass scale;
G	≡	Gauge boson and Gaugino;	k	≡	quadrimomentum;
H	≡	Higgs and Shiggs;	C	≡	Charge Conjugation;
RGEs	≡	Renormalization Group Equations;	P	≡	Parity;
GUT	≡	Grand Unified Theory;	T	≡	Time Reversal;
SUSY	≡	Supersymmetry;	≠	≡	Breakdown of Symmetry Operators.
RQST	≡	Relativistic Quantum String Theory;			
SSB	≡	Spontaneous Symmetry Breaking.			

The five basic steps in our understanding of nature. ① The renormalization group equations (RGEs) imply that the gauge couplings (α_i) and the masses (m_j) all run with k^2. It is this running which allows GUT, suggests SUSY and produces the need for a non point-like description (RQST) of physics processes, thus opening the way to quantize gravity. ② All forces originate in the same way: the gauge principle. ③ Imaginary masses play a central role in describing nature. ④ The mass-eigenstates are mixed when the Fermi forces come in. ⑤ The Abelian force QED has lost its role of being the guide for all fundamental forces. The non-Abelian gauge forces dominate and have features which are not present in QED.

Figure 11

9 COMPLEXITY EXISTS AT ALL SCALES

The conclusion is that we do have AFB *phenomenology* in Science ≡ EBUS, and *Sarajevo-type events* (UEEC) in Science ≡ EBUS. Therefore: *Sarajevo-type events* and *Beethoven-type phenomenology* both exist in History ≡ EWRL as well as in Science ≡ EBUS. There is no question that nature shows structures which are considered complex on the basis of AFB and UEEC events. What should be considered as not granted is that in order to unreveal the roots of AFB and UEEC events, we must go from Reductionism to Holism. In fact, as shown in chapters 7 and 8 these two basic elements, AFB and UEEC – which are at the origin of Complexity, with its consequences permeating all our existence, from molecular biology to life in all its innumerable forms up to our own, including History – do exist at the fundamental level [1–4]. It turns out that Complexity in the real world exists, no matter the mass-energy and space-time scales considered [1–4], as illustrated in figure 22.

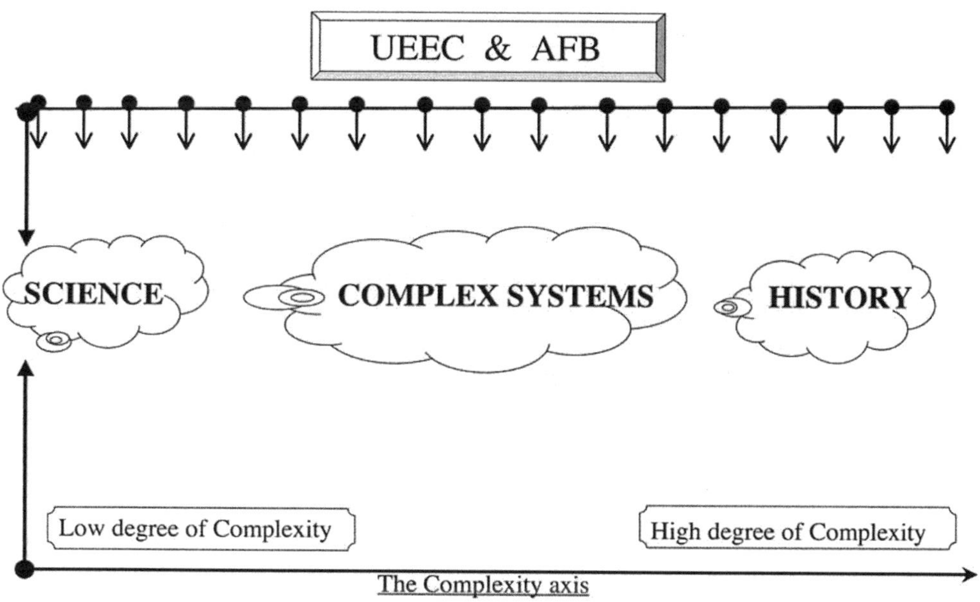

Figure 22

At this point it is interesting to analyse how progress is made in advanced research, where UEEC events do take place. This is done in the following chapter 10.

10 THIS IS HOW PROGRESS IS MADE IN ADVANCED RESEARCH *'Logical Reasoning is Difficult'*

On page 209 of my book (whose front-page is reported below) '*Subnuclear Physics – the first 50 years*' [12] the following sentence is quoted: '*Copying is easy, logical reasoning is difficult.*' The author being Gerardus 't Hooft, Erice 1997.

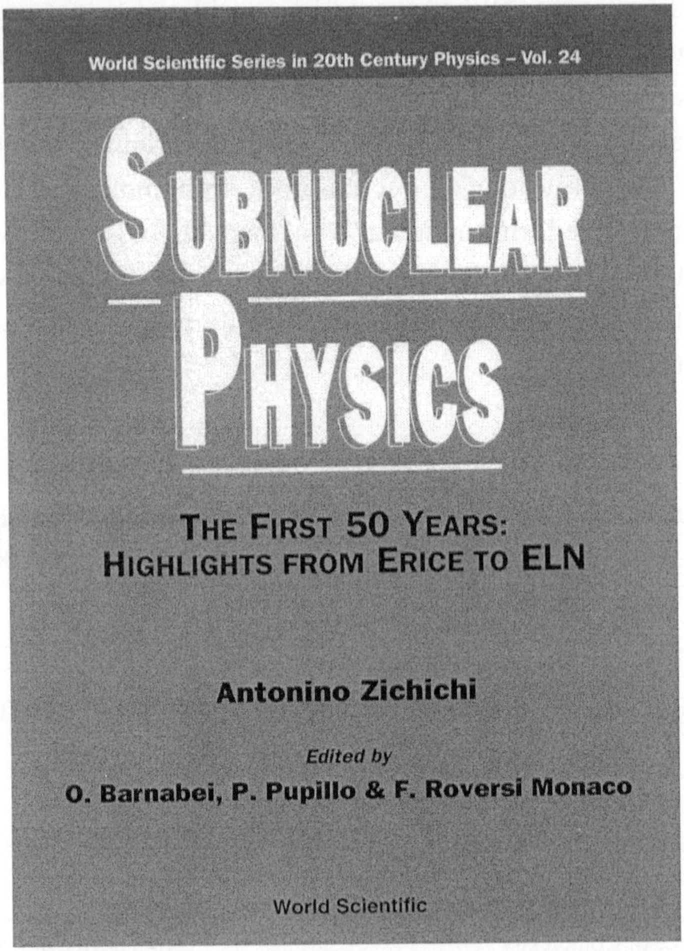

Figure 23

With his statement, my great friend Gerardus, worded his view on the progress made, and progress still to be made, in theoretical physics.

On the occasion of his 60th Anniversary Celebrations I gave my own testimony on the validity of 't Hooft's statement in experimental physics [13].

In fact, in experimental physics as well, it is not enough to have an original idea. My great teacher, Patrick Blackett, used to tell us, young fellows of his group: '*We experimentalists are not like theorists: the originality of an idea is not for being printed in a paper, but for being shown in the implementation of an original experiment.*' To reach the level of performing an original experiment corresponds to bring 'logical reasoning' to its asymptotic limit of reality. Thus, both in theory and in experiment, the progress of physics is due to those who have the perseverance of not only having an original idea, but of investigating its logical structure in terms of its consequences. At the end of this series of creative steps what is always found is an UEEC event. Here a few examples from my own past experience follow.

The third lepton. *This is UEEC event no. 1 in chapter 13.*

In the late fifties, I realized that if the pion mass was not what it was, the muon had very little chance of being so obviously present everywhere; and if a new lepton of 1 GeV mass (or heavier) would have been there, no one would have seen it; I did not limit myself to discussing this topic with a few colleagues; I followed Blackett's teaching. And this is how I realized that the best 'signature' for a heavy lepton would have been '$e\mu$' acoplanar pairs; this is how I invented the 'preshower' to improve electron identification by many orders of magnitude; this is why I studied how to improve muon identification; this is how I experimentally established that the best production mechanism could not be ($p\,\overline{p}$), but (e^+e^-) annihilation. We will return to the third lepton in chapter 13.

Matter-Antimatter Symmetry. *This is UEEC event no. 2 in chapter 13.*

In the sixties, the need to check the symmetry between nuclear matter and antimatter came to the limelight. The reason being the apparent triumph of the S-matrix theory to describe strong interactions and the violation of the 'well-established' symmetry operators (C, P, CP, T) in weak interactions and in the K-meson decay physics. When the discovery of scaling in Deep Inelastic Scattering (DIS) and the non-breaking of the protons in high energy collisions come in the late sixties, the basic structure of all Relativistic Quantum Field Theories (RQFT) were put in serious difficulties, and therefore the validity of the celebrated CPT theorem. On the other hand, the basic reason why nuclear antimatter had to exist was CPT.

In the early sixties the first example of nuclear antimatter, the antideuteron, had been searched for and found not to be there at the level of one antideuteron per 10^7 pions produced. I did not limit myself to saying that it would have been important to build a beam of negatively charged 'partially separated' particles in order to have a very high intensity. I did not limit myself to suggesting a very advanced electronic device in order to increase, by an order of magnitude, the accuracy for time-of-flight (TOF) measurements. I did bring all my ideas to the point of full implementation in a detailed experiment, where the antideuteron was found, thus proving nuclear matter-antimatter symmetry. Therefore credence could be given to CPT and to RQFT. We will return to matter-antimatter symmetry in chapter 13.

The mesonic mixings: $\theta_{PS} \neq \theta_V \neq 0$. *This is UEEC event no. 5 in chapter 13.*

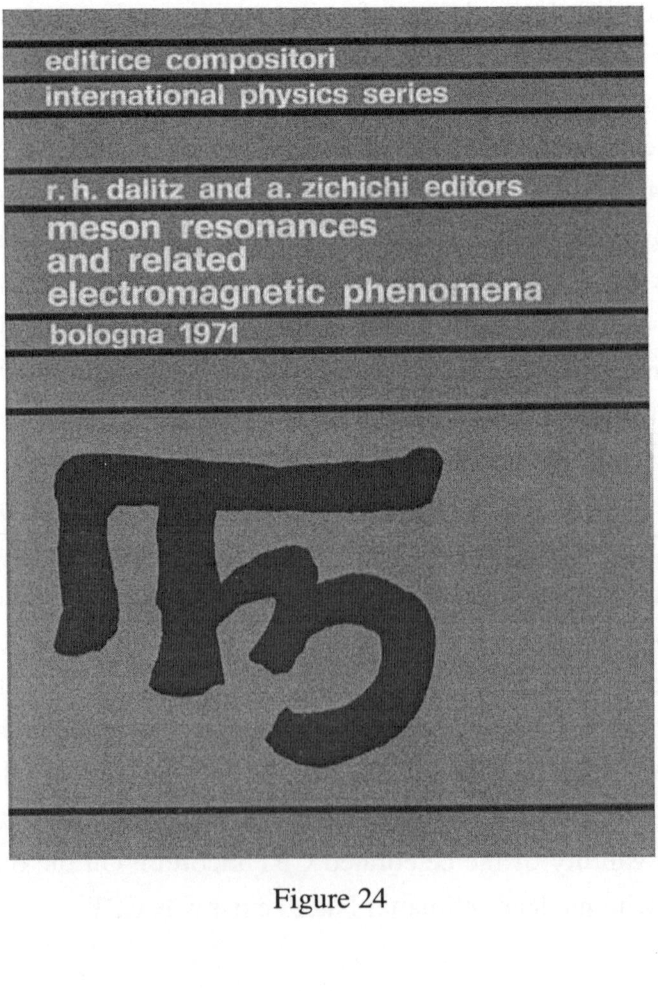

Figure 24

The problem of concern in the physics of strong interactions was the 'mixing' in meson physics. It was necessary to know why this mixing was there and why the vector mesons (ρ, ω, ϕ) did not show the same behaviour as the pseudoscalar mesons (π, η, η'). At the end of the 'logical reasoning' in terms of experimental searches never conducted before (figure 24 is the cover-page of a volume dedicated to this topic), the result was that the mesonic mixing was there and the two mixing angles were drastically different: $\theta_{PS} \neq \theta_V$. This is what Dick Dalitz defined the most significant results from all mesonic physics [14]. We will return to this topic in chapter 13.

In homage to Richard Dalitz let me show figure 25.

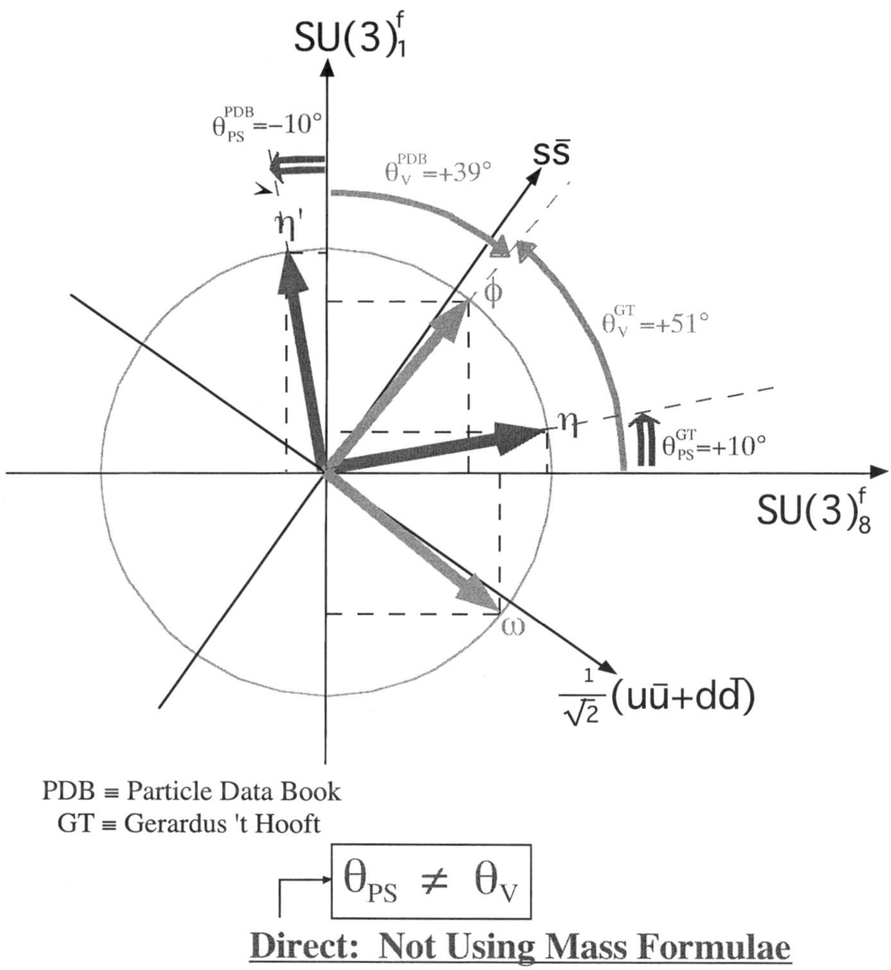

Figure 25

Let me go back to 'logical reasoning' in experimental physics. I did not limit myself to saying that the most appropriate way to study this problem [(e^+e^-) colliders did not yet exist], was to measure with the best possible accuracy the electromagnetic decay rates of the vector mesons

$$(\rho \to e^+e^-),$$
$$(\omega \to e^+e^-),$$
$$(\phi \to e^+e^-),$$

and to see if the heaviest meson (known at that time with the symbol X^0) was decaying into two γ's ($X^0 \to \gamma\gamma$).

These were times when experimental physics was dominated by bubble chambers. I designed and built a non-bubble-chamber detector, NBC; it consisted of an original neutron missing mass spectrometer coupled with a powerful electromagnetic detector which allowed to clearly identify all final states of the decaying mesons into (e^+e^-) or $(\gamma\gamma)$ pairs. The mass of the meson (be it pseudoscalar or vector) was measured by the neutron missing mass spectrometer. The two 'mixing angles', the pseudoscalar θ_{PS} and the vector θ_V, where directly measured (without using the masses) to be, not as expected by $SU(3)_{uds}$, i.e. $\theta_{PS} = \theta_V = 0$, but, $\theta_{PS} \neq 0$, $\theta_V \neq 0$ and totally different $\theta_{PS} \neq \theta_V$. Many years were needed and the Gerard's instantons to explain why $\theta_{PS} \simeq 10°$ and $\theta_V \simeq 51°$.

The proton does not break into 3 quarks despite 1968 Panofsky. *This is UEEC event no. 4 in chapter 13.*

When in 1968 I heard Pif (W.K.H.) Panofsky reporting in Vienna on (ep) deep-inelastic-scattering, whose immediate consequence was that 'partons' inside a proton behaved as 'free' particles, I did not limit myself to saying that it would have been interesting to check if, in violent (pp) collisions, 'free' partons were produced. Since the 'partons' were suspected to be the quarks earlier suggested by M. Gell-Mann and G. Zweig (we now know that partons can also be gluons), the experiment needed was a search for fractionally charged particles in the final states of violent (pp) interactions at the CERN ISR. To perform the experiment, a new type of plastic scintillator was needed, with very long attenuation length since the counters had to be put inside a very big magnet. These scintillators did not exist on the market. We studied the problem and built the most powerful and sensitive scintillators. The result was that free quarks were not produced, despite the violent (pp) collisions.

The Gribov QCD light. *This is UEEC event no. 6 in chapter 13.*

When the physics of strong interactions finally became the physics of quarks and gluons, QCD had a problem, defined by Gribov as being its 'hidden side': i.e., the large number of different final states produced by different pairs of interacting particles, such as (πp, pp, $\bar{p}$p, Kp, e^+e^-, νp, μp, ep, etc.). I did not limit myself to suggesting that a totally different approach was needed to put all these final states on the same basis. I found what this basis could be and this is how the 'Effective Energy' became the correct quantity to be measured in each interaction. The 'Effective Energy' was not predicted by QCD. To perform this study, it was necessary to analyze tens of thousands of (pp) interactions at the ISR. This was done despite all the difficulties to be overcome. And this is how QCD light was discovered (figures 26 and 27). We will return to the 'QCD light' in chapter 13.

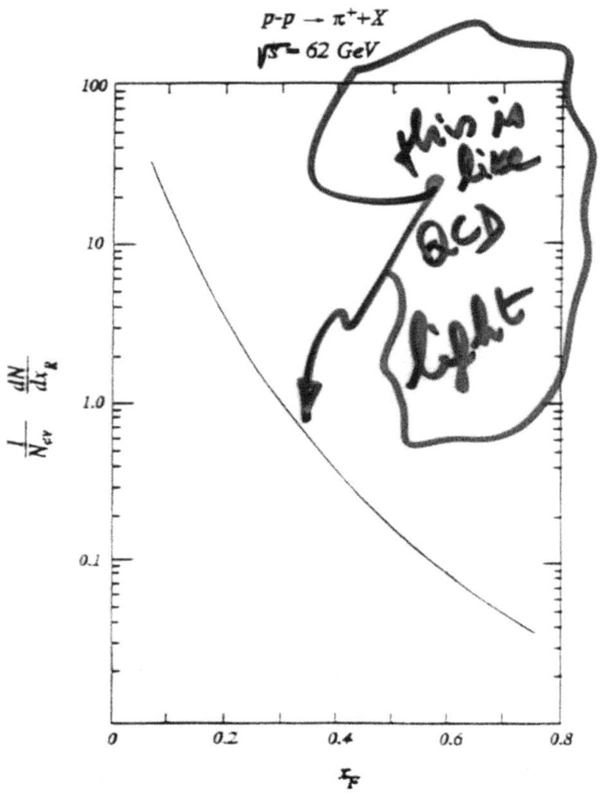

Figure 26

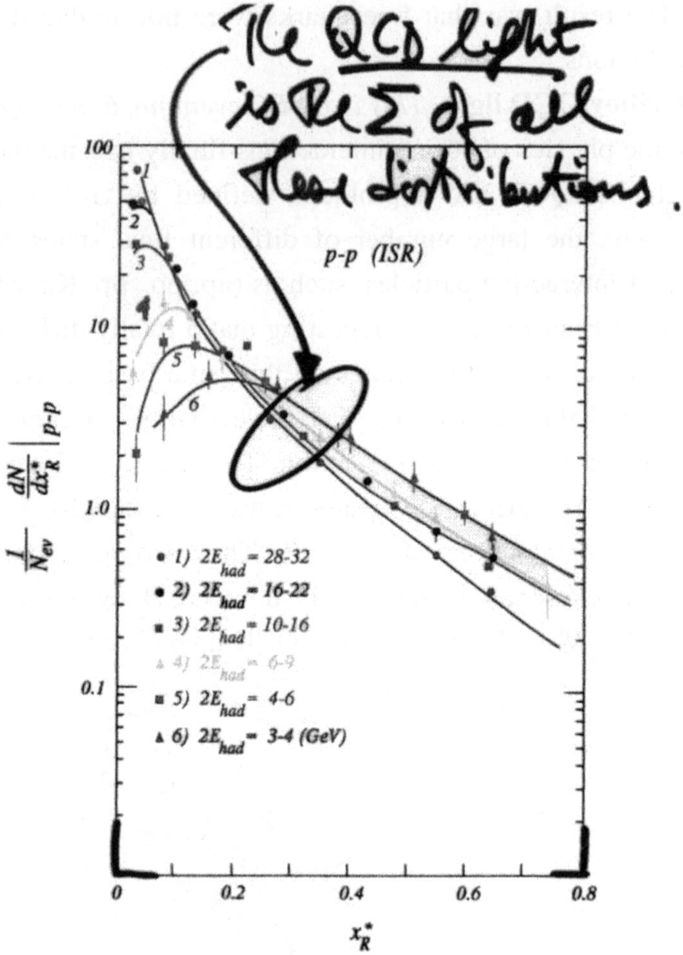

Figure 27

So, when a new problem appears, the only way out is to bring the logical reasoning – be it of experimental, theoretical or technical nature – to the deepest level of consequences. At the very end of this 'logical reasoning' what is found is a UEEC event: i.e. a result which was totally unexpected.

This is how progress is made in the most advanced frontier of reductionism.

And now we move to the platonic Simplicity. This is related to the simplest definition of Complexity discussed in chapter 3.

We will first discuss the Platonic Grand Unification (chapter 11) and then the platonic supersymmetry (chapter 12).

11 THE PLATONIC GRAND UNIFICATION

Let us now move towards the Platonic structure of a Grand Unification, taking as basic points the gauge principle and the SSB (Spontaneous Symmetry Breaking) mechanism, which represent the conceptual structure of the Standard Model.

The simplest way to have a Platonic Grand Unification is to have one and only one basic fundamental particle, B. This particle must obey the very simple symmetry law which puts fermions and bosons on the same basis. This basic fundamental particle B can therefore exist either as being a boson B_B or as being a fermion B_F.

Let us consider first B_B. The fundamental forces exist because a basic fundamental boson B_B exists. Figure 28 illustrates the simple sequence which generates all known forces of nature.

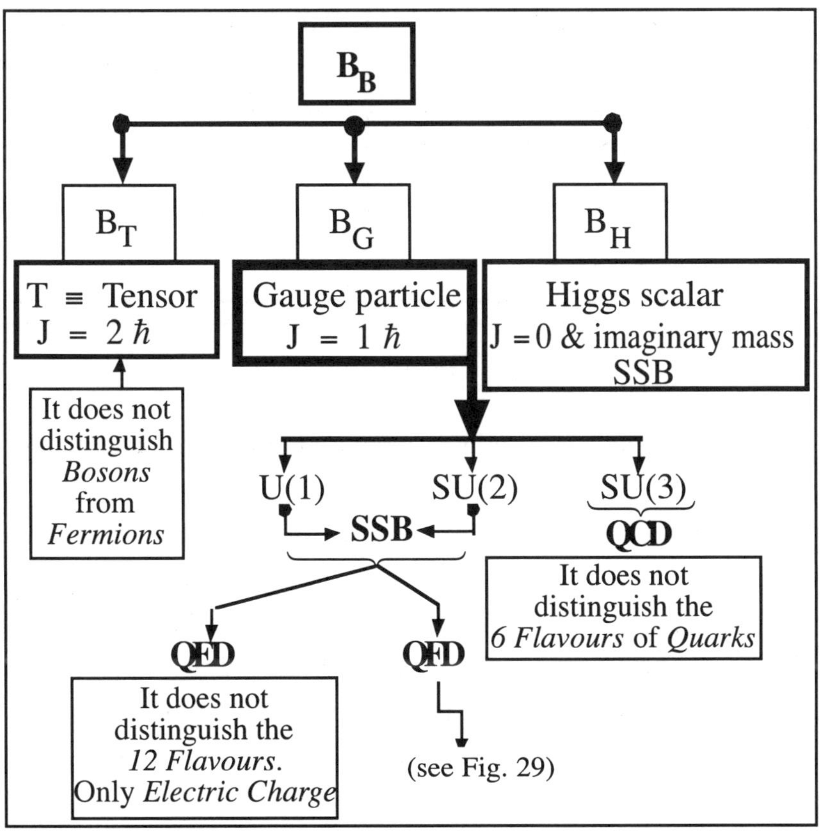

Figure 28

At the bottom of figure 28 there is the QFD force, illustrated in figure 29. The 'platonic' Simplicity suffers a further deviation.

QFD

$(q^u_m)_{1,2,3}$ $(q^d_m)_{1,2,3}$	$(\ell^u_m)_{1,2,3}$ $(\ell^d_m)_{1,2,3}$
Experimentally proved	*Experimentally proved*
mixing & C ≠; P ≠; CP≠	?

Figure 29

In fact, we need to introduce many complications. The quarks and the leptons are 'mixed'.

This mixing is indicated by the index m, while the indices 'u' and 'd' refer to the two types of flavours (up-type) and (down-type) which are present in each of the three families: 1, 2, 3. There is a further complication.

The two mixings for the 'up' and the 'down' flavours must be different. In the case of the quark, this mixing is experimentally measured. In the case of the leptons, the experimental results are with nearly half a century of delay, compared with the quark case.

Mixing and violation of symmetry laws (for charge conjugation, C, parity, P, and the product of the two, CP) are well-established in the quark case. In the leptonic sector, only future experiments will tell us if the same symmetry laws are violated.

There is no known reason why all these details – mixing of states and symmetry law violations – are needed. They have been experimentally discovered and show how many deviations from the simple 'platonic' structure are needed. So far we have developed the sequence of Platonic Deviations from *Simplicity*, starting from the basic fundamental boson B_B.

We show in figure 30 the deviations needed from the Platonic Simplicity, when we start from the basic fundamental fermion B_F. It has to be with 'quark' and 'lepton' flavours and have two flavours in each class (called family). The total number of flavours is *12*: *6 for quarks*, *6 for leptons*.

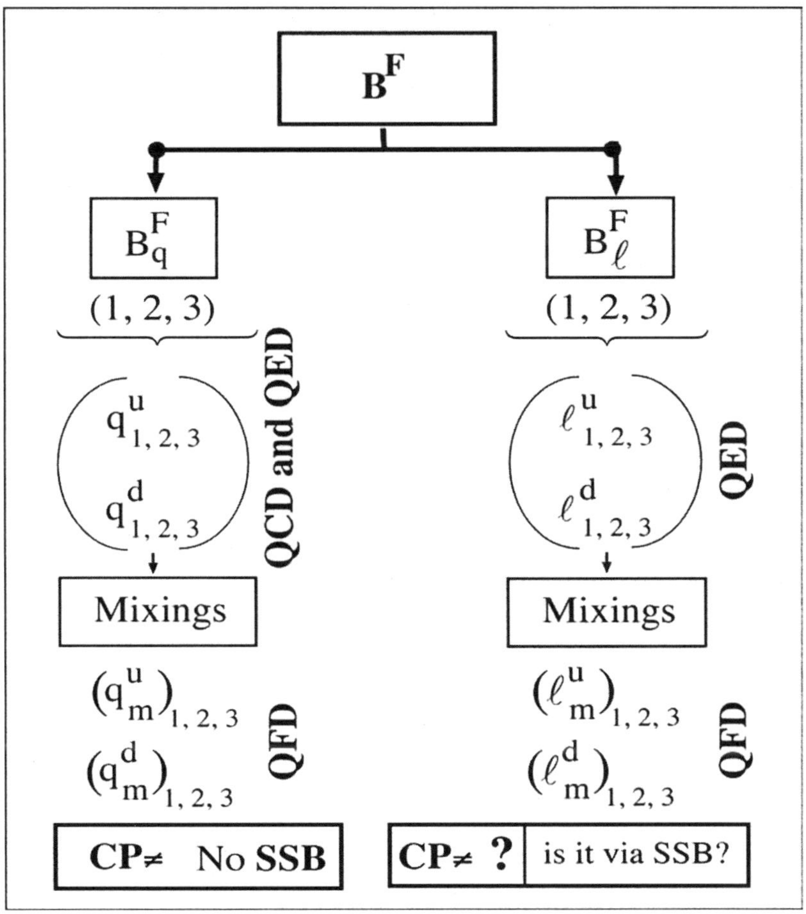

Figure 30

Why so many? The answer will probably come from the superspace with 43 dimensions compactified into (3+1).

The quark sector interacts with two forces, QCD and QED, while the lepton sector interacts using only QED. The QFD force comes into play only after all the mixings come in. No one knows why all these deviations from the Platonic Simplicity are needed.

The bold symbols, QCD, QED in the column

$$B_q^F$$

indicate that the 6 quark flavours interact via these two forces. In the lower part of the same column, the 'mixing' indicates that the quark states are no longer 'pure' states. They are 'mixed'; only these mixed states

$$\left(q_m^u\right)_{1,\,2,\,3} \quad \text{and} \quad \left(q_m^d\right)_{1,\,2,\,3}$$

interact via the QFD forces.

The column below

$$B_\ell^F$$

has the same structure, but the 'mixings' are not the same as in the 'quark' column.

Furthermore, no one knows at present if the symmetry CP is violated as it is in the quark case. This is why in the box CP ≠ there is a question mark. Another detail needs to be specified.

In the quark case, the CP symmetry breaking, CP ≠ , has been experimentally established not to be via the basic Standard Model mechanism, SSB.

A further deviation from simplicity.

In the leptonic case, we do not know if the CP symmetry is violated. It could be it is. In this case it will be interesting to know if it follows the SSB mechanism.

All these question marks are evidence of further deviations from the simple Platonic descriptions of natural phenomena.

A synthesis of the Platonic Grand Unification and the deviations needed is shown in figure 31.

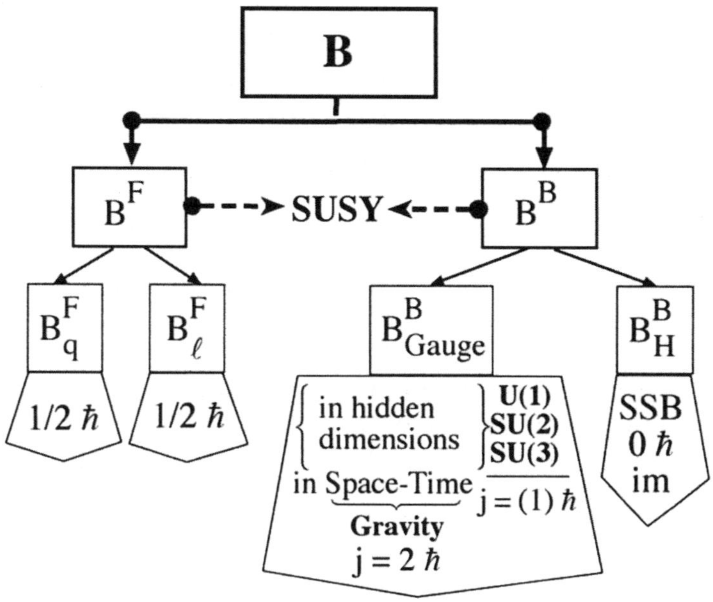

- The fundamental **forces** exist because a Basic Fundamental Boson **exists** ≡ B^B
- The fundamental **fermions** exist because a Basic Fundamental Fermion **exists** ≡ B^F
- The SSBs originate from the existence of many scalars with imaginary masses (im).

Figure 31

12 THE PLATONIC SUPERSYMMETRY

The Platonic concept of Supersymmetry is schematically reported in figure 30, where the basic point for a Platonic concept of Supersymmetry is given; i.e. the only fermions with spin (*1/2 h*) allowed to exist would be the 'gauginos'.

THE PLATONIC CONCEPT OF SUPERSYMMETRY

The **Gauge Principle** should generate a
Gauge Force ⟶ **Gauge Bosons**
If *NATURE* was platonically *SUPERSYMMETRIC*
Supersymmetry Transformation should generate **Gauginos**

1st **DEVIATION FROM PLATONIC SIMPLICITY**
OUR FERMIONS ARE NOT THE GAUGINOS

2nd **DEVIATION FROM PLATONIC SIMPLICITY**
THE FUNDAMENTAL FERMIONS ARE OF TWO DIFFERENT CLASSES: LEPTONS AND QUARKS

3rd **DEVIATION FROM PLATONIC SIMPLICITY**
THERE IS NOT ONLY ONE BUT THREE FAMILIES OF FUNDAMENTAL FERMIONS

4th **DEVIATION FROM PLATONIC SIMPLICITY**
THE FUNDAMENTAL FERMIONS BECAME MIXED WHEN THE WEAK FORCES ARE SWITCHED ON: MIXINGS EXIST

5th **DEVIATION FROM PLATONIC SIMPLICITY**
THERE ARE DIFFERENT MIXINGS

Figure 32

If the only allowed fermions would be the 'gauginos', there would be no way to have quarks and leptons.

Our fermions are not the gauginos. A deviation is needed. And this is the first one. Our fermions are in fact of two classes: quarks and leptons. Another deviation is needed to introduce quarks and leptons. And this is not enough: one family would not suffice. We need another deviation, the third one, in order to produce three families. Once again this is not enough.

We need a further deviation: the fundamental fermions became mixed when the weak forces are switched on. This fourth deviation is followed by another one, the fifth: the mixing of states in the quark sector and in the leptonic sector are different. Having proved that Platonic Simplicity is not at work in the Grand Unification and in the law of supersymmetry we go on illustrating a few examples of other deviations from Platonic Simplicity in the detailed construction of the Standard Model (SM). These deviations are coupled to UEEC events.

13 EXAMPLES OF UEEC EVENTS IN THE CONSTRUCTION OF THE STANDARD MODEL AND BEYOND

The Standard Model (SM) is the greatest synthesis of all times in the study of fundamental phenomena governing the Universe in its microscopic structure. We will see that the basic achievements of the SM have been obtained via UEEC events; moreover the SM could not care less about the existence of Platonic Simplicity. Platonic Simplicity is violated at every corner in the process of construction of the SM.

We will devote our analysis to this point. The conclusion is that Complexity exists at the elementary level. In fact, starting from Platonic Simplicity, the Standard Model needs a series of 'ad hoc' inputs. These inputs are the proof that at the fundamental level of scientific knowledge there is experimental evidence for the existence of the AFB phenomena and the UEEC events. A few cases where I have been directly involved are summarised in figure 33. Each UEEC event is coupled with a **despite**, in order to emphasize the reason why the event is unexpected.

UEEC EVENTS IN THE CONSTRUCTION OF THE SM&B

① **The 3rd lepton**
despite the abundance of neutrinos: ν_e (ν_μ) *another* ***one*** (ν_{HL}).

② **Antimatter**
despite *S-matrix and C, P, CP breaking.*

③ **Nucleon Time-like** EM structure
despite *S-matrix.*

④ **No quarks** in violent (pp) collisions
despite *scaling.*

⑤ **Meson mixings**
$\theta_V \neq \theta_{PS} : (51^o) \neq (10^o) \neq 0$ ***despite*** $SU(3)_{uds}$.

⑥ **Effective energy:** *the QCD-light*
despite *QCD.*

⑦ **The running** of α_1 α_2 α_3 versus **energy** at ***a point*** $\boldsymbol{E_{GU}}$.
(1979) (1991) ***despite*** *straight line convergence.*

EGM

Figure 33

The UEEC event no.1, the third lepton and the UEEC events no. 2, 4, 5, 6, have been discussed in chapter 10 in the contest of how progress is made in advanced research. Here, we further discuss the UEEC events no. 2, 5, 6 in the logic of corroborating their UEEC characteristic. Needless to say that there is a strong correlation between UEEC events and how progress is made at the frontier of our knowledge. Chapter 14 will be devoted to the UEEC no. 7, which as we will see, corresponds to a set of UEEC events.

UEEC event no. 2. In figure 34 the case of mass ≠ matter is shown in a very synthetic form. From the Greeks who associated 'stability' of matter with 'heaviness' to our present understanding, the number of Sarajevo-type events is really impressive.

Mass ≠ Matter

$$|m_i\rangle \equiv \text{Mass} \equiv \text{Antimass} \equiv |\overline{m}_i\rangle$$

i ≡ 1 (Intrinsic); **i ≡ 2** (Confinement); **i ≡ 3** (Binding)

$$C|m_i\rangle = |m_i\rangle \qquad \boxed{***}$$

i ≡ 1, 2, 3

$$|m_i Q_j\rangle \equiv \text{Matter} \neq \text{Antimatter} \equiv |m_i \overline{Q}_j\rangle$$

Q_j ° Flavour Charges

$$\mathbf{j} \equiv (u\ d\ c\ s\ t\ b) \equiv (1, 2, 3, 4, 5, 6)$$

$$(\nu_e\ e^-\ \nu_\mu\ \mu^-\ \nu_{HL}\ HL^-) \equiv (7, 8, 9, 10, 11, 12)$$

$$HL^- \downarrow \tau^-$$

$$C|m_i Q_j\rangle = |m_i \overline{Q}_j\rangle \qquad \boxed{***}$$

i ≡ 1, 2, 3 ; J ≡ 1, 2, 3, 4, 5, 6, 7, 8, 9, 10, 11, 12.

Figure 34

There are in fact seven decades of developments which started from the antielectron and C-invariance and brought us to the discovery of nuclear antimatter and to the unification of all gauge forces with all deviations from simplicity. These steps are reported in figure 35, which looks as complex and full of deviations from simplicity as a page of History (EWRL), despite being a page of Science (EBUS).

THE INCREDIBLE STORY
TO DISENTANGLE THE ORIGIN OF THE STABILITY OF MATTER
SEVEN DECADES FROM THE ANTIELECTRON TO ANTIMATTER AND THE UNIFICATION OF ALL GAUGE FORCES

- **The validity of C invariance from 1927 to 1957.**

After the discovery by Thomson in 1897 of the first example of an elementary particle, the Electron, it took the genius of Dirac to theoretically discover the Antielectron thirty years after Thomson.

1927 → Dirac equation [15]; the existence of the antielectron is, soon after, theoretically predicted. Only a few years were needed, after Dirac's theoretical discovery, to experimentally confirm (Anderson, Blackett and Occhialini [16]) the existence of the Dirac antielectron.

1930-1957 → **Discovery of the C operator** [(charge conjugation) H. Weyl and P.A.M. Dirac [17]]; discovery of the P Symmetry Operator [E.P. Wigner, G.C. Wick and A.S. Wightman [18, 19]]; discovery of the T operator (time reversal) [E.P. Wigner, J. Schwinger and J.S. Bell [20, 21, 22, 23]]; discovery of the CPT Symmetry Operator from RQFT (1955-57) [24].

1927-1957 → Validity of C invariance: e^+ [16]; $\bar{p}$ [25]; $\bar{n}$ [26]; $K_2^0 \to 3\pi$ [27] but see LOY [28].

- **The new era starts: C ≠ ; P ≠ ; CP ≠ (*).**

1956 → Lee & Yang P ≠; C ≠ [29].

1957 → Before the experimental discovery of P ≠ & C ≠, Lee, Oehme, Yang (LOY) [28] point out that the existence of the second neutral K-meson, $K_2^0 \to 3\pi$, is proof neither of C invariance nor of CP invariance. Flavour antiflavour mixing does not imply CP invariance.

1957 → C.S. Wu et al. P ≠; C ≠ [30]; CP ok [31].

1964 → $K_2^0 \to 2\pi \equiv K_L$: CP ≠ [32].

1947-1967 → QED divergences & Landau poles.

1950-1970 → The crisis of RQFT & the triumph of S-matrix theory (i.e. the negation of RQFT).

1965 → Nuclear antimatter is (experimentally) discovered [33]. See also [34].

1968 → The discovery [35] at SLAC of Scaling (free quarks inside a nucleon at very high q^2) but in violent (pp) collisions no free quarks at the ISR are experimentally found [36]. Theorists consider Scaling as being evidence for RQFT not to be able to describe the Physics of Strong Interactions. The only exception is G. 't Hooft who discovers in 1971 that the β-function has negative sign for non-Abelian theories [12].

1971-1973 → $\beta = -$; 't Hooft; Politzer; Gross & Wilczek. The discovery of **non-Abelian** gauge theories. Asymptotic freedom in the interaction between quarks and gluons [12].

1974 → All gauge couplings $\alpha_1\ \alpha_2\ \alpha_3$ run with q^2 but they do not converge towards a unique point.

1979 → A.P. & A.Z. point out that the new degree of freedom due to SUSY allows the three couplings $\boldsymbol{\alpha_1\ \alpha_2\ \alpha_3}$, **to converge towards a unique point** [37].

1980 → QCD has a 'hidden' side: the multitude of final states for each pair of interacting particles: (e^+e^-; $p\bar{p}$; πp; Kp; νp; pp; etc.)
The introduction of the Effective Energy allows to discover the Universality properties [38] in the multihadronic final states.

1992 → All gauge couplings converge towards a unique point at the gauge unification energy: $E_{GU} \cong 10^{16}$ GeV with $\alpha_{GU} \cong 1/24$ [39, 40].

1994 → The Gap [41] between E_{GU} & the String Unification Energy: $E_{SU} \cong E_{Planck}$.

1995 → **CPT loses its foundations at the Planck scale (T.D. Lee)** [42].

1995-1999 → **No CPT theorem from M-theory (B. Greene)** [43].

1995-2000 → A.Z. points out the need for new experiments to establish if matter-antimatter symmetry or asymmetry are at work.

(*) The symbol ≠ stands for 'Symmetry Breakdown'.

Figure 35

UEEC event no. 3. The nucleon time-like electromagnetic structure (form factors).

For a long time Quantum ElectroDynamic (QED) was taken as the ideal model to describe a fundamental force of nature, such as the nuclear forces (proposed by Yukawa) and the weak forces (proposed by Fermi). The mathematical description of these forces had to be like QED, i.e. a Relativistic Quantum Field Theory (RQFT). Many unexpected experimental discoveries started to create difficulties; these discoveries included the violation of the symmetry operators (parity **P**, charge conjugation **C**, and time reversal **T**), mentioned in UEEC no. 2. This is how the scattering matrix, S, become the way out. We have discussed this problem in chapter 10 with reference to the matter-antimatter symmetry. The S-matrix was the negation of RQFT.

In fact, the field concept involves a larger set of functions than those derived by the analytic continuation of the S-matrix. But no one knew (and even now knows) how to construct fields purely in terms of analytic scattering amplitudes. Scattering amplitudes are 'on the mass shell' while fields imply extension to 'off the mass shell'. Form factors are not scattering amplitudes, nevertheless they do exist and they are due to strong interactions. The conjectured analyticity properties of the nuclear scattering matrix is a very restricted concept, if compared with the concept of a 'field'.

S-matrix theory is not designed to describe experiments in which interactions between particle states do take place while momentum measurements are being performed. In other words all the physics due to '**virtual processes**' fell outside the physics described by the S-matrix theory, which had a period of large success in the description of strong interactions [44]. This apparent triumph of the S-matrix [44] put in serious troubles the foundations of Relativistic Quantum Field Theories (RQFT). On the other hand these theories were restricted to be of Abelian nature, since the non-Abelian ones were shrouded by even more mystifying problems. The 'prediction' was that the 'time-like' electromagnetic structure of the nucleon had not be there. A totally unexpected result [45, 46] came with the experiment performed at CERN to study the annihilation process between a proton and an antiproton ($\bar{p}p$) giving rise to a 'virtual photon' (γ)

transforming into an electron-antielectron pair (e^+e^-). The corresponding reaction is $\bar{p}p \to \gamma \to e^+e^-$.

The experimental results [45, 46] proved that the nucleon had a very large 'time-like' electromagnetic form factor: totally unexpected.

UEEC event no. 5. In the physics of mesons the totally unexpected result was the difference existing between the two mesonic mixing angles, pseudoscalar and vector: $\theta_{PS} \neq \theta_V$. They should both be zero if $SU(3)_{uds}$ was a good Symmetry. The existence of instantons was not known. They came after the discovery that $\theta_{PS} \neq \theta_V$. A strong supporter of my experiment was Richard Dalitz, to whom I would like to dedicate the results reported in figure 25.

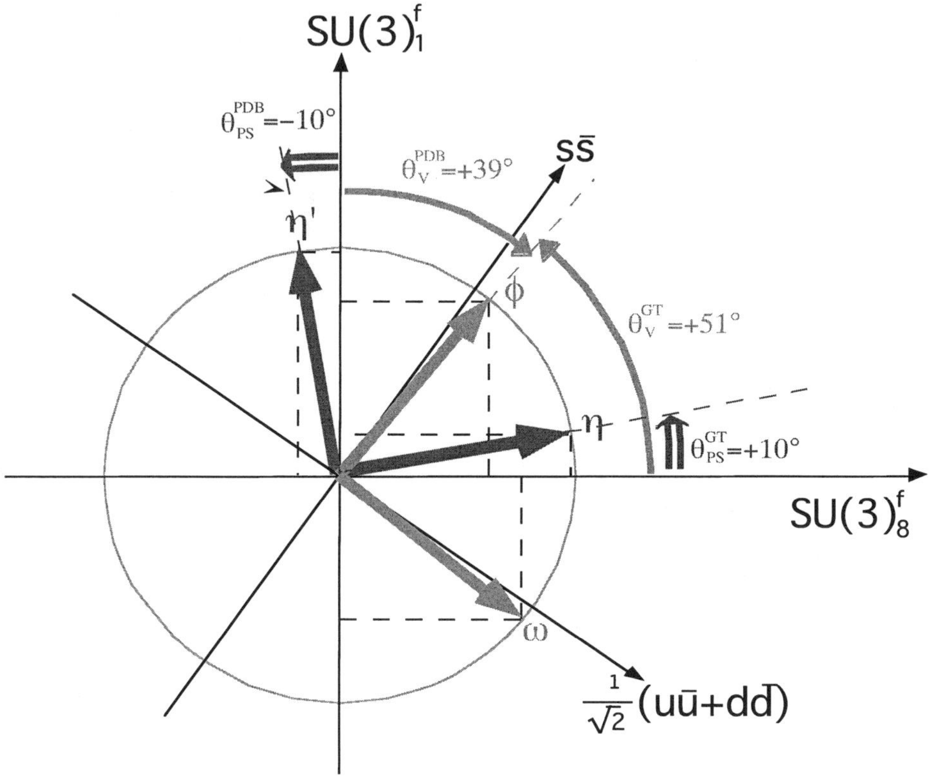

PDB ≡ Particle Data Book
GT ≡ Gerardus 't Hooft

$$\theta_{PS} \neq \theta_V$$

Direct: Not Using Mass Formulae

Figure 25

UEEC event no. 6. The 'effective energy'. Another great physicists I would like to express my gratitude is Vladimir Gribov who pointed out what follows. Newton discovered that light is the sum of different colours: this is QED light.

In QCD we have quarks and gluons interacting and producing Jets made of many pions, as for example in the (pp) reaction pp → π + X whose spectrum is shown in figure 26 (from chapter 10). The horizontal axis is for the fractional energy of the pion (also called Feynman x), while the vertical axis is for the number of pions having fractional energy x_F.

p-p → π^+ + X
Nominal Energy of the (pp) collision = √s= 24 GeV

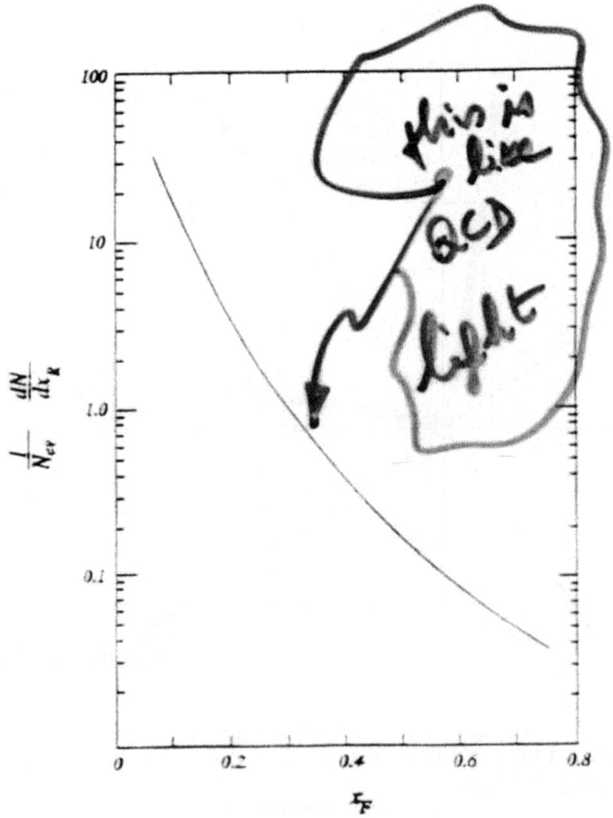

Figure 26

The spectrum reported in figure 26 is an example of what Gribov defined: *the QCD light.*

The 'effective energy' shown in figure 27 (from chapter 10) is at its origin, despite being totally unexpected. In fact the same initial state (pp) produces many final states each one having different 'effective energy', called E_{had} in figure 27.

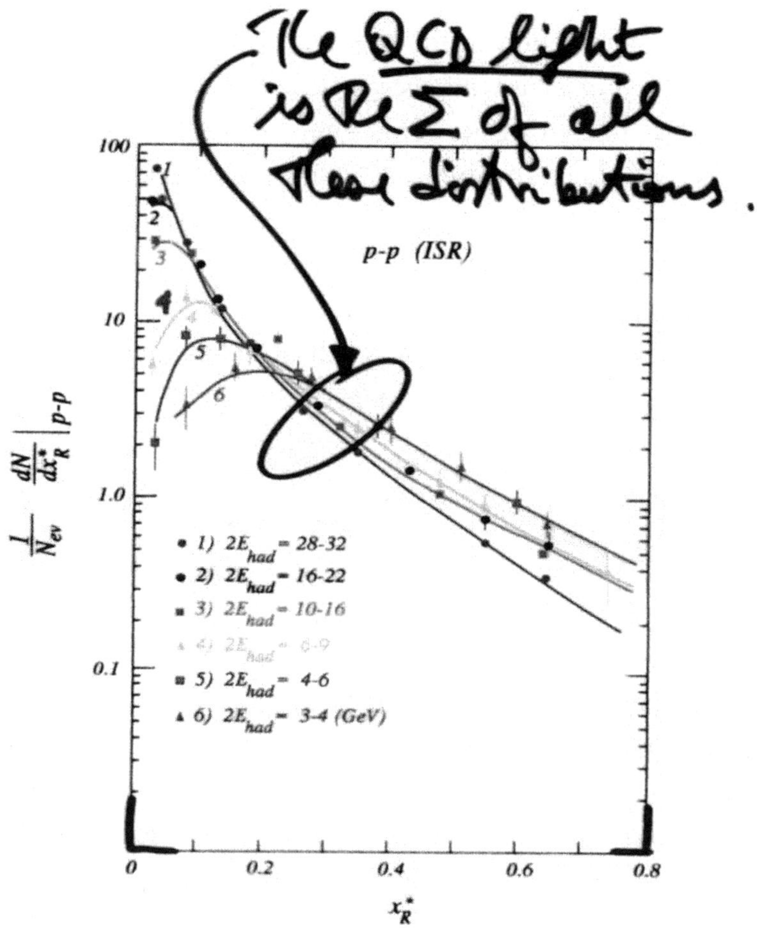

Figure 27

Each E_{had} has a given π–spectrum. The sum of all these spectra gives the total spectrum of figure 26. If, instead of (pp) we study other initial states, purely hadronic (kp), ($\bar{p}$p) or mixed (lepton–hadron) (ep) (νp) (μp) and even (γp), no matter what the initial states is, the only relevant quantity is the 'effective energy'. Each **'effective energy'** produces the same π–spectrum in analogy with the QED light of given colour.

We now move towards the unification of all fundamental forces, another source of UEEC events listed as UEEC n. 7 in figure 33. It is really a set of UEEC events which we discuss in the next chapter 14.

14 THE GRAND UNIFICATION IN THE REAL WORLD

The grand unification in the real world depends on how the gauge couplings (α_1 α_2 α_3) change with energy. It is this change which allows the fundamental forces to converge towards a unique origin.

The mathematical structure describing this 'change' is a system of coupled differential non linear equations, called the 'renormalization group equations', RGEs, reported below.

The Renormalization Group Equations

The lines in figure 36 are the result of calculations executed with a supercomputer using the following system of equations:

$$\mu \frac{d\alpha_i}{d\mu} = \frac{b_i}{2\pi} \alpha_i^2 + \sum_j \frac{bij}{8\pi^2} \alpha_j \alpha_i^2 . \qquad (1)$$

This is a system of coupled non-linear differential equations (RGEs) that describes the superworld, from the maximum level of energy (Planck scale) to our world at the minimum of energy.

The results reported in figure 36 are the most exact use of the renormalization group equations for the running of the three gauge couplings α_1 α_2 α_3. The unification of all forces and the threshold, where to find the first particle of the superworld, with its problems are reported in figures 36 and 37 respectively.

During more than ten years (from 1979 to 1991), no one had realized that the energy threshold for the existence of the Superworld was strongly dependent on the 'running' of the masses. This is now called: the EGM effect (from the initials of Evolution of Gaugino Masses).

To compute the energy threshold using only the 'running' of the gauge couplings (α_1, α_2, α_3) corresponds to neglecting nearly three orders of magnitude in the energy threshold for the discovery of the first particle (the lightest) of the superworld [47], as illustrated in figure 37.

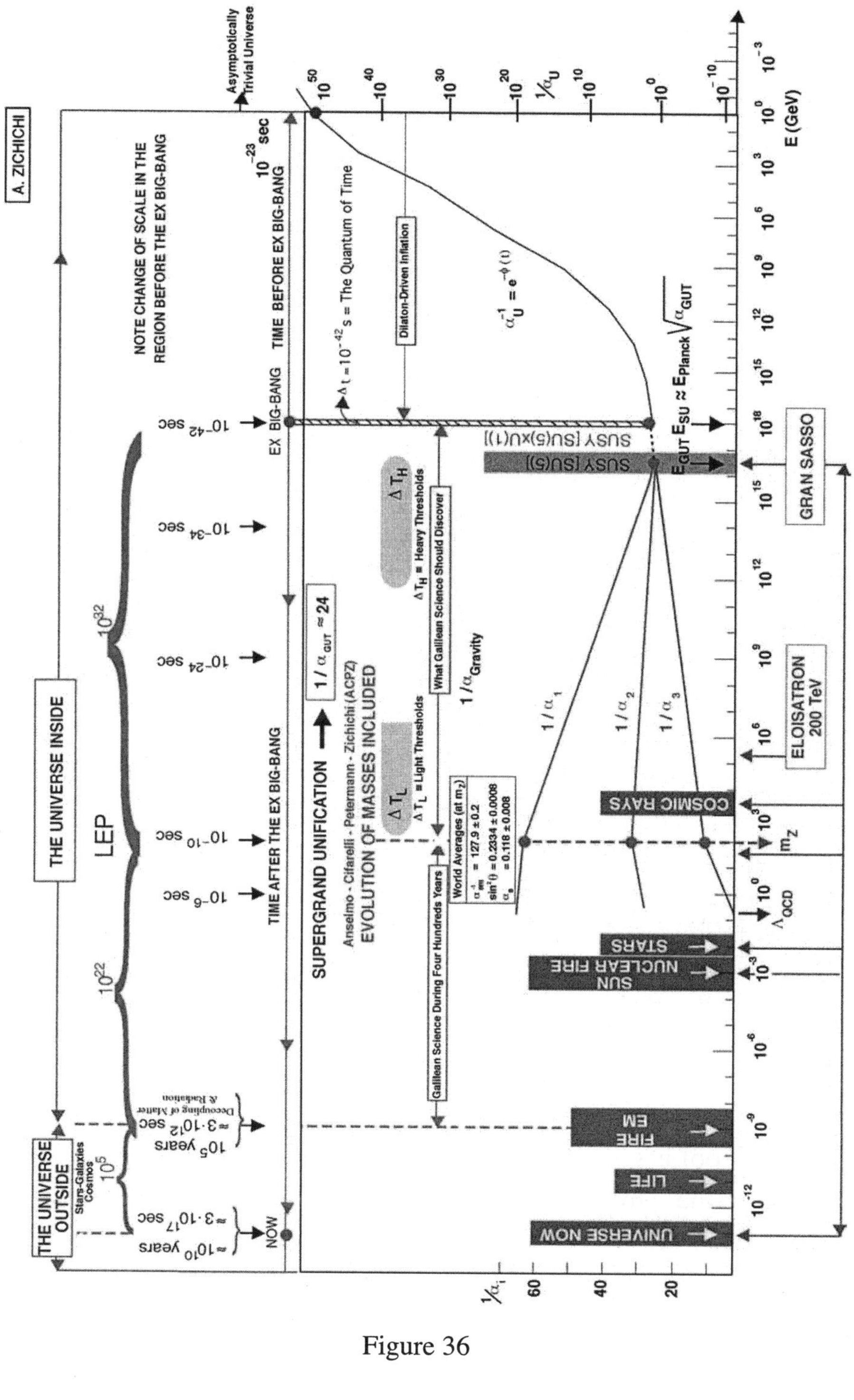
A. ZICHICHI
THE UNIVERSE OUTSIDE
THE UNIVERSE INSIDE
LEP
NOTE CHANGE OF SCALE IN THE REGION BEFORE THE EX BIG-BANG
TIME AFTER THE EX BIG-BANG
TIME BEFORE EX BIG-BANG
EX BIG-BANG
NOW
Asymptotically Trivial Universe
SUPERGRAND UNIFICATION
Anselmo - Cifarelli - Petermann - Zichichi (ACPZ)
EVOLUTION OF MASSES INCLUDED
Dilaton-Driven Inflation
What Galilean Science Should Discover
Galilean Science During Four Hundreds Years
UNIVERSE NOW
LIFE
FIRE EM
SUN NUCLEAR FIRE
STARS
COSMIC RAYS
SUSY [SU(5)]
SUSY [SU(5)xU(1)]
ELOISATRON 200 TeV
GRAN SASSO
E (GeV)

Figure 36

This is just a further example of comparison between the 'Platonic' Simplicity and the 'real world', when we deal with the Grand Unification.

Talking about supersymmetry, there is another important step: how we go from pure theoretical speculations to phenomenology. This is not an easy task.

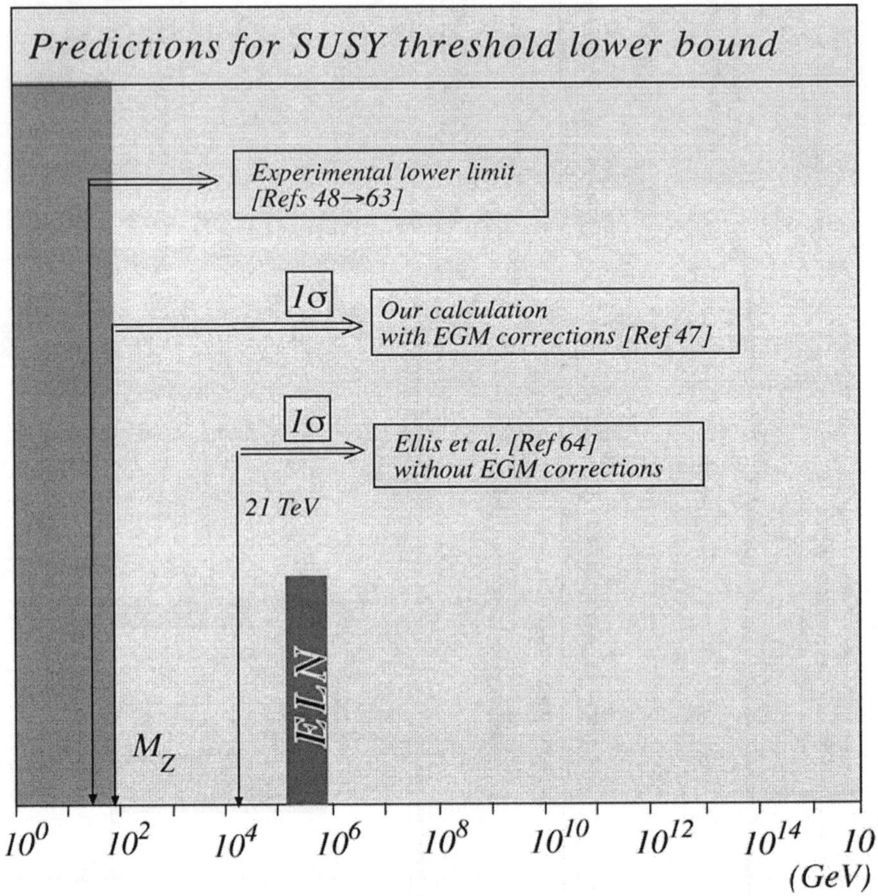

Figure 37

The proof is given in figure 38 where it is shown how many important properties of the physics to be described have been neglected by some authors (AdBF) whose claim was to 'predict' the energy scale at which supersymmetry is broken.

In order to attempt to give such a prediction, there are at least five 'details' to be taken into account, as reported in the last five columns (6–10) of figure 38.

●	①	②	③	④	⑤	⑥	⑦	⑧	⑨	⑩
Authors	*Input data*	*Errors*	*EC*	M_{SUSY}	*CC*	*UC*	ΔT_L	M_X	ΔT_H	*EGM*
ACPZ [47, 65→70]	WA	$\pm 2\,\sigma$	all possible solutions (24)	Yes	physical	Yes	Yes	Yes	Yes	Yes

Authors	*Input data*	*Errors*	*EC*	M_{SUSY}	*CC*	*UC*	ΔT_L	M_X	ΔT_H	*EGM*
AdBF [71]	only one experiment	$\pm 1\,\sigma$	only one solution	Yes	Geometrical	No	No	No	No	No

① *WA* ≡ *World Average*
② *Errors* ≡ *Uncertainty taken from all data (World Average) or from a single experiment*
③ *EC* ≡ *Evolution of Couplings*
④ M_{SUSY} ≡ *Mass Scale assumed to represent the Supersymmetry Scale breaking*
⑤ *CC* ≡ *Convergence of Couplings*
⑥ *UC* ≡ *Unification of Couplings above* E_{GUT}
⑦ ΔT_L ≡ *Low Energy threshold*
⑧ M_X ≡ *Mass Scale at the breaking of the Grand Unified Theory to the SU(3)xSU(2)xU(1)*
⑨ ΔT_H ≡ *High Energy threshold*
⑩ *EGM* ≡ *Evolution of Gaugino Masses*

Figure 38

It is interesting to study the point where the three gauge couplings meet (the GUT point). This is neither at the 'String Unification Point', E_{SU}, nor at the Planck scale, as reported in figure 39.

There is in fact a 'gap' of few orders of magnitudes between these points. A detailed study of this gap has been performed by ACPZ [69].

In figure 40 there is a different way of reporting the results obtained using the same (1) mathematical structure (RGEs). The three axis are the gauge couplings α_1 α_2 α_3 and the other details are given in the figure caption.

After we have published these results [37], the (α_1 α_2 α_3) graph, has been given the name of 'action space'. In this space we have emphasized the 'straight' line as being the one which would naively be considered the 'platonic' way of imagining the changes of α_1 α_2 α_3 in order to meet at the same point E_{GUT}.

The 'Platonic' Simplicity would indicate the series of points making up the straight line as the platonic ideally simple solution. The real solution is the sequence of points which totally deviate from the straight line.

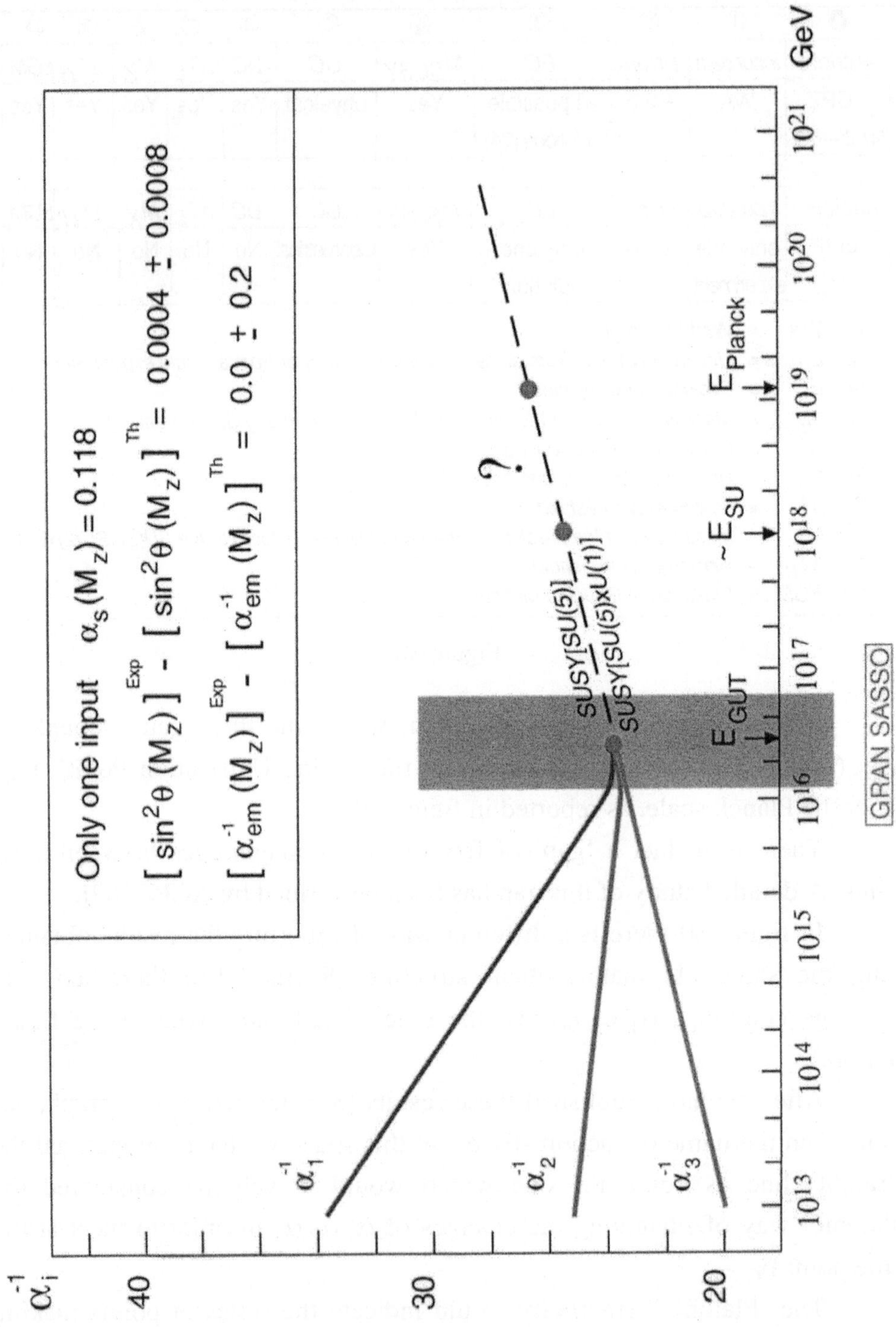
Only one input $\alpha_s(M_Z) = 0.118$
$[\sin^2\theta(M_Z)]^{Exp} - [\sin^2\theta(M_Z)]^{Th} = 0.0004 \pm 0.0008$
$[\alpha_{em}^{-1}(M_Z)]^{Exp} - [\alpha_{em}^{-1}(M_Z)]^{Th} = 0.0 \pm 0.2$
α_i^{-1}
40
30
20
α_1^{-1}
α_2^{-1}
α_3^{-1}
SUSY[SU(5)]
SUSY[SU(5)xU(1)]
?
E_{GUT}
$\sim E_{SU}$
E_{Planck}
10^{13}
10^{14}
10^{15}
10^{16}
10^{17}
10^{18}
10^{19}
10^{20}
10^{21}
GeV
GRAN SASSO

Figure 39

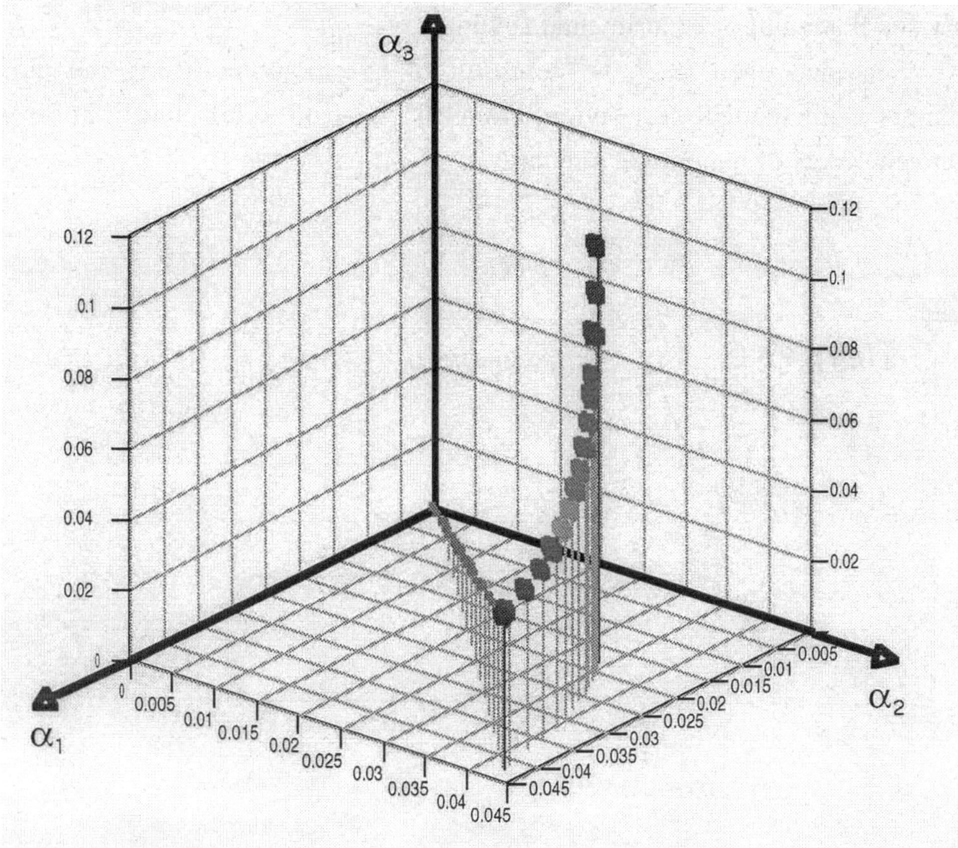

The points have a sequence of 100 GeV in energy. The last point where the 'ideal' platonic straight line intercepts the theoretical prediction is at the energy of the Grand Unification. This corresponds to $E_{GU} = 10^{16.2}$ GeV. Other detailed information on the theoretical inputs: the number of fermionic families, N_F, is 3; the number of Higgs particles, N_H, is 2. The input values of the gauge couplings at the Z^0-mass is α_3 $(M_Z) = 0.118 \pm 0.008$; the other input is the ratio of weak and electromagnetic couplings also measured at the Z^0-mass value: $\sin^2 \theta_W (M_Z) = 0.2334 \pm 0.0008$.

Figure 40

Finally, in figure 41 we give a simplified version of our analysis with the ELN energy at (500+500) TeV level and the LHC at its present value.

The 'origin of space-time' is indicated immediately above E_{GUT} since it is there where all theoretical speculations stop to be coupled with even a very small amount of experimental finding.

In fact, even E_{GUT} is the result of extrapolation (using the most accurate mathematical description) from 10^2 GeV up to 10^{16} GeV, i.e. over fourteen orders of magnitude in energy.

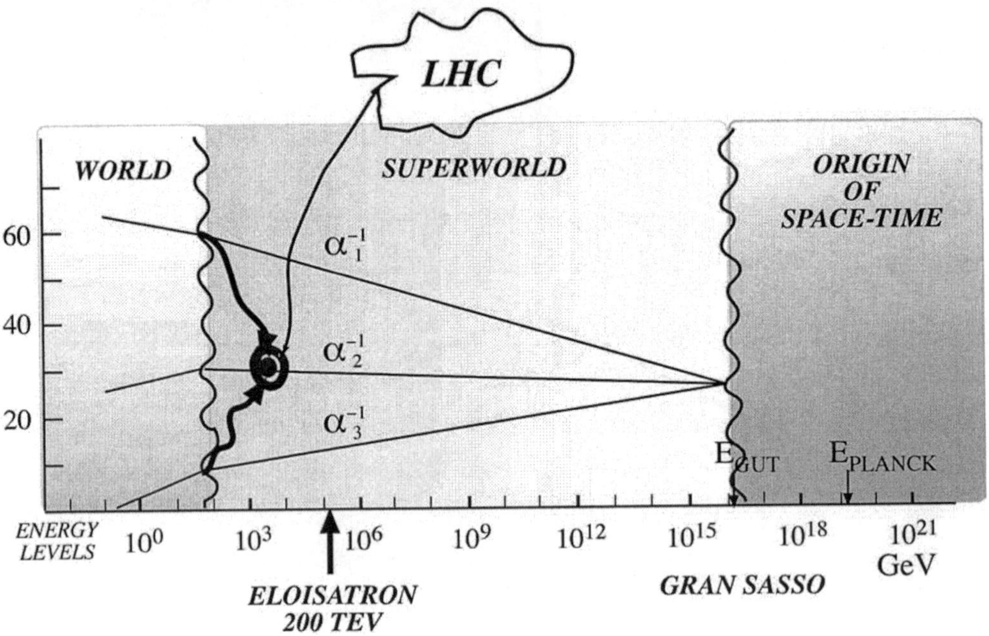

Figure 41

15 LET ME SUMMARIZE THE MAIN POINTS

Let me summarize what has been the contest of this lecture.

I have reviewed the present status of all we know in the Reductionistic achievements together with our present understanding of the rigorous attempts towards the basic features which allow Complexity to exist, i.e. AFB phenomena and UEEC events.

The conclusion is that Complexity exists at the fundamental level.

We do not know what will be the final outcome of String Theory.

What we do know is that all these attempts to understand the Logic of Nature appear to be very '*simple*' when compared with the '*Complexity*' of the **world** we are part of.

This has been shown in figure 1 (from chapter 4) where the various components of our knowledge which make up the apparently very different domains of our world are specified.

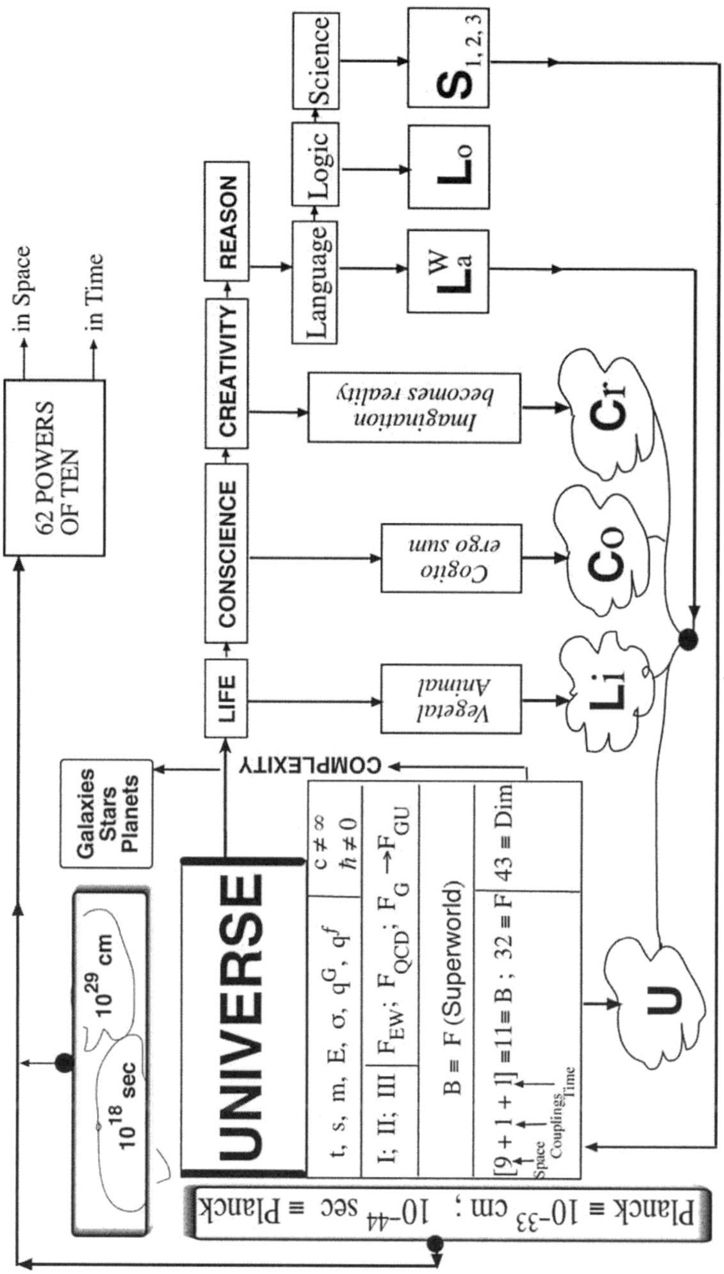

Figure 1

The conceptual understanding of the constituents of the Universe after 400 years of Reductionism are shown in figure 42.

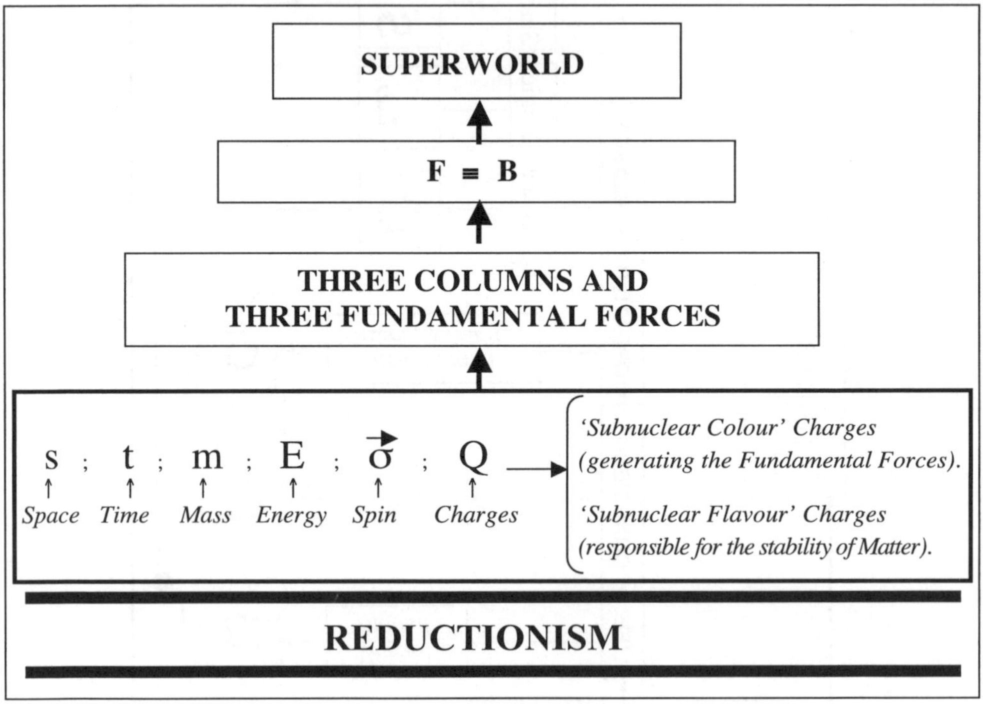

Figure 42

The key point is that the vacuum we are looking for must have the properties shown in figure 42.

Furthermore it is important to emphasize that we have reached this extremely 'simple' understanding of the properties needed for our vacuum via an incredible series of **ABF** and **UEEC events (Sarajevo-type)**.

Furthermore it should be pointed out that the apparently simple structure of the hypothetical goal, the superworld, is full of problems to be solved and, once again, it has been obtained with a remarkable set of **AFB** and **UEEC** events.

We have also seen that if we try to formulate the simplest version of our rigorous logic to describe Nature, i.e. the platonic GUT and the platonic supersymmetry, the result is that nothing in the real world looks like it, not even the Standard Model.

Present trend: from Reductionism to Holism.

Is Holism the way to understand the origin of Complexity?

Is the study of the Big Numbers the way to understand the origin of Complexity?

We have a few fundamental sources of Big Numbers.

Big Numbers come from the ratio of Space, Time, Mass-Energy and Action.

Space $\Rightarrow$ $$\frac{S^{Universe}}{Planck\ \ Length} \cong 10^{62}$$

Time $\Rightarrow$ $$\frac{\tau_p}{Planck\ \ Time} = \frac{10^{33} \cdot \pi \times 10^{7}\ \sec}{10^{-43}\ \sec} = \frac{10^{40}\ \sec}{10^{-43}\ \sec} \cong 10^{83}$$

Mass-Energy $\Rightarrow$ $$\frac{M^{Universe}}{M_{\nu}} \cong \frac{10^{80}\, GeV \cdot 10^{2}}{10^{-4}\, eV} = \frac{10^{91}\, eV}{10^{-4}\, eV} = 10^{95}$$

Action $\Rightarrow$ $$\frac{h^{Universe}}{h_{Planck}} = 10^{123}$$

With these Big Numbers compete those coming from the Mathematical Model of Brain. According to some models, in order to produce new ideas the number of possible combinations is much larger than 10^{123}.

What about the Number of Constituents?

There is a recent experimental finding where the number seems not to be relevant. The decodification of the Human Genome has shown that it contains only about 25,000 genes. Where is the seat of human species Complexity since the number of genes is similar to that of many other living organisms?

The number of genes does not seem to play a role as far as Complexity levels are concerned.

What seems to play an important role is the *network of interactions between Genes and Proteins*. It should be noticed that the interaction with the environment stops at the level of molecules, atoms and all components of the real world down to the fundamental constituents and the **SM&B**, as shown in the table below.

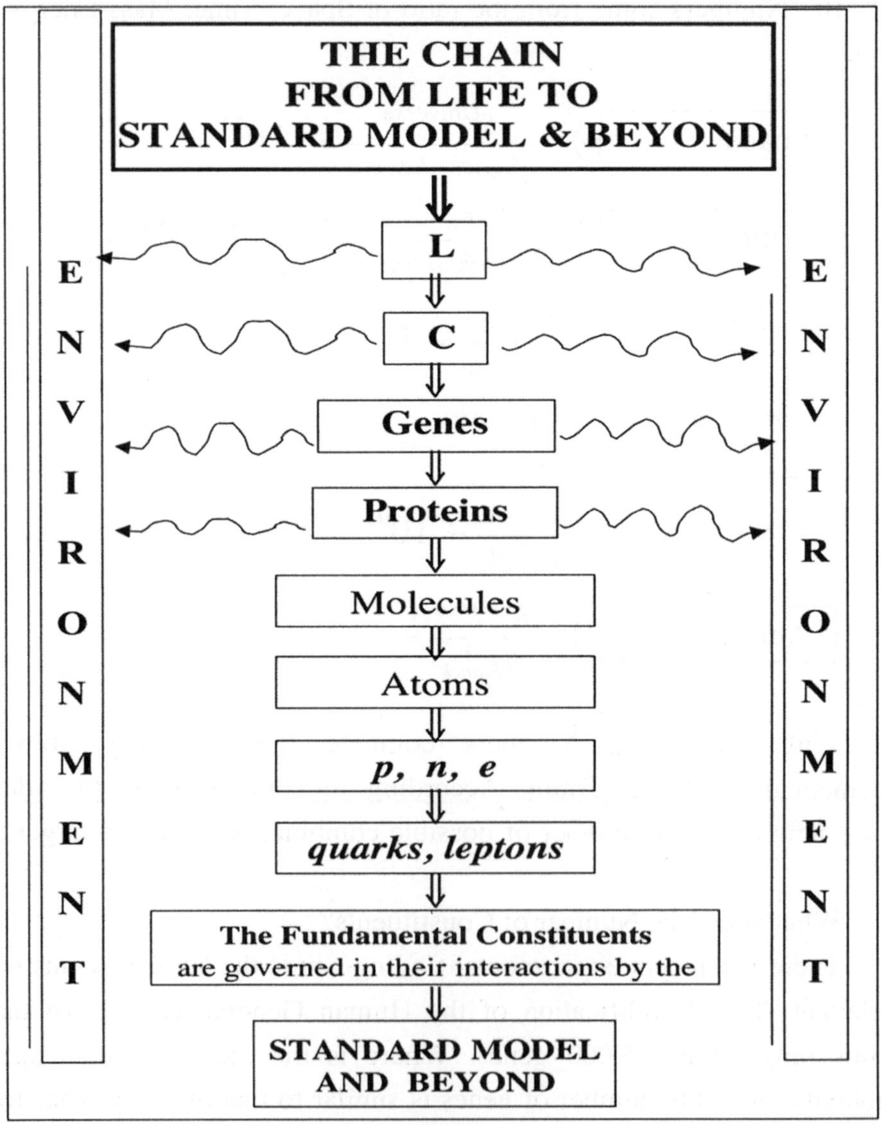

Table

<u>So far, no new fundamental forces</u> have been discovered by those fellows engaged in the study of Complex Systems.

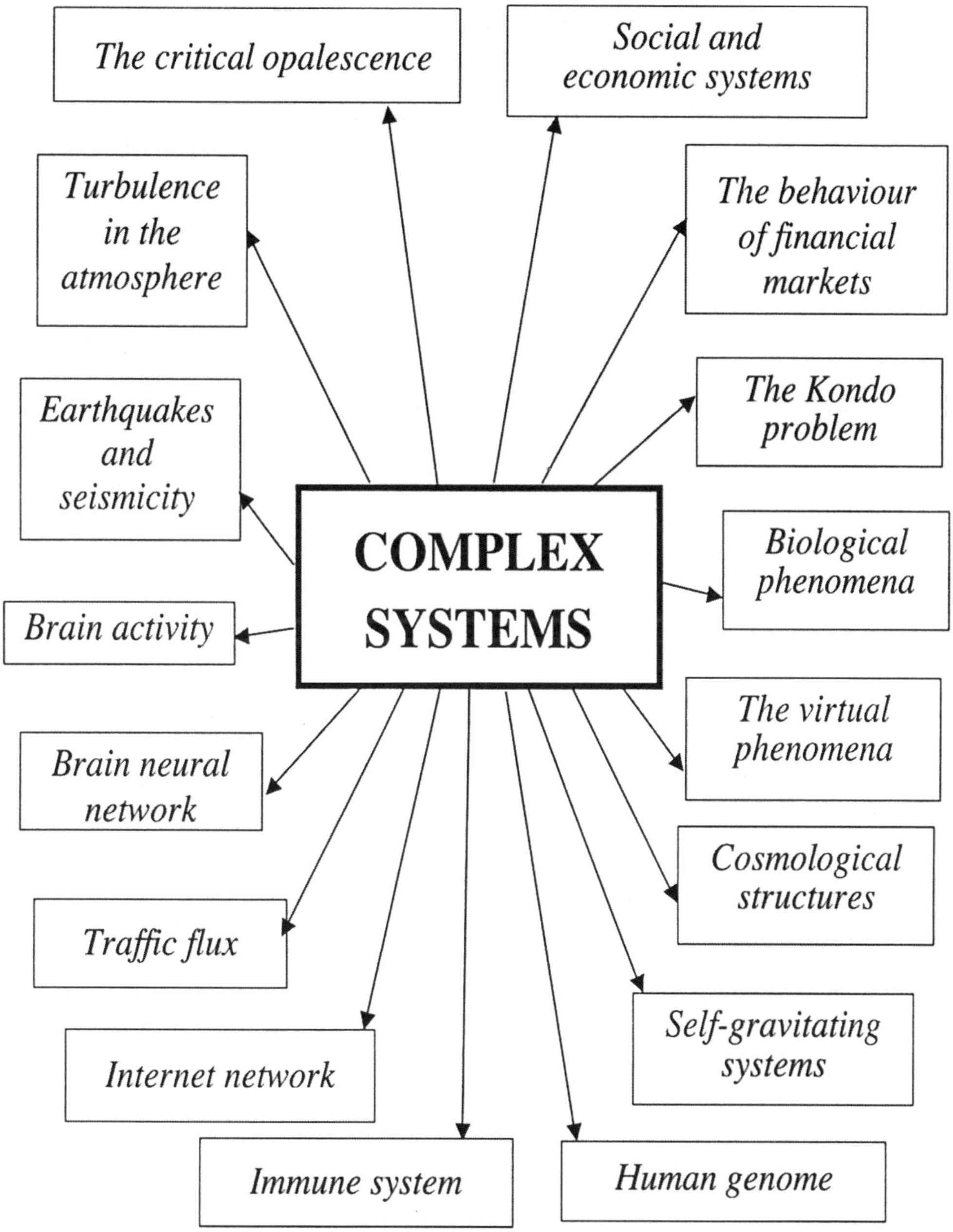

Figure 9 (from chapter 5)

<u>We have shown</u> that the experimental foundations for the existence of Complexity are found in the Galilean study of the Logic of Nature. It is therefore not correct to claim that holism is the right way to go in order to understand Complexity. In fact Complexity exists at the fundamental level.

Since Complexity exists at the fundamental level then:

- **Totally Unexpected Effects** should **show up**.
- **Effects**, which are impossible to be predicted on the basis of **present knowledge**.
- Where these effects are most likely to be, no one knows.

16 FROM PLANCK TO COMPLEXITY: TWO GREAT NOVELTIES AND A CONCLUSION

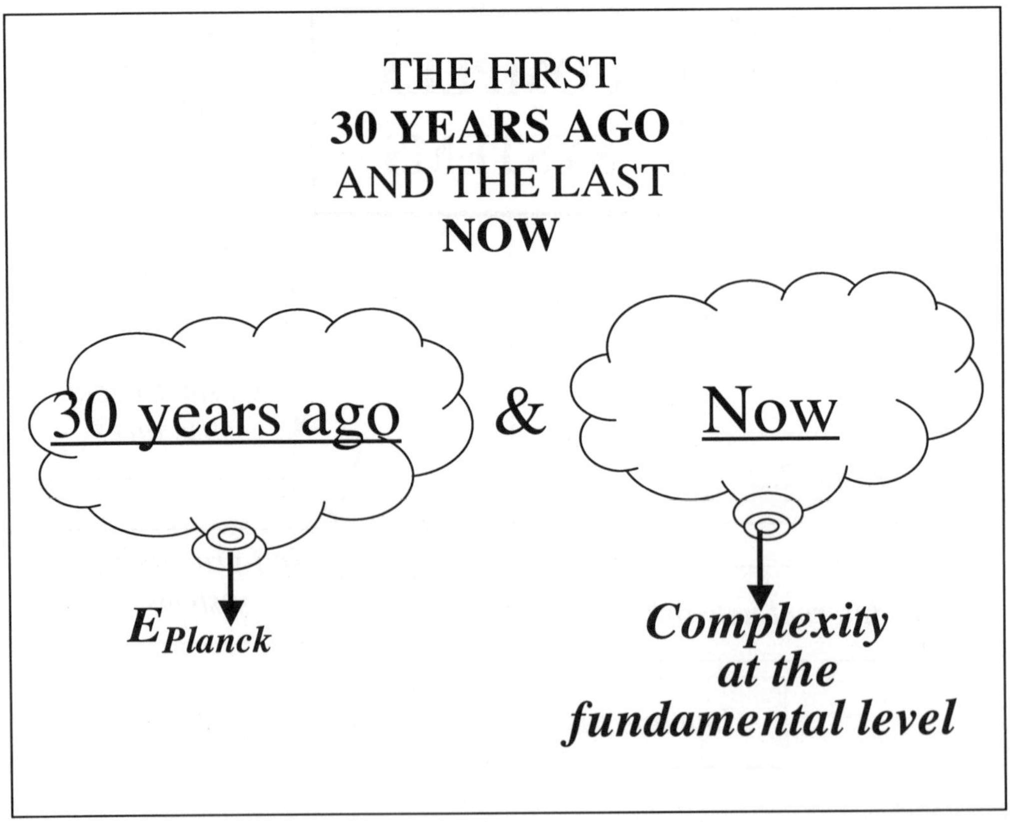

Thirty years ago a great scientific novelty came; all experimental discoveries obtained with our powerful accelerators were to be considered only matters of extremely low energy.

The scale of energy on which to direct the attention to understand the Logic that rules the world, from the tiniest structures to the galactic ones, had to be shifted at a much higher level: to the mass-energy named after Planck, E_{Planck}, something like seventeen powers of ten above the Fermi scale, E_{Fermi}, that already seemed to be an extremely high level of energy.

Now, after thirty years, it comes about the novelty of our time: Complexity exists at the fundamental level.

This discovery has very interesting consequences, as we will see in the conclusion.

For the time being let me call your attention on what happens with the only mathematical structure to be in a position of describing all that happens at the Planck scale: the Relativistic Quantum String Theory (**RQST**).

This mathematical structure produces innumerable minima of energy, named **Landscape**. One of this minima is the vacuum needed for our world, with 3 space plus one time dimensions (not being both real) plus all other properties described in chapter 6.

The theoretical discovery of the **Landscape** (Leonard Susskind) [72], has been followed by another formidable discovery in mathematical physics: the most rigorous model of RQST (Raphael Bousso and Joseph Polchinski) is **NP-complete** (Michael R. Douglas and Frederik Denef) [73].

This discovery corroborates all that we have put in evidence during the last five years [1–4]: **Complexity exists at the fundamental level**.

In fact, UEEC events and AFB phenomena exist at all scales, as illustrated in the Figure 22 (from chapter 9) of page 68.

The figure 41 (from chapter 14) on page 68 illustrates the extrapolation of our present knowledge, from the energy level where facilities are being implemented, such as the LHC, to the Gran Sasso, which is operative since more than a decade and will be upgraded with other detectors.

Other facilities exist the world over, but they are all planned to search for events which can be predicted on the basis of our knowledge.

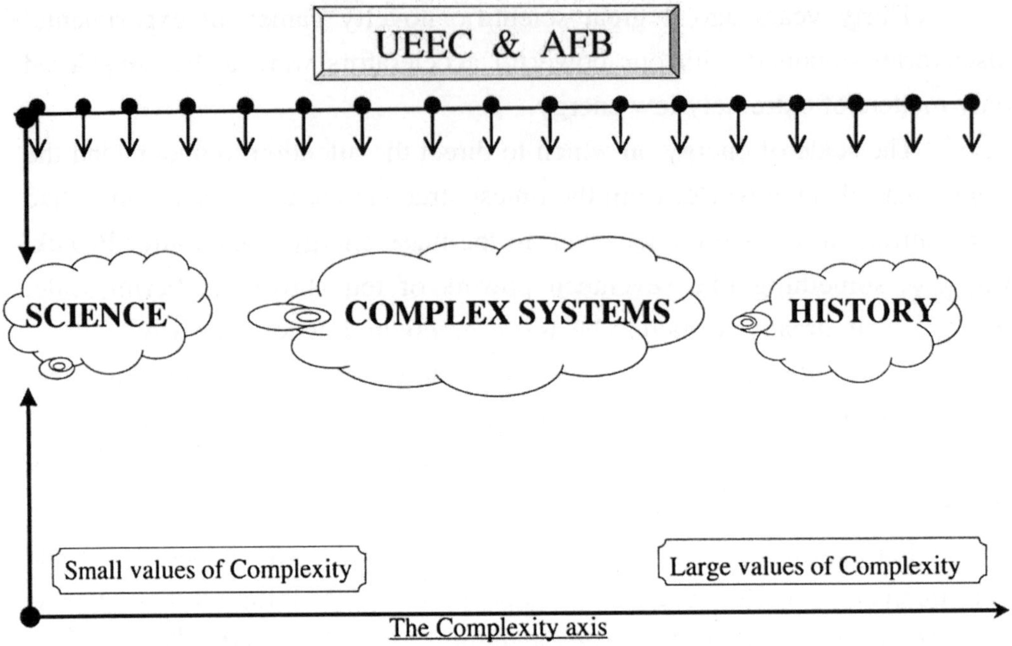

Figure 22

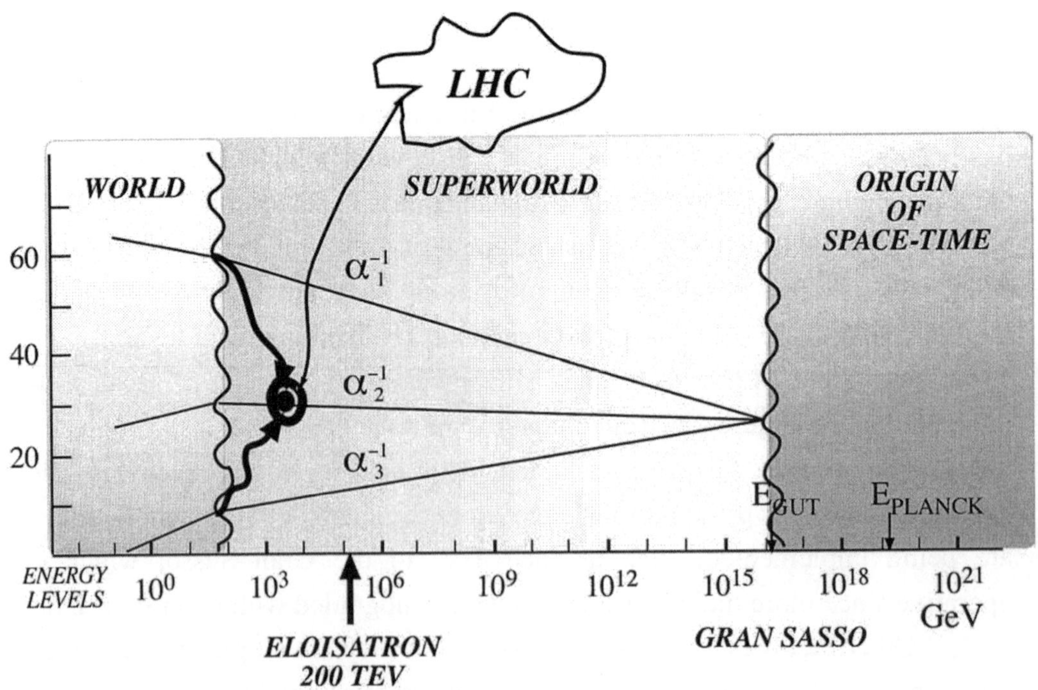

Figure 41

All these extrapolations ignore the fact that, the status of our knowledge is the one reported in the figure 43 below.

In fact, after 400 years of Reductionism, the present trend is to go from **Reductionism** to **Holism** and the Future could be with dominated by the **Science being NP-complete**.

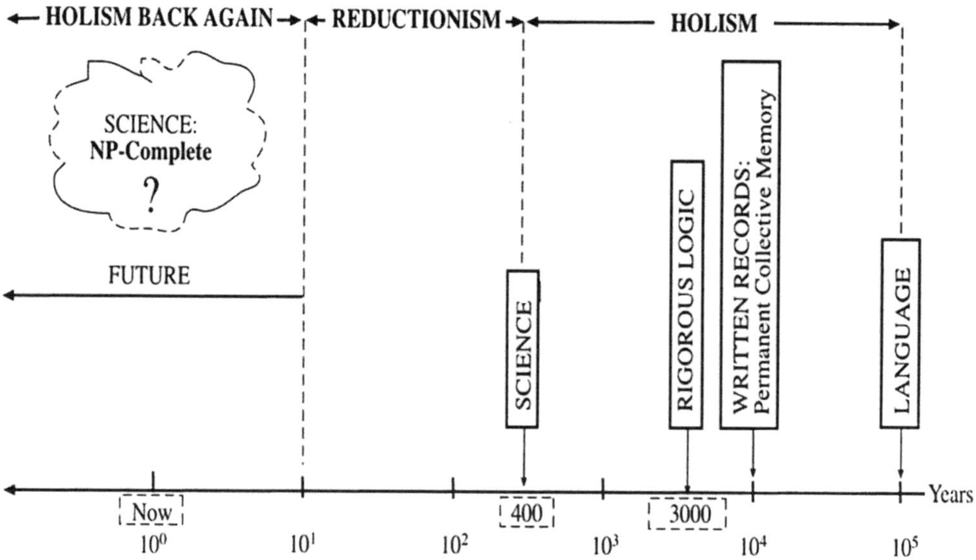

Figure 43

We do not know what will be the final outcome of String Theory.

What we know is that: '*The world appears to be complex at every scale. **Therefore** we must expect a continued series of surprises that we cannot easily predict*'.

CONCLUSION

With reference to advanced physics experiments [74] it is of great importance the outcome of our analysis which shows that **Complexity exists at the fundamental level**.

- Therefore **Totally Unexpected Effects** should **show up**.
- **Effects**, which are impossible to be predicted on the basis of **present knowledge**.

We should be prepared with **powerful experimental instruments, technologically at the frontier of our knowledge**, to discover Totally Unexpected Events at LHC, at Gran Sasso and / or at other laboratories.

RECALL:

- The **discoveries in Electricity**, **Magnetism** and **Optics** (UEEC).
- **Radioactivity** (UEEC).
- The **Strange Particles** (UEEC).
- The **3 Columns** (UEEC).
- The **origin of the Fundamental Forces** (UEEC).

17 REFERENCES

The References [1–4] refer to the various occasions where I have presented papers on highly specialized topics and discussed the connection of these topics with Complexity. The title on the upper part refers to the connection with Complexity while the specialized topic is reported in the detailed references.

[1] *Complexity at the Fundamental Level*
A. Zichichi
presented at:

- *International Conference on 'Quantum [un]speakables' in Commemoration of John S. Bell, International Erwin Schrödinger Institut (ESI), Universität Wien (Austria), November 2000,* 'John Bell and the Ten Challenges of Subnuclear Physics'.
- 40th Course of the International School of Subnuclear Physics, Erice (Italy), September 2002, '*Language Logic and Science*'.
- 31st, 32nd and 33th Course of the International School of Solid State Physics, Erice (Italy), July 2004, '*Complexity at the Elementary Level*'.
- 42nd International School of Subnuclear Physics, Erice (Italy), August - September 2004, '*Complexity at the Elementary Level*'.
- Trinity College, Dublin (Ireland), February 2005, '*Complexity at the Elementary Level*'.
- Department of Physics, University of Padova (Italy), March 2005, '*Complexity at the Elementary Level*'.
- 43th Course of the International School of Subnuclear Physics, Erice (Italy), September 2005, '*Complexity at the Elementary Level*'.

- Italian Physics Society (SIF) XCI Annual National Congress, University of Catania (Italy), September 2005, '*Complexity at the Elementary Level*'.
- Desy, Hamburg, November 2005, '*Complexity at the Fundamental Level*'.
- 44th Course of the International School of Subnuclear Physics, Erice (Italy), September 2006, '*Complexity at the Fundamental Level*'.

[2] *The Logic of Nature and Complexity*
A. Zichichi
presented at:

- Pontificia Academia Scientiarum, The Vatican, Rome (Italy), November 2002, '*Scientific Culture and the Ten Statements of John Paul II*'; '*Elements of Rigour in the Theory of Evolution*'.
- The joint Session of:
6th Course of the International School of Biological Magnetic Resonance; Erice (Italy), July 2003, '*Language Logic and Science*'.
- 2nd Workshop on Science and Religion of the Advanced School of History of Physics; Erice (Italy), July 2003, '*Language Logic and Science*'.
- 10th Workshop of the International School of Liquid Crystals; Erice (Italy), July 2003, '*Language Logic and Science*'.
- International School on Complexity, 1st Workshop on Minimal Life, Erice (Italy), December 2004, '*Evolution and Complexity at the Elementary Level*'.

[3] *Complexity and New Physics*
A. Zichichi
presented at:

- INFN-Alice Meeting, University of Catania (Italy), January 2005, '*Complexity at the Elementary Level*'.
- INFN Eloisatron Project 'The 1st Physics ALICE Week', Erice (Italy), December 2005, '*Complexity and New Physics with ALICE*'.
- 50th Anniversary of INFN Bologna - ALICE Week, Bologna (Italy), June 2006, '*Complexity at the Fundamental Level*'.

[4] *Complexity and Planetary Emergencies*
A. Zichichi
presented at:

- 27th Sessions of the International Seminars on Planetary Emergencies, Erice (Italy), August 2002, '*Language, Logic and Science*'.

- 28th Sessions of the International Seminars on Planetary Emergencies, Erice (Italy), August 2003, '*Language Logic and Science, Evolution and Planetary Emergencies*'.
- 36th Sessions of the International Seminars on Planetary Emergencies, Erice (Italy), August 2006, '*Complexity and Planetary Emergencies*'.

[5] *Language, Logic and Science*
A. Zichichi, proceedings of the 26th Session of the International Seminar on Nuclear War and Planetary Emergencies, Erice 18-26 August 2002 (World Scientific, 2003).

[6] *The value of Statistical Laws in Physics and Social Sciences*
Ettore Majorana
Quantitative Finance 5, 133 (2005).

[7] *Ettore Majorana: Genius and Mystery*
A. Zichichi
Review Lecture presented at the Centenary Celebrations in Rome, Catania, Bologna, Academy of Sciences (2006), World Scientific (2006).

[8] *A Proposal to Search for Leptonic Quarks and Heavy Leptons Produced by ADONE*
M. Bernardini, D. Bollini, E. Fiorentino, F. Mainardi, T. Massam, L. Monari, F. Palmonari and A. Zichichi
INFN/AE-67/3, 20 March 1967.

[9] *Limits on the Electromagnetic Production of Heavy Leptons*
V. Alles-Borelli, M. Bernardini, D. Bollini, P.L. Brunini, T. Massam, L. Monari, F. Palmonari and A. Zichichi
Nuovo Cimento Letters 4, 1156 (1970).

[10] *Limits on the Mass of Heavy Leptons*
M. Bernardini, D. Bollini, P.L. Brunini, E. Fiorentino, T. Massam, L. Monari, F. Palmonari, F. Rimondi and A. Zichichi
Nuovo Cimento 17A, 383 (1973).

[11] *The Origin of the Third Family*
by C.S. Wu, T.D. Lee, N. Cabibbo, V.F. Weisskopf, S.C.C. Ting, C. Villi, M. Conversi, A. Petermann, B.H. Wiik and G. Wolf, in honour of A. Zichichi on the 'XXX Anniversary of the Proposal to search for the Third Lepton at ADONE'; O. Barnabei, L. Maiani, R.A. Ricci and F. Roversi Monaco (eds), Rome (1997); and World Scientific (1998)

[12] A. Zichichi in '*Subnuclear Physics - The first fifty years*', O. Barnabei, P. Pupillo and F. Roversi Monaco (eds), a joint publication by University and Academy of Sciences of Bologna, Italy (1998); World Scientific Series in 20th Century Physics, Vol. 24 (2000).

[13] *Logical Reasoning in Experimental Physics: Past and Future*
A. Zichichi
in *Gerardus 't Hooft Liber Amicorum to celebrate his 60th anniversary* (2006).

[14] *Meson Resonances and Related Electromagnetic Phenomena*
Proceedings of the 1st International Conference organized by the High Energy and Particle Physics Division of the EPS at Bologna 14-16 April 1971, R.H. Dalitz and A. Zichichi (eds), International Physics Series, Editrice Compositori Publishers, Bologna, Italy (1971).

[15] P.A.M. Dirac, '*The Quantum Theory of the Electron*', *Proc. Roy. Soc.* (London) A117, 610 (1928); '*The Quantum Theory of the Electron, Part II*', *Proc. Roy. Soc.* (London) A118, 351 (1928).

[16] *The Positive Electron*
C.D. Anderson, *Phys. Rev.* 43, 491 (1933);

Some Photographs of the Tracks of Penetrating Radiation
P.M.S. Blackett and G.P.S. Occhialini, *Proc. Roy. Soc.* A139, 699 (1933).

[17] H. Weyl, '*Gruppentheorie und Quantenmechanik*', 2nd ed., 234 (1931).

[18] E.P. Wigner, '*Unitary Representations of the Inhomogeneous Lorentz Group*', *Ann. Math.*, 40, 149 (1939).

[19] G.C. Wick, E.P. Wigner, and A.S. Wightman, '*Intrinsic Parity of Elementary Particles*', *Phys. Rev.* 88, 101 (1952).

[20] E.P. Wigner, '*Über die Operation der Zeitumkehr in der Quantenmechanik*', *Gött. Nach.* 546-559 (1931). Here for the first time an anti-unitary symmetry appears.

[21] E.P. Wigner, *Ann. Math.* 40, 149 (1939).

[22] J. Schwinger, *Phys. Rev.* 82, 914 (1951).

[23] J.S. Bell, '*Time Reversal in Field Theory*', *Proc. Roy. Soc.* (London) A231, 479-495 (1955).

[24] To the best of my knowledge, the CPT theorem was first proved by W. Pauli in his article '*Exclusion Principle, Lorentz Group and Reflection of Space-Time and Charge*', in 'Niels Bohr and the Development of Physics' [Pergamon Press, London, page 30 (1955)], which in turn is an extension of the work of J. Schwinger [*Phys. Rev.* 82, 914 (1951); '*The Theory of Quantized Fields. II.*', *Phys. Rev.* 91, 713 (1953); '*The Theory of Quantized Fields. III.*', *Phys. Rev.* 91, 728 (1953); '*The Theory of Quantized Fields. VI.*', *Phys. Rev.* 94, 1362 (1954)] and G. Lüders, '*On the Equivalence of Invariance under Time Reversal and under Particle-Anti-particle Conjugation for Relativistic Field Theories*' [*Dansk. Mat. Fys.*

Medd. 28, 5 (1954)], which referred to an unpublished remark by B. Zumino. The final contribution to the CPT theorem was given by R. Jost, in '*Eine Bemerkung zum CPT Theorem*' [*Helv. Phys. Acta* 30, 409 (1957)], who showed that a weaker condition, called 'weak local commutativity' was sufficient for the validity of the CPT theorem.

[25] *Observation of Antiprotons*
O. Chamberlain, E. Segrè, C. Wiegand, and T. Ypsilantis, *Physical Review* 100, 947 (1955).

[26] *Anti-Neutrons Produced from Anti-Protons in Charge Exchange Collisions*
B. Cork, G.R. Lambertson, O. Piccioni, W.A. Wenzel, *Physical Review* 104, 1193 (1957).

[27] *Observation of Long-Lived Neutral V Particles*
K. Lande, E.T. Booth, J. Impeduglia, L.M. Lederman, and W. Chinowski, *Physical Review* 103, 1901 (1956).

[28] *Remarks on Possible Noninvariance under Time Reversal and Charge Conjugation*
T.D. Lee, R. Oehme, and C.N. Yang, *Physical Review* 106, 340 (1957).

[29] *Question of Parity Conservation in Weak Interactions*
T.D. Lee and C.N. Yang, *Phys. Rev.* 104, 254 (1956).

[30] *Experimental Test of Parity Conservation in Beta Decay*
C.S. Wu, E. Ambler, R.W. Hayward, D.D. Hoppes, *Phys. Rev.* 105, 1413 (1957);

Observation of the Failure of Conservation of Parity and Charge Conjugation in Meson Decays: The Magnetic Moment of the Free Muon
R. Garwin, L. Lederman, and M. Weinrich, *Phys. Rev.* 105, 1415 (1957);

Nuclear Emulsion Evidence for Parity Non-Conservation in the Decay Chain $\pi^+\mu^+e^+$
J.I. Friedman and V.L. Telegdi, *Phys. Rev.* 105, 1681 (1957).

[31] *On the Conservation Laws for Weak Interactions*
L.D. Landau, *Zh. Éksp. Teor. Fiz.* 32, 405 (1957).

[32] *Evidence for the* 2π *Decay of the* K_2^0 *Meson*
J. Christenson, J.W. Cronin, V.L. Fitch, and R. Turlay, *Physical Review Letters* 113, 138 (1964).

[33] *Experimental Observation of Antideuteron Production*
T. Massam, Th. Muller, B. Righini, M. Schneegans, and A. Zichichi, *Nuovo Cimento* 39, 10 (1965).

[34] *The Discovery of Nuclear Antimatter*
L. Maiani and R.A. Ricci (eds), Conference Proceedings 53, Italian Physical Society, Bologna, Italy (1995); see also A. Zichichi in '*Subnuclear Physics - The first fifty years*', O. Barnabei, P. Pupillo and F. Roversi Monaco (eds), a joint publication by University and Academy of Sciences of Bologna, Italy (1998); World Scientific Series in 20th Century Physics, Vol. 24 (2000).

[35] The first report on 'scaling' was presented by J.I. Friedman at the 14th International Conference on *High Energy Physics* in Vienna, 28 August-5 September 1968. The report was presented as paper n. 563 but not published in the Conference Proceedings. It was published as a SLAC preprint. The SLAC data on scaling were included in the Panofsky general report to the Conference where he says «... the apparent success of the parametrization of the cross-sections in the variable ν / q^2 in addition to the large cross-section itself is at least indicative that point-like interactions are becoming involved». '*Low q^2 Electrodynamics, Elastic and Inelastic Electron (and Muon) Scattering*', W.K.H. Panofsky in Proceedings of 14th International Conference on *High Energy Physics* in Vienna 1968, J. Prentki and J. Steinberger (eds), page 23, published by CERN (1968). The following physicists participated in the inelastic electron scattering experiments: W.B. Atwood, E. Bloom, A. Bodek, M. Breidenbach, G. Buschhorn, R. Cottrell, D. Coward, H. DeStaebler, R. Ditzler, J. Drees, J. Elias, G. Hartmann, C. Jordan, M. Mestayer, G. Miller, L. Mo, H. Piel, J. Poucher, C. Prescott, M. Riordan, L. Rochester, D. Sherden, M. Sogard, S. Stein, D. Trines, and R. Verdier. For additional acknowledgements see J.I. Friedman, H.W. Kendall and R.E. Taylor, '*Deep Inelastic Scattering: Acknowledgements*', *Les Prix Nobel 1990*, (Almqvist and Wiksell, Stockholm/Uppsala 1991), also *Rev. Mod. Phys.* 63, 629 (1991). For a detailed reconstruction of the events see J.I. Friedman '*Deep Inelastic Scattering Evidence for the Reality of Quarks*' in '*History of Original Ideas and Basic Discoveries in Particle Physics*', H.B. Newman and T. Ypsilantis (eds), Plenum Press, New York and London, 725 (1994).

[36] *Quark Search at the ISR*
T. Massam and A. Zichichi, *CERN preprint*, June 1968;

Search for Fractionally Charged Particles Produced in Proton-Proton Collisions at the Highest ISR Energy
M. Basile, G. Cara Romeo, L. Cifarelli, P. Giusti, T. Massam, F. Palmonari, G. Valenti and A. Zichichi, *Nuovo Cimento* 40A, 41 (1977); and
A Search for quarks in the CERN SPS Neutrino Beam
M. Basile, G. Cara Romeo, L. Cifarelli, A. Contin, G. D'Alì, P. Giusti, T. Massam, F. Palmonari, G. Sartorelli, G. Valenti and A. Zichichi, *Nuovo Cimento* 45A, 281 (1978).

[37] *New Developments in Elementary Particle Physics*
A. Zichichi, *Rivista del Nuovo Cimento* 2, n. 14, 1 (1979). The statement on page 2 of this paper, «*Unification of all forces needs first a Supersymmetry. This can be broken later, thus generating the sequence of the various forces of nature as we observe them*», was based on a work by A. Petermann and A. Zichichi in which the renormalization group running of the couplings using supersymmetry was studied with the result that the convergence of the three couplings improved. This work was not published, but perhaps known to a few. The statement quoted is the first instance in which it was pointed out that supersymmetry might play an important role in the convergence of the gauge couplings. In fact, the convergence of three straight lines ($\alpha_1^{-1}\alpha_2^{-1}\alpha_3^{-1}$) with a change in slope is guaranteed by the Euclidean geometry, as long as the point where the slope changes is tuned appropriately. What is incorrect about the convergence of the couplings is that, with the initial conditions given by the LEP results, the change in slope needs to be at $M_{SUSY} \sim 1$ TeV as claimed by some authors not aware in 1991 of what was known in 1979 to A. Petermann and A. Zichichi.

[38] V.N. Gribov, G. 't Hooft, G. Veneziano and V.F. Weisskopf '*The Creation of Quantum ChromoDynamics and the Effective Energy*', L.N. Lipatov (ed), a joint publication by the University and the Academy of Sciences of Bologna, Italy (1998); World Scientific Series in 20th Century Physics, Vol. 25 (2000).

[39] *The Effective Experimental Constraints on M_{SUSY} and M_{GUT}*
F. Anselmo, L. Cifarelli, A. Petermann and A. Zichichi, *Nuovo Cimento* 104A, 1817 (1991).

[40] *The Simultaneous Evolution of Masses and Couplings: Consequences on Supersymmetry Spectra and Thresholds*
F. Anselmo, L. Cifarelli, A. Petermann and A. Zichichi, *Nuovo Cimento* 105A, 1179 (1992).

[41] *A Study of the Various Approaches to M_{GUT} and α_{GUT}*
F. Anselmo, L. Cifarelli and A. Zichichi, *Nuovo Cimento* 105A, 1335 (1992).

[42] *Are Matter and Antimatter Symmetric?*
T.D. Lee, in Proceedings of the '*Symposium to celebrate the 30th anniversary of the Discovery of Nuclear Antimatter*', L. Maiani and R.A. Ricci (eds), Conference Proceedings 53, page 1, Italian Physical Society, Bologna, Italy (1995).

[43] *String Theory: the Basic Ideas*
B. Greene, Erice Lectures - Discussion 1999 in '*Basics and Highlights in Fundamental Physics*', A. Zichichi (ed), World Scientific (to be published).

[44] *The Analytic S-Matrix*
G.F. Chew, W.A. Benjamin Inc., New York, Amsterdam (1966).

[45] *Search for the Time-Like Structure of the Proton*
M. Conversi, T. Massam, Th Muller and A. Zichichi, *Phys. Lett.* 5, 195 (1963).

[46] *The Leptonic Annihilation Modes of the Proton-Antiproton System at 6.8 (GeV/c)2*
M. Conversi, T. Massam, Th Muller and A. Zichichi, *Nuovo Cimento 40,* 690 (1965).

[47] *The Evolution of Gaugino Masses and the SUSY Threshold*
F. Anselmo, L. Cifarelli, A. Peterman and A. Zichichi, *Nuovo Cimento* 105A, 581 (1992).

[48] *Search for Supersymmetric Particles using Acoplanar Charged Particle Pairs from Z^0 decays*
ALEPH Collab., D. Decamp et al., *Phys. Lett.* B236, 86 (1990).

[49] *Search for Neutral Higgs Bosons from Supersymmetry in Z decays*
ALEPH Collab., D. Decamp et al., *Phys. Lett.* B237, 291 (1990).

[50] *Search for Neutralino Production in Z decays*
ALEPH Collab., D. Decamp et al., *Phys. Lett.* B244, 541 (1990).

[51] *Search for the Neutral Higgs Bosons of the MSSM and other two Doublet Models*
ALEPH Collab., D. Decamp et al., *Phys. Lett.* B265, 475 (1991).

[52] *Search for Heavy Charged Scalars in Z^0 decays*
DELPHI Collab., P. Abreu et al., *Phys. Lett.* B241, 449 (1990).

[53] *Search for Pair Production of Neutral Higgs Bosons in Z^0 decays*
DELPHI Collab., P. Abreu et al., *Phys. Lett.* B245, 276 (1990).

[54] *Search for Scalar Quarks in Z^0 decays*
DELPHI Collab., P. Abreu et al., *Phys. Lett.* B247, 148 (1990).

[55] *A Search for Sleptons and Gauginos in Z^0 Decays*
DELPHI Collab., P. Abreu et al., *Phys. Lett.* B247, 157 (1990).

[56] *Mass Limits for Scalar Muons, Scalar Electrons and Winos from e^+e^- Collisions near S**(1/2) = 91-GeV*
L3 Collab., B. Adeva et al., *Phys. Lett.* B233, 530 (1989).

[57] *Search for the Neutral Higgs Bosons of the Minimal Supersymmetric Standard Model from Z^0 Decays*
L3 Collab., B. Adeva et al., *Phys. Lett.* B251, 311 (1990).

[58] *Search for the Charged Higgs Boson in Z^0 decay*
L3 Collab., B. Adeva et al., *Phys. Lett.* B252, 511 (1990).

[59] *A Search for Acoplanar Pairs of Leptons or Jets in Z^0 decays: Mass Limits on Supersymmetric Particles*
OPAL Collab., M.Z. Akrawy et al., *Phys. Lett.* B240, 261 (1990).

[60] *A Search for Technipions and Charged Higgs Bosons at LEP*
OPAL Collab., M.Z. Akrawy et al., *Phys. Lett.* B242, 299 (1990).

[61] *A Direct Search for Neutralino Production at LEP*
OPAL Collab., M.Z. Akrawy et al., *Phys. Lett.* B248, 211 (1990); P.D. Acton et al., preprint CERN-PPE/91-115, 22 July 1991.

[62] *Searches for Supersymmetric Particles Produced in Z Boson decay*
MARK II Collab., T. Barklow et al., *Phys. Rev. Lett.* 64, 2984 (1990).

[63] *Searches for New Particles at LEP*
M. Davier, LP-HEP 91 Conference, Geneva, CH, Preprint LAL 91-48, December 1991.

[64] *A Detailed Comparison of LEP Data with the Predictions of the Minimal Supersymmetric SU(5) GUT*
J.R. Ellis, S. Kelley, D.V. Nanopoulos, preprint CERN-TH/6140-91, *Nucl. Phys.* B373, 55 (1992).

[65] *The Effective Experimental Constraints on M_{SUSY} and M_{GUT}*
F. Anselmo, L. Cifarelli, A. Peterman and A. Zichichi, *Nuovo Cimento* 104A, 1817 (1991).

[66] *The Convergence of the Gauge Couplings at E_{GUT} and above: Consequences for $\alpha_3(M_Z)$ and SUSY Breaking*
F. Anselmo, L. Cifarelli, A. Peterman and A. Zichichi, *Nuovo Cimento* 105A, 1025 (1992).

[67] *The Simultaneous Evolution of Masses and Couplings: Consequences on Supersymmetry Spectra and Thresholds*
F. Anselmo, L. Cifarelli, A. Peterman and A. Zichichi, *Nuovo Cimento* 105A, 1179 (1992).

[68] *Analytic Study of the Supersymmetry-Breaking Scale at Two Loops*
F. Anselmo, L. Cifarelli, A. Peterman and A. Zichichi, *Nuovo Cimento* 105A, 1201 (1992).

[69] *A Study of the Various Approaches to M_{GUT} and α_{GUT}*
F. Anselmo, L. Cifarelli and A. Zichichi, *Nuovo Cimento* 105A, 1335 (1992).

[70] *A χ^2-Test to Study the α_1, α_2, α_3 Convergence for High-Precision LEP Data, Having in Mind the SUSY Threshold*
F. Anselmo, L. Cifarelli and A. Zichichi, *Nuovo Cimento* 105A, 1357 (1992).

[71] U. Amaldi, W. de Boer and H. Furstenau, *Phys. Lett.* B260, 447 (1991).

[72] *The Landscape and its Physics Foundations*
How String Theory Generates the Landscape
L. Susskind
Proceedings of the 2006–Erice Subnuclear Physics School to be published by World Scientific.

[73] *Complexity and Landscape in String Theory*
M.R. Douglas and F. Denef
Proceedings of the 2006–Erice Subnuclear Physics School to be published by World Scientific.

[74] *The Logic of Nature, Complexity and New Physics: From Quark-Gluon Plasma to Superstrings, Quantum Gravity and Beyond*
Proceedings of the 2006–Erice Subnuclear Physics School to be published by World Scientific.

CHAIRMAN: Antonino Zichichi

Scientific Secretaries: Jacob Bourjaily and Diego Casadei

DISCUSSION I

— *Sevil Salur*:

Why do we have three families of quarks? What is special about the number three?

— *Antonino Zichichi*:

Nobody knows. There is no theoretical explanation; but possibly, this arises from the compactification from (10+1)-dimensions to (3+1)-dimensions in string theory. Witten and Horava, for example, have developed a model in which this occurs. Compactifying (10+1)-dimensional supergravity on a line segment, they obtain the desired SU(3)xSU(2)xU(1) gauge group with three families. However, there are many features of the subnuclear world still missing, for example the mixing of states needed for the weak interactions. But we at least have a way to explain the origin of three families.

— *Sergio Ferrara*:

Theories of gravity in higher dimensions can have a mechanism different from Kaluza-Klein for explaining forces. For example, there can be topological effects called branes—which do not need to be four dimensional in the low energy limit—in which interactions do not arise from the compactifications to lower dimensions but from the degrees of freedom of the topological defect.

— *Antonino Zichichi*:

I'm not presenting the current theoretical status of our understanding. It could be that we're in a brane of four dimensions, itself embedded in higher dimensions to which we are gravitationally coupled. Hence, our world could be immersed in a fantastic superworld. But the introduction of branes is a further complication that I did not want to include in my lecture since it is already too complicated.

— *Johan Alwall*:

Are there any examples of the Anderson-Feynman-Beethoven phenomena that you described where it is *impossible* to explain physics at higher scales from the fundamental laws?

— *Antonino Zichichi*:

There is no such 'no-go theorem'. However, I invite all of you to try to go from QCD to nuclear physics. You will become famous in no-time.

— *Jernej Kamenik*:

If it is not possible to derive the laws of physics at higher levels from those describing lower levels, then why would we expect a quantum theory of gravity to be useful for our description of the standard model?

— *Antonino Zichichi*:

If you say this, then all the RQST theorists go home. Not one knows the end of this very interesting story. Probably Professor Ferrara or Professor Veneziano will speak about this. I hope that RQST will be the source of our further understanding of the problem.

I remember several times when everyone thought that everything was understood in physics. The last time was in 1947. In my book *Subnuclear Physics: the Fist 50 Years,* I describe many examples. With what was known, people believed you could explain everything. The little details would eventually be described. But this was never the case.

— *Giovanni Imponente*:

There has been a great deal of interest in the cosmological implications of brane-worlds. Is there any way to experimentally determine if the bulk exists? Would it be possible for interactions from extra dimensions to effect physics on the brane via SUSY-particles? Could these particles live in the bulk?

—*Antonino Zichichi*:

Supersymmetry particles are in the brane. The entire universe exists only in the brane and the only interaction that exists in the bulk is gravitation. How can we discover this? Good question. Study gravitational interactions. How? We should take advice from Galileo: ask only questions that can be tested experimentally. At present we are in the midst of a great theoretical development: relativistic quantum string theory. Unfortunately, so far this theory has produced no falsifiable prediction for experiment. String theory is not yet able to describe reality.

Professor Ferrara may have more to say.

— *Sergio Ferrara*:

I believe that we should have more modest goals. I should note that you can have supersymmetry without strings and strings without supersymmetry. We should not put everything together. Rather, we should first try to find supersymmetry at low energies; then, the quantization of gravity might bring everything together. But in principle they can be separate.

—*Antonino Zichichi*:

We don't know the end of the story. We really have to hope that nature is going to be kind with us. If the desert announced in 1979 is true, we're stuck. If the desert is there, there is nothing from 100 GeV to the Planck mass, and we go home.

—*Oleg Loktik*:

How many future discoveries do you think will have a significant impact on society? Will any TeV events have relevance for real life? Is the future of high energy physics still relevant to the real world? How can these discoveries have an impact similar to that of quantum mechanics? Do you think that superstring theory can have similar consequences for society?

—*Antonino Zichichi*:

Our enemies say this: "we couldn't care less." In reality, we are working most rigorously on complexity. Throughout the history of science, we have observed totally unexpected events. Strange particles, radioactivity, the unification of electricity and magnetism, and Lorentz invariance were all unexpected discoveries. You cannot imagine what will be discovered tomorrow at the next collider: maybe Higgs; maybe supersymmetry; or maybe quark gluon plasma, will open new horizons.

—*Luciano Maiani*:

We should be more optimistic about the possibility to describe large scale physics from the laws of underlying levels: for example, we can explain the laws of thermodynamics in terms of the collisions of atoms, and phase transitions in terms of thermodynamic quantities. There are enormously different scales involved in these examples. So it may be difficult, but it can be done. Some of us believe that such descriptions are always possible.

I have a somewhat different question: you said that it is difficult to defend our discipline to those who fund it. How then do you think we should proceed?

—*Antonino Zichichi*:

The strategy is multilateral. Our detractors say that we are stuck on reductionism and for this reason we are not able to address the complexity arising from economics, social activities, the internet, traffic, *etc.* They say the new frontier is holism. You want to convince decision making leaders that we *do* study complex phenomena. We wish to describe complexity at a fundamental level in order to be able to understand complexity at the macroscopic level. How can our complexity be translated into the other fields? We need to do something that proves to the people in the streets that we are doing something very relevant for the world we live in. If, for example, you could prove that the arrow of time derives from the violation of time reversal invariance at the microscopic scale then physics would get a tremendous boost. Your link with thermodynamics is very important.

CHAIRMAN: A. ZICHICHI

Scientific Secretaries: D. Baldassari, V. Bindi

DISCUSSION II

- *G. Veneziano*:

Can you make similar tests of universality using LEP data?

- *A. Zichichi*:

The LEP data fit our extrapolation, there is no hint for anything deviating. We have analyzed the leading effect for η' and we found that in jets with η' leading the universality feature are indeed present.

- *L. Cifarelli:*

Following the answer of Prof. Zichichi: there are presently again studies concerning the higher energy regimes (from LEP, HERA, Tevatron, RHIC) in particular in terms of average charged multiplicity versus the corrected energy parameters, indicating that the "universality features" still hold true.

- *S. Ferrara:*

What are the low energy limits of the "effective energy" in which your hypothesis has been tested?

- *A. Zichichi*:

The lowest possible effective energy investigated at CERN-ISR was 3GeV and the data well compared with the e^+e^- collision at SPEAR. The maximum energy investigated was 30GeV and the (pp) data were compared with TASSO, the most high energy e^+e^- machine at that time. The ISR collider was at three nominal energies of the proton beams: 30, 44 and 62GeV. Notice that the same effective energy can be obtained using different values of the "nominal" (pp) energy.

- *M. Gyulassy:*

The universality of the effective energy phenomena is specific to hadronization in the cold physical confinement/vacuum. This is why Gribov said he has to first understand confinement to understand the universal features that you found.

However, in proton-nucleus and especially nucleus-nucleus reaction, final state interactions strongly changed feature that are universal to the confining vacuum. However, a new type of universality emerges where, for example $K/\pi, \Lambda/\pi, \Omega/\pi$, are very strongly enhanced, relative to e^+e^- and pp, and infact well approximated by statistical thermodynamics applied just below the confinement temperature $T \approx 170$ MeV.

The total entropy production in AA does conform to the effective energy hypothesis (as shown at RHIC by PHOBOS, etc.). However, the hadronic composition, the collective elliptic flow, the jet quenching, the correlation between dynamical observables are drammatically different than in pp collisions.

I believe one could extend the effective energy hypothesis to a new effective energy density that enables one to relate observables, not only at T=0 vacuum, but also to $T \geq T_C$ QGP production in AA at SPS, RHIC and soon LHC.

- *A. Zichichi*:

The universality exists also in QGP, but when you introduce the universality in a medium, what happends to the universality features is due to the enormous complexity of the medium, where these universality features indeed are present. In other words, the universality is exactly there but it is forced by the medium and disturbed by the huge number of quarks and gluons it contains.

I think that we asymptotically go to the same interpretation. It is important that you start from the universality features; in the collisions all the interactions do respect universality features, and each one has its effective energy in the QCD world, full of gluons and quarks, and therefore, the jets which need to interact with the QCD world get totally changed. But it would be very interesting that we go on interacting, in order to see what could be done, if there is a key feature which can be disentangled in this QCD dense atmosphere which is the opposite of our clean vacuum state.

- *G. Veneziano:*

I think that the problem with proving universality in QCD is that one expects much larger transverse momenta in e^+e^- than in pp soft collisions. In the case of e^+e^- a large $p_\perp$ event is only suppressed by $\alpha(p_\perp^2) \approx 1/\log p_\perp^2$ while in pp the suppression is much stronger.

- *A. Zichichi*:

If you speak about transverse momenta, where we have been working, in all the analysis we have done, for the jets in proton-proton collisions and equivalently in e^+e^- collisions, the p_T distributions look the same up to the limit of 1GeV; what you ask me is at the level of 1% of the distribution; what is most important is all the rest.

People at present are mostly interested in other things like Higgs, SUSY, etc. It is important to show if QCD can theoretically "predict" the existence of the effective energy.

- *S. Salur:*

For example, for different energies of pp collision, we observe the same $\langle p_T \rangle$ in π, K, p for 25GeV and 200GeV. Is there a threshold value in effective energy?

- *A. Zichichi*

No. If you start with 100 GeV protons beam, and this collision uses 3GeV effective energy, this is exactly the analogous of the proton-proton of 10GeV producing 3GeV effective energy. No matter what the nominal energy of the incoming beam is. If the effective energy is the same, the physics is the same. This has been done just on purpose to prove that the physics which corresponds to a given effective energy remains the same when you change the nominal energy. The threshold is just technical.

- *C. Klein-Bösing*:

Is there a general dependence of the effective energy on the nominal energy?

- *A. Zichichi*:

The effective energy does not care about what is the machine energy of the collider, the effective energy range is due to energy momentum conservation. The nominal energy produces a range of effective energies; for example if we consider a beam with energy 15GeV against 15GeV, you produce an effective energy of 10GeV; the same effective energy can be produced by beams of 22GeV against 22GeV and 31GeV against 31GeV. If you change the primary energies, you produce the same physics when the effective energy is the same.

- *J. Kamenik*:

What is the relationship between the mean effective energy of reactions and the nominal energy at which the experiment is performed?

- *A. Zichichi*:

The average value of the effective energy is always roughly one half of the nominal.

NEW TALENTS SESSIONS

A series of special sessions was devoted to young talents: those fellows who cannot be easily identified in the constantly increasing dimensions of experimental groups. The problem is less acute in the field of theoretical research. The results of the presentations follow.

Exploring the Proton's Spin at PHENIX

C. Aidala for the PHENIX Collaboration
Columbia University, New York, New York 10027, USA

Abstract

In late 2001 the first polarized proton collisions at the Relativistic Heavy Ion Collider (RHIC) took place. The PHENIX experiment at RHIC has a broad program to investigate the spin structure of the proton. This program will be described, and first results will be presented.

1 Introduction

Far from the point particle it was once believed to be, the proton has proven to be an extremely complex entity. A very rich structure has gradually been uncovered over the past 40 years of research. A thorough comprehension of proton structure, in particular its spin structure, remains the goal of extensive ongoing study. The PHENIX experiment at the Relativistic Heavy Ion Collider (RHIC) at Brookhaven National Laboratory is in a position to make significant contributions to further understanding the origin of the proton's spin.

2 History of Proton Structure

2.1 The Quark-Parton Model

In the 1960s, in deep-inelastic scattering (DIS) experiments at SLAC analogous to the famous Rutherford scattering experiment that led to the discovery of the atom's hard core, it was found that protons also had "hard" subcomponents [1, 2]. These hard subcomponents came to be known as partons. It took some time before the experimentally observed partons inside the proton came to be identified as the theoretically hypothesized quarks, but eventually the quark-parton model of the proton came into being. As experimental work progressed and higher-energy lepton beams were used as probes, the proton came to reveal a much more intricate structure than that of the three so-called "valence" quarks. These other subcomponents are now known to be sea quarks and gluons.

2.2 The Spin Structure of the Proton

For many years it was assumed that the proton's spin of $\frac{1}{2}\hbar$ was due to the spins of the three spin-$\frac{1}{2}$ valence quarks, with two oriented in one direction and one in the other. In the late 1980s, however, the EMC experiment at CERN [3] discovered that only $12 \pm 16\%$ of the proton's spin was carried by quarks. This surprising result became known as the "proton spin crisis". Subsequent experimental work, mostly through DIS, has continued to explore this problem for more than 25 years, yet there remains much to be understood. In particular, the magnitude and even sign of the gluon spin's contribution to the spin of the proton remains to be determined, the flavor breakdown of the sea quarks' contributions is largely unknown, and the contribution from orbital angular momentum of both quarks and gluons has yet to be probed.

3 The Relativistic Heavy Ion Collider (RHIC)

3.1 RHIC Physics

RHIC is the most versatile hadronic collider in the world. It is capable of colliding heavy ions up to $\sqrt{s} = 200$ GeV/nucleon and polarized protons up to $\sqrt{s} = 500$ GeV, as well as asymmetric species. Collision of asymmetric species is possible due to independent rings with independent steering magnets. In the first four years of running, RHIC has provided gold collisions at four different energies, deuteron-gold collisions, and polarized proton-proton collisions, with plans for copper collisions in the upcoming year. The flexibility of RHIC allows for an extremely diverse physics program. The heavy-ion physics program investigates strongly-interacting matter under extreme conditions of density and temperature. Systematic variations of nuclear matter with collision species and energy are being examined, and nucleon structure in a nuclear environment is being studied.

The polarized proton program seeks a better understanding of the proton's spin structure, in particular contributions from the gluons and sea quarks. By studying hadronic collisions rather than deep-inleastic scattering, RHIC experiments may directly observe gluon-scattering processes. As a collider, RHIC can provide collisions at much higher energy than can be achieved in fixed-target measurements. As a result hard processes, describable by perturbative QCD (pQCD), can be studied, and new probes such as W bosons will eventually become available. The application of factorization to pQCD processes is of particular importance because it allows one to separate out parton distribution functions (pdf's), partonic hard-scattering cross sections, and fragmentation functions (FF's). Partonic hard-scattering cross sections are directly calculable in pQCD, while pdf's and FF's must be determined experimentally. RHIC experiments have access to a wealth of data from other experiments on pdf's and FF's, allowing them to check the applicability of pQCD calculations to their unpolarized data and subsequently utilize factorized pQCD to determine various polarized pdf's more accurately.

The unpolarized cross sections for mid-rapidity production (see Figure 1) as well as forward production of neutral pions have been measured in 200-GeV proton-proton collisions at RHIC and have been found to agree well with next-to-

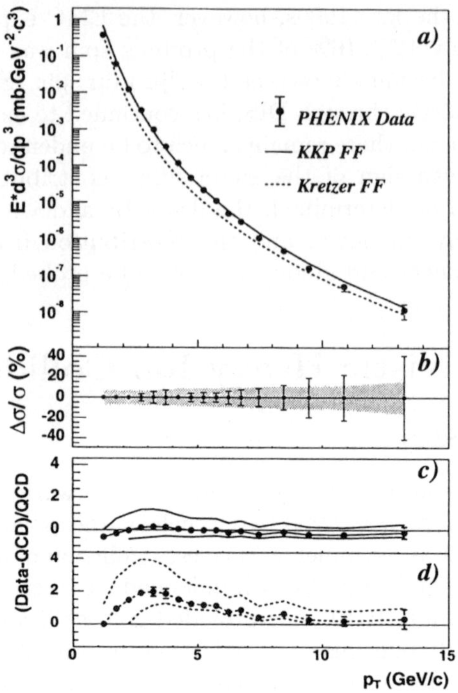

Figure 1: PHENIX results (points) for the invariant differential cross section for inclusive neutral pion production at 200 GeV. In panel *b)* the relative statistical (points) and point-to-point systematic (band) errors are shown.The curves are the results from NLO pQCD calculations using two different sets of fragmentation functions. See [4] for more details.

leading order (NLO) pQCD calculations [4, 5]. In addition, as shown in Figure 2, there are preliminary results from PHENIX for mid-rapidity production of inclusive charged hadrons which also demonstrate consistency with NLO pQCQ. This agreement indicates that NLO pQCD will be applicable in interpreting polarized data from RHIC as well and provides a solid theoretical foundation for the spin physics program.

3.2 RHIC as a Polarized Proton Collider

RHIC is the first high-energy polarized proton collider in the world. This achievement is possible due to the development of a variety of technologies to create, maintain, and measure the beam polarization throughout acceleration and storage.

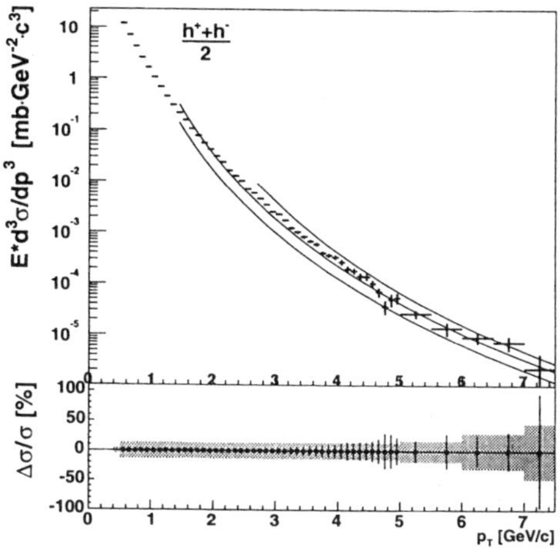

Figure 2: Preliminary PHENIX results (points) for the invariant differential cross section for inclusive charged hadron production at 200 GeV. In the bottom panel the relative statistical (points) and point-to-point systematic (band) errors are shown. The curves indicate NLO pQCD calculations by W. Vogelsang, using renormalization scales of $\frac{p_T}{2}$, p_T, and $2p_T$.

RHIC-AGS complex For proton-proton running, the path traveled by the protons is through a linac, a booster, the Alternating Gradient Synchrotron (AGS), and finally RHIC. The polarized source reliably provides a polarization of approximately 80%. The polarization in the AGS is maintained via careful tuning to avoid depolarizing resonances during acceleration and a partial Siberian snake. Siberian snakes, helical magnets developed at Novosibirsk, rotate the spin vector of the proton 180° such that any effects of depolarizing resonances will effectively be averaged out on the next turn around the ring. In 2005 a full-length, superconducting Siberian snake will be installed, completing the array of equipment related to running polarized protons in RHIC. Once the superconducting snake is installed, RHIC should be capable of reaching its design beam polarization of 70% at a beam energy of 250 GeV. In RHIC, there are two Siberian snakes installed in each ring. Very little polarization loss has been observed in the RHIC rings through acceleration and storage.

Polarimetry The RHIC-AGS complex utilizes various polarimeters to determine the beam polarization at different points along its path. In particular, there are proton-carbon (pC) polarimeters in both the AGS and RHIC which make use of Coulomb nuclear interference (CNI) [6]. A filament of carbon is inserted into the proton beam, and the left-right (azimuthal) asymmetry of recoil carbon atoms from $p^{\uparrow}C \rightarrow p^{\uparrow}C$ elastic scattering is measured. The analyzing power, $A_N \approx 0.015$, originating from the anomalous magnetic moment of the proton is exploited. The polarization of the beam can be determined from the following set of equations, in which N_L (N_R) is the number of recoil carbon atoms observed to the left (right) of the beam :

$$P_{Beam} = \frac{\varepsilon_{LR}}{A_N}, \varepsilon_{LR} = \frac{N_L - N_R}{N_L + N_R}$$

The uncertainty on the polarization measurement from the CNI polarimeters is currently approximately 30%, and the analyzing power must be further calibrated for improved polarization measurements.

In the spring of 2004, a hydrogen-jet polarimeter [7] was commissioned. A polarized hydrogen gas jet target is inserted into the beam, and the left-right asymmetry in p-p elastic scattering is measured in the CNI regime. The hydrogen-jet polarimeter will be used to calibrate the pC polarimeters and is expected to reduce the uncertainty on the polarization from ~30% to 5%.

Spin direction The naturally stable spin direction is transverse to the proton's momentum, in the vertical direction. Spin-rotator magnets immediately outside the STAR and PHENIX interaction regions are used to achieve longitudinal spin. These magnets were not commissioned until 2003, so during the 2001-2 run only data with transverse spin were taken. A detector which exploits previously measured forward-neutron azimuthal asymmetries in transverse-spin collisions is used to confirm the longitudinal component of the spin at the PHENIX interaction region.

3.3 The PHENIX Experiment

There are four major experiments at RHIC: two larger experiments, PHENIX and STAR, and two smaller ones, BRAHMS and PHOBOS. PHENIX, STAR, and BRAHMS all have spin physics programs. The four experiments have capabilities that overlap in many areas, making it possible to corroborate new results, but also areas of specialization which make the experiments complementary.

The PHENIX collaboration is comprised of approximately 480 participants from 12 different nations. The PHENIX detector [8] consists of two central spectrometer arms to track charged particles and detect electromagnetic processes, two forward spectrometer arms to identify and track muons, and three global detectors to determine when a collision occurs. The central arms cover a pseudo-rapidity range of $|\eta| < 0.35$ and 90° in azimuth each, while the forward arms cover $1.2 < |\eta| <$ ~2.2 and 2π in azimuth. PHENIX was specifically designed to have a high rate capability and high granularity as well as good mass resolution and particle-identification capabilities.

3.4 The Spin Physics Program at PHENIX

The first polarized proton collisions at RHIC were achieved in December 2001. In the 2001-2002 run, an average beam polarization of 15% was achieved, and

150 nb^{-1} of transverse-spin data were collected by PHENIX. In 2003, the average polarization reached 27%, and 220 nb^{-1} of longitudinal-spin data were taken. 2004 was principally a commissioning run to improve the polarization in the AGS and to commission the hydrogen jet polarimeter. During four days of data taking at the end of the commissioning period 75 nb^{-1} with an average polarization of approximately 40% were collected. There has been tremendous progress in machine performance over the first three years of running polarized protons at RHIC, and an extensive spin run of approximately 10 weeks with close to 50% polarization is anticipated in 2005. Proton running up until this point has been at $\sqrt{s} = 200$ GeV; 500-GeV runs are planned for the future.

PHENIX has a broad spin physics program. The principal areas of investigation are the gluon polarization (ΔG), flavor separation of the sea quark polarization ($\Delta\overline{u}$, $\Delta\overline{d}$), and transverse spin physics. PHENIX will be able to access a number of channels which probe ΔG through double longitudinal-spin asymmetries. These channels include pion production, for which results have already been published (see below), prompt photon production, dominated by gluon Compton scattering, heavy flavor production, mainly from gluon-gluon fusion, and jet production. When RHIC begins running 500-GeV protons, PHENIX will have access to W bosons, which will be identified via their leptonic decay mode. Because W^+ (W^-) production will be almost entirely from $u+\overline{d}$ ($\overline{u}+d$) and Δu and Δd are already well known, it will be possible to single out $\Delta\overline{u}$ and $\Delta\overline{d}$ from measurement of the single longitudinal-spin asymmetry of W production. The transverse spin physics program at PHENIX seeks to understand the transverse spin structure of the proton. This structure will be explored through a variety of measurements, including single transverse-spin asymmetries, for which there are already results (see below), jet correlations, the double transverse-spin asymmetry of the Drell-Yan process, and the interference fragmentation of pion pairs.

3.5 Recent Spin Physics Results

From the first two years of polarized proton collisions at RHIC, PHENIX has results on the single transverse-spin asymmetry of neutral pions and charged hadrons as well as the double longitudinal-spin asymmetry of neutral pions [9], [10].

3.5.1 Single Transverse-Spin Asymmetry of Neutral Pions and Charged Hadrons

The single transverse-spin asymmetry in the yield of a particular particle is given by

$$A_N = \frac{1}{P_{beam}}\left(\frac{N_L - N_R}{N_L + N_R}\right)$$

where N_L (N_R) is the particle yield to the left (right) of the polarized beam.

Large single transverse-spin asymmetries on the order of 20-30% have been observed in a number of experiments [11, 12, 5], ranging in energy from $\sqrt{s} = 20$-

200 GeV. The large asymmetries seen have stimulated more careful study by the theoretical community of polarized cross sections, in particular their dependence on the intrinsic transverse momentum of the partons (k_T) (see e.g. [13]).

Over the years, a number of models based on pQCD have been developed to predict these k_T dependencies and to explain the observed asymmetries. Among these models are the Sivers effect [14], transversity and the Collins effect [15], and various models which attribute the observed asymmetries to higher-twist contributions (see e.g. [16]). The Sivers effect hypothesizes that the asymmetries are due to spin-dependent intrinsic partonic momentum; the Collins effect suggests that they stem from a spin-dependent transverse momentum kick in the fragmentation process. The Collins effect requires transversity, the degree to which quarks in a transversely polarized proton are transversely polarized, to be non-zero in order to produce a non-zero asymmetry.

In Figure 3 preliminary PHENIX data on the transverse single-spin asymmetry of inclusive charged hadrons as well as neutral pions is shown as a function of p_T. The data were taken with the central arms and thus represent a pseudorapidity coverage of $|\eta| < 0.35$, corresponding to $x_F = \frac{p_L}{\sqrt{s}/2} \approx 0$. The single transverse-spin asymmetries observed for production of both neutral pions and inclusive charged hadrons at $x_F \approx 0$ are consistent with zero over the measured transverse momentum range. A small asymmetry in this kinematic region follows the trend of previous results, which indicate a decreasing asymmetry with decreasing x_F [11, 17, 5]. As a significant fraction of neutral pion production in this kinematic region comes from gluon scattering, any contribution to the asymmetry from transversity and the Collins effect, requiring a scattered quark, would be suppressed, while contributions from the Sivers effect or other mechanisms would remain a possibility. Further theoretical study of the results will have to be performed in order to interpret their full implications for the transverse spin structure of the proton. Future measurements reaching higher p_T will be dominated instead by quark scattering and a better probe of transversity and the Collins effect. See [9] for further discussion of these results.

3.5.2 Double Longitudinal-Spin Asymmetry of Neutral Pions

From the 2003 data-taking period, PHENIX obtained its first results probing ΔG, the gluon spin contribution to the spin of the proton. The double longitudinal-spin asymmetry of neutral pions was measured at mid-rapidity. The double longitudinal-spin asymmetry is given by

$$A_{LL} = \frac{1}{|\langle P_1 P_2 \rangle|} \left(\frac{N_{++} - RN_{+-}}{N_{++} + RN_{+-}} \right)$$

where P_1 and P_2 are the beam polarizations, N_{++} (N_{+-}) is the particle yield from same-helicity (opposite-helicity) bunch crossings and R is the relative luminosity between same- and opposite-helicity crossings. In Figure 4, the double-longitudinal asymmetry of neutral pions is shown as a function of p_T. The curves indicate two theoretical calculations based on NLO pQCD. The data

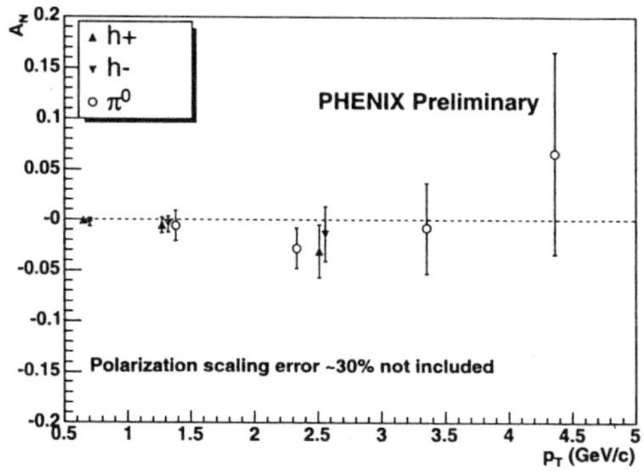

Figure 3: Preliminary PHENIX results as a function of transverse momentum for the single transverse-spin asymmetry of inclusive charged hadrons and neutral pions at mid-rapidity ($x_F \approx 0$). The positive charged hadron points are shifted 50 MeV/c to the left for readability. See [9] for more details.

points do not suggest a large contribution from gluon spin. For further details regarding the analysis and these results, see [10].

As mentioned above, in the current kinematic range π^0 production has a significant contribution from g-g scattering. This gluon dominance makes the $A_{LL}^{\pi^0}$ measurement quite sensitive to the polarized gluon pdf; however, because the polarized gluon pdf enters the factorized cross section twice at approximately equal values of x_{Bj}, it is not straightforward to determine the sign of ΔG from this measurement. Further theoretical discussion of these results and the sign of ΔG can be found in [18]. Future measurements of the double longitudinal-spin asymmetry of charged pions, produced largely via g-q scattering, will provide an additional handle on the magnitude of ΔG and allow determination of its sign.

4 Conclusions

RHIC, as a polarized hadron collider, provides a wealth of new opportunities to study the spin structure of the proton. The accelerator has already demonstrated success, and the RHIC community is looking forward to many more years of running with further improvements in luminosity and polarization as well as at higher energy. The PHENIX experiment has a broad program to investigate this structure, with particular focus on the gluon's contribution to the spin of the proton, the flavor decomposition of the sea quarks' contributions, and the transverse spin structure of the proton. First results are already available, indicating that the small single transverse-spin asymmetries seen at $x_F \approx 0$ at lower energies remain small at RHIC energies and that ΔG is not

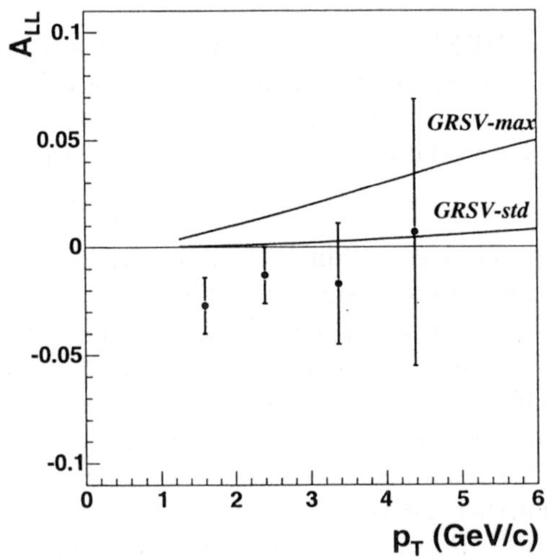

Figure 4: PHENIX results for the double longitudinal-spin asymmetry of inclusive neutral pions at mid-rapidity. A scale uncertainty of ±65% is not included. Two theoretical calculations based on NLO pQCD are shown for comparison with the data. See [10] for more details.

large. The spin structure of the proton continues to be a field of study of great interest with much still to be explored.

5 Acknowledgements

The author wishes to express her deep appreciation to Miriam Kartch-Hughes for making it possible to attend ISSP 2004 via a scholarship in memory of her husband, Vernon W. Hughes. PHENIX acknowledges support from the Department of Energy and NSF (U.S.A.), MEXT and JSPS (Japan), CNPq and FAPESP (Brazil), NSFC (China), CNRS-IN2P3, CEA, and ARMINES (France), BMBF, DAAD, and AvH (Germany), OTKA (Hungary), DAE and DST (India), ISF (Israel), KRF and CHEP (Korea), RAS, RMAE, and RMS (Russia), VR and KAW (Sweden), U.S.CRDF for the FSU, US-Hungarian NSF-OTKA-MTA, and US-Israel BSF.

References

[1] E. D. Bloom et al., Phys. Rev. Lett. **23**:930-934 (1969).

[2] M. Breidenbach et al., Phys. Rev. Lett. **23**:935-939 (1969).

[3] J. Ashman et al., Phys. Lett. **B206**, 364 (1988); Nucl. Phys. **B328**, 1 (1989).

[4] S. S. Adler et al., Phys. Rev. Lett. **91**, 182301 (2003).

[5] J. Adams et al., Phys. Rev. Lett. **92**, 171801 (2004).

[6] O. Jinnouchi et al., 15th Int. Spin Physics Symposium (SPIN 2002), AIP Conf. Proc. **675**:817-825 (2003).

[7] A. Zelenski et al., 15th Int. Spin Physics Symposium (SPIN 2002), AIP Conf. Proc. **675**:954-960 (2003).

[8] K. Adcox et al., NIM **A499**, 469-479 (2003)

[9] C. Aidala for the PHENIX Collaboration, Proceedings of DIS 2004, hep-ex/0410003.

[10] S. S. Adler et al., hep-ex/0404027, accepted to Phys. Rev. Lett.

[11] D. L. Adams et al., Phys. Lett. **B261**, 201 (1991); **B264**, 462 (1991).

[12] A. Airapetian et al., Phys. Rev. Lett. **84**:4047-4051 (2000).

[13] P. J. Mulders and R. D. Tangerman, Nucl. Phys. **B461**:197-237 (1996); Erratum Nucl. Phys. **B484**:538-540 (1997).

[14] D. Sivers, Phys. Rev. **D41**, 83 (1990); **D43**, 261 (1991).

[15] J. Collins, Nucl. Phys. **B396**, 161 (1993).

[16] J. Qiu and G. Sterman, Phys. Rev. **D59**, 014004 (1998).

[17] D. L. Adams et al., Phys. Rev. **D53**:4747-4755 (1996).

[18] B. Jaeger et al., Phys. Rev. Lett. **92**, 121803 (2004).

An improved description of charged Higgs boson production*

Johan Alwall
High Energy Physics, Uppsala Univ., Box 535, S-751 21 Uppsala, Sweden
E-mail: Johan.Alwall@tsl.uu.se

October 11, 2004

Abstract:
Many extensions of the Standard Model predict the existence of charged Higgs bosons. In order to be able to find those particles, an accurate description of their production is needed. In Monte Carlo simulations of charged Higgs boson production at hadron colliders, the two tree-level processes $gb \to H^{\pm}t$ and $gg \to H^{\pm}tb$ are used. Since those processes overlap in the collinear region of the phase-space of the outgoing b-quark, care must be taken not to introduce double-counting if both processes are to be used together. In this talk I present a method for matching these processes, developed by Johan Rathsman and myself. The method also allows for investigations of the factorization scale dependence of the processes and a better understanding of which factorization scale to choose to get a reliable description of charged Higgs production.

1 Introduction

The existence of a charged Higgs boson is a common feature of many extensions of the Standard Model of particle physics, most notably supersymmetric extensions such as the MSSM. In the Standard Model, the fermions get their mass from their interaction with the Higgs field, which gets a non-zero vacuum expectation value from spontaneous symmetry breaking of the isospin $SU(2)_L$ symmetry. The vector bosons $W^{\pm}$ and Z^0 get mass by absorbing three of the originally four scalar degrees of freedom from the Higgs doublet. In supersymmetric extensions of the Standard Model, one Higgs doublet is not enough - at least two are necessary for mainly two reasons (see *e.g.* [1]):

1. The fermionic superpartner of the Higgs boson, so called higgsino, destroys the cancellation of the ABJ anomaly (see *e.g.* [2]) in the Standard Model. In order for the anomaly to cancel, two Higgs doublets (and thus two higgsinos) with opposite hypercharge are needed.

2. With supersymmetry, the same Higgs doublet cannot interact with (and hence give mass to) both the up-type (u,c,t) and the down-type fermions (d,s,b and the charged leptons).

*Talk given at the 42nd International School of Subnuclear Physics at Erice, Sicily, 1 September 2004

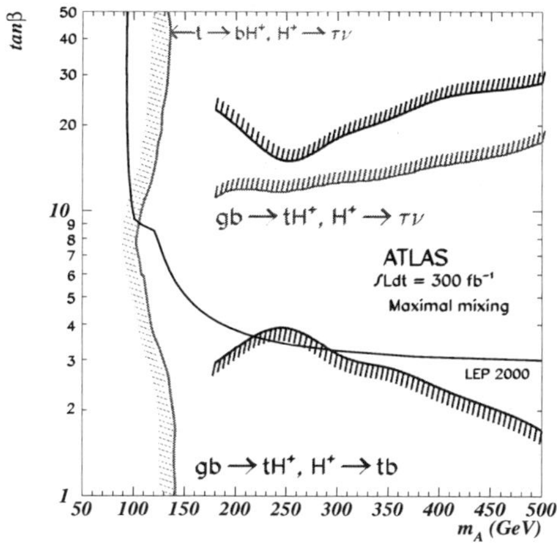

Figure 1: The ATLAS 5-σ discovery contour for charged Higgs. The gap in the region around the top mass could be bridged using a properly matched sum of the $gb \to H^{\pm}t$ process and the $gg \to H^{\pm}tb$ process. Figure taken from [3].

With two Higgs doublets (8 real fields) but still only three massive vector fields, we get five surviving Higgs fields:

$$h,\ H,\ H^+,\ H^-,\ A \text{ (pseudoscalar)}$$

Such a theory is called a (type II) two Higgs doublet model (2HDM).

In the MSSM there are only two parameters determining the masses and interactions of these fields: the first is $\tan\beta = \frac{v_1}{v_2}$, the ratio of the vacuum expectation values for the two Higgs doublets, the second is one of the masses, *e.g.* the pseudoscalar mass M_A. In a general 2HDM, however, there are seven (or more) parameters.

Needless to say, the discovery of a charged scalar particle would be a clear signal of physics beyond the Standard Model. In order to search for such a particle, we need an accurate description of the production mechanisms and phase-space distributions. Using Monte Carlo programs, the production of charged Higgs can then be simulated, and one can optimize search strategies (*i.e.* minimize the Standard Model background) using different cuts on the data from the collider. Even before any actual experiment is done, one can in this way put limits on the parameter-space regions where different experiments will be able to find a signal (see fig. 1).

In Monte Carlo generators such as PYTHIA[4] and HERWIG[5], the production channels used to simulate single charged Higgs boson production (as opposed to pair-production, which is not discussed here) are $g\bar{b} \to H^+\bar{t}$ and $gg \to H^+\bar{t}b$ and their charge conjugates. (There is also a process $q\bar{q} \to H^{\pm}tb$, which gives a large contribution in a $p\bar{p}$-collider such as Tevatron, but a very small contribution at the high energies of the LHC.) Here (as usual) g stands for gluon and q ($\bar{q}$) for an arbitrary quark (antiquark). The $gg \to H^{\pm}tb$ process gives a better description of the part of phase-space where the outgoing b-quark

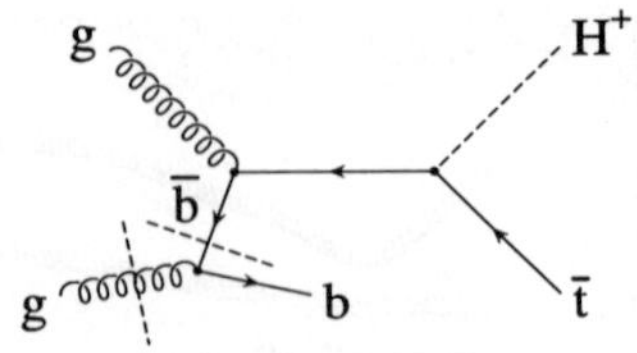

Figure 2: Illustration of the relation between the $gb \to H^{\pm}t$ and $gg \to H^{\pm}tb$ processes. If the factorization between the parton densities and the hard scattering is done at the gluon line we get the $gg \to H^{+}\bar{t}b$ process, while if instead this factorization is done at the $\bar{b}$ line, we get the $g\bar{b} \to H^{+}\bar{t}$ process. They can therefore be viewed as the same process in two different approximations.

has a large transverse momentum (p_T), while the $gb \to H^{\pm}t$ process resums potentially large logarithms $(\alpha_s \log(\mu_F/m_b))^n$ and hence give a better description in the rest of the phase space, as will be discussed later. In the region where the outgoing b-quark has small transverse momentum, the two processes overlap. Therefore, if both processes would be used and summed naively we would get double-counting in this region of phase-space. Together with Johan Rathsman, I have developed a method to remove this double-counting by generating events from a distribution corresponding to the double-counted part of phase space, and subtracting these events from the sum. Our work is presented in [7], where also more references are found.

2 The twin-processes and their double-counting

As discussed in the introduction, the two tree-level processes (*i.e.* no-loop processes) used in Monte Carlo simulation of single production of charged Higgs at hadron colliders are

$$g\bar{b}(b) \;\to\; \bar{t}H^{+}\;(tH^{-}) \tag{1}$$
$$gg \;\to\; \bar{t}bH^{+}\;(t\bar{b}H^{-}) \tag{2}$$

The first one (1), which I will denote the leading order (LO) or $2 \to 2$ process, includes the b-quark density, $b(\mu_F^2) \sim \sum(\alpha_s \log(\mu_F/m_b))^n$, which comes from the logarithmic DGLAP resummation of gluon splitting to $b\bar{b}$ pairs. This means that the b-quark going into the process is accompanied by a $\bar{b}$ (or vice versa) which is not explicitly shown in the equation. Due to the approximation made in the DGLAP expansion, this accompanying b-quark is nearly collinear with the beam. The second production process, (2), which I will denote the $2 \to 3$ process, gives the correct treatment of the kinematics of the accompanying b-quark to order α_s^2. The relation between the two processes is illustrated in fig. 2. Since the processes really have the same initial and final states, they can be viewed as the same process in two different approximations, hence the term "twin-processes".

As suggested by fig. 2, there is an overlap between the two processes: When the transverse momentum of the outgoing b-quark in the $2 \to 3$ process is small, there is no distinction between the full $2 \to 3$ matrix element and a gluon splitting to $b\bar{b}$ convoluted with the $gb \to H^{\pm}t$ matrix element. Therefore there is a double-counting between the processes, which can be expressed as [6]

$$\sigma_{\rm DC} = \int dx_1 dx_2 \left[g(x_1,\mu_F) b'(x_2,\mu_F) \frac{d\hat{\sigma}_{2\to 2}}{dx_1 dx_2}(x_1,x_2) + x_1 \leftrightarrow x_2 \right] \tag{3}$$

where

$$b'(x,\mu_F^2) = \frac{\alpha_s(\mu_R^2)}{2\pi} \int \frac{dQ^2}{Q^2+m_b^2} \int \frac{dz}{z} P_{g\to q\bar{q}}(z)\, g\left(\frac{x}{z},Q^2\right) \tag{4}$$

$$\approx \frac{\alpha_s(\mu_R^2)}{2\pi} \log\frac{\mu_F^2}{m_b^2} \int \frac{dz}{z} P_{g\to q\bar{q}}(z)\, g\left(\frac{x}{z},\mu_F^2\right) \tag{5}$$

This is just the leading logarithmic contribution to the b-quark density included in the $2 \to 2$ process. Here $P_{g\to q\bar{q}}(z) = \frac{1}{2}\left[z^2+(1-z)^2\right]$ is the splitting function for g going to $q\bar{q}$, μ_F is the factorization scale (*i.e.* the scale where the parton densities are evaluated) and μ_R is the renormalization scale used in evaluating α_s, and $Q^2 = -k^2$, the 4-momentum of the incoming b-quark squared. We need to take care to include kinematic constraints due to the non-zero b-quark mass in our calculation of the integration limits, since such constraints are implicitly included in the $2 \to 3$ matrix element. This is done in detail in our paper [7].

The matched integrated cross-section is then given by

$$\sigma = \sigma_{2\to 2} + \sigma_{2\to 3} - \sigma_{\rm DC} \tag{6}$$

The matched integrated cross-section and its components are shown as a function of the charged Higgs boson mass in fig. 3. For charged Higgs masses below the top mass the cross-section can be well approximated by top pair production with subsequent decay of one of the top quarks to $H^\pm b$, *i.e.* $gg \to t\bar{t} \to tbH^\pm$ (for a comparison between this process and the $2 \to 3$ process, see [8]). Our matching procedure works for all charged Higgs masses, but is of greatest interest for $m_{H^\pm} \gtrsim m_t$. In the following I will use $m_{H^\pm} = 250$ GeV and $\tan\beta = 30$ as a case study.

3 Matching the differential cross-sections

As we saw in the last section, the cancellation of the double-counting between the $2 \to 2$ and $2 \to 3$ processes on the integrated cross-section level was simple enough. But how do we do it for the differential cross-sections? Whatever approach we take, we need to make sure some basic requirements are fulfilled:

1. The integrated cross-section should equal the correct one given by eq. (6).
2. All differential cross-sections should be smooth after matching.
3. The matched p_T-distribution for the outgoing b-quark should be given by the $2 \to 2$ process for small transverse momenta, and by the $2 \to 3$ process for large transverse momenta, with a smooth interpolation between those regions.

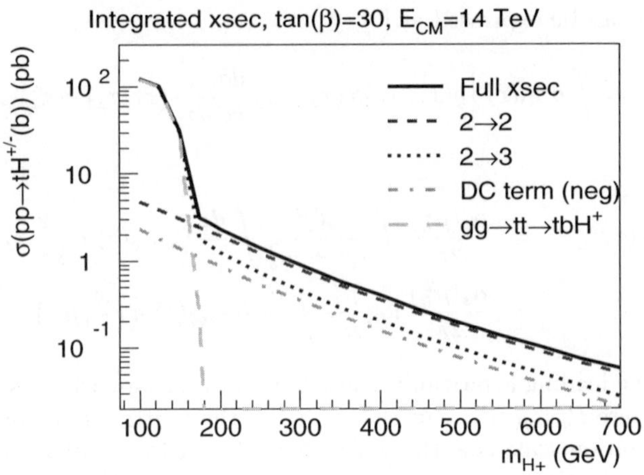

Figure 3: Integrated cross-section components (leading order process, $2 \to 3$ process and double-counting term) and matched total as a function of the $H^{\pm}$ mass at LHC, with $\tan\beta = 30$ and $\mu_F = (m_t + m_{H^\pm})/4$. Note that the double-counting term contribution (DC) is subtracted from the sum. At $m_{H^\pm} < m_t$ the $2 \to 3$ process can be approximated by $gg \to t\bar{t} \to tbH^{\pm}$.

A Monte Carlo event generator simulates collision events by picking kinematical variables from a probability distribution given by the folding of parton density functions with the matrix element of the hard scattering. In this way the events will be distributed in all kinematic variables according to the differential cross-section distribution of the process in question. This is true for the $2 \to 2$ process as well as for the $2 \to 3$ process. The number of events generated for each process is proportional to the integrated cross-section of the process. In a process where we have an incoming b ($\bar{b}$), such as the $2 \to 2$ process, we get an accompanying outgoing $\bar{b}$ (b) from the gluon splitting as explained above. The kinematic distribution of this accompanying $\bar{b}$ is given by the DGLAP evolution and generated through so called parton showers (PS) (see *e.g.* [4]).

Now, our solution to the matching problem is simple: We view the double-counting term (given by eq. (3)) as a probability distribution in kinematic variables and pick events from this distribution. Then we subtract this contribution (*i.e.* add with negative weight) from the sum of the two processes (1) and (2) in the final data analysis, *i.e.* the histograms.[1]

One might worry that this procedure could give a negative number of events in some phase-space region. However, the leading-logarithmic part of the b density used in the double-counting term is always smaller then the full b density used in the $2 \to 2$ process, ensuring that if only a sufficiently large number of events is generated, the sum of the events from the $2 \to 2$ and the $2 \to 3$ process will always be larger than the number of

[1]The code for the double-counting term, implemented as an external process to PYTHIA, will be available for download at `http://www3.tsl.uu.se/thep/MC/pytbh`, including a manual in preparation.

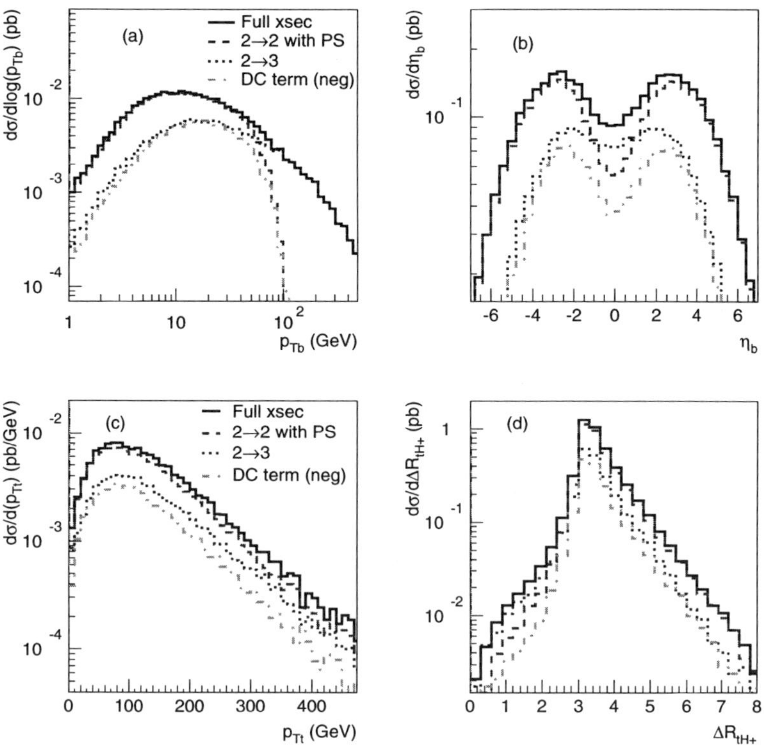

Figure 4: Differential cross-sections in (a) $p_{T,b}$, (b) η_b, (c) $p_{T,t}$ and (d) $\Delta R_{H^\pm t}$ for the cross-section components and the resulting matched cross-section with $\tan\beta = 30$, $m_{H^\pm} = 250$ GeV and $\mu_F = (m_t + m_{H^\pm})/4$. Note that the double-counting term contribution (DC) is subtracted from the sum.

events from the double-counting term.

Some resulting differential cross-sections from our matching are shown in fig. 4. Looking at fig. 4a we see that the matched differential cross-section in p_T for the outgoing b-quark looks exactly as expected (and wanted): for small p_T it follows the $2 \to 2$ process distribution, while for large p_T it follows the $2 \to 3$ process. However, there is a rather large intermediate region, from about 30 GeV to about 100 GeV, where the whole matching procedure is really necessary to get the correct differential cross-section. For comparison one can note that at LHC, the region where b-quarks can be tagged is $p_{T,b} \gtrsim 20$ GeV. In fig. 4b we see large differences in the rapidity distribution for the outgoing b-quark between the matched cross section and each of the two contributing processes, up to a factor $\simeq 2$ in the experimentally interesting region $|\eta_b| < 2.5$, indicating that matching is really necessary to get a correct description of the b-quark kinematics..

Even if the outgoing b is not observed, we still see a difference between the matched cross-section and the $2 \to 2$ process, which is usually used in this case. In the transverse momentum distribution of the top quark (fig. 4c) we get a small enhancement for large $p_{T,t}$, amounting to $\sim 20\%$ compared to the $2 \to 2$ process. There is a similar effect in the distribution of the charged Higgs boson (not shown). In fig. 4d, showing the distance measure $\Delta R_{H^\pm t} = \sqrt{\Delta\varphi^2_{H^\pm t} + \Delta\eta^2_{H^\pm t}}$, we see an effect especially for small separations.

All in all, we see that unless we make very limiting cuts on the phase space, we need to use the matched differential cross-sections to get reliable predictions.

4 Factorization scale dependence

The factorization scale is the scale where the factorization is done between the parton description of the proton and the hard scattering of the partons, *i.e.* where the parton distributions are evaluated. This scale also determines the maximum transverse momentum of the parton shower, *e.g.* of the outgoing b-quark of the $2 \to 2$ process. Of course, the factorization scale, just like the renormalization scale, is a fictitious entity in the sense that if all orders of perturbation theory was included in the cross-section calculation, the scale would never show up in the result. However, in real life we can only use a few orders in perturbation theory, and then it is necessary to introduce such scales. Often the factorization scale is chosen to be some hard scale of the process, *e.g.* the invariant mass $\hat{s}$ of the partonic system or the transverse mass of one of the outgoing particles. In the process we are discussing, $gb \to H^\pm t$, often $m_{H^\pm} + m_t \approx \hat{s}$ is used as the factorization scale. What scale we should use in the lower-order calculations can not really be known until all orders of perturbation theory have been calculated, but the factorization scale dependence should be weaker the more orders we include. Therefore we get an indication of which scale to use in the first-order expression already by doing the next-to-leading order calculation.

The double-counting term of eq. (3) is designed to describe the part of the $2 \to 3$ process which is already included in the $2 \to 2$ process through the resummation of logarithmic contributions to the b-quark density. This means that the double-counting term should not exceed the $2 \to 3$ process contribution in any part of the phase-space. However, the upper integration limit of the transverse momentum of the outgoing b-quark in the double-counting term is determined by the factorization scale. Up to this limit the transverse momentum-distribution is almost flat, which means that the integrated value of the double-counting term is nearly proportional to the factorization scale (although for large scales the kinematic constraints modify this behaviour). This is illustrated in figs. 5 and 6.

Fig. 5 shows the integrated cross-section components, as well as the matched total, as a function of the factorization scale μ_F parametrized by $\rho = \mu_F/\overline{m}$, where $\overline{m} = (m_{H^\pm} + m_t)/2$ is the average of the charged Higgs and top masses. Here two things can be noted. The first is that the double-counting term exceeds the $2 \to 3$ process term already near $\rho = 1$, *i.e.* $\mu_F \gtrsim (m_{H^\pm} + m_t)/2$. This indicates that the factorization scale should definitely not be chosen above this value. The other thing to be noted is that the matched integrated cross-section shows a significantly smaller dependence on the factorization scale than any one of the component cross-sections, indicating that the matching also stabilizes the cross-section. This is not surprising, since the $2 \to 3$ process is a part of the next-to-leading order calculation of the $2 \to 2$ process. Hence including it in a correct way should reduce the factorization scale dependence, as is the case when calculating the full

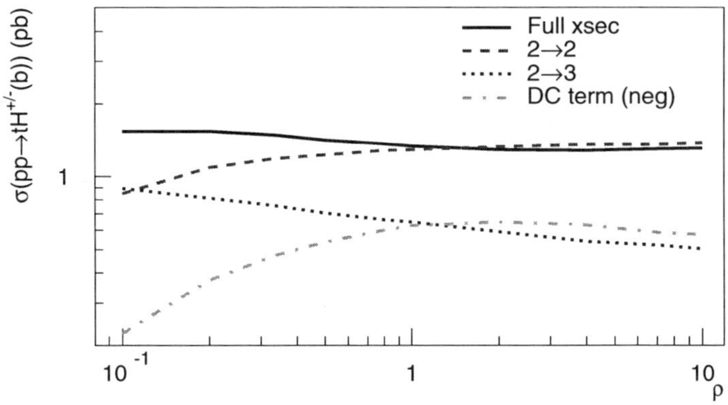

Figure 5: Integrated cross-sections at $m_{H^\pm} = 250$ GeV and $\tan\beta = 30$ as a function of the factorization scale parametrized by $\rho = 2\mu_F/(m_t + m_{H^\pm})$. Note that for $\rho \gtrsim 1$ the double-counting term exceeds the $2 \to 3$ term. Also note that the factorization scale dependence is smaller for the matched total cross-section than for any one of the components.

next-to-leading order expression.

In fig. 6 we see the distribution in p_T of the outgoing b-quark for the $2 \to 3$ process (without parton showers) and the double-counting term at different factorization scales. Here we see that already for $\rho = 1$, the double-counting term exceeds the $2 \to 3$ term around $p_{T,b} \approx 50$ GeV. $\rho = 0.5$, *i.e.* $\mu_F = (m_{H^\pm} + m_t)/4$, seems to be a limiting case where the double-counting distribution just touches the $2 \to 3$-distribution. Therefore we have chosen to use this unconventionally small value of the factorization scale in our plots in the previous sections. Similar results for the size of the factorization scale has been achieved in next-to-leading order calculations of the $2 \to 2$ process, see [9, 10].

5 Conclusions

The discovery of a charged scalar particle would be a clear signal of physics beyond the Standard Model of particle physics. The detailed study of the properties of such a particle would give valuable insight into the nature of this new physics. But to be able to find the charged Higgs boson, we need to devise search strategies and to reduce the Standard Model background. For this, an appropriate description of charged Higgs boson production, using Monte Carlo events generators, is necessary.

For single charged Higgs boson production, mainly two processes are used in Monte Carlos, the bottom-gluon fusion $gb \to H^\pm t$ production channel and the gluon-gluon fusion $gg \to H^\pm tb$ channel. These have different virtues: $gb \to H^\pm t$ resums large logarithms describing the b-quark density, why it gives the major contribution to the total cross-section, and also gives the best description of the differential cross-section for small values of the transverse momentum of the associated b-quark. On the other hand, the $gg \to H^\pm tb$

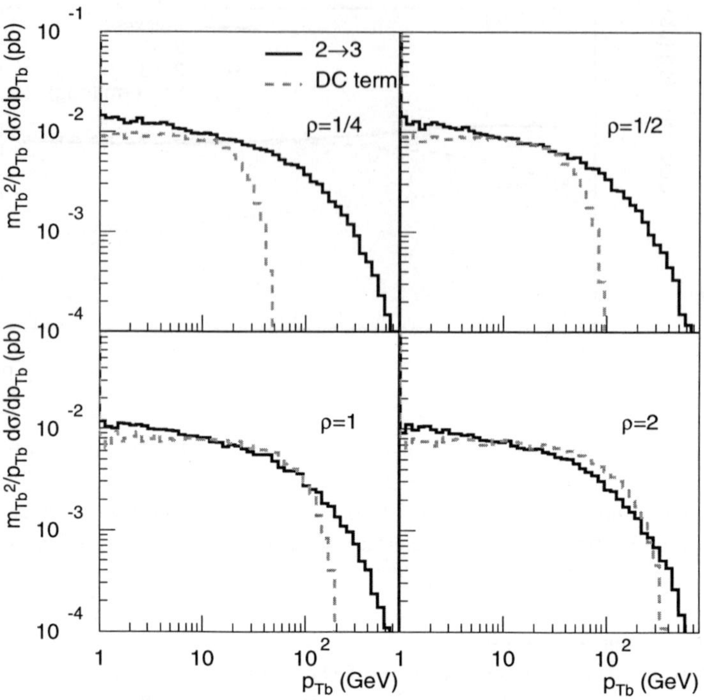

Figure 6: The differential cross-section $d\sigma/dp_{T,b}$ multiplied by $m_{T,b}^2/p_{T,b}$ for the $2 \to 3$ matrix element and the double-counting term for different factorization scales parameterized by $\rho = 2\mu_F/(m_t + m_{H^\pm})$. Note that the double-counting term overshoots the $2 \to 3$ term already for $\rho = 1$.

process gives a correct description to order α_s^2 of the outgoing b-quark for large values of the transverse momentum. However, the two processes overlap in the small transverse momentum region, so in order to use them both we must compensate for this double-counting.

In this talk I have presented our algorithm for the matching of the two processes. This matching is done by summing the events from the two processes (as is usually done in Monte Carlo generators), and then subtract events generated from a double-counting distribution term from the sum. In this way we are able to combine the virtues of the two processes to get a good description of the full differential cross-section.

This method also allows us to get a better understanding of what choice of factorization scale is appropriate, by comparing the transverse momentum distributions for the double-counting term with the distribution for the $2 \to 3$ process matrix element. Since the double-counting term should remove the part of the $2 \to 3$ distribution already contained in the $2 \to 2$ distribution, the double-counting term should not overshoot the $2 \to 3$ term. The result is that the appropriate factorization scale is significantly smaller than the conventionally used value.

Acknowledgements:
I would like to thank the organizers of ISSP42 for giving me the opportunity to present this talk. I would also like to thank all participants of the school for giving me a great time.

References

[1] S. Dawson, "SUSY and such," arXiv:hep-ph/9612229.

[2] T. P. Cheng and L. F. Li, *Gauge Theory Of Elementary Particle Physics*, Oxford science publications, 1984

[3] K. A. Assamagan and Y. Coadou, Acta Phys. Polon. B **33** (2002) 1347.

[4] T. Sjöstrand *et al.*, "High-energy-physics event generation with PYTHIA 6.1," Comput. Phys. Commun. **135** (2001) 238 [arXiv:hep-ph/0010017].

[5] G. Corcella *et al.*, "HERWIG 6: An event generator for hadron emission reactions with interfering gluons (including supersymmetric processes)," JHEP **0101** (2001) 010, [arXiv:hep-ph/0011363].

[6] F. Borzumati, J. L. Kneur and N. Polonsky, "Higgs-strahlung and R-parity violating slepton-strahlung at hadron colliders," Phys. Rev. D **60** (1999) 115011 [arXiv:hep-ph/9905443].

[7] J. Alwall and J. Rathsman, "Improved description of charged Higgs boson production at hadron colliders," arXiv:hep-ph/0409094.

[8] J. Alwall, C. Biscarat, S. Moretti, J. Rathsman and A. Sopczak, "The p anti-p $\to$ t b H+- process at the Tevatron in HERWIG and PYTHIA simulations," in *The Higgs working group: Summary report 2003*, arXiv:hep-ph/0312301.

[9] T. Plehn, "Charged Higgs boson production in bottom gluon fusion," Phys. Rev. D **67** (2003) 014018 [arXiv:hep-ph/0206121].

[10] E. L. Berger, T. Han, J. Jiang and T. Plehn, "Associated production of a top quark and a charged Higgs boson," arXiv:hep-ph/0312286.

New Issues in the Inflationary Scenario

Giovanni Imponente
School of Mathematical Sciences
Queen Mary, University of London
*Mile End Road, London E1 4NS, UK**

We briefly review the arising of an inflationary phase in the Universe evolution in order to discuss an inhomogeneous cosmological solution in presence of a real self interacting scalar field minimally coupled to gravity in the region of a slow rolling potential plateau. During the inhomogeneous de Sitter phase the scalar field dominant term is a function of the spatial coordinates only. We apply this generic solution to the Coleman-Weinberg potential and to the Lemaitre-Tolman metric. This framework specialized nearby the FLRW model allows a classical origin for the inhomogeneous perturbations spectrum.

Keywords: inhomogeneous inflation, perturbation spectrum

I. GENERAL STATEMENTS

The homogeneous and isotropic Universe so far observed at sufficient large scales (of the order of 100 Mpc) inevitably breaks down when considering galaxies, clusters, etc. The early Universe, as testified by the extreme uniformity of the Cosmic Microwave Background Radiation (CMBR), exhibits a level of isotropy and homogeneity of order of 10^{-4}; this is well tested up to $10^{-3} - 10^{-2}$ seconds of its life by the very good agreement between the abundances of light elements as predicted by the Standard Cosmological Model (SCM) and the one observed [1].

The theoretical framework of the Einstein equations applied to cosmology when following the evolution towards the initial singularity shows instability for density perturbations [2], eventually giving space to cosmological models differing from the SCM. Furthermore, reliable indications support the idea that the Universe evolved through an inflationary scenario and since that age it reached isotropy and homogeneity on the horizon scale at least [3].

Despite of these evidences favourable to the large scale homogeneity, many relevant features suggest that in the very early stage of evolution it had to be described by much more general inhomogeneous solutions of the Einstein equations. With respect to this we remark the following two issues:

(*i*) the backward instability of the density perturbations to an isotropic and homogeneous Universe [2] allows us to infer that, when approaching the initial singularity, the dynamics evolved as more general and complex models; thus the appearance of an oscillatory regime is expected in view of its general nature, i.e. because its perturbations correspond simply to redefine the spatial gradients involved in the Cauchy problem.

(*ii*) Since the Universe dynamics during the Planckian era underwent a quantum regime, then no symmetry restrictions can be *a priori* imposed on the cosmological model; in fact, the wave functional of the Universe can provide information on the casual scale at most, and therefore the requirement for a global symmetry to hold would imply a large scale correlation of different horizons. Thus, the quantum evolution of the Universe is appropriately described only in terms of a generic inhomogeneous cosmological model.

A peculiar feature of the inflationary scenario consists of the violent expansion the Universe underwent during the de Sitter phase [4, 5]-[7]; indeed via such a mechanism the inflationary model provides a satisfactory explanation of the so-called *horizons* and *flatness* paradoxes by stretching the inhomogeneities at a very large scale [3, 8]. When referred to a (homogeneous and isotropic) Friedmann–Lemaitre–Robertson–Walker (FLRW) model [1], the de Sitter phase of the inflationary scenario rules out the small inhomogeneous perturbations so strongly, that it makes them unable to become seeds for the later structures formation [9, 10]. This picture emerges sharply within the inflationary paradigm and it is at the ground level of the statements according to which the cosmological perturbations arise from the scalar field quantum fluctuations [11].

Though this argument is well settled down and results very attractive even because the predicted quantum spectrum of inhomogeneities takes the Harrison-Zeldovich form, nevertheless the question whether it is possible, in a more general context, that classical inhomogeneities survive up to a level relevant for the origin of the actual Universe

*Electronic address: g.imponente@qmul.ac.uk

large scale structures remains open, as recently presented [12].

Indeed here we show the behaviour of an inhomogeneous cosmological model [13] which undergoes a de Sitter phase [14] and show how such a general scheme allows the scalar field to retain, at the end of the exponential expansion, a generic inhomogeneous term to leading order (for connected topic see [14]).
Thus our analysis provides relevant information either with respect to the morphology of an inhomogeneous inflationary model, either stating that the scalar field is characterized by an arbitrary spatial function which plays the role of its leading order.
The paper is structured as follows: in Section II we review the basic equations of the SCM and the arising of the typical inflationary behaviour of the metric scale factor. In Section III we introduce the formalism of [7], in terms of the invariant function ζ, to treat density perturbations in an inflationary scenario. In Section IV we will show the dynamics of an inhomogeneous cosmological model coupled with a real self-interacting scalar field. The solution concerns the phase when the scalar field slow rolls on a potential plateau and the Universe evolution is dominated by the effective cosmological constant related to the energy level over the true vacuum state of the theory. In Section V.A we specify such a framework to the Coleman-Weinberg expression for the potential, while in Section V.B we consider also the explicit spherically symmetric Tolman-Bondi metric. Then we apply in Section V.C the density perturbations estimate to the FLRW Universe and in Section VI we draw some conclusions.
In the overall work, we can neglect the contribution of the ultra-relativistic matter which is relevant only for higher order terms and becomes more and more negligible as the exponential expansion develops (for a discussion of an inflationary scenario with relevant ultra-relativistic matter and different outcoming behaviour, see [9, 10]).

II. FROM THE SCM TO INFLATION

A. The Friedmann Equation

The FLRW metric is the most general spatially homogeneous and isotropic one which, in terms of the *comoving coordinates* (t, r, θ, ϕ), has a line element reading as

$$ds^2 = dt^2 - R^2(t)\left(\frac{dr^2}{1-\kappa r^2} + r^2 d\theta^2 + r^2 \sin^2\theta d\phi^2\right) \tag{1}$$

where the scale factor $R(t)$ is a generic function of time only and, for an appropriate rescaling of the coordinates, the factor $\kappa = 0, \pm 1$ distinguishes the sign of constant spatial curvature.
The FLRW dynamics is reduced to the time dependence of the scale factor $R(t)$, solving the Einstein equations with a diagonal stress-energy tensor $T_{\mu\nu}$ for all the fields present (matter, radiation, etc.); for a perfect fluid it is characterized by a space-independent energy density $\rho(t)$ and pressure $p(t)$ as given by $T^{\mu}{}_{\nu} = \mathrm{diag}(\rho, -p, -p, -p)$; in this case the $0-0$ component of the Einstein equations (the *Friedmann* one) and the $i-i$ components read as

$$\frac{\dot{R}^2}{R^2} + \frac{\kappa}{R^2} = \frac{8\pi G}{3}\rho\,, \qquad 2\frac{\ddot{R}}{R} + \frac{\dot{R}^2}{R^2} + \frac{\kappa}{R^2} = -8\pi G p\,, \tag{2}$$

respectively. The difference between them leads to

$$\frac{\ddot{R}}{R} = -\frac{4\pi G}{3}(\rho + 3p)\,, \tag{3}$$

which can be solved for $R(t)$ once provided an equation of state, i.e. a relation between ρ and p.
When the Universe was radiation-dominated, as in the early period, the radiation component provided the greatest contribution to its energy density and for a photon gas we have $p = \rho/3$. The present-time Universe is, on the contrary, matter-dominated: the "particles" (i.e. the galaxies) have only long-range (gravitational) interactions and can be treated as a pressureless gas ("dust"): the equation of state is $p = 0$.
The energy tensor appearing in the Einstein field equations describes the complete local energy due to all non-gravitational fields, while gravitational energy has a non-local contribution. An unambiguous formulation for such a non-local expression is found only in the expressions used at infinity for an asymptotically flat space-time [17]. This is due to the property of the mass-energy term to be only one component of the energy-momentum tensor which can be reduced only in a peculiar case to a four-vector expression which can not be summed in a natural way.
Bearing in mind such difficulties, the conservation law $T^{\mu\nu}{}_{;\nu} = 0$ leads to $d(\rho R^3) = -p d(R^3)$, i.e. the first law of

thermodynamics in an expanding Universe which, via the equation of state, leads to a differential relation for $\rho(R)$. A solution $\rho(R)$ for it in the Friedmann equation (2) corresponding to $\kappa = 0$ provides the following behaviours

$$\text{Radiation: } \rho \propto R^{-4}\,, \quad R \propto t^{1/2}\,; \qquad \text{Matter: } \rho \propto R^{-3}\,, \quad R \propto t^{2/3} \tag{4}$$

where as long as the Universe is not curvature-dominated, i.e. for sufficiently small values of R, the choice of $\kappa = 0$ is not relevant. Let us define the Hubble parameter $H \equiv \dot{R}/R$ and the critical density $\rho_c \equiv 3H^2/8\pi G$ in order to rewrite (2) as

$$\frac{\kappa}{H^2R^2} = \frac{\rho}{3H^2/8\pi G} - 1 \equiv \Omega - 1, \tag{5}$$

where Ω is the ratio of the density to the critical one $\Omega \equiv \rho/\rho_c$; since H^2R^2 is always positive, the relation between the sign of κ and the sign of $(\Omega - 1)$ reads

$$\text{Closed}: \ \kappa = +1 \Rightarrow \Omega > 1 \quad ; \quad \text{Flat}: \ \kappa = 0 \Rightarrow \Omega = 1 \quad ; \quad \text{Open}: \ \kappa = -1 \Rightarrow \Omega < 1 \tag{6}$$

B. Shortcomings of the SM: Horizon and Flatness Paradoxes

Despite the simplicity of the Friedmann solution some paradoxes occur when taking into account the problem of *initial conditions*. The observed Universe has to match very specific physical conditions in the very early epoch, but [18] showed that the set of initial data that can evolve to a Universe similar to the present one is of zero measure and the standard model tells *nothing* about initial conditions.

Flatness.
The value of Ω at present time Ω_0 is related to the radius of curvature and to the density by Eq. (5). The observational data restrict Ω_0 to be of order of few units and consequently $R_{curv} \sim H_0^{-1}$ and $\rho_0 \sim \rho_c$. By eliminating H^2 via the two equations in (5) one finds the dependence of Ω on time as approaching unity for decreasing time as

$$|\,\Omega(10^{-43}\ \text{sec}) - 1\,| \lesssim \mathcal{O}(10^{-60})\,, \qquad |\,\Omega(1\ \text{sec}) - 1\,| \lesssim \mathcal{O}(10^{-16})\,, \tag{7}$$

and for the radius of curvature this would imply

$$R_{\text{curv}}(10^{-43}\ \text{sec})\,| \gtrsim 10^{30}H^{-1}\,, \qquad R_{\text{curv}}(1\ \text{sec})\,| \gtrsim 10^{8}H^{-1}\,. \tag{8}$$

These estimates imply that the FLRW model was characterized at the beginning by very special initial data in order to evolve in the Universe we observe today as, for example, a primeval deviation from the critical density when the temperature was $T = 10^{17}$ GeV (at the Grand Unification epoch) given by

$$\left|\frac{\rho - \rho_c}{\rho}\right|_{T=10^{17}\text{GeV}} < 10^{-55}\,. \tag{9}$$

A flat Universe today requires Ω at ancient time close to unity up to a part in 10^{55}. A little displacement from flatness at the beginning – for example 10^{-30} – would produce an actual Universe either very open or very closed, so that $\Omega = 1$ is a very unstable condition: this is the *flatness problem.*
The natural time scale for cosmology is the Planck time ($\sim 10^{-43}$ sec): in a time of this order a typical closed Universe would reach the maximum size while an open one would become curvature dominated. The actual Universe has survived 10^{60} Planck times without neither recollapsing nor becoming curvature dominated.

Horizon.
The other important problem arising from the SCM regards the explanation of the smoothness of the CMBR: the entire observable Universe could not have been in causal contact at remote time and eventually smoothed out, as a consequence of particle horizon. In fact, a light signal (as a sample of the fastest possible interaction) emitted at $t = 0$ travelled during a time t the physical distance

$$l(t) = R(t)\int_0^t R(t')^{-1}dt' = 2\,t \tag{10}$$

in a radiation-dominated Universe with $R \propto t^{1/2}$, measuring the physical horizon size, i.e. the linear size of the greatest region causally connected at time t. The distance (10) has to be compared with the radius $L(t)$ of the region which will evolve in our currently observed part of the Universe. Conservation of entropy for $s \propto T^3$ gives

$$L(t) = (s_0/s(t))^{1/3} L_0 , \tag{11}$$

where s_0 is the present entropy density and $L_0 \sim H^{-1} \simeq 10^{10}$ years is the radius of the observed Universe. An estimate of the ratio of the volumes provides, as $T \sim 10^{17}$GeV,

$$l^3/L^3|_{T=10^{17}\text{GeV}} \sim 10^{-83} . \tag{12}$$

The actual observable Universe is composed of several regions which have *not* been in causal contact for the most part of their evolution, preventing an explanation about the present days Universe smoothness. In particular, the spectrum of the CMBR is uniform up to 10^{-4}. Moreover, we have at the time of recombination, i.e. when the photons of the CMBR last scattered, the ratio $l^3/L^3 \sim 10^5$: the present Hubble volume consists of about 10^5 causally disconnected regions at recombination and no process could have smoothed out the temperature differences between these regions without violating causality. The particle horizon at recombination subtends an angle of only 0.8° in the sky today, while the CMBR is uniform across the sky.

C. The Inflationary Paradigm

The basic ideas of the theory of inflation rely firstly on the original work [4], i.e. the *old inflation*, which provides a phase in the Universe evolution of exponential expansion; then the formulation of *new inflation* by [15] introduced the slow-rolling phase in inflationary dynamics; finally, many models have sprung from the original theory (for a criticism, see [16].

In [4] is described a scenario capable of avoiding the horizon and flatness problems: both paradoxes would disappear dropping the assumption of adiabaticity. In such a case, the entropy per comoving volume s would be related as $s_0 = Z^3 s_{\text{early}}$, where s_0 and s_{early} refer to the values at present and at very early times, for example at $T = T_0 = 10^{17}$ GeV, and Z is some large factor. The estimate of the value of $|\rho - \rho_c|/\rho$ is then multiplied by a factor Z^2 and would be of the order of unity if $Z > 3 \cdot 10^{27}$, getting rid of the flatness problem.

The right-hand side of (11) is multiplied by Z^{-1}: for any given temperature, the length scale of the early Universe is smaller by a factor Z than previously evaluated, and for Z sufficiently large the initial region which has evolved in our observed one would have been smaller than the horizon size at that time; for $Z > 5 \cdot 10^{27}$ the horizon problem disappears.

Making some *ad hoc* assumptions, the model accounts for the horizon and flatness paradoxes while a suitable theory needs a physical process capable of such a large entropy production. A simple solution relies on the assumption that at very early times the energy density of the Universe was dominated by a scalar field $\phi(\vec{x}, t)$, i.e. $\rho = \rho_\phi + \rho_{\text{rad}} + \rho_{\text{mat}} + \ldots$ with $\rho_\phi \gg \rho_{\text{rad, mat, etc}}$ and hence $\rho \simeq \rho_\phi$.

The quantum field theory Lagrangian density in this case and the corresponding stress-energy tensor read as

$$\mathcal{L} = \partial^\mu \phi \, \partial_\mu \phi/2 - V(\phi) , \quad T^\mu_{\ \nu} = \partial^\mu \phi \, \partial_\nu \phi - \mathcal{L} \delta^\mu_{\ \nu} , \tag{13}$$

resepctively, and hence for a perfect fluid leads to

$$\rho_\phi = \dot{\phi}/2^2 + V(\phi) + R^{-2}/2\nabla^2\phi , \qquad p_\phi = \dot{\phi}/2^2 - V(\phi) - R^{-2}/6\nabla^2\phi . \tag{14}$$

Since spatial homogeneity implies a slow variation of ϕ with position, the spatial gradients are negligible and the ratio $\omega = p/\rho$ reads

$$\frac{p_\phi}{\rho_\phi} \simeq \frac{\dot{\phi}^2/2 + V(\phi)}{\dot{\phi}^2/2 - V(\phi)} . \tag{15}$$

For a field at a minimum of the potential $\dot{\phi} = 0$ and (15) becomes an equation of state as $p_\phi = -\rho_\phi$, giving rise to a phase of exponential growth of $R \propto e^{Ht}$, where the Hubble parameter H remains constant: this is the *inflationary* or *de Sitter phase.*

A very different evolution arises for a field in a thermal bath, in which case the coupling can be summarized by adding a term $-(1/2)\lambda T^2 \phi^2$ to the Lagrangian. The potential $V(\phi)$ is replaced by the *finite-temperature* effective potential

$$V_T(\phi) = V(\phi) + \lambda T^2 \phi^2/2 . \tag{16}$$

This model can recorver the Standard Cosmology via a phase transition of the scalar field between a metastable state (false minumum) and the true vacuum; the reheating due to oscillations around this state are damped by particle decay and, when the corresponding products thermalize, the Universe is reheated, and inflation comes to an end. This process (we will not discuss here the details) nevertheless leaves some problems open as, schematically: (i) inflation never ends, due to smallness of the tunnelling transition rate between the two minima; (ii) the phase transition is never completed; (iii) the discontinuous process of bubble nucleation (exponential expansion of vacuum phases) via quantum tunnelling should produce a lot of inhomogeneities which aren't actually observed.

D. New Inflation: the Slow Rolling Model

In 1982, both [15] proposed a variant of Guth's model, now referred to as *new inflation* or *slow-rolling inflation*, in order to avoid the shortcomings of the old inflation. Their original idea considered a different mechanism of symmetry breaking, the so-called Coleman-Weinberg (CW) one, based on the gauge boson potential with a finite-temperature effective mass $m_T \equiv \sqrt{-m^2 + \lambda T^2}$ reading as

$$V_T(\phi) = \frac{B\sigma^4}{2} + B\phi^4 \left[\ln\left(\frac{\phi^2}{\sigma^2}\right) - \frac{1}{2}\right] + \frac{1}{2}m_T^2\phi^2\,, \tag{17}$$

where $B \simeq 10^{-3}$ is connected to the fundamental constants of the theory, while $\sigma \simeq 2 \cdot 10^{15}$ GeV gives the energy associated to the symmetry breaking process.
The quantity ${m_T}^2$ can be used to parametrize the potential (17):

1. when ${m_T}^2 > 0$, the point $\phi = 0$ is a minimum of the potential, while when $m_T^2 < 0$ it is a maximum;
2. when ${m_T}^2 < 4\sigma^2/e \simeq 1.5\sigma^2$, a second minimum develops for some $\bar{\phi} > 0$; initially this minimum is higher than the one at 0, but when m_T becomes lower than a certain value m_T^* $(0 < m_T^* < 1.5\sigma^2)$ it will eventually become the global minimum of the potential.

If at some initial time the ϕ-field is trapped in the minimum at $\phi = 0$, the true minimum can disappear as the temperature lowers: as m_T approaches 0, the potential barrier becomes low and can be overcome by *thermal* (*weakly* first order process) tunnelling, i.e. due to classical fluctuations of the ϕ field around its minimum; the barrier can disappear completely when $m_T = 0$ (second order process). The phase transition doesn't proceed via a quantum tunnelling – a very discontinuous and a strongly first order process. The transition occurs rather smoothly, avoiding the formation of undesired inhomogeneities.
When the ϕ-field has passed the barrier (if any), it begins to evolve towards its true minimum. The model (17) has the feature that if the coefficient of the logarithmic term is sufficiently high, the potential is very flat around 0 and the field ϕ "slow rolls" in the true vacuum state, rather than falling abruptly: during this phase the inflation takes place, lasting enough to produce the required supercooling. When the field reaches the minimum, it begins to oscillate around it thus originating the reheating.
The problems of Guth's originary model are skipped moving the inflationary phase *after* the field has escaped the false vacuum state, by adding the slow-rolling phase.
Virtually all models of inflation are based upon this principle.

III. EVOLUTION OF DENSITY PERTURBATIONS

During the de Sitter phase the Hubble radius H^{-1} is roughly constant while it increases during the FLRW phase. The scale factor R undergoes an e-folding in either case so that microphysics (and then interaction between different close points) can operate only at scale less then $\mathcal{O}(H^{-1})$. Hence, during the late inflationary epoch the evolution of the perturbations is essentially scale independent, since nothing can alter the amplitude of a real physical perturbation when its scale is larger then H^{-1}.
The perturbed metric tensor gives place to the equations [7]

$$[k^2 + 12\pi G(\rho_0 + p_0)R^2]a = -k^2(1 + 3c_s^2)h\,, \qquad (\dot{h} - Ha)\dot{} + 3H(\dot{h} - Ha) = 0\,, \tag{18}$$

where h and a denote two functions measuring the metric perturbations independent of the choice of the spatial coordinates, ρ_0 and p_0 have to be considered as the backgroungd energy density and pressure and $c_s^2 = dp_0/d\rho_0$ is

formally the "speed of sound", while k is the wave-number corresponding to the comoving wavelength $\lambda = 2\pi/k$ (which is constant as the Universe expands).
The equations (18) permit to identify, in the uniform Hubble constant gauge, the amplitude

$$\zeta \equiv h\left[1+\frac{k^2}{12\pi G(\rho_0+p_0)R^2}\right] \tag{19}$$

as a nearly time-independent quantity. Hence, by the second of (18), there are two independent modes: a decaying one for which $\dot{h} - Ha \sim \exp(-3Ht)$, clearly negligible, and a constant one for which $\dot{h} = Ha$. This second mode, with a few calculations, during inflation is expressed by the time-independence of the function (19) $\zeta = \zeta_f = const.$ and has the form $\zeta = \delta\rho/(\rho+p)$ [6, 7]; its value has to match the final one ζ_f of the FLRW epoch.

IV. INHOMOGENEOUS INFLATIONARY MODEL

In a synchronous reference, the generic line element of a cosmological model takes the form (in units $c = \hbar = 1$)

$$ds^2 = dt^2 - \gamma_{\alpha\beta}(t,x^\mu)dx^\alpha dx^\beta\,, \quad \alpha,\beta,\mu = 1,2,3 \tag{20}$$

where $\gamma_{\alpha\beta}(t,x^\mu)$ is the three-dimensional metric tensor describing the geometry of the spatial slices. The Einstein equations in the presence of a self interacting scalar field $\{\phi(t,x^\mu), V(\phi)\}$ read explicitly [2]

$$\frac{1}{2}\partial_t k^\alpha_\alpha + \frac{1}{4}k^\beta_\alpha k^\alpha_\beta = \chi\left[-(\partial_t\phi)^2 + V(\phi)\right] \tag{21a}$$

$$\frac{1}{2}(k^\beta_{\alpha;\beta} - k^\beta_{\beta;\alpha}) = \chi\,(\partial_\alpha\phi\,\partial_t\phi) \tag{21b}$$

$$\frac{1}{2\sqrt{\gamma}}\partial_t(\sqrt{\gamma}\,k^\beta_\alpha) + P^\beta_\alpha = \chi\left[\gamma^{\beta\mu}\,\partial_\alpha\phi\,\partial_\mu\phi + V(\phi)\,\delta^\beta_\alpha\right]\,, \tag{21c}$$

where $\chi = 8\pi G$, the three-dimensional Ricci tensor P^β_α is constructed via $\gamma_{\alpha\beta}$, $\gamma \equiv det\ \gamma_{\alpha\beta}$, $k_{\alpha\beta} \equiv \partial_t\gamma_{\alpha\beta}$. The dynamics of the scalar field $\phi(t,x^\gamma)$ is coupled to the system (21) and is described by the equation

$$\partial_{tt}\phi + \frac{1}{2}k^\alpha_\alpha\partial_t\phi - \gamma^{\alpha\beta}\phi_{;\alpha;\beta} + \frac{dV}{d\phi} = 0\,. \tag{22}$$

In what follows we will consider the three fundamental statements:

(*i*) the three metric tensor is taken in the general factorized form $\gamma_{\alpha\beta}(t,x^\mu) = \Gamma^2(t,x^\mu)\xi_{\alpha\beta}(x^\mu)$, where $\xi_{\alpha\beta}$ is a generic symmetric three-tensor and therefore contains six arbitrary functions of the spatial coordinates, while Γ is to be determined by the dynamics;

(*ii*) the self interacting scalar field dynamics is described by a potential term which satisfies all the features of an inflationary one, i.e. a symmetry breaking configuration characterized by a relevant plateau region;

(*iii*) the inflationary solution is constructed under the assumptions

$$1/2\,(\partial_t\phi)^2 \ll V(\phi) \tag{23a}$$

$$\mid \partial_{tt}\phi \mid \ll \mid k^\alpha_\alpha\,\partial_t\phi \mid\,. \tag{23b}$$

Our analysis, following [12], concerns the evolution of the cosmological model when the scalar field slow rolls on the plateau and the corresponding potential term is described as

$$V(\phi) = \Lambda_0 - \lambda U(\phi)\,, \tag{24}$$

where Λ_0 behaves as an effective cosmological constant of the order $10^{15} - 10^{16}$ GeV and λ ($\ll 1$) is a coupling constant associated to the perturbation $U(\phi)$.
Since the scalar field moves on a plateau almost flat, we infer that to lowest order of approximation $\phi(t,x^\gamma) \sim \alpha(x^\gamma)$ (see below (29)) and therefore the potential reduces to a space-dependent effective cosmological constant

$$\Lambda(x^\gamma) \equiv \Lambda_0 - \lambda U(\alpha(x^\gamma))\,. \tag{25}$$

In this scheme the $0-0$ (21a) and $\alpha - \beta$ (21c) components of the Einstein equations reduce, under condition *(iii)* and neglecting all the spatial gradients, to the simple ones

$$3\,\partial_{tt}\ln\Gamma + 3\,(\partial_t \ln\Gamma)^2 = \chi\Lambda(x^\gamma)\,, \qquad (\partial_{tt}\ln\Gamma)\delta^\alpha_\beta + 3\,(\partial_t\ln\Gamma)^2\delta^\alpha_\beta = \chi\Lambda(x^\gamma)\delta^\alpha_\beta\,, \tag{26}$$

respectively. A simultaneous solution for Γ of both equations (26) takes the form

$$\Gamma(x^\gamma) = \Gamma_0(x^\gamma)\exp\left[\sqrt{\chi\Lambda(x^\gamma)/3}(t-t_0)\right]\,, \tag{27}$$

where $\Gamma_0(x^\gamma)$ is an integration function while t_0 a given initial instant of time for the inflationary scenario. Under the same assumptions and taking into account (27) for Γ, the scalar field equation (22) rewrites as

$$3H(x^\gamma)\partial_t\phi - \lambda W(\phi) = 0\,, \qquad H(x^\gamma) = \partial_t\ln\Gamma = \sqrt{\chi\Lambda(x^\gamma)/3}\,, \quad W(\phi) = dU/d\phi\,. \tag{28}$$

We search for a solution of the dynamical equation (28) in the form

$$\phi(t,x^\gamma) = \alpha(x^\gamma) + \beta(x^\gamma)\,(t-t_0)\,. \tag{29}$$

Inserting expression (29) in (28) and considering it to the lowest order, it is possible to express β in terms of α as

$$\beta = \frac{\lambda W(\alpha)}{\sqrt{3\chi\Lambda_0 - \lambda U(\alpha)}}\,. \tag{30}$$

Of course the validity of solution (30) takes place in the limit

$$t - t_0 \ll \left|\frac{\alpha}{\beta}\right| = \left|\frac{\alpha}{W(\alpha)}\sqrt{3\chi\frac{\Lambda_0}{\lambda^2} - \frac{U(\alpha)}{\lambda}}\right| \tag{31}$$

where the ratio Λ_0/λ^2 takes in general very large values.
The $0-\alpha$ component (21b), in view of (27) and (29) through (30), reduces to $\partial_\gamma\,(\Lambda + \lambda U) = 0$, which for $\Lambda(x^\gamma)$ is an identity as (25).
The spatial gradients, either of the three-metric field either of the scalar one, behave as Γ^{-2} and decay exponentially.

If we take into account the coordinate characteristic lengths L and l for the inhomogeneity scales regarding the functions Γ_0 and $\xi_{\alpha\beta}$, i.e.

$$\partial_\gamma\Gamma_0 \sim \Gamma_0/L\,, \qquad \partial_\gamma\,\xi_{\alpha\beta} \sim \xi_{\alpha\beta}/l\,, \tag{32}$$

respectively, negligibility of the spatial gradients at the beginning leads to the inequalities for the physical quantities

$$\Gamma_0 l = l_{\rm phys} \gg H^{-1}\,, \qquad \Gamma_0 L = L_{\rm phys} \gg H^{-1}\,. \tag{33}$$

These conditions state that all the inhomogeneities have to be much greater then the physical horizon H^{-1}.
Negligibility of the spatial gradients at the beginning of inflation is required (as well known) by the existence of the de Sitter phase itself; however, spatial gradients having a passive dynamical role allow to deal with a fully inhomogeneous solution: space point dynamically decouple to leading order.

The condition (23a) is naturally satisfied since states that the dominant contribution in $\Lambda(x^\gamma)$ is provided by Λ_0, i.e. $\lambda U(\alpha) \ll \Lambda_0$; the same can be said for (23b).
The only important restriction on the spatial function $\alpha(x^\gamma)$ is $|\alpha| \ll \left|U^{-1}\,(\Lambda_0/\lambda)\right|$.
A satisfactory exponential expansion able to overcome the SCM shortcomings, i.e. to freeze the density fluctuations between the beginning and the end of the inflation, requires that in each space point the condition for the *e*-folding $H(t_f - t_i) \sim \mathcal{O}(10^2)$ holds, where t_i and t_f denote the instants when the de Sitter phase starts and ends, respectively. We may take $t_i \equiv t_0$ and hence for t_f we have, by (28) and (30),

$$H(t_f - t_i) \ll (\Lambda_0/\lambda)\;|\alpha/W(\alpha)|\,. \tag{34}$$

Since Λ_0/λ is a very large quantity, no serious restrictions appear for the e-folding of the model.

The obtained solution possesses a very general festure: in fact, once satisfied all the dynamical equations, still eight arbitrary spatial functions remain, i.e. six for $\xi_{\alpha\beta}(x^\gamma)$, and then $\Gamma_0(x^\gamma)$, $\alpha(x^\gamma)$. However, taking into account the possibility to choose an arbitrary gauge via the set of the spatial coordinates, we have to kill three degrees of freedom; hence five physically arbitrary functions finally remain: four corresponding to gravity degrees of freedom and one related to the scalar field.
This picture corresponds exactly to the allowance of specifying a generic Cauchy problem for the gravitational field, on a spatial non-singular hypersurface, nevertheless one degree of freedom of the scalar field is lost against the full generality.

V. APPLICATIONS

A. Coleman–Weinberg Model

Let us specify this solution in the case of the vanishing mass zero-temperature CW potential [20]

$$V(\phi)=\frac{B\sigma^4}{2}+B\phi^4\left[\ln\left(\frac{\phi^2}{\sigma^2}\right)-\frac{1}{2}\right]. \tag{35}$$

In the region $|\phi| \ll |\sigma|$ the potential (35) approaches a plateau profile similar to (24) and acquires the form

$$V(\phi)\simeq B\sigma^4/2-\lambda\phi^4/4\,,\qquad \lambda\simeq 80B\simeq 0.1\,. \tag{36}$$

This is reducible to (24) by setting $\Lambda_0=B\sigma^4/2$, $U(\phi)=\phi^4/4$ and $W(\phi)=\phi^3$. Hence the relations (30) rewrites as $\beta=\lambda\alpha^3/3H$ while the inequality for α is equivalent to fulfil the initial assumption $\Lambda_0\gg\lambda U(\alpha)\sim\lambda\alpha^4/4$, that is

$$|\alpha|\ll\sqrt[4]{\Lambda_0/\lambda}\sim\sigma\,. \tag{37}$$

B. Lemaitre–Tolman Spherically Symmetric Metric

It is interesting to consider the framework developed so far when in presence of a spherically symmetric line element written as

$$ds^2=dt^2-e^{2\pi}dr^2-e^{2\psi}d\Omega \tag{38}$$

where π and ψ are functions of t and r only. The corresponding Einstein equations read as

$$(\dot\psi)^2+2\dot\pi\dot\psi+e^{-2\psi}-e^{-2\pi}\left[2\psi''+3(\psi')^2-2\pi'\psi'\right]=\frac{\chi}{2}\left[(\partial_t\phi)^2+(\partial_r\phi)^2e^{-2\pi}+V(\phi)\right] \tag{39a}$$

$$2\ddot\psi+3(\dot\psi)^2+e^{-2\psi}-(\psi')^2e^{-2\pi}=-\frac{\chi}{2}\left[(\partial_t\phi)^2+(\partial_r\phi)^2e^{-2\pi}-V(\phi)\right] \tag{39b}$$

$$2\dot\psi'+2\psi'\dot\psi-2\dot\pi\psi'=\chi\dot\phi\phi'\,, \tag{39c}$$

while the scalar field $\phi(t,x^\gamma)$ is described by the equation

$$\partial_{tt}\phi+\left[\dot\pi+2\dot\psi\right]\partial_t\phi-\left[\phi''+\phi'\left(-\pi'+2\psi'\right)\right]e^{-2\pi}+\frac{dV}{d\phi}=0\,. \tag{40}$$

Once defined $e^\psi\equiv rm(r,t)$, without entering in the details of the derivation [21], we can find again for large values of m

$$\sqrt{\frac{2\chi\Lambda}{3}}m=e^{\sqrt{\chi\Lambda/6}\,(t-t_0)}\,, \tag{41}$$

which admits linear solution for the scalar field similar to (29), satisfying the set of Einstein equations under the same assumptions of the previous discussion.

C. Towards FLRW Universe

The conditions (33) state that the validity of the inhomogeneous inflationary scenario discussed so far requires the inhomogeneous scales to be out of the horizon when inflation starts. The situation is different when treating the small perturbations to the FLRW case; in fact, the negligibility of the spatial curvature corresponds to require the radius of curvature of the Universe to be much greater than the physical horizon, the inhomogeneous terms being small in amplitude. To this end, let us consider the three-metric

$$\gamma_{\alpha\beta}=\Gamma(t,\varphi^\mu)^2\left[h_{\alpha\beta}+(t-t_0)\delta\theta_{\alpha\beta}(\varphi^\mu)\right]\,, \tag{42}$$

where $h_{\alpha\beta}$ denotes the FLRW spatial part of the three metric ($\{\varphi^\mu\}$ are the three usual angular coordinates) and $\delta\theta_{\alpha\beta}$ denote a small inhomogeneous perturbation. The Einstein equations (21) coupled to the scalar field dynamics (22) on the plateau (36) admit, to leading order in the inhomogeneities, the solution

$$\Gamma = \Gamma_0 e^{H(t-t_0)}\,, \qquad H = H_0 - \frac{\delta\theta}{6}\,, \qquad H_0 = \sigma^2\sqrt{\frac{\chi B}{6}}\,, \tag{43a}$$

$$\phi = \alpha_0\left[1 + \frac{\lambda\alpha_0^2}{3H_0}(t-t_0)\right] + \frac{\delta\theta}{3\chi}\left[1 + \frac{\lambda\alpha_0^2}{H_0}(t-t_0)\right]\,, \qquad \delta\theta_{\alpha\beta} = \frac{\delta\theta}{3}h_{\alpha\beta}\,, \tag{43b}$$

where t_0 and Γ_0 are constants. The solution (43) holds and provides the correct e-folding of order $\mathcal{O}(10^2)$ when the inequalities

$$t - t_0 \ll 3H_0/\lambda\alpha_0^2\,, \tag{44a}$$

$$\alpha_0 \ll \mathcal{O}\left(10^{-1}\sqrt{\Lambda_0/\lambda}\right)\,, \tag{44b}$$

take place, the first one ensuring that the dominant term of the scalar field remains the time-independent one during the de Sitter phase, while the second one accounts for the e-folding; the spatial gradients in the Einstein field equations behave as (see, for instance, the RHS of Eq. (21c)) $\Gamma^{-2}\partial_\mu\alpha\partial_\nu\alpha \sim \Gamma^{-2}\delta^2/l^2$ which have to be compared to the potential term $V \sim \lambda\alpha^4$ and therefore are negligible, while the case for the scalar field is similar. Hence we get

$$\Gamma_0 l = l_{\text{phys}} \gg H_0^{-1}\delta\,, \qquad l_{\text{phys}} \gg \delta/\sqrt{\lambda\alpha_0^3}\,, \tag{44c}$$

respectively, where δ ($\ll H_0/100$) and l in (44c) denote the characteristic amplitude and length, of the *arbitrary* function $\delta\theta$, trace of the tensor $\delta\theta_{\alpha\beta}$. When inflation starts the inhomogeneous scales can be inside the physical horizon H_0^{-1}.

The physical implications on the density perturbation spectrum of such a nearly homogeneous model rely on the dominant behaviour of the potential term over the energy density ρ_ϕ associated to the scalar field during the de Sitter phase and therefore

$$\Delta \equiv \left|\frac{\delta\rho_\phi}{\rho_\phi}\right| \sim \left|\frac{d\ln V}{d\phi}\delta\phi\right| \simeq \left|\frac{\lambda}{\Lambda_0}W(\alpha_0)\,\delta\alpha\right|\,, \tag{45}$$

where $\delta\alpha = \delta\theta/(3\chi)$ for our scalar field solution (43). In particular, in the CW case, Equation (45) reduces to

$$\Delta_{CW} \simeq \frac{50}{\sigma^4}{\alpha_0}^3\frac{\delta\theta}{\chi}\,. \tag{46}$$

However, to compute the physically relevant perturbations after the scales re-entry in the horizon, let us evaluate the gauge invariant quantity ζ which has now the form (19)

$$\zeta = \frac{\delta\rho}{\rho + p} \cong \frac{\delta\theta}{W(\alpha_0)}\frac{\Lambda_0}{\lambda} \tag{47}$$

when the perturbations leave the horizon and $\rho + p = (\partial_t\phi)^2$ (see Eq. 14); in the CW case it reads as

$$\zeta_{CW} = \frac{\sigma^4}{160}\frac{\delta\theta}{{\alpha_0}^3}\,. \tag{48}$$

Since ζ remains constant during the super-horizon evolution of the perturbations, then at the re-entry to the causal scale in the matter-dominated era we get $\zeta_{MD} \sim \delta\rho/\rho \sim \zeta_{CW}$.
By restoring physical units and assuming $\alpha_0 \lesssim 10^{-4}\sigma/\sqrt{\hbar c}$ in agreement with (37), then it is required $\delta\alpha/\alpha_0 \lesssim 10^{-2}$ in order to obtain perturbations $\delta\rho/\rho \sim 10^{-4}$ at the horizon re-entry during the matter-dominated age.
Hence the expression (48) explains how the perturbation spectrum after the de Sitter phase can still arise from classical inhomogeneous terms. Indeed, the function $\delta\theta(\varphi^\mu)$ is an arbitrary one and can be chosen for it a Harrison–Zeldovich spectrum by assigning its Fourier transform as

$$|\delta\alpha(k)|^2 \propto \text{const.}/k^3\,; \tag{49}$$

such a spectrum has to hold for $k \ll \Gamma_0/(H_0^{-1}\delta)$.

Thus, the pre-inflationary inhomogeneities of the scalar field remain almost of the same amplitude during the de Sitter phase as a consequence of the linear form of the scalar field solution (29). We get that the Harrison–Zeldovich spectrum can be a pre-inflationary picture of the density perturbations and it survives to the de Sitter phase, becoming a classical seed for structure formation. The existence of such a classical spectrum is not related with the quantum fluctuations of the scalar field, whose effect is an independent contribution to the classical one.

VI. CONCLUDING REMARKS

The merit of our analysis relies on having provided a dynamical framework within which classical inhomogeneous perturbations to a real scalar field minimally coupled with gravity can survive even after that the de Sitter expansion of the Universe stretched the geometry; the key feature underlying this result consists *(i)* of constructing an inhomogeneous model for which the leading order of the scalar field is provided by a spatial function and then *(ii)* of showing how the very general case contains as a limit a model close to the FLRW one. Moreover such results have been extended to the spherically symmetric Lemaitre–Tolman metric.

It is relevant to remark that the metric tensor (42) seems of the same form as the one considered in [10]; however in the present paper the function $\eta(t)$ appearing in the previous work is linear in time and does not decay exponentially. The different behaviour relies on the negligibility of the matter with respect to the scalar field which is at the ground of the present analysis. We are here assuming the dynamics of $\eta(t)$ to be driven by the scalar field alone, instead of by the ultra-relativistic matter. This situation corresponds to an initial conditions for which the scalar field dominates over the ultra-relativistic matter when inflation starts and this is the reason for the resulting different issues of the two analyses.

Acknowledgements

This work has been done when visiting the Maths Department at Queen Mary, University of London, benefiting of the kind hospitality of the General Relativity and Cosmology group, under grant funded partially by "Accademia Nazionale dei Lincei - Royal Society" and partially by Fondazione "Angelo Della Riccia".

[1] E.W. Kolb and M.S. Turner, *The Early Universe*, (Adison-Wesley, Reading) (1990).
[2] E.M.Lifshitz and I.M.Khalatnikov, *Adv.Phys.* **12**, 185 (1963).
[3] P. de Bernardis et al., *Astrophys.J.* **564**, 559, (2002); D.N. Spergel, *Astrophys.J. Suppl. Series* (2003) **148**, 175
[4] A.H. Guth, *Phys. Rev. D*, **23**, 347 (1981).
[5] A.D. Linde, *Phys. Lett.*, **108B**, 389 (1982); A.A. Starobinsky, *JETP Lett.*, **30**, 682 (1979); A.A. Starobinsky, *Phys. Lett. B*, **91**, 99 (1980); A.D. Linde, *Phys. Lett.*, **129B**, 177 (1983); J. Silk and M.S. Turner, *Phys. Rev. D*, **35**, 419 (1987).
[6] A.H. Guth and S. Pi *Phys. Rev. Lett* **49**, n.15 1110 (1982)
[7] J.M. Bardeen, P.J. Steinhardt and M.S.Turner *Phys.Rev.D* **28**, n.4, 679 (1983)
[8] C.P.Ma and E. Bertschinger, *Astrophys.J.*, **455**, 7 (1995).
[9] K. Tomita and N. Deruelle, *Phys.Rev.D*, **50**, 7216 (1994).
[10] G. Imponente and G. Montani *Int. Journ. Mod. Phys. D*, **12**, n.10, 1845 (available gr-qc/0307048) (2003).
[11] D. Polarski and A.A. Starobinsky, *Class. Quant. Grav.*, **13**, 377 (1996).
[12] G. Imponente and G. Montani *Mod. Phys. Lett. A*, **19**, n.17, 1281-1290 (2004) (available gr-qc/0404031).
[13] M.A.H. MacCallum, In "General relativity: an Einstein centenary survey", ed. S.W. Hawking and W. Israel, p.533. Cambridge University Press (1979); V.A. Belinski, I.M. Khalatnikov and E.M. Lifshitz, *Adv. Phys.* **31**, 639 (1982).
[14] A.A. Starobinsky, *Pis'ma Zh.Eksp.Teor.Fiz.* **37**, 55 (1983); A.Kirillov and G. Montani, *Phys.Rev.D* **66**, 064010 (2002); J.D. Barrow, *Phys. Lett. B* **187**, 12 (1987).
[15] A.D. Linde, *Phys. Lett.*, **108B**, (1982) 389; A. Albrecht and P. J. Steinhardt, *Phys. Rev. Lett.*, **48**, (1982) 1220.
[16] A. Albrecht and L. Sorbo, to appear in *Phys. Rev. D* (2004), (available hep-th/0405270),
[17] R. Penrose (1982), *Proc.R.Soc.Lond.* **A 381**, 53; R. Arnowitt, S. Deser and C.W. Misner, in *Gravitation: an introduction to current research* (1962), eds. I. Witten and J. Wiley, New York.
[18] C.B. Collins and S.W. Hawking, (1973) *Astrophys. Journ.* **180**, 317
[19] G. Montani, *Class. and Quantum Grav.* **17**, 2205 (2000).
[20] S.Coleman and E. Weinberg, *Phys.Rev.D* **7**, 1888 (1973).
[21] G. Imponente, in preparation

The Ghost-Gluon Vertex in Landau Gauge Yang-Mills Theory

Wolfgang Schleifenbaum

Institute for Nuclear Physics, Darmstadt University of Technology, Schloßgartenstraße 9, D-64289 Darmstadt, Germany

Abstract

Possible mechanisms for gluon confinement in Yang-Mills theory in covariant gauges are discussed briefly. Implications for the infrared behavior of the Green's functions are outlined. Results for the gluon and Faddeev-Popov ghost propagators from Dyson-Schwinger studies and lattice calculations are presented. The importance of the ghost-gluon vertex in these studies is discussed. Numerical results for a non-perturbatively constructed ghost-gluon vertex in four and three space-time dimensions are presented.

1 Confinement in Covariant Gauges

The strong interactions governing hadron and ultimately nuclear dynamics are described by quantum chromodynamics, the theory of quarks and gluons. This theory has several genuine non-perturbative features, such as *e.g.* dynamical chiral symmetry breaking. The most prominent phenomenon is, however, confinement. Strong evidence exists that it is generated in the Yang-Mills subsector, *i.e.* by the gluons. Naïvely, a Yang-Mills theory is defined in terms of the generating functional [1]

$$\begin{aligned} Z[J] &= \int \mathcal{D}A^a_\mu \, \exp\left(-\int d^dx\, \mathcal{L} + \int d^dx\, J^a_\mu A^a_\mu\right) , \qquad (1) \\ \mathcal{L} &= \frac{1}{4} F^a_{\mu\nu} F^a_{\mu\nu} , \\ F^a_{\mu\nu} &= \partial_\mu A^a_\nu - \partial_\nu A^a_\mu + g_d f^{abc} A^b_\mu A^c_\nu . \end{aligned}$$

Here we have already chosen a description in Euclidean space-time, which is used throughout this work. $A^a_\mu(x)$ is the gluon field, $F^a_{\mu\nu}(x)$ is the field strength tensor, g_d is the gauge coupling constant in d dimensions, and f^{abc} are the structure constants of the gauge group. The measure of the functional integral (1) extends over the complete configuration space of the gluon field.

However, the functional integral (1) is ill-defined. The Lagrangian is invariant under local gauge transformations and therefore the complete configuration space contains infinite sets of mutually equivalent field configurations. They are said to lie on the same gauge orbit. To avoid this over-counting in the functional integral, it is necessary to select only one copy on each gauge orbit. One tries to achieve this by gauge-fixing [1]. Implementing the linear covariant gauge fixing condition $\partial^\mu A_\mu^a = 0$ leads to the generating functional

$$\begin{aligned} Z[J] &= \int \mathcal{D}A_\mu^a \, M \, \exp\left(-\int d^dx \, (\mathcal{L}+\mathcal{L}_{\text{gf}}) + \int d^dx J_\mu^a A_\mu^a\right) , \\ M &= \det(-\partial_\mu D_\mu^{ab}) , \\ D_\mu^{ab} &= \partial_\mu \delta^{ab} + g_d f^{abc} A_\mu^c , \end{aligned} \tag{2}$$

where M is the Faddeev-Popov determinant, and $\mathcal{L}_{\text{gf}}$ the gauge-fixing term. The determinant is then rewritten as a functional integral over scalar Grassmann fields, the Faddeev-Popov ghosts. The generating functional then reads

$$Z[J=0] = \int \mathcal{D}A_\mu^a \mathcal{D}\bar{c}^b \mathcal{D}c^c \, \exp\left(-\int d^dx \, (\mathcal{L}+\mathcal{L}_{\text{gf}} + \bar{c}^d \partial_\nu D_\nu^{de} c^e)\right) . \tag{3}$$

The ghost and anti-ghost fields, $c^a(x)$ and $\bar{c}^a(x)$, respectively, encode part of the quantum fluctuations of the gluon field. Their appearance is a generic feature of covariant gauges. As will be detailed below, this procedure is in general not sufficient to completely specify the gauge beyond the use in perturbation theory.

One of the most intriguing features of QCD is confinement. Mechanisms for gluon confinement have been related to the infrared behaviour of propagators, see *e.g.* ref. [2]. We will focus here on aspects which have been substantiated recently, *c.f.* section 2.

First we discuss a sufficient but not necessary condition for confinement. If a propagator D vanishes sufficiently fast in the infrared,

$$\lim_{p^2 \to 0} D(p^2) = 0 , \tag{4}$$

then the "particle" represented by D cannot have a positive semi-definite spectral norm. Therefore it violates the Osterwalder-Schrader axiom of reflection positivity [3], and the propagator D cannot have a Källen-Lehmann representation. Thus this "particle" is not part of the physical spectrum, and is hence confined [4].

For a massless excitation such as the gluon this confinement criterion is rather intuitive. Then, the condition (4) is just the statement that the on-shell propagator vanishes. Therefore, such a "particle" does not propagate, *i.e.* it is confined.

The criterion (4) is related to a scenario which was introduced by Kugo and Ojima [5]. In perturbation theory, ghosts and longitudinal gluons are not part of the physical spectrum, since they form doublets with respect to BRST charge. As these doublets are conjugate pairs by ghost number, this is called the quartet mechanism. The conjecture of Kugo and Ojima states that also transverse gluons form together with ghost-gluon bound states a quartet, and are thus not part of the physical spectrum. This does not imply that transverse gluons do not have asymptotic states. However, these states are not in the physical subspace and therefore do not contribute to S-matrix elements.

This scenario is based on a proof which requires three preconditions. The first is an unbroken BRST charge. Whether this is the case is not known. The second is the failure of cluster decomposition in the complete state space. This requires the existence of a massless excitation in the unphysical subsector.

However, at the same time a mass gap in the physical subsector is required. The third precondition is an unbroken global color charge. This condition can be cast in Landau gauge into a condition for the ghost propagator D_G [6]

$$\lim_{p^2 \to 0} p^2 D_G(p^2) \to \infty . \tag{5}$$

As will be seen below this scenario also entails condition (4) for the gluon propagator.

Another scenario has been proposed by Zwanziger [7] based on work by Gribov [8]. Only one gauge copy on each gauge orbit should be included in the functional integral constituting the generating functional. However, no local gauge condition is known which is able to specify the gauge completely also for large gauge field fluctuations. This is known as the Gribov problem [8]. The origin of this problem actually provides a possible handle to understand confinement. Therefore a closer look at the properties of the field configuration space is necessary.

It is possible to partition configuration space by the zeros of the Faddeev-Popov determinant (2). Concentric regions are found and each gauge orbit intersects each region at least once [8]. The region containing the origin, and thus relevant for perturbation theory, is known as the first Gribov horizon. It can be shown to be compact and convex. It has been found, that this region also contains gauge copies, see *e.g.* [9], and it is necessary to further restrict to the so-called fundamental modular region. This region is specified by minimizing an appropriate functional along each gauge orbit [10]. The fundamental modular region is also convex and compact around the origin [7].

The latter properties lead to an interesting observation. As configuration space is infinite-dimensional, any volume is dominated by its boundary. It has therefore been argued that only the common boundary of the first Gribov horizon and the fundamental modular region contribute to any correlation function built from a finite number of field operators, especially Green's functions [7]. Therefore, restricting to the first Gribov horizon, would be sufficient to calculate Green's functions.

This has a further important consequence [7]. The Faddeev-Popov determinant (2) vanishes by definition on this boundary. Therefore, the condition (5), termed in this context the Zwanziger horizon condition, is fulfilled as the determinant is directly linked to the inverse self-energy of the ghost. This also entails that the action is dominated by the gauge fixing term alone. Thus, in the extreme infrared all Green's functions are determined by contributions involving ghosts. This leads to the notion of ghost dominance.

Assuming the horizon condition (5), it can be shown that an infrared tree-level ghost-gluon vertex complies with the renormalization group in leading order. This leads to the hypothesis of a bare ghost-gluon vertex also in the infrared. It has then been shown that, for the Landau gauge gluon propagator, the condition (4) follows and gluons are confined [7].

Whether the identical infrared behavior of the ghost and gluon propagators within the Kugo-Ojima and the Zwanziger-Gribov scenarios, respectively, is more than a formal coincidence is not known. However, a ghost propagator with a stronger infrared divergence than a massless pole is also intuitively seen to be able to mediate confinement. After Fourier transformation such a singularity

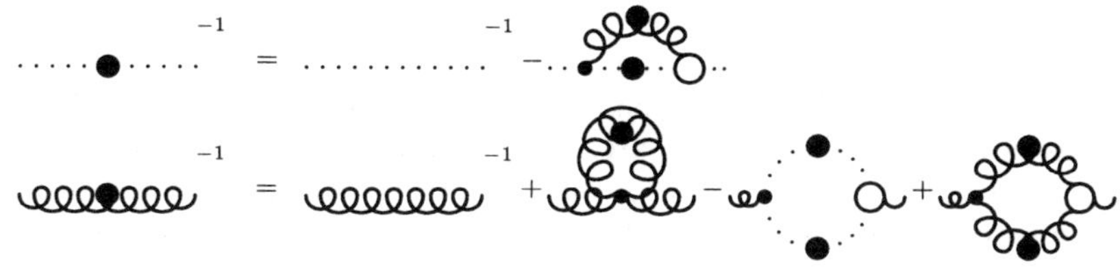

Figure 1: The Dyson-Schwinger equations for the 2-point functions of Yang-Mills theory. Dotted lines are ghosts and wiggly lines are gluons. Lines with a dot are full propagators, vertices with a small black dot are bare vertices and white circles are full vertices. The equation for the gluon has already been truncated to the one-loop level, see text.

corresponds to a correlation being more long-range than that of a photon, and thus mediates a stronger force than a Coulomb force.

2 Propagators in Yang-Mills Theory

In both confinement scenarios described above the infrared behavior of propagators plays a central role. As infrared singularities are anticipated, *c.f.* eq. (5), a non-perturbative method formulated in space-time continuum is desirable. Such a formulation is provided by the Dyson-Schwinger equations (DSEs), determining the Green's function of a theory, see *e.g.* [1, 2]. A graphical representation of the equations for the propagators is shown in fig. 1.

The Landau gauge will be used throughout, as it is advantageous in the study of confinement for various reasons. First, the Landau gauge is a fixed point of the renormalization group. Second, the conditions for the Kugo-Ojima and Zwanziger-Gribov scenario take the simple forms (4) and (5) in Landau gauge. Last but not least, it is less singular than other gauges. This manifests itself in the bareness of the ghost-gluon vertex for a vanishing incoming ghost momentum [11]

$$\lim_{p\to 0}\Gamma_\mu^{abc}(k;q,p)=ig_d f^{abc}q_\mu. \tag{6}$$

Here p, q, and k are the momenta of the incoming ghost, the outgoing ghost, and the gluon, respectively.

Employing the ghost DSE and quite general assumptions, it can then be shown [12] that the ghost propagator necessarily fulfills condition (5). It is not possible in this generality to show that the gluon propagator satisfies (4). This requires a solution of the DSEs. However, it is not possible to solve these equations as they form an infinite set of coupled non-linear integral equations. Therefore, it is necessary to truncate the system.

During the last years, the truncation scheme depicted in fig. 1 has been established [13]. This scheme has been successfully applied to Yang-Mills theory in four dimensions [14], three dimensions [15] and at finite temperature [16] as well as to QCD including dynamical quarks [17]. In this scheme, all equations beyond those of the propagators are neglected, as well as genuine two-loop contributions in the equations for the propagators. Furthermore, a perturbative color structure of the Green's functions is assumed. It then only remains to specify the two remaining full vertices, the ghost-gluon and the 3-gluon vertex. The 3-gluon vertex is chosen such that leading-order resummed perturbation theory is reproduced in the ultraviolet. Due to the property (6) the ghost-gluon vertex is usually taken to be bare. The results for the ghost and gluon propagators are shown in figs. 2 and 3 compared to lattice calculations. Both methods agree remarkably well. The largest deviations are found at intermediate momenta where the truncation scheme is expected to be least reliable. Although this scheme imposes several problems, especially concerning gauge invariance, in a systematic analysis only small quantitative effects due to truncation errors have been found, for a detailed discussion of this issue see *e.g.* refs. [14, 15].

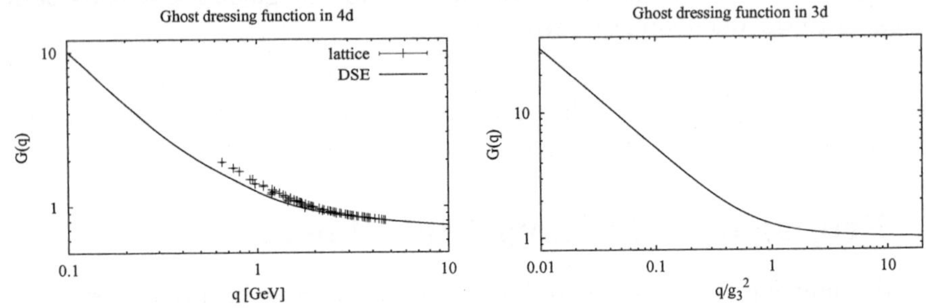

Figure 2: The ghost dressing function from DSEs in four dimensions [14] compared to lattice results [18] (left) and from DSEs in three dimensions [15] (right). g_3 is the dimensionful coupling constant in three-dimensional Yang-Mills theory.

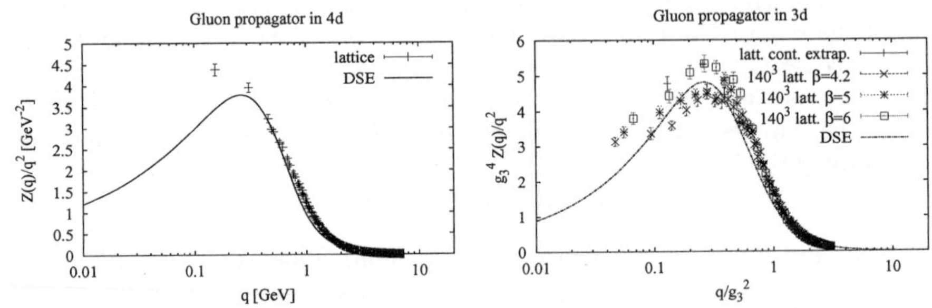

Figure 3: The gluon propagator from DSEs in four dimensions [14] compared to lattice data [19] (left) and from DSEs in three dimensions [15] compared to corresponding lattice data [20, 21] (right). Note that the lattice data at low momenta are likely to be strongly influenced by finite volume effects.

The gluon propagator obeys condition (4), and the gluon is thus confined. As the ghost propagator also satisfies the Zwanziger horizon condition (5), the Kugo-Ojima and the Zwanziger-Gribov scenarios are both fulfilled. These findings have also been confirmed recently by exact renormalization group methods [22]. Furthermore, such propagators are in sharp contrast to the older concept of infrared slavery, which expected a gluon dressing function diverging as $1/p^2$ [23]. These earlier attempts neglected the ghost contribution [24]. This so-called

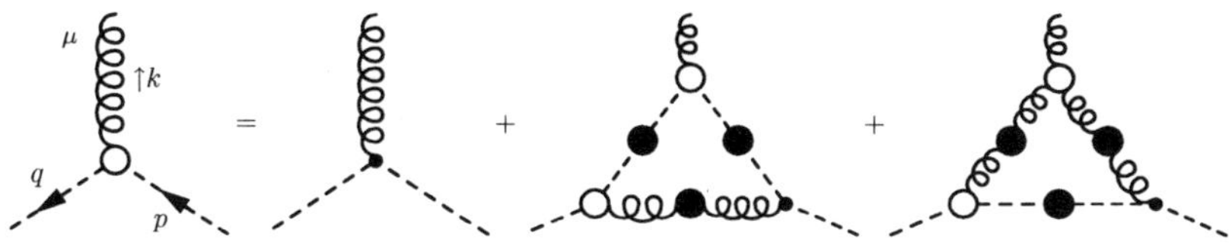

Figure 4: The DSE for the ghost-gluon vertex. Dotted lines are ghosts and wiggly lines are gluons. Lines with a dot are full propagators, vertices with a small black dot are bare vertices and white circles are full vertices. Contributions from the ghost-gluon scattering kernel have been removed, see text.

Mandelstam approximation leads to results contradicting current lattice results and can now be dismissed.

In the infrared, ghost dominance is found. Thus, besides the full gluon propagator, only diagrams with at least one ghost line in fig. 1 contribute. This is in accordance with the Zwanziger-Gribov scenario. The single crucial point in this derivation is then the structure of the ghost-gluon vertex [25]. It is therefore of great importance to understand it. The infrared behavior of the propagators has been studied using several ansätze, motivated by the Slavnov-Taylor identity (STIs) for the ghost-gluon vertex. The results are found to depend only quantitatively on the input vertex [25]. However, a more extensive analysis is highly desirable and will be presented below.

3 Ghost-Gluon Vertex

The most general tensor structure of the ghost-gluon vertex in Landau gauge is given by

$$\Gamma_\mu^{abc}(k;q,p) = ig_d\left(q_\mu\left(f^{abc} + A^{abc}(k;q,p)\right) + k_\mu B^{abc}(k;q,p)\right) , \tag{7}$$

where A^{abc} and B^{abc} are scalar functions. The tree-level contribution is left explicit. As there is no indication for a color structure different from perturbation theory [2], the structure constant is factored out, $A^{abc} =: f^{abc}A$ and $B^{abc} =: f^{abc}B$. Due to the transversality of the gluon propagator, the on-shell ghost-gluon vertex is transverse with respect to the gluon momentum. Therefore, the function B is only relevant when not contracting with a transverse projector, *e.g.* in the corresponding STI. Hence, only the function A will be discussed here. A detailed account on B is given elsewhere [26].

In principle, the ghost-gluon vertex should be determined self-consistently by solving the DSEs for the propagators and the vertex simultaneously. A truncated DSE for the vertex neglecting four-point functions is shown in fig. 4. The completely self-consistent solution of such an equation is of significant technical complexity and has not been achieved yet. However, due to the hypothesis of Zwanziger, it is most likely sufficient to perform a semi-perturbative calculation, *i.e.* to do one iteration step in the ghost-gluon vertex DSE. If the hypothesis is correct then the resulting vertex should not significantly deviate from the input tree-level vertex.

Hence, the evaluation scheme neglects the ghost-gluon scattering kernel in the ghost-gluon vertex DSE. Further, as this is the first iteration step, all vertices

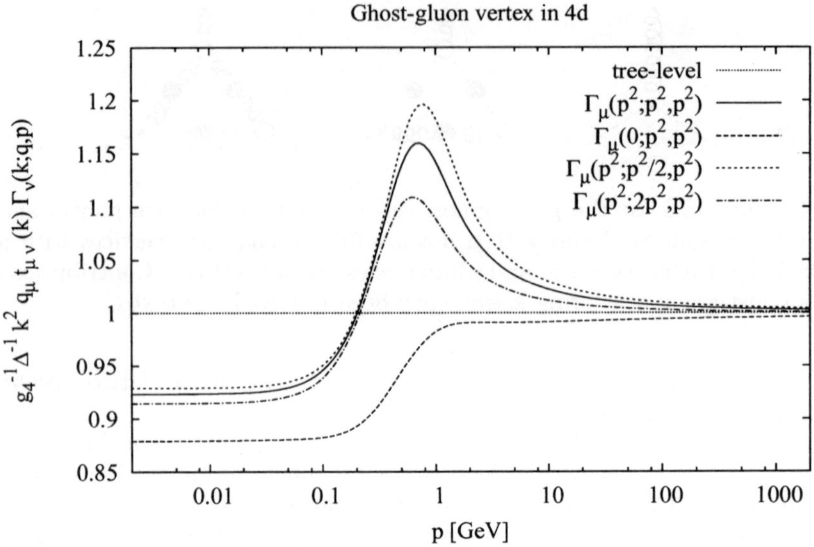

Figure 5: The normalized transverse part (8) of the ghost-gluon vertex in four dimensions in various kinematical regions.

are left bare and only the propagators are dressed with the results presented in the previous section. The results are displayed in fig. 5 for four dimensions and in fig. 6 for three dimensions. Shown is the transverse part of the ghost-gluon vertex, $1 + A$, defined by

$$\frac{k^2}{g_d\Delta}q_\nu\left(\delta_{\mu\nu}-\frac{k_\mu k_\nu}{k^2}\right)\Gamma_\mu^{abc}(k;q,p)=f^{abc}(1+A(k;q,p)) \tag{8}$$

where $\Delta = q^2k^2 - (q\cdot k)^2$ is a Gram determinant.

Several observations can be made [26]. First, although not shown in the plots, the results comply with the infrared limit (6) and thus satisfy the respective STI in this specific kinematical limit. Second, they have the proper ghost-antighost symmetry [25], $A(k;q,p) = A(k;p,q)$. Therefore, they fulfill all required conditions. In addition, in three dimensions, asymptotic freedom is manifest and the vertex reduces to the bare one in the ultraviolet. However, in four dimensions the anomalous dimensions is incorrect, as the vertex is not resummed yet.

As can be seen, the result is a vertex which is very close to the tree-level one at all momenta, especially in the infrared. This does not change in other kinematical conditions [26] and thus confirms Zwanziger's hypothesis in an impressive manner. The origin for the small deviations from the exact bare vertex in the infrared may be twofold. On the one hand, certainly iteration to self-consistency and truncation artifacts change the results quantitatively. As the ghost-gluon vertex DSE is already quite well fulfilled, this will most likely not lead to a qualitative change. On the other hand, in Zwanziger's hypothesis only agreement with leading-order renormalization group has been required. Subleading effects may generate a deviation of the order observed.

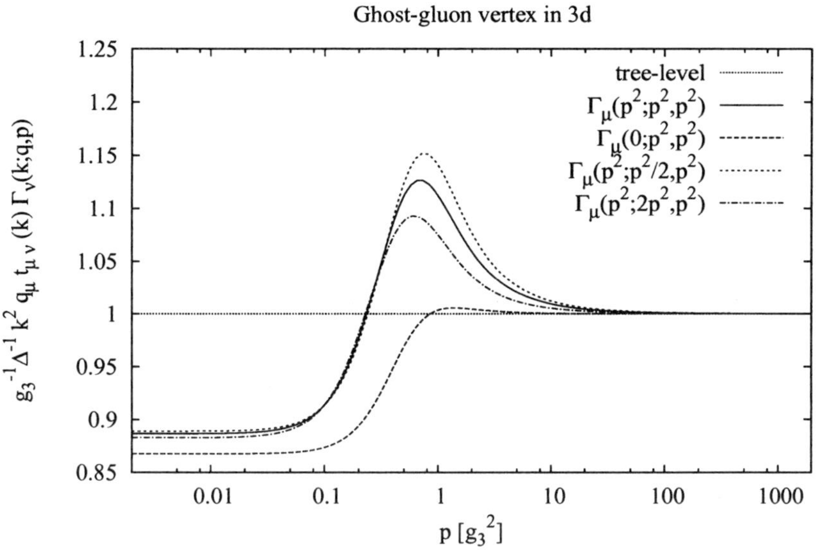

Figure 6: The normalized transverse part (8) of the ghost-gluon vertex in three dimensions in various kinematical regions.

In addition, the vertex is nearly tree-level for all momenta. This supports the assumption of a bare ghost-gluon vertex in the truncation scheme for the propagators. It especially provides additional evidence for ghost dominance and therefore for the Zwanziger-Gribov confinement mechanism.

Similar results have also been reported recently in lattice calculations in four dimensions [27]. They confirm the results found independently here and indicate also a nearly tree-level ghost-gluon vertex.

To assess the error of the approximations adopted, the full ghost-gluon vertices in the ansatz shown in fig. 4 have been replaced by the vertices motivated by the corresponding STI [25]. It has been found [26] that also these vertices are mapped to a nearly tree-level like vertex by the truncated DSE, as long as no singularity in an external momentum is introduced. (As there is no iteration in the current ansatz, these divergencies would survive the calculation.) Moreover, the results have been found to be slightly more sensitive to the propagators than to the input vertices. All these variations are quantitatively small. This indicates that the solution of the ghost-gluon vertex DSE is a vertex which is nearly tree-level.

It is probable that the influence of quarks, *c.f.* ref. [17], or of an adjoint Higgs field, *c.f.* ref. [15], on these findings is negligible. Neither couple directly to the ghost and therefore their presence does not give rise to new terms in the ghost-gluon vertex DSE. Hence, they can only contribute indirectly via their effect on the Yang-Mills propagators, which are very robust against coupling a small number of quark flavours [17], or scalars [15].

The ghost-gluon vertex may also be interesting with respect to extracting the running coupling constant [27]. This issue has to be investigated further.

4 Conclusions

In Landau-gauge Yang-Mills theory we have calculated the ghost-gluon vertex employing non-perturbative propagators. The results show quite conclusively that deviations from the tree-level value are very small, even in the infrared. On the one hand, this supports the truncation scheme used to calculate the propagators of the Yang-Mills theory. The results found in these calculations give strong evidence for a confinement mechanism of the Kugo-Ojima or Zwanziger-Gribov type. Especially, no sign of an infrared enhanced gluon propagator in Landau gauge is found. More importantly, the results on the ghost-gluon vertex alone, provide independent support for the Zwanziger-Gribov scenario as they fulfill Zwanziger's hypothesis of a bare ghost-gluon vertex in the infrared. The results presented here are thus a further indication for a confinement mechanism driven by the dynamics on or near the Gribov horizon. This coincides with recent findings in Coulomb gauge [28]. All these evidences combined promise a new era of quantitative confinement physics.

Acknowledgments

This work has been done in collaboration with Axel Maas and Jochen Wambach from the Darmstadt University of Technology and with Reinhard Alkofer from the University of Tübingen.

W. S. thanks the organizers of the International School of Subnuclear Physics for the very interesting meeting and for the opportunity to present this talk. The collaboration is grateful to C. S. Fischer, J. M. Pawlowski and L. von Smekal for helpful discussions. This work is supported by the BMBF under grant numbers 06DA917 and 06DA116, and by the Helmholtz association (Virtual Theory Institute VH-VI-041).

References

[1] R. J. Rivers, "Path Integral Methods In Quantum Field Theory," Cambridge, UK: Univ. Pr. (1987) 339 p. (Cambridge monographs on mathematical physics).

[2] R. Alkofer and L. von Smekal, Phys. Rept. **353** (2001) 281 [arXiv:hep-ph/0007355] and references therein.

[3] K. Osterwalder and R. Schrader, Commun. Math. Phys. **31** (1973) 83. K. Osterwalder and R. Schrader, Commun. Math. Phys. **42** (1975) 281.

[4] R. Oehme and W. Zimmermann, Phys. Rev. D **21** (1980) 1661. R. Oehme and W. Zimmermann, Phys. Rev. D **21** (1980) 471.

[5] T. Kugo and I. Ojima, Prog. Theor. Phys. Suppl. **66** (1979) 1.

[6] T. Kugo, arXiv:hep-th/9511033.

[7] D. Zwanziger, Phys. Rev. D **69** (2004) 016002 [arXiv:hep-ph/0303028] and references therein.

[8] V. N. Gribov, Nucl. Phys. B **139** (1978) 1.

[9] P. van Baal, arXiv:hep-th/9711070.

[10] D. Zwanziger, Nucl. Phys. B **412** (1994) 657.

[11] J. C. Taylor, Nucl. Phys. B **33** (1971) 436. W. J. Marciano and H. Pagels, Phys. Rept. **36** (1978) 137.

[12] P. Watson and R. Alkofer, Phys. Rev. Lett. **86** (2001) 5239 [arXiv:hep-ph/0102332]; R. Alkofer, L. von Smekal and P. Watson, arXiv:hep-ph/0105142.

[13] L. von Smekal, R. Alkofer and A. Hauck, Phys. Rev. Lett. **79** (1997) 3591 [arXiv:hep-ph/9705242]. L. von Smekal, A. Hauck and R. Alkofer, Annals Phys. **267** (1998) 1 [Erratum-ibid. **269** (1998) 182] [arXiv:hep-ph/9707327].

[14] C. S. Fischer and R. Alkofer, Phys. Lett. **B536** (2002) 177 [arXiv:hep-ph/0202202];

[15] A. Maas, J. Wambach, B. Grüter and R. Alkofer, Eur. Phys. J. **C37** (2004) 335 [arXiv:hep-ph/0408074].

[16] B. Grüter, R. Alkofer, A. Maas and J. Wambach, arXiv:hep-ph/0408282.

[17] C. S. Fischer and R. Alkofer, Phys. Rev. D **67** (2003) 094020 [arXiv:hep-ph/0301094].

[18] K. Langfeld, J. C. R. Bloch, J. Gattnar, H. Reinhardt, A. Cucchieri and T. Mendes, arXiv:hep-th/0209173.

[19] P. O. Bowman, U. M. Heller, D. B. Leinweber, M. B. Parappilly and A. G. Williams, arXiv:hep-lat/0402032.

[20] A. Cucchieri, F. Karsch and P. Petreczky, Phys. Rev. D **64** (2001) 036001 [arXiv:hep-lat/0103009].

[21] A. Cucchieri, T. Mendes and A. R. Taurines, Phys. Rev. D **67** (2003) 091502 [arXiv:hep-lat/0302022].

[22] J. M. Pawlowski, D. F. Litim, S. Nedelko and L. von Smekal, Phys. Rev. Lett. **93** (2004) 152002 [arXiv:hep-th/0312324]; C. S. Fischer and H. Gies, arXiv:hep-ph/0408089.

[23] N. Brown and M. R. Pennington, Phys. Rev. D **38** (1988) 2266.

[24] S. Mandelstam, Phys. Rev. D **20** (1979) 3223.

[25] C. Lerche and L. von Smekal, Phys. Rev. D **65** (2002) 125006 [arXiv:hep-ph/0202194].

[26] W. Schleifenbaum, A. Maas, J. Wambach, R. Alkofer, to be published; W. Schleifenbaum, diploma thesis, TU Darmstadt, 2004.

[27] A. Cucchieri, T. Mendes and A. Mihara, arXiv:hep-lat/0408034.

[28] J. Greensite, S. Olejnik and D. Zwanziger, arXiv:hep-lat/0407032.

Little Supersymmetry with heavy sfermions[1].

Jakub Wagner
Institute of Theoretical Physics, Warsaw University, Hoża 69, 00-681, Warsaw, Poland

Abstract

I present a mechanism of electroweak symmetry breaking in supersymmetry, inspired by Little Higgs theories. The mechanism works for scalar superpartner masses of order 10 TeV and the cut-off scale $\Lambda \leq 250$ TeV. The necessary fine tuning is two orders of magnitude smaller than in the MSSM with soft scalar masses of the same order of magnitude. The physical Higgs boson mass is in the range 120 – 135 GeV.

1. It is well known that although the Standard Model provides a succesfull description of fundamental interactions at currently accesible energies, it suffers from the hierarchy problem. If it is not valid to arbitrarily high energies, and we know that new physics must eventually enter at some energy scale, because the Standars Model does not include for example gravity, we have dangerous quantum corrections to the higgs mass parameter m_H^2. If the cut-off scale of SM is equal Λ the one loop corrections give:

$$\delta m_H^2 = -\frac{3\lambda_t^2}{8\pi^2}\Lambda^2 \tag{1}$$

$$\delta m_H^2 = \frac{g^2}{16\pi^2}\Lambda^2 \tag{2}$$

[1]Talk is based on ref.[1]

$$\delta m_H^2 = \frac{\lambda^2}{16\pi^2}\Lambda^2 \tag{3}$$

from the loops involving top quark, gauge bosons, and Higgs field respectively. For $\Lambda = 10$ TeV it means that we have to fine tune the bare higgs mass with precision of the order of 1%. If the cut-off scale is the GUT or the Planck scale, the fine tuning is worse then 10^{-30}.

Most explored solution to the hierarchy problem is the Minimal Supersymmetric extension of the Standard Model (MSSM) . However, the present LEP limits on the lightest Higgs boson mass push the masses of the scalar partners of the top quark into the range 500 GeV- 1 TeV. Also a natural suppresion of the flavour changing neutral currents (FCNC) in the MSSM require squark masses to be at least O(10 TeV). This implies a big fine tuning because the logarithmic divergent correction to the Higgs mass parameter is proportional to the stop mass squared.

Because it is not easy to quantify the problem of fine tuning, in ref.[1] we have asked a different question: is it possible to avoid quadratic dependence of the Fermi scale on the stop masses? To answer this question we have used the model of Birkedal, Chacko and Gaillard [2] inspired by the Little Higgs models [3] which we briefly review below.

2.In the little Higgs models, the SM Higgs doublet is the pseudo-Goldston boson of the spontanuosly broken approximate symmetry. Lagrangians of such theories has the following structure:

$$L = L_0 + \epsilon_1 L_1 + \epsilon_2 L_2 \tag{4}$$

The L_0 part has a big symmetry group G which is spontanously broken in such a way that Higgs dublet becomes a Goldstone boson. This symmetry is explicitly broken by L_1 and L_2, but in such a way that if $\epsilon_1 \to 0$ or $\epsilon_2 \to 0$ Higgs is still an exact, massless to all orders of perturbation calculation, Goldston boson. So all corections to its mass, must be proportional to $\epsilon_1\epsilon_2$. At one loop level this implies the absence of quadratic divergence.

The cut-off scale Λ is constrained by the magnitude of the dominant quadratically divergent two loop corrections:

$$\delta m^2 = \frac{\epsilon_1}{16\pi^2}\frac{\epsilon_2}{16\pi^2}\Lambda^2 \tag{5}$$

If $\epsilon_i \sim 1$, then Λ can be pushed up to about 10 TeV.

There are diffrent Little Higgs models based on diffrent pattern of symmetry breaking ($SU(5)/SO(5)$, $SU(3)^2/SU(3)$, $SU(3)/SU(2)$) and all of them predict some additional particles with masses of the order of 1 TeV. These particles are partners of the same spin of the SM particles. Enlarging symmetry of the top Yukawa interactions implies existence of the additional heavy top quark, enlarging gauge group - additional heavy gauge bosons, and enlarging symmetry of Higgs sector - additional heavy scalar fields. Extended symmetry of Yukawa interactions ensures the absence at one-loop of the corrections $\sim Y_t^2\Lambda^2$, where Y_t is the top quark Yukawa coupling, to the Higgs mass parameter m^2. Extending the gauge group does the same for the $g^2\Lambda^2$ contribution.

3.The model we consider has been originally proposed in ref.[2]. The Yukawa sector has the form similar to the one in the Little Higgs models. The MSSM Higgs dublets $\hat{H}_u$ and $\hat{H}_d$ and third generation quark superfields $\hat{Q}_3$ are extended to fit fundamental and anti-fundamental representations of a global SU(3) symmetry:

$$\hat{\mathcal{H}}_u = \begin{pmatrix} \hat{H}_u \\ \hat{S}_u \end{pmatrix}, \quad \hat{\mathcal{H}}_d^T = \begin{pmatrix} \hat{H}_d \\ \hat{S}_d \end{pmatrix}, \quad \hat{\mathcal{Q}}^T = \begin{pmatrix} \hat{Q}_3 \\ \hat{T} \end{pmatrix} . \tag{6}$$

In addition there is a new SU(3) singlet quark supermultiplet $\hat{T}^c$. At some high scale Λ_S the relevant part of the superpotential takes the form:

$$W = Y_T \hat{\mathcal{Q}} \hat{\mathcal{H}}_u \hat{t}^c + \mu \hat{\mathcal{H}}_d \hat{\mathcal{H}}_u + \mu_T \hat{T}^c \hat{T} \tag{7}$$

The Yukawa coupling as well as the μ term respect the global $SU(3)$ symmetry. In the superpotential it is broken explicitly only by the μ_T term. We also assume that at the scale Λ_S the soft symmetry breaking terms respect $SU(3)$ symmetry. The gauge symmetry of the model is $SU(3) \times SU(2) \times U(1)_Y \times U(1)_E$. The $U(1)_E$ and the ordinary strong gauge interactions respect the global $SU(3)$ whereas the electroweak gauge interactions break it explicitly. Hence, the Higgs potential consist of the symmetric part:

$$V_{\text{symm}} = m_1^2|\mathcal{H}_d|^2 + m_2^2|\mathcal{H}_u|^2 - (m_3^2\mathcal{H}_d\mathcal{H}_u + \text{h.c.}) + \frac{1}{2}g_E^2(|\mathcal{H}_u|^2 - |\mathcal{H}_d|^2)^2 , \tag{8}$$

and the $SU(3)$ symmetry breaking part:

$$V_{EW} = \frac{g_2^2 + g_y^2}{8}(|H_u|^2 - |H_d|^2)^2 - \frac{g_2^2}{2}(|H_dH_u|^2 - |H_u|^2|H_d|^2) . \tag{9}$$

$U(1)_E$ gauge and $SU(3)$ global symmetry breaking occurs simultanously in a similar way as the electroweak symmetry breaking in the MSSM. In this mechanism the Higgs triplets in (6) aquire VEVs in the S_u and S_d directions:

$$s_u \equiv \langle S_u \rangle = f \sin\beta \ , \qquad s_d \equiv \langle S_d \rangle = f \cos\beta \ , \tag{10}$$

Spontanous breaking of the $SU(3)$ down to $SU(2)$ produces five Goldstone bosons , of which one is „eaten" in the Higgs mechanism by the gauge boson of the broken $U(1)_E$. The remaining four physical degrees of freedom can be identified with the SM Higgs field. Below the $SU(3)$ symmetry breaking scale we can work with the non-linear sigma model. The Higgs triplets are then parametrized as:

$$\mathcal{H}_u = e^{i\Pi/f} \begin{pmatrix} 0 \\ s_u \end{pmatrix}, \qquad \mathcal{H}_d = (0,\ s_d)\, e^{-i\Pi/f} \tag{11}$$

where $i\Pi$ is a matrix of the four Goldstone bosons:

$$i\Pi = \begin{pmatrix} \mathbf{0}_{2\times 2} & H \\ -H^\dagger & 0 \end{pmatrix} \tag{12}$$

which leads to

$$\begin{pmatrix} H_u \\ S_u \end{pmatrix} = \sin\beta \begin{pmatrix} H \frac{\sin(|H|/f)}{|H|/f} \\ f\cos(|H|/f) \end{pmatrix} \qquad \begin{pmatrix} H_d \\ S_d \end{pmatrix}^T = \cos\beta \begin{pmatrix} H^\dagger \frac{\sin(|H|/f)}{|H|/f} \\ f\cos(|H|/f) \end{pmatrix}^T \tag{13}$$

The tree level SM Higgs potential is obtained by using (13) in (8) and (9). It has only the quartic part arising from the electroweak D-terms (9):

$$V_h = \frac{g_2^2 + g_y^2}{8} \cos^2(2\beta) f^4 \sin^4(h/f) \ . \tag{14}$$

where h is the neutral component of H.

4. The SM higgs mass parameter m_H^2 is generated already at the one loop level, because $SU(3)$ global symmetry is only approximate. In order to estimate m_H^2 we calculate the one loop effective potential, which is given by the standard formula [4]:

$$V_{1-\text{loop}} = \frac{1}{64\pi^2} \mathrm{STr} \left\{ \mathcal{M}^4 \left(\log \frac{\mathcal{M}^2}{\Lambda^2} - \frac{3}{2} \right) \right\} \tag{15}$$

By supersymmetry m_H^2 can not be $\sim \Lambda^2$. Due to the global $SU(3)$ symmetry at one loop m_H^2 can not be generated by the top Yukawa interactions and therefore can not be proportional to the stop mass squared. The contribution coming from top and stop mass matrices could only be proportional to the only $SU(3)$ breaking parameter μ_T:

$$\mathrm{Tr}\mathcal{M}^4_{top} \sim \mu_T^2 h^2$$
$$\mathrm{Tr}\mathcal{M}^4_{stop} \sim \mu_T^2 h^2 \tag{16}$$

but μ_T does not break supersymmetry, hence one has:

$$\mathrm{STr}\mathcal{M}^4 = 0 \tag{17}$$

The only possibile higgs mass term coming from Yukawa interactions is:

$$\mathrm{STr}\mathcal{M}^4 \log \mathcal{M}^2 \sim (\mu_T^2 \log m_{\tilde{q}}^2) h^2 \tag{18}$$

It is finite and only logarithmically dependent on squark masses. We can check it directly calculating the h dependent top and stop mass matrices. The top mass matrix is:

$$\mathcal{L}_{\text{mass}} = -(t^c, T^c)\mathcal{M}_{\text{tops}}(t,T)^T + \text{h.c.} \qquad \mathcal{M}_{\text{tops}} = \begin{pmatrix} Y_T f_u s_h & Y_T f_u c_h \\ 0 & \mu_T \end{pmatrix} \tag{19}$$

where we have used te abbreviations $s_h = \sin(h/f)$, $c_h = \cos(h/f)$. Hence,

$$\mathcal{M}^2_{\text{tops}} \equiv \mathcal{M}^\dagger_{\text{tops}}\mathcal{M}_{\text{tops}} = \begin{pmatrix} Y_T^2 f_u^2 s_h^2 & Y_T^2 f_u^2 s_h c_h \\ Y_T^2 f_u^2 s_h c_h & \mu_T^2 + Y_T^2 f_u^2 c_h^2 \end{pmatrix} . \tag{20}$$

In the stop sector

$$\mathcal{L}_{\text{mass}} = -(\tilde{t}^*, \tilde{T}^*, \tilde{t}^c, \tilde{T}^c)\mathcal{M}^2_{\text{stops}}(\tilde{t}, \tilde{T}, \tilde{t}^{c*}, \tilde{T}^{c*})^T \tag{21}$$

where $\mathcal{M}^2_{\text{stops}}$ takes the form

$$\begin{pmatrix} m_{Q_3}^2 + Y_T^2 f_u^2 s_h^2 & Y_T^2 f_u^2 s_h c_h & (Y_T f_d \mu - A_T f_u) s_h & 0 \\ Y_T^2 f_u^2 s_h c_h & m_{Q_3}^2 + \mu_T^2 + Y_T^2 f_u^2 c_h^2 & (Y_T f_d \mu - A_T f_u) c_h & 0 \\ (Y_T f_d \mu - A_T f_u) s_h & (Y_T f_d \mu - A_T f_u) c_h & m_{U_3}^2 + Y_T^2 f_u^2 & Y_T f_u \mu_T c_h \\ 0 & 0 & Y_T f_u \mu_T c_h & m_{T^c}^2 + \mu_T^2 \end{pmatrix} \tag{22}$$

With the help of (20), (22) from (15) we find the Y_T contribution to the higgs potential:

$$V_{1\text{-loop}} = -\frac{3}{8\pi^2} Y_T^2 \mu_T^2 f^2 \sin^2\beta \left(\log \frac{m_{\tilde{q}}^2}{m_T^2} + 1\right) \sin^2(h/f)$$
$$-\frac{3}{8\pi^2} \frac{Y_T^4 \mu_T^4 \sin^4\beta}{m_T^4} f^4 \left[\log\left(\frac{Y_T \mu_T f \sin\beta}{m_T^2} \sin(h/f)\right) + \frac{1}{4}\right] \sin^4(h/f) \tag{23}$$

where m_T is the mass of the heavy fermionic partner of the top quark. The first term gives the Higgs mass squared which is negative allowing electroweak breaking, and does not depend quadraticaly on stop mass. This allows for squark masses much larger then in the MSSM without increasing the fine tuning. In addition to (23) there are also other contributions to $V_{1-\mathrm{loop}}$ coming from the gauge and scalar interactions, which we will discuss in the next paragraph. Here let us stress only the that due to supersymmetry, there is no contribution proportional to the $g_2^2 m_{\tilde{q}}^2$, and we do not have to introduce complicated gauge group structure like those in the Little Higgs models in order to eliminate such contribution at one loop.

5. The global $SU(3)$ symmetry is broken radiativly by quantum correction induced by the large Yukawa coupling Y_T. For $\tan\beta \geq 5$ the expression for $M_{Z'}$ can be estimated as:

$$M_{Z'}^2 = g_E f^2 \approx -\mu^2 + 0.1 m_{\tilde{q}}^2 \log\frac{\Lambda_S}{m_{\tilde{q}}} \tag{24}$$

Assuming $g_E \approx 1$, and minimal allowed value $f \sim 2.5$ TeV(from limits on $M_{Z'}$ [5]) we have practically no fine tuning for $m_{\tilde{q}} \leq 10 \div 50$ TeV , and $\Lambda \sim 100 \div 250$ TeV. Relating Y_T to the physical top quark Yukawa coupling y_t one finds that μ_T and m_T have to be both of order ~ 4 TeV. The Higgs mass parameter obtained from (23) is therefore too large: $m_H^2 \sim -1.6\,\mathrm{TeV}^2$ whereas for succesfull electroweak symmetry breaking one needs $m_H^2 \sim -0.01\,\mathrm{TeV}^2$. Thus, the electroweak breaking in our model requires fine tuning of order 1%. The neccesary positive contribution to m_H^2 can be provided by the logarithmically divergent contributions of gauginos and higgsinos to $V_{1-\mathrm{loop}}$. The complete one loop m_H^2 reads:

$$\begin{aligned} m_H^2 \approx & -\frac{3}{4\pi^2} y_t^2 m_T^2 \log\frac{m_{\tilde{q}}}{m_T} + \frac{g_y^2}{16\pi^2}\cos(2\beta)\mathrm{Tr}[Y m^2]\log\frac{\Lambda_S}{m_{\tilde{q}}} \\ & + \frac{1}{32\pi^2}\Big[3(3g_2^2 + g_y^2)\mu^2 + 12 g_2^2 M_2^2 + 4 g_y^2 M_y^2 \\ & \qquad -4\sin 2\beta(3g_2^2 M_2 + g_y^2 M_y)\,\mu \\ & \qquad -g_E^2(3g_2^2 + 4g_y^2)\cos^2 2\beta f^2\Big]\log\frac{\Lambda_S}{\mu} \end{aligned} \tag{25}$$

and for sufficiently large μ and gaugino masses, m_H^2 can be of the right order of magnitude. Moreover, the requirement that $m_H^2 < 0$ set the upper limit on $\Lambda_S \leq 250$ TeV.

The full one loop SM Higgs potential obtained from (15) has the following form:

$$V = m_H^2 h^2 + \lambda h^4 + \kappa h^4 \log \frac{h^2}{m_T^2} \tag{26}$$

where:

$$\lambda = \frac{1}{8}(g_2^2 + g_y^2)\cos^2(2\beta) - \frac{3}{8\pi^2} y_t^4 \left[\log(y_t) + \frac{1}{4}\right]$$

$$-\frac{3}{64\pi^2}(g_2^2 + g_y^2)\cos(2\beta)\sin^2\beta Y_T^2 \log\frac{\Lambda_S^2}{m_{\tilde{q}}^2} - \frac{1}{64\pi^2} g_E^2 (g_2^2 + g_y^2) \log\frac{\Lambda_S^2}{\mu^2}$$

$$\kappa = -\frac{3}{16\pi^2} y_t^4 \tag{27}$$

The physical higgs mass is predicted by the model to be in the range:

$$M_h \approx 120 - 135\,\text{GeV}\,. \tag{28}$$

6.To conclude: in the presented model the Higgs field is a pseudo Goldstone boson of an approximate $SU(3)$ global symmetry. This ensures that the corrections to the Fermi scale do not depend quadraticaly on squark masses, which can therefore be as heavy as $\mathrm{O}(10 \div 50)$ TeV. The fine tuning in such model is of the order 1 % - two orders of magnitude smaller then in MSSM with soft masses of the similar order of magnitude. There is another fundamental scale $\Lambda_S \sim 100 \div 250$ TeV at which supersymmetry breaking is mediated to the visible sector.

The main phenomenological consequence of this model is a natural suppresion of flavour changing neutral currents due to havy soft scalar masses. Sfermions are too heavy to be seen in the LHC. The additional heavy top with mass at least 4 TeV is also out of the LHC reach [6]. There is a Z' boson with mass $\sim$ 3 TeV which should be discovered at the LHC. The prediction on Higgs boson mass is 120 – 135 GeV.

References

[1] P. H. Chankowski, A. Falkowski, S. Pokorski and J. Wagner, Phys. Lett. B **598** (2004) 252 [arXiv:hep-ph/0407242].

[2] A. Birkedal, Z. Chacko and M. K. Gaillard, arXiv:hep-ph/0404197.

[3] N. Arkani-Hamed, A. G. Cohen and H. Georgi, Phys. Lett. B **513** (2001) 232 [arXiv:hep-ph/0105239]. N. Arkani-Hamed, A. G. Cohen, E. Katz, A. E. Nelson, T. Gregoire and J. G. Wacker, JHEP **0208** (2002) 021 [arXiv:hep-ph/0206020]. N. Arkani-Hamed, A. G. Cohen, E. Katz and A. E. Nelson, JHEP **0207** (2002) 034 [arXiv:hep-ph/0206021]. D. E. Kaplan and M. Schmaltz, JHEP **0310**, 039 (2003) [arXiv:hep-ph/0302049];

[4] S. R. Coleman and E. Weinberg, Phys. Rev. D **7** (1973) 1888.

[5] T. Appelquist, B. A. Dobrescu and A. R. Hopper, Phys. Rev. D **68** (2003) 035012 [arXiv:hep-ph/0212073].

[6] G. Azuelos *et al.*, arXiv:hep-ph/0402037.

(DOI: 10.1140/epj/s2004-02053-6) of The European Physical Journal (EPJ) C

Determining the Actual Local Density of Dark Matter Particles

Jacob L. Bourjaily

Michigan Center for Theoretical Physics, University of Michigan, Ann Arbor, MI 48109-1120

Abstract. Even if dark matter particles were unambiguously discovered in experiments, there is no clear reason to expect that this would resolve the dark matter problem. It is very easy to provide examples of dark matter scenarios (*e.g.* in supersymmetric models) in which nearly identical detector signals correspond to extremely different relic densities. The density of the discovered particles, therefore, must be determined before their cosmological relevance can be established. In this paper, I present a general method that utilizes both dark matter and hadron collider experimental data (once they become available) to estimate the local density of dark matter particles. These results were obtained in collaboration with Gordon Kane at the University of Michigan.

PACS: not given

1. Introduction

We are now confident that our universe contains a large amount of cold dark matter. The most popular particle candidates for dark matter are weakly interacting massive particles (wimps). These particles are being searched for directly and indirectly by dozens of experimental groups throughout the world. If we are fortunate, wimps may soon be discovered experimentally.

Although the discovery of wimps in the galactic halo would have enormous implications for our understanding of elementary particle physics, it would infact contribute very little to our understanding of dark matter. It is unreasonable to expect that the particle discovered would represent all dark matter. This issue was first raised technically, though not resolved, in [1]. Indeed, current (and future) experiments readily provide examples of particles that in fact constitute less than 1% of all dark matter.

Even if weakly interacting massive particles were produced at colliders, it would still be very important to directly determine their local density in the galactic halo. This can only be done with experimental data. In this paper, I describe how this can be determined in the context of supersymmetry, in which the dark matter is the lightest supersymmetric particle (LSP). Similar analyses could be done for any dark matter candidate.

I start out by illustrating why a discovery of dark matter particles in the halo would be insufficient for addressing the dark matter problem, and describing some of the uncertainties in relating the local and relic densities of dark matter. I then describe how dark matter is directly detected in experiments and present the general form of the interaction rate. This shows what is required for determining the local density of wimps. I present a very useful method for improving on these expressions by using data from different detector materials and at different recoil energies.

In order to deduce the local density of wimps, it is essential to know the wimp's mass. I present two methods for determining the mass of a wimp using direct detection data alone. The first is based on a well known relationship, and the second presents the preliminary results of the author. All of this work is then combined in the framework of the most general minimally-supersymmetric standard model, in which the neutralino is the wimp seen in dark matter experiments. A general procedure is presented for estimating the local density, and explicit bounds are given.

2. Discovering (Some of?) the Dark Matter

Let us imagine that a weakly interacting massive particle χ has been unambiguously observed in direct detection experiments. Such a discovery would represent an enormous triumph for theoretical and experimental particle cosmology; it would have deep implications for our understanding of the universe, it may herald the existence of supersymmetry, and it would account for (at least) some of the dark matter in the universe. However, a discovery of dark matter particles is far from a solution to the dark matter problem: there is no reason to suspect that χ is *all* the dark matter.

What fraction of dark matter is represented by χ is a question that cannot be answered by experiment or theory alone. Furthermore, the answer will depend crucially on dark matter detection experiments. The purpose of this paper is to describe how this question may be answered.

It should be possible to determine the local density

of χ using direct detection experiments. This is because, in a rough sense, they measure the local wimp density times its scattering cross section. Unfortunately, there are very few constraints on the scattering cross sections of most wimp candidates.

However, it is not true that the signal rate depends on the cross section and density independently, because these are somewhat related. This is because the relic density Ω_χ is related to thermal production and freeze-out in the early universe. The rate of wimp-annihilation affects the relic density and depends on the wimp annihilation cross section, which is in turn somewhat related to the scattering cross section by crossing.

As a result, there is less freedom in the observed signal rate than one may have naïvely suspected. This can be illustrated qualitatively as follows. If the cross section is large, then most wimps annihilate in the early universe and local density is small. Alternatively, if the cross section is small, then thermal freeze-out occurs very early and the density is higher. In either case, the cross section and density tend to compensate each other.

The crude arguments above suggest that even a very small component of dark matter may be detectable because it may have a higher cross section. This has been referred to as the 'no-lose theorem' in recent conferences. If even tiny fractions of dark matter may be detectable, in other words, experimentalists cannot 'lose' on making a discovery [2].

This is seen in many realistic dark matter scenarios. In figure 1, we have plotted the relic density against the direct detection signal for some six thousand randomly generated, constrained minimally supersymmetric standard models (without assuming any specific supersymmetry breaking scenario). By "constrained", we mean that all of the models are allowable under the current experimental constraints on supersymmetry. These models were generated and analyzed using the DarkSUSY code [3].

Notice that for any particular signal strength, the relic density fluctuates over at least two orders of magnitude. In accordance with the 'no-lose theorem,' experiments in the near future may detect even 1% of the dark matter or less.

However, the no-lose theorem also implies, unfortunately, that the discovery of wimps in the galactic halo tells us very little about how much of the dark matter they compose. A wimp discovery could easily represent a negligible fraction of the dark matter.

Therefore, although wimps may be discovered in the near future, the dark matter problem will not be addressed until the density of wimps has been directly determined.

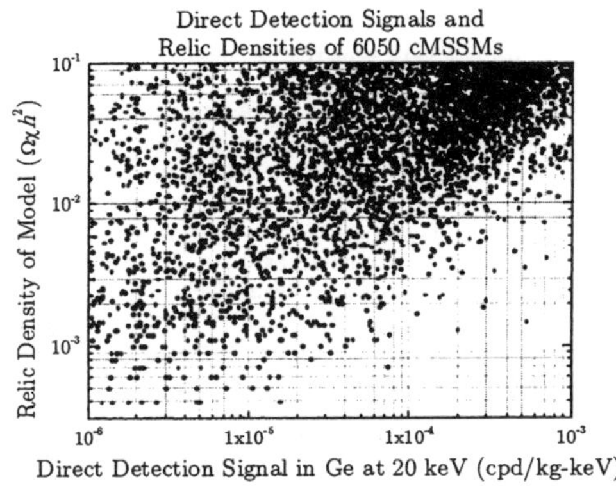

Fig. 1. The relic densities of 6050 constrained MSSMs as a function of direct detection signal strength in germanium. Experiments currently in planning or under construction may be able to observe signals of the order 10^{-4} cpd/kg·keV.

3. Local and Relic Densities

From studies of the cosmic microwave background, large scale structure formation, and big bang nucleosynthesis, we know the cosmic-scale relic density of cold dark matter to be approximately $\Omega_{\rm cdm}h^2 \sim 0.11$ [4]. From our knowledge of the rotation of the Milky Way galaxy, the local dark matter halo density is known to be approximately $\rho_{\rm cdm} \sim 0.3$ GeV/cm^3 [5]. It is obvious that any dark matter experiment on Earth is only sensitive to the local density and not the relic density.

Unfortunately, the relationship between local and relic densities involves many details of galaxy formation and structure that are still not understood. Even if we were able to demonstrate that χ has a local density of precisely 0.3 GeV/cm^3, there remain important subtleties about our understanding of dark matter in the universe as a whole.

Acknowledging these shortcomings, it nevertheless remains extremely important to determine the local density of any wimp discovered in direct detection experiments. It would be very promising if the entire local dark matter halo could be accounted for by wimps discovered in these experiments.

Because direct detection experiments are sensitive to small-scale structure in the local halo density, a knowledge of the ambient halo density, $\rho_{\rm cdm} \sim 0.3$ GeV/cm^3 may not be sufficient. Our knowledge of the local halo density is based on large scale surveys of star velocities in the Milky Way, and these measurements are not very sensitive to small-scale structure in the halo.

There are several types of small-scale halo structure that may affect direct detection experiments. For example, the earth may be within a stream of dark matter. This situation has been suggested in studies of the Sgr A stream and it has been estimated that our local halo density could be $0.3 - 23\%$ higher than the ambient density [6]. Alternatively, some authors have proposed that the halo may be clumpy or contain caustic structures [7]. These small-scale perturbations in dark matter density could have significant effects on direct detection rates.

Most of the small-scale structure considered in the literature involves local, high-density regions of dark matter within the halo. Although these structures may make it easier to discover dark matter, they make it nearly impossible to assess what fraction of the halo is composed of χ.

Fortunately, there exist ways to check the smoothness of the halo profile. For example, dark matter streams or caustics may be identified or excluded using directional dark matter experiments like DRIFT [8]. Also, it may be possible to identify a clumpy dark matter halo by studying the time-dependence of a wimp signal. These questions, therefore, may find answers in the foreseeable future.

These ambiguities must be addressed before the dark matter problem can be put to rest. For the purposes of this paper, however, we assume that the halo is locally smooth and that if $\rho_\chi = \rho_{\rm cdm} = 0.3$ GeV/cm^3, then χ represents all of the dark matter.

4. Dark Matter Direct Detection

Dark matter particles in the halo can be observed directly through their interaction with ordinary matter [13]. Although wimps interact only weakly, occasionally they will scatter off matter in detectors, depositing a small amount of energy. Because wimps typically have masses on the order of a hundred GeV and move relatively slowly in the halo (on the order of a few hundred km/s) they typically deposit recoil energies from $\sim 1 - 200$ keV. Using very sensitive detectors, experiments can observe signals as low as a few keV. Using sophisticated coincidence algorithms, most experiments can remove virtually all background noise except scattering from neutrons.

In essence, direct detection experiments measure the χ-nucleus scattering rate as a function of recoil energy and time. In general, the signal rate is a function of the cross section for χ-nucleon scattering, the nuclear physics describing the nuclei in a particular detector, and the local velocity profile of the wimp fraction of the dark matter halo.

A. Elastic Scattering Rate

It is helpful here to state the explicit form of the differential interaction rate for a particular detector at the recoil energy q. Let the detector in question be composed of nuclei labeled with index j, each with mass fraction c_j. Thus the differential rate of wimp scattering at recoil energy q is,

$$\left.\frac{dR}{dQ}\right|_{Q=q} = \frac{2\rho_\chi}{\pi m_\chi}\sum_j c_j \int_{v_{\min_j}(q)}^{\infty} \frac{f(v,t)}{v} dv \left\{ F_j^2(q)[Z_j f_p + (A_j - Z_j) f_n]^2 + \frac{4\pi}{(2J_j+1)}\left[a_1^2 S_{j00}(q) + a_0^2 S_{j11}(q) + a_1 a_0 S_{j01}(q)\right]\right\}, \quad (1)$$

where $v_{\min_j}(q)$ is the minimum velocity kinematically capable of depositing energy q into the $j^{\rm th}$ nucleus, $f(v,t)$ is the local velocity distribution function for the galactic halo, $F_j^2(q)$ and $S_{j_{mn}}(q)$ are nuclear form factors for coherent and incoherent scattering, respectively, Z_j and A_j are atomic and mass numbers, J_j is the nuclear spin, $a_1 \equiv a_p + a_n$ and $a_0 \equiv a_p - a_n$, and the constant parameters $f_{p,n}$ and $a_{p,n}$ describe the coherent and incoherent wimp-nucleon scattering cross sections, respectively [14]. For a more detailed discussion of equation 1, please refer to any modern review of dark matter (see, *e.g.*, [5]).

It is important to note that equation 1 depends on several unknown parameters: (1) the wimp's mass, m_χ; (2) the particle physics of χ which determines the interaction parameters $f_{p,n}$ and $a_{p,n}$; (3) the velocity distribution of the halo, $f(v,t)$; (4) the local density of wimps, ρ_χ. These details will not be known when wimps are first discovered and may take many years to resolve.

B. Prerequisites to Determine ρ_χ

From the discussion above, it is clear that to determine the density ρ_χ, one must first: (1) identify the particle χ; determine m_χ; estimate the halo profile; calculate the interaction parameters from the theory describing χ. Each of these will require enormous efforts from both dark matter and collider physics experiments.

The most important and perhaps most difficult requirement is the identification of χ. This is not possible through dark matter experiments alone. This is because these experiments observe only a few of χ's quantum numbers. For example, it is unlikely that any amount of direct detection data would enable us to differentiate between the lightest supersymmetric particle and the lightest Kaluza-Klein particle; if possible at all, this would probably require very precise data from several different nuclei. Therefore, while direct detection experiments may unequivocally discover dark matter wimps, they would not be able to explain the dark matter itself.

Perhaps the most important parameter describing χ is its mass. This determines all of its kinematics and is crucially linked to the local density. Fortunately, m_χ may be calculable from direct detection experiments alone. The known methods of calculating m_χ from dark matter experiments are described below.

Because the mass may be observable, it may prove to be the key to the identification of χ. If a neutral, stable particle is observed at colliders with the same mass as that observed in dark matter experiments, then we might suspect that they are the same particle. Although this association is imprecise, it appears to be one of the best methods of identification.

We should note, however, that determining the mass of χ may be very difficult at hadron colliders. For example, if χ is the LSP, it could take several years and an enormous effort to determine m_χ in a model-independent way. Most of the known techniques for determining the mass of the LSP rely on either the framework of mSUGRA or specific assumptions about the relative masses of squarks and sleptons. Therefore, it should be stressed that χ may not be identified until long after it is discovered.

Also, the halo profile must be known sufficiently well. As described above, any small-scale structure in the halo will dramatically alter the analysis of the local density. It is imperative that these issues be sufficiently resolved.

Lastly, to compute the local density ρ_χ, one must know the interaction parameters $f_{p,n}$ and $a_{p,n}$. To compute these parameters, one must know a great deal about the theory describing χ. It is clear that these parameters cannot be obtained from direct detection data alone: they depend on the many parameters of whichever extended standard model χ is a part of. For example, if χ is the LSP, then these parameters will be functions of the squark masses, mixing angles, gauge-content of the LSP, and higgs parameters. It is extremely unlikely that all of these will be known when χ is discovered in direct detection experiments.

C. Combining Data

All of the required analysis can be strengthened and empowered by combining data from different detectors over a range of recoil energies. There are many important insights and results that derive from the following framework.

In general, the expression for the scattering rate (equation 1) is a second order polynomial in the four unknown interaction parameters $f_{p,n}$ and $a_{p,n}$. To highlight this, it can be recast in the suggestive form:

$$\begin{aligned}\left.\frac{dR}{dQ}\right|_{Q=q} = \frac{2\rho_\chi}{\pi m_\chi}\Bigg\{ & f_p^2\left(\sum_j c_j \int_{v_{\min_j}(q)}^{\infty} \frac{f(v,t)}{v} dv F_j^2(q) Z_j^2\right) + a_p^2\left(4\pi\sum_j c_j \int_{v_{\min_j}(q)}^{\infty} \frac{f(v,t)}{v} dv \frac{[S_{j00}(q)+S_{j11}(q)+S_{j01}(q)]}{2J_j+1}\right) \\ & + f_n^2\left(\sum_j c_j \int_{v_{\min_j}(q)}^{\infty} \frac{f(v,t)}{v} dv F_j^2(q)(A_j-Z_j)^2\right) + a_n^2\left(4\pi\sum_j c_j \int_{v_{\min_j}(q)}^{\infty} \frac{f(v,t)}{v} dv \frac{[S_{j00}(q)+S_{j11}(q)-S_{j01}(q)]}{2J_j+1}\right) \\ & + f_p f_n\left(2\sum_j c_j \int_{v_{\min_j}(q)}^{\infty} \frac{f(v,t)}{v} dv F_j^2(q) Z_j(A_j-Z_j)\right) + a_p a_n\left(8\pi\sum_j c_j \int_{v_{\min_j}(q)}^{\infty} \frac{f(v,t)}{v} dv \frac{[S_{j00}(q)-S_{j11}(q)]}{2J_j+1}\right)\Bigg\}. \end{aligned} \tag{2}$$

It is clear from the expressions above that by using data from

1. different detector materials (varying the mass fractions, nuclear form factors, nuclear spins, and minimum velocities),
2. different recoil energies (varying the nuclear form factors and minimum velocities),

one could invert equation 2 to solve for $\sqrt{\rho_\chi} f_{p,n}$ and $\sqrt{\rho_\chi} a_{p,n}$ if the halo velocity distribution and m_χ were known. Given a halo model and the wimp mass, that is, the data from different detector materials and different recoil energies are sufficient for determining $\sqrt{\rho_\chi} f_{p,n}$ and $\sqrt{\rho_\chi} a_{p,n}$ (up to quadratic ambiguities).

We should mention that there are many important situations in which the above analysis can be simplified. For example, because $a_{p,n}$ are already scaled by linearly independent combinations of the incoherent nuclear form factors, $S_{j_{mn}}(q)$, it is not necessary to use different detector materials to solve for $\sqrt{\rho_\chi} a_{p,n}$. However, this will only work if there is data available from a detector with nuclei that have non-zero spin, which is sufficiently sensitive to incoherent scattering.

Although knowing the scaled interaction parameters will not directly determine the local density, it can give enormous insight into the particle physics of χ. For example, if χ is the LSP, then the ratios a_p/a_n or a_p/f_n could possibly lead to important insights on $\tan\beta$, the degeneracy of the squark masses, mixing, and perhaps other information as well. This could be very important for collider physics and disentangling the MSSM.

5. Determining the Wimp Mass

Of all the factors required to interpret the observed signal rate, perhaps the most important is m_χ. Not only is the mass required to compute the density, but it also plays a critical role in the identification of χ as described above. Fortunately, it may be possible to determine m_χ from direct detection experiments alone.

There are roughly two ways to determine m_χ from direct detection data. One method, that using the annual modulation crossing energy, was described in the dark matter review article by Primack *et. al.* in 1988 [9]. Although it seems unlikely to have originated in a review article, we have been unable to find any previous work mentioning this effect. The other method has been developed by the author in collaboration with Gordon Kane and represents work still in preparation. Both of these methods rely solely on the kinematics of the halo, although neither are particularly sensitive to the precise halo model [15] (although, see caveats in [10]).

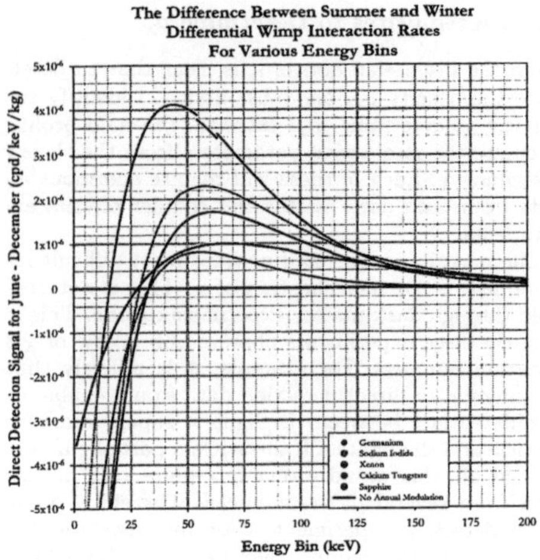

Fig. 2. The difference between direct detection signals in germanium in June and December as a function of recoil energy. This plot was generated for an MSSM with $m_\chi \sim 161$ GeV.

A. Annual Modulation Crossing Energy

As the earth orbits the sun, its velocity through the galactic dark matter halo varies between roughly 250 and 190 km/s [8]. This in turn results in annual modulation in the scattering rate. However, the amplitude of this modulation varies as a function of recoil energy and changes sign. For a more detailed description, see, for example, [10].

In figure 2, we plot the difference between the scattering rates in June and December as a function of recoil energy for several detector materials. Notice that there is a particular energy, called the 'crossing energy,' at which no annual modulation is observed. As pointed out by Primack *et. al.* [9], the crossing energy is an explicit function of the masses of the wimp and detector nuclei that can be derived easily from kinematics. Therefore, if crossing is observed, one can determine the mass of the wimp explicitly.

This method is moderately robust. Specifically, if there is an energy at which the annual modulation amplitude changes sign, then one can confidently determine the wimp mass to within approximately 10%. There are, however, some important subtleties and caveats to this analysis as described in, for example, [10]. These include the effects of bin-sizes and small-scale halo structure. Furthermore, if the wimp is very light, then the crossing energy may be well below the threshold of the detector and therefore not observed at all.

B. Kinematical Consistency

Recall that if the halo velocity profile and m_χ are known, then direct detection data from different detector materials and different energies can be used to solve for $\sqrt{\rho_\chi} f_{p,n}$ and $\sqrt{\rho_\chi} a_{p,n}$. If the halo velocity distribution is known, then only the wimp mass is required to determine these.

Let us assume that the local halo velocity profile can be adequately approximated and that there exists enough data to solve for $\sqrt{\rho_\chi} f_{p,n}$ and $\sqrt{\rho_\chi} a_{p,n}$ as long as the mass is known. (If, for example, there was enough data to solve for $a_{p,n}$, it would be clear how to proceed along similar lines). Because many direct detection experiments observe scattering rates in a large number of recoil energy bins, we can generally expect to have many more measurements than the minimum required to solve the system of equations.

Because the interaction parameters are absolute constants, all minimal, linearly independent combination of measurements used to solve for the scaled interaction parameters will agree as long as the correct mass is used in the derivation. However, if an arbitrary m'_χ was used to solve for these parameters, different calculations will tend to disagree.

This motivates us to define a 'kinematical consistency' function, $\zeta(m'_\chi)$, which compares the values of $\sqrt{\rho_\chi} f_{p,n}, \sqrt{\rho_\chi} a_{p,n}$ obtained using different independent subsets of the data as a function of the m'_χ used. Specifically, let $\zeta(m'_\chi)$ be given by

$$\zeta(m'_\chi) \equiv \sum_{i \neq j} \sqrt{\rho_\chi} \left\{ (a_p(i) - a_p(j))^2 + \text{similar terms} \right\},$$

where the indices i, j represent a minimal set of data used to compute the constants given the particular m'_χ. It is necessary that $\zeta(m'_\chi) = 0$ when $m'_\chi = m_\chi$, but this is not a sufficient condition. Specifically, we have not found any way to demonstrate that m_χ is the unique root of $\zeta(m'_\chi)$, although we have found no example where it has multiple roots.

To determine the wimp mass, one varies m'_χ until $\zeta(m'_\chi) = 0$. To test how useful this technique is, we applied it to some six thousand random, constrained MSSMs. In every single model tested, the correct mass was determined to near-arbitrary precision. Figure 3 illustrates a typical plot of $\zeta(m'_\chi)$. Notice that ζ has an extremely sharp minimum, decreasing many orders of magnitude within a few GeV of the true mass of the LSP. It should be stressed, however, that experimental uncertainties and resolutions were not considered in these calculations.

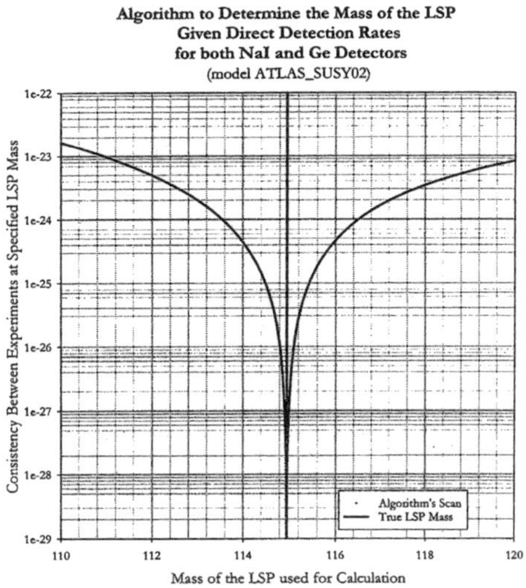

Fig. 3. The function $\zeta(m'_\chi)$, where the wimp corresponds to the neutralino in the MSSM specified by ATLAS SUSY point 2 [11]. The models and data were generated within the framework of the DarkSUSY package, [3].

Although this method appears quite promising, these results should be considered preliminary. Several questions remain regarding how robust this calculation would be upon the introduction of experimental uncertainties and halo model ambiguities. Without these uncertainties, the algorithm yields the correct mass to seemingly arbitrary precision. It will be interesting to see how this changes in more realistic circumstances.

6. Bounds on the Local Density

Because m_χ can be determined, in principle, using the methods described above and because we are working under the assumption that the local halo velocity profile can be adequately approximated, we can determine $\sqrt{\rho_\chi} f_{p,n}$ and/or $\sqrt{\rho_\chi} a_{p,n}$ independently of the identification of χ. Therefore, to determine ρ_χ it is sufficient to know any one of the interaction parameters. This is an enormous improvement over the general case, for which all of the interaction parameters were required.

Therefore, any bounds on the interaction parameters will translate into bounds on the local density. Unfortunately, these parameters can only be computed in the framework of a very explicit model for the wimp. Furthermore, these parameters are typically very poorly constrained.

In order to address the question of the local density ρ_χ, one must specialize on a particular candidate particle in a specific extension of the standard model. In other words, we cannot proceed further without losing some generality.

7. Neutralino Dark Matter

The most popular and perhaps best-justified candidate for cold dark matter is the lightest supersymmetric particle (LSP), as predicted by supersymmetric extensions of the standard model that conserve R-parity. Indeed, the existence of supersymmetric dark matter was predicted before it was known that non-baryonic dark matter would be needed. In most MSSMs allowed by experimental constraints, the LSP is the neutralino, χ, which is the supersymmetric partner of the neutral gauge and higgs bosons. For an extensive and authoritative review of supersymmetric dark matter, see Jungman *et. al.* [5].

A. Interaction Parameters

Given a completely specified supersymmetric standard model, it is rather straightforward to calculate the interaction parameters. It should be noted, however, that one does not compute $f_{p,n}$ or $a_{p,n}$ directly. Rather, χ-quark (an χ-gluon) interaction parameters are calculated and these are used to determine the χ-nucleon parameters.

Even at tree level, it is clear that the χ-quark interaction parameters will depend on many of the parameters in the model. Specifically, they are functions of the

1. gauge content of the lightest neutralino,
2. most of the squark masses and mixing angles,
3. $\tan\beta$, the ratio of the vacuum expectation value of the two higgs bosons,
4. higgs mass parameters (only for the coherent interactions).

It should be emphasized that most of these parameters will be extraordinarily difficult to measure in practice (especially at hadron colliders). Currently, no general, model-independent methods exist for determining most of these parameters.

To illustrate the dependence on each of these parameters (although we do not derive them here), the incoherent scattering of χ with a u-quark is given by:

$$
\begin{aligned}
a_u = &-\frac{g^2}{16m_W^2}(N_{\tilde{H}_1}^2 - N_{\tilde{H}_2}^2) + \frac{g^2}{8}\sum_{\tilde{q}_j}\frac{1}{m_{\tilde{q}_j}^2-(m_\chi+m_u)^2}\Bigg\{\; 2\left(\frac{1}{2}N_{\tilde{W}}^* + \frac{1}{6}\tan\theta_W N_{\tilde{B}}^*\right)^2 (\Pi_L\Theta_u)_{1j}^2 \\
&+\frac{m_u}{m_W\sin\beta}\Re\left[\left(N_{\tilde{H}_2}N_{\tilde{W}}^* + \frac{1}{3}\tan\theta_W N_{\tilde{H}_2}N_{\tilde{B}}^*\right)(\Pi_R\Theta_u)_{1j}^*(\Pi_L\Theta_u)_{1j}\right] \\
&+\frac{m_u^2}{2m_W^2\sin^2\beta}N_{\tilde{H}_2}^2(\Pi_R\Theta_u)_{1j}^2 + \frac{8}{9}\tan^2\theta_W N_{\tilde{B}}^2(\Pi_R\Theta_u)_{1j}^2 \\
&-\frac{4m_u\tan\theta_W}{3m_W\sin\beta}\Re\left[N_{\tilde{H}_2}N_{\tilde{B}}^*(\Pi_L\Theta_u)_{1j}^*(\Pi_R\Theta_u)_{1j}\right] + \frac{m_u^2}{2m_W^2\sin^2\beta}N_{\tilde{H}_2}^2(\Pi_L\Theta_u)_{1j}^2 \;\Bigg\}.
\end{aligned}
$$

In the expression above, the matrices $\Pi_{L,R}$ are 3×6 projection matrices given in the basis $(\tilde{u}_L,\tilde{c}_L,\tilde{t}_L,\tilde{u}_R,\tilde{c}_R,\tilde{t}_R)$; Θ_u is a unitary matrix which diagonalizes $\tilde{M}_u^2$ so that $\tilde{M}_u^{2\ \mathrm{diag}} = \Theta_u^\dagger \tilde{M}_u^2\Theta_u$ [16]; the subscript j on $\tilde{q}_j$ corresponds to the flavor and handedness of the quarks so that $j=1,\ldots,6$ corresponds to $(\tilde{u}_L,\tilde{c}_L,\tilde{t}_L,\tilde{u}_R,\tilde{c}_R,\tilde{t}_R)$; and gauge content of χ is given by $\chi = N_{\tilde{B}}|\tilde{B}\rangle + N_{\tilde{W}}|\tilde{W}\rangle + N_{\tilde{H}_1}|\tilde{H}_1\rangle + N_{\tilde{H}_2}|\tilde{H}_2\rangle$. A similar expression describes scattering with d,s-quarks.

The coherent parameters are similar to the incoherent ones except that they also contain higgs exchange at tree-level. This implies that in addition to squark masses, mixing angles, $\tan\beta$, and the gauge content of χ, one must also know the higgs masses. Therefore, in general, less knowledge of the MSSM is required to compute $a_{p,n}$ than $f_{p,n}$.

B. Limits on Scattering Parameters

As we have shown, the interaction parameters depend on a very detailed knowledge of the MSSM. Unfortunately, these may not be known until well after dark matter particles have been directly observed in experiments. We should try to estimate them somehow using partial information and any bounds on the MSSM that are available.

It should be clear that even if all of the squark masses and mixing angles are unknown, we can still place limits on the interaction parameters by using exclusion bounds. In general, one can typically find a way to make use of what is known and constrain what is not known in order to estimate and limit the interaction parameters.

Beginning with almost none of the MSSM parameters determined, we find that we can still place rather strong limits on $a_{p,n}$. For example, we have found that given only upper and lower bounds on $\tan\beta$ and a lower bound on the lightest squark mass, there is a strict upper bound for the incoherent χ-quark scattering parameters. If there is a strict lower bound on the lightest squark mass, such as $m_{\tilde{q}_\ell}$, and $\tan\beta$ is bounded so that $\sin\beta \geq \sin\beta_\ell$ [17], then there is a strict upper bound on $a_{p,n}$. It should be mentioned that these types of bounds already exist today, at least in the framework of particular supersymmetry breaking scenarios. In this case, it can be shown that the magnitude of a_u is strictly bounded by

$$a_u \leq -\frac{g^2}{16m_W^2}(N_{\tilde{H}_1}^2 - N_{\tilde{H}_2}^2) + \frac{g^2}{8}\frac{1}{(m_{\tilde{q}_\ell}^2 - (m_\chi + m_u)^2}\left\{\frac{17}{18}\tan^2\theta_W N_{\tilde{B}}^2 + \frac{1}{2}N_{\tilde{W}}^2 + \frac{m_u^2}{m_W^2\sin^2\beta}N_{\tilde{H}_2}^2\right.$$
$$\left. + \frac{1}{3}\tan\theta_W|N_{\tilde{B}}||N_{\tilde{W}}|\cos(\alpha_{\tilde{W}}) + \frac{m_u}{m_W\sin\beta}|N_{\tilde{W}}||N_{\tilde{H}_2}|\cos(\alpha_{\tilde{H}_2} - \alpha_{\tilde{W}}) - \frac{m_u}{m_W\sin\beta}\tan\theta_W|N_{\tilde{B}}||N_{\tilde{H}_2}|\cos(\alpha_{\tilde{H}_2})\right\},$$

where $\alpha_{\tilde{H}_2}$, and $\alpha_{\tilde{W}}$ are the relative phases between $N_{\tilde{H}_2}, N_{\tilde{W}}$ and $N_{\tilde{B}}$, respectively.

This expression has six real unknowns. Notice that by the normalization of the neutralino wave function, $|N_{\tilde{B}}|^2 + |N_{\tilde{W}}|^2 + |N_{\tilde{H}_1}|^2 + |N_{\tilde{H}_2}|^2 = 1$, the parameter space is compact. Therefore, a_u can be absolutely maximized with respect to all six unknowns. Specifically, although all of the gauge content of the neutralino may be unknown, one can limit the χ-quark and hence the χ-nucleon interaction parameters absolutely.

It should be emphasized that the analysis used to derive the above bound was for the most general softly-broken supersymmetric standard model; no *ad hoc* supersymmetry breaking scenarios such as mSUGRA were assumed. It is obvious that if a particular supersymmetry breaking scenario were assumed, the above expressions would be enormously simplified. However, these types of assumptions are very difficult to justify (theoretically or experimentally) and therefore greatly limit the generality of the work.

It is important to note the flexibility of the derivation involved in computing these bounds. If, for example, the masses of several light squarks were known, one could greatly improve the above bounds by including these in the explicit expression for a_q and then maximizing it relative to the parameters that remain unknown. In this manner, almost any additional knowledge can be incorporated to arrive at stronger statements. Therefore, not only do these bounds grow more restrictive with increasing knowledge of the MSSM, but they continue to approach a realistic estimate of the interaction parameters.

C. Strong Lower Bound on ρ_χ

From the work above it is clear that, given adequate bounds on $\tan\beta$ and a lower bound on the lightest squark mass, there exist strong, model-independent upper bounds on $a_{p,n}$. These in turn can be used to place a very strong lower bound on the neutralino relic density, because we know $\sqrt{\rho_\chi}a_{p,n}$.

To test this method, we considered some six thousand randomly generated MSSMs that are consistent with all known bounds on supersymmetry. For each of

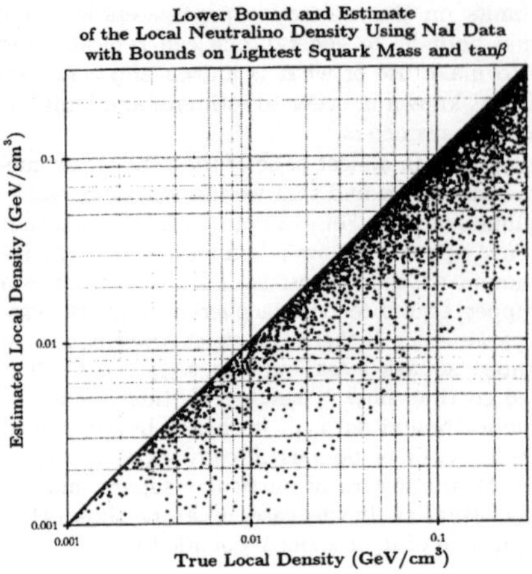

Fig. 4. This plot compares the lower bound and estimate of the local density computed using the strong upper bound for $a_{p,n}$ to the true local density for each model. The red line indicates perfect agreement. Notice that the procedure correctly determined a lower bound for the local density for every model.

these models, we calculated the interaction rates for a NaI detector in twelve recoil energy bins. This (idealized) data was used to compute the mass of the LSP - using the kinematical consistency function - and to solve for $\sqrt{\rho_\chi}a_{p,n}$. Upper bounds were calculated for $a_{p,n}$ assuming 10% uncertainty in $\tan\beta$ and a lower bound on the lowest squark mass of either 200 GeV or the actual mass of the lightest squark, whichever was less. For the sake of computational simplicity, however, the specific gauge content of the neutralino was taken to be known for each model[18]. Using the upper bounds for $a_{p,n}$, we obtained a lower bound on the local density ρ_χ.

Figure 4 illustrates the results of this algorithm for each of the randomly generated MSSMs. Notice that the estimated local density is always strictly less than the true local density—as required for a lower bound. Notice also that for many models the lower bound was not such a poor estimate. This will be the case, for example, when the lightest squark mass is near or below the 200 GeV bound.

8. Conclusions

We have seen that, by itself, a discovery of dark matter particles in our galactic halo cannot address directly the dark matter problem of the universe. However, when combined with data from colliders in order to identify the particle and limit its interaction parameters, we can give a general estimate of its local density.

In the framework of supersymmetry, we have presented a robust and iteratively improvable method for estimating, using any information available about the MSSM, the local density of neutralino LSPs observed in direct detection experiments .

Thus we can conclude that even though the observation of wimps is unlikely to resolve the dark matter problem immediately, there do exist clear and general methods for addressing their cosmological significance.

Acknowledgements

This research was done in collaboration with Gordon Kane of the University of Michigan and was supported by the Michigan Center for Theoretical Physics and the National Science Foundation's 2004 REU program.

I would like to thank the organizers of the 42nd *International School of Subnuclear Physics*, Antonino Zichichi and Gerardus 't Hooft, for the opportunity to present this work and for organizing such a wonderful school.

References

1. M. Brhlik, D. J. H. Chung, and G. L. Kane, "Weighing the Universe with Accelerators and Detectors," *Int. J. Mod. Phys.*, vol. D10, p. 367, 2001, hep-ph/0005158.
2. G. Dūda, G. Gelmini, P. Gondolo, J. Edsjö, and J. Silk, "Indirect Detection of a Subdominant Density Component of Cold Dark Matter," *Phys. Rev.*, vol. D67, p. 023505, 2003, hep-ph/0209266.
3. P. Gondolo *et al.*, "DarkSUSY: Computing Supersymmetric Dark Matter Properties Numerically," *JCAP*, vol. 0407, p. 008, 2004, astro-ph/0406204.
4. C. L. Bennett *et al.*, "First year Wilkinson Microwave Anisotropy Probe (WMAP) Observations: Preliminary Maps and Basic Results," *Astrophys. J. Suppl.*, vol. 148, p. 1, 2003, astro-ph/0302207.
5. G. Jungman, M. Kamionkowski, and K. Griest, "Supersymmetric Dark Matter," *Phys. Rept.*, vol. 267, pp. 195–373, 1996, hep-ph/9506380.
6. K. Freese, P. Gondolo, and H. J. Newberg, "Detectability of Weakly Interacting Massive Particles in the Sagittarius Dwarf Tidal Stream," 2003, astro-ph/0309279.

7. P. Sikivie, "Caustic Rings of Dark Matter," *Phys. Lett.*, vol. B432, pp. 139–144, 1998, astro-ph/9705038.
8. C. Muñoz, "Dark Matter Detection in the Light of Recent Experimental Results," 2003, hep-ph/0309346.
9. J. R. Primack, D. Seckel, and B. Sadoulet, "Detection of Cosmic Dark Matter," *Ann. Rev. Nucl. Part. Sci.*, vol. 38, p. 751, 1988.
10. P. Ullio and M. Kamionkowski, "Velocity Distributions and Annual-Modulation Signatures of Weakly Interacting Massive Particles," *JHEP*, vol. 03, p. 049, 2001, hep-ph/0006183.
11. "ATLAS Detector and Physics Performance. Technical Design Report. vol. 2," CERN-LHCC-99-15.
12. M. W. Goodman and E. Witten, "Detectability of Certain Dark-Matter Candidates," *Phys. Rev.*, vol. D31, p. 3059, 1985.
13. This was first proposed by Goodman and Witten in [12].
14. Informally, coherent scattering is sometimes called 'spin-independent' and incoherent scattering 'spin-dependent.'
15. Specifically, these calculations are insensitive to which isothermal halo model is assumed if the halo is in fact locally isothermal. Any extra structure in the halo (e.g. streams, caustics, etc.), however, can disturb these calculations significantly.
16. We have chosen the basis for the squark-mass matrices so that M_u is diagonal.
17. Both the upper and lower bounds of $\tan\beta$ are important and it is not sufficient to have only a lower bound on $\sin\beta$. This is because the nucleon scattering parameters $a_{p,n}$ will depend on u- d- and s-quark scattering. In particular, a lower bound on $\cos\beta$ is needed to place bounds on $a_{d,s}$.
18. If the gauge content of the neutralino was unknown, the interaction parameters could have been maximized with respect to these parameters as described earlier. In general, therefore, the upper bounds obtained were more restrictive than they would be in practice.

Reprinted with permission from the electronic-only supplement "Young Scientists Papers" (DOI: 10.1140/epj/s2004-02053-6) of The European Physical Journal (EPJ) C

MSSM Higgs Bosons at a Future Linear Collider. Heavy Quark Production at HERA using the H1 Detector

T. Klimkovich

DESY, Notkestraße 85, D-22607 Hamburg, Germany.; e-mail: tklimk@mail.desy.de

Abstract. We study the potential of the future e^+e^- linear collider operated at a center-of-mass energy of 500 to 1000 GeV for the measurement of the neutral Higgs boson properties within the framework of the MSSM. The process of associated Higgs boson production with subsequent decays of Higgs bosons into b-quarks is considered. An integrated luminosity of 500 fb^{-1} is assumed at each energy. The Higgs boson masses and production cross sections are measured by reconstructing the $\mathrm{b\bar{b}b\bar{b}}$ final state. The precision of these measurements, depending on the Higgs boson masses, is evaluated. Under the assumed experimental conditions, a statistical accuracy ranging from 0.1 to 1.0 GeV is achievable on the Higgs boson mass. The topological cross section $\sigma(\mathrm{e^+e^-} \to \mathrm{HA} \to \mathrm{b\bar{b}b\bar{b}})$ can be determined with the relative precision of 1.5 – 6.6 %. The 5σ discovery limit corresponds to the Higgs mass of around 385 GeV for the degenerate Higgs boson masses in the $\mathrm{HA} \to \mathrm{b\bar{b}b\bar{b}}$ channel at $\sqrt{s} = 800$ GeV with integrated luminosity of 500 fb^{-1}. The potential of the Higgs mass determination for the benchmark point SPS 1a for the process $\mathrm{HA} \to \mathrm{b\bar{b}b\bar{b}}$ at $\sqrt{s} = 1$ TeV and luminosity 1000 fb^{-1} is investigated. The possibility of parameter measurements in a CP-violating MSSM scenario is shown.

In the second part the inclusive method of charm and beauty cross section measurements in e^+p collisions at H1 experiment for values of $\mathrm{Q}^2 > 12$ GeV^2 is explained. Preliminary plots are shown.

PACS: not given

1 MSSM Higgs Bosons at a Future Linear Collider

1.1 Introduction

A future e^+e^- linear collider operating at a center-of-mass energy of 500 to 1000 GeV can become an excellent tool for investigating the mechanism of electroweak symmetry breaking. In the minimal Standard Model (SM), breaking of electroweak symmetry is provided by one Higgs doublet, which introduces an additional spin-0 particle, the Higgs boson [1]. A distinct feature of the supersymmetric models is the extended Higgs sector. In the CP-conserving MSSM (Minimal Supersymmetric extension of the Standard Model) the Higgs sector consists of five physical states: two CP-even Higgs bosons, the lighter of which is denoted as h and heavier H, one CP-odd Higgs Boson A and two charged bosons $\mathrm{H}^\pm$. The mass eigenstates in this scenario are equal to CP eigenstates. The Higgs sector can be parametrized by $\mathrm{m_A}$ and $\tan\beta$, where $\tan\beta$ is the ratio of the vacuum expectation values of the two Higgs doublets.

In the MSSM it is possible to explicitly or spontaneously break CP symmetry by radiative corrections [2, 3]. The SM fails to provide enough CP-violation to explain the cosmological matter-antimatter asymmetry. CP-violating effects can help to reduce this crisis [4]. CP-violation is introduced in the MSSM Higgs potential via loop effects and manifests itself in complex phases of $\mathcal{L}_{MSSM}$; in particular, the phases of $A_{t,b}$, the soft SUSY-breaking trilinear couplings of the top (bottom) squark to the Higgs boson, and $\mathrm{m_{\tilde{g}}}$, the mass of the gluino. The phase of $A_{t,b}$ is the only parameter to introduce the CP-violation in the Higgs potential on one-loop level. On two-loop level the phase of $\mathrm{m_{\tilde{g}}}$ enters as an additional parameter, which can provide CP-violation. If CP is broken in the Higgs sector, then the Higgs boson mass eigenstates no longer correspond to CP eigenstates. There are three neutral Higgs bosons $\mathrm{H_1}$, $\mathrm{H_2}$, $\mathrm{H_3}$ ($\mathrm{m_{H_1}} \leq \mathrm{m_{H_2}} \leq \mathrm{m_{H_3}}$) that have mixed CP parities. The parametrization of the Higgs sector is chosen to be in $\mathrm{m_{H^+}}$ and $\tan\beta$, because the Higgs boson A is no longer a mass eigenstate.

The present analysis is motivated by the existing LHC "wedge" region for MSSM Higgs boson production (Fig. 1). There is a large part of the parameter space where ATLAS will be able to see only h^0. A future linear collider can check the MSSM Higgs boson production in that region. In this paper, the study of Higgs boson pair production at a future linear e^+e^- collider is presented. This analysis is an extension of previous studies [6, 7]. The analyzed topology is the $b\bar{b}b\bar{b}$ final state. In the following we will assume the $e^+e^- \to$ HA production process. However, this assumption does not restrict the generality of our study and the analysis developed in this paper is applicable to the process $e^+e^- \to$ hA as well as to the processes of associated Higgs pair production in CP-violating MSSM scenarios.

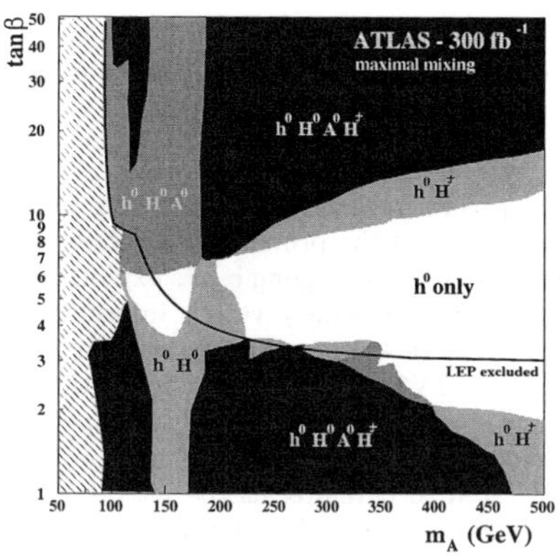

Fig. 1. ATLAS Higgs boson production possibility as a function of parameters m_A and $\tan\beta$.

The study of the Higgs pair production process is motivated by a so-called 'decoupling limit' of the MSSM, in which the h boson approaches the properties of the SM Higgs boson. The closer the MSSM scenario moves towards the decoupling limit the more difficult it becomes to distinguish the Higgs sector from the SM. In such a scenario the detection of heavier neutral Higgs bosons would be crucial for establishing an extended Higgs sector. The decoupling limit is approached relatively fast for large values of the H and A boson masses, $m_A, m_H >$ 200 GeV, in a large portion of the MSSM parameter space. The most distinct feature of this scenario is the vanishing coupling of heavy CP-even Higgs boson to weak bosons, $\cos(\beta - \alpha) \to 0$:

$$g_{HZZ,HWW} \sim \cos(\beta - \alpha). \quad (1)$$

As a consequence, the H boson production via the fusion and Higgs-strahlung processes is significantly suppressed, whereas the cross section of the $e^+e^- \to$ HA process reaches its maximal value, making associated heavy Higgs pair production a promising channel for the detection of the H and A bosons at a future linear e^+e^- collider:

$$g_{HAZ} \sim \sin(\beta - \alpha). \quad (2)$$

It should also be emphasized that in the decoupling limit the H and A bosons are almost degenerate in mass and have similar decay properties.

1.2 Detector Simulations and Monte Carlo Samples

The study presented here [5] was performed for a linear collider operating at center-of-mass energies of 500 GeV to 1000 GeV and an event sample corresponding to an integrated luminosity of 500 fb^{-1}.

The detector used in the simulation followed the proposal for the TESLA collider as presented in the TDR [6]. Studies have also been performed in Asia and North America [8, 9]. The detector response was simulated with the parametric fast simulation program SIMDET [10]. Beamstrahlung was accounted for using CIRCE [11].

Samples of $e^+e^- \to$ HA events were generated for several Higgs boson mass hypotheses with PYTHIA 6.2 [12], including initial state radiation. The tree level cross section of the HA production was related to the Higgs-strahlung cross section in the SM, σ_{HZ}^{SM}, in the following way [13]:

$$\sigma_{HA} = g_{HAZ}^2 \bar{\lambda}_{HA} \sigma_{HZ}^{SM}, \quad (3)$$

where

$$\bar{\lambda}_{HA} = \frac{\lambda_{HA}^{3/2}}{\lambda_{HZ}^{1/2}[12m_Z^2/s + \lambda_{HZ}]} \quad (4)$$

accounts for the suppression of the P-wave cross section near the kinematic threshold. The quantity $\lambda_{ij} = [1 - (m_i + m_j)^2/s][1 - (m_i - m_j)^2/s]$ is the usual momentum factor of the two particle phase space. The Higgs-strahlung cross section in the SM was computed using the program HPROD [14].

For the case of maximal allowed g_{HAZ} coupling, $\sin^2(\beta - \alpha) = 1$, which was assumed for these studies, signal cross sections are given in Table 1. The branching fractions of the Higgs bosons into b-quarks were set to their typical values in the MSSM: $Br(H \to b\bar{b}) = Br(A \to b\bar{b}) = 90\%$. For

$\sqrt{s}$ [GeV]	(m_H,m_A) [GeV]	σ_{HA} [fb]
500	(150,100)	33.62
	(200,100)	25.30
	(250,100)	16.61
	(150,140)	28.39
	(150,150)	26.90
	(200,150)	18.85
	(250,150)	10.67
	(200,200)	11.35
800	(300,150)	10.55
	(290,200)	9.54
	(300,250)	7.49
	(300,300)	5.70
	(350,350)	2.23
	(400,150)	6.46
	(400,200)	5.17
	(400,250)	3.70

Table 1. Tree level cross sections σ_{HA} for $e^+e^- \to HA$ expected for the Higgs boson mass hypotheses (m_H,m_A) considered in the study. Numbers are given for $\sin^2(\beta-\alpha)$=1. Cross sections were calculated including ISR at center-of-mass energies of $\sqrt{s}$ = 500, 800 GeV.

the background estimation, the following processes were generated using PYTHIA 6.2: $e^+e^- \to WW$, $e^+e^- \to ZZ$, $e^+e^- \to q\bar{q}$. The cross sections for the most important background processes are given in Table 2.

1.3 Analysis tools

Identification of b-quarks plays a crucial role in this analysis. The efficient tagging of jets containing heavy flavour hadrons will be achieved with a highly granular micro-vertex detector, which allows

Process	$\sigma \times BR$ [fb]	
	500 GeV	800 GeV
$t\bar{t} \to W^+bW^-\bar{b}$	$6.69 \cdot 10^2$	$1.65 \cdot 10^2$
$WW \to q\bar{q}q\bar{q}$	$4.13 \cdot 10^3$	$2.34 \cdot 10^3$
$ZZ \to q\bar{q}q\bar{q}$	$3.11 \cdot 10^2$	$1.74 \cdot 10^2$
$q\bar{q}(q = u, d, s, c, b)$	$1.29 \cdot 10^4$	$5.45 \cdot 10^3$

Table 2. Topological cross sections $\sigma \times BR$ of background processes at $\sqrt{s}$ = 500, 800 GeV.

for the precise reconstruction of track parameters in the vicinity of the primary interaction point.

The procedure of tagging b-jets exploits both single track and secondary vertex information. Secondary vertices are searched for within jets using the package ZVTOP [15] developed for the SLD experiment. A neural network is developed [16] for jet flavour separation. The analysis uses a jet-wise tag, referred to hereafter as the jet b-tag variable. It is defined as follows for a jet with neural network output x:

$$B(x) = \frac{f_b(x)}{f_b(x) + f_{udsc}(x)}, \tag{5}$$

where $f_b(x)$ and $f_{udsc}(x)$ are probability density functions of neural network output in samples of b– and udsc–jets, respectively. Tagging of c-jets proceeds in a similar way.

The mass resolution of the reconstructed Higgs bosons is improved by means of a kinematic fit. In the $b\bar{b}b\bar{b}$ analysis, conservation of four-momentum is required, leading to a total of four constraints. The kinematic fit is performed using the code developed by DELPHI [17].

1.4 Analysis Procedure and Results

Events of the $b\bar{b}b\bar{b}$ topology are characterized by four high multiplicity hadronic jets, which contain the decay products of b-hadrons. A cut-based technique is employed to separate signal from background. Selection criteria are optimized separately for 500 GeV and 800 GeV center-of-mass energies. Each event is required to pass the following cuts:

1. Total energy deposited in the detector, the visible energy E_{vis}, must be greater than 340 GeV

(600 GeV) for 500 GeV (800 GeV) center-of-mass energies.

2. Each event is forced into four jets using the DURHAM algorithm [18] and the number of tracks per jet is required to be greater than three.

3. To separate centrally produced H and A bosons from the WW and ZZ events, peaking in forward/backward direction, we apply a cut on the polar angle of the thrust vector [12], $|\cos\theta_T| < 0.8$.

4. Further suppression of the WW and ZZ backgrounds is achieved by requiring the event thrust value to be less than 0.88.

5. Two-fermion background is suppressed by applying a cut on the DURHAM jet resolution parameter, for which the event changes from four to three jets, $\log_{10} y_{34} \geq -2.9$.

6. High multiplicity six-jet events originating from $e^+e^- \to t\bar{t}$ production are reduced by requiring the number of energy flow objects[1] in the event to be less than 130. This cut is applied only at $\sqrt{s} = 500$ GeV.

7. The background from $e^+e^- \to t\bar{t}$ events is further reduced by applying a cut on the jet resolution parameter, for which the event changes from six to five jets, $\log_{10} y_{56} \leq$ -3.1 (-2.8) at $\sqrt{s} = 500$ GeV (800 GeV).

8. Finally, we make use of the b-tag information to enhance the purity of the selected event sample. First, the b-tag variable for each jet is calculated as described in Section 1.3. The four b-tag variables are sorted in descending order, $B_1 > B_2 > B_3 > B_4$. Two quantities B_{12}, B_{34} are then defined as

$$B_{12} = \frac{B_1 B_2}{B_1 B_2 + (1 - B_1)(1 - B_2)},$$

$$B_{34} = \frac{B_3 B_4}{B_3 B_4 + (1 - B_3)(1 - B_4)}.$$

The value of B_{12} must be greater than 0.75 (0.6) at $\sqrt{s} = 500$ GeV (800 GeV). The value of B_{34} must be greater than 0.05 independent of the center-of-mass energy.

Events accepted in the final sample are subjected to a 4C kinematic fit. For each of the three possible di-jet pairings, the di-jet mass sum and the di-jet mass difference are reconstructed. In the final step of the analysis, the spectra of the di-jet mass sum and difference, obtained in the HA $\to b\bar{b}b\bar{b}$ channel are used to determine Higgs boson properties.

First, the analysis is performed assuming that the natural widths of Higgs bosons are small relative to detector resolution. As an example Fig. 2 shows the distributions of the di-jet mass sum and di-jet mass difference obtained after selection cuts and kinematic fit in the HA $\to b\bar{b}b\bar{b}$ channel for the Higgs boson mass hypothesis of $(m_H, m_A) =$ (300,250) GeV at $\sqrt{s} = 800$ GeV. Three entries per event contribute to these distributions, corresponding to three possible pairings of jets in the four-jet events. Two entries form a so-called combinatorial background. Fig. 3 demonstrates the final di-jet mass sum and di-jet mass difference after the cut on the di-jet mass difference sum, respectively, as indicated by arrows in Fig. 2. Higgs boson masses can be measured with an accuracy ranging from 0.1 to 1 GeV for Higgs pair production above and close to the kinematic threshold. The topological cross section $\sigma(e^+e^- \to HA \to b\bar{b}b\bar{b})$ can be measured with a relative precision varying between 1.5 and 6.6%.

A large part of the SUSY parameter space leads to degenerate H and A Higgs boson masses. For this case the discovery significance as a function of m_H ($=m_A$) (Fig. 4) is calculated for the HA $\to b\bar{b}b\bar{b}$ channel at $\sqrt{s} = 800$ GeV. Approaching the kinematic limit, the significance drops below 5σ between 380 to 390 GeV. The whole range of the significances for the Higgs masses is from 28.2 to 3.4. Fig. 4 also shows η^2 as a function of m_H ($=m_A$), where η^2 is the assumed $e^+e^- \to$ HA cross section relative to that for $\sin^2(\beta - \alpha) = 1$.

[1]The event reconstruction is done in terms of energy flow objects. First, tracks are measured with tracking system and associated to calorimetric clusters to define charged energy flow objects (electrons, muons and charged hadrons). Calorimetric clusters with no associated tracks are regarded as neutral energy flow objects (photons and neutral hadrons).

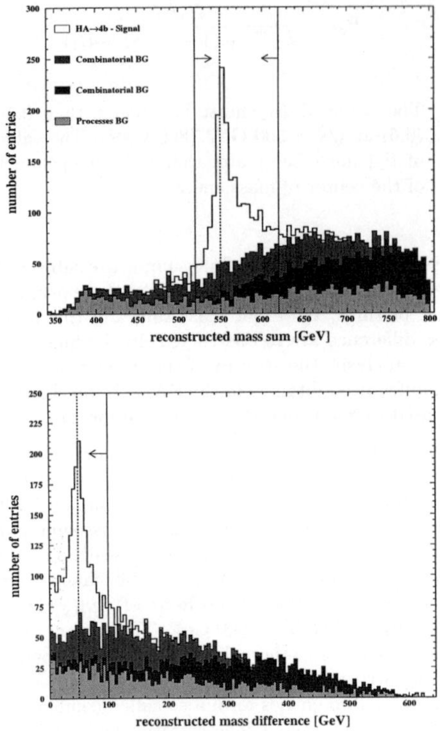

Fig. 2. Distributions of the di-jet mass sum (upper figure) and di-jet mass difference (lower figure) in the HA → $b\bar{b}b\bar{b}$ channel for Higgs boson mass hypothesis $(m_{\rm H},m_{\rm A})$ = (300,250) GeV at $\sqrt{s}$ = 800 GeV after selection cuts and kinematic fit.

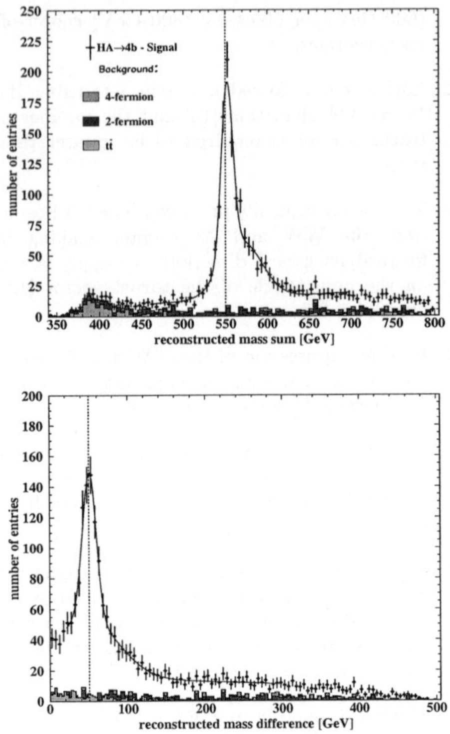

Fig. 3. Upper figure: distribution of the di-jet mass sum after selection cuts, kinematic fit and cut on di-jet mass difference. Lower figure: distribution of the di-jet mass difference after selection cuts, kinematic fit and cut on di-jet mass sum.

1.5 SPS 1a

The present analysis was applied to one of the so-called benchmark points SPS 1a for SUSY searches [19]. SPS 1 is a typical mSUGRA scenario, which consists of a point with an intermediate value of $\tan\beta$ and a model line attached to it (SPS 1a) and of a "typical" mSUGRA point with relatively high $\tan\beta$ (SPS 1b). The parameters for the SPS 1a point are m_0 = 100 GeV, $m_{1/2}$ = 250 GeV, A_0 = -100 GeV, $\tan\beta = 10$, $\mu > 0$. For this point the Higgs masses are m_{h^0} = 113.7 GeV, m_{A^0} = 394.65 GeV, m_{H^0} = 394.9 GeV, $m_{H^\pm}$ = 403.6 GeV according to the Hdecay and Feynhiggsfast programs [20, 21].

The analysis is carried out for the center-of-mass energy of $\sqrt{s}$ = 1 TeV, at which the cross section for the process $e^+e^- \to$ HA is 2.5 fb. The luminosity assumed is 1000 fb^{-1}. The branching ratio for the H (A) Higgs boson to $b\bar{b}$ is 0.64 (0.40), $\Gamma_{\rm tot}$ =

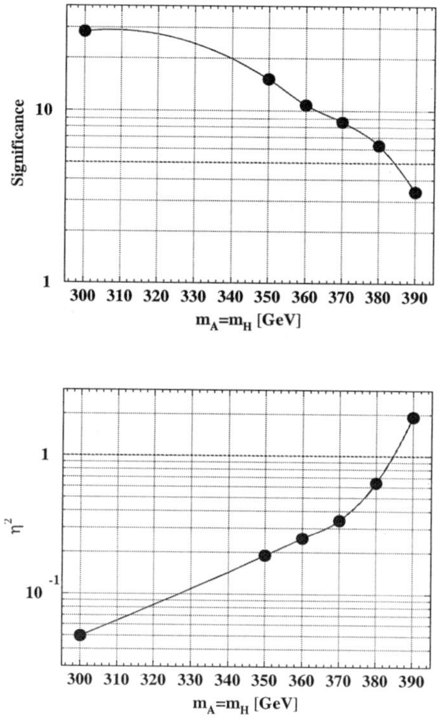

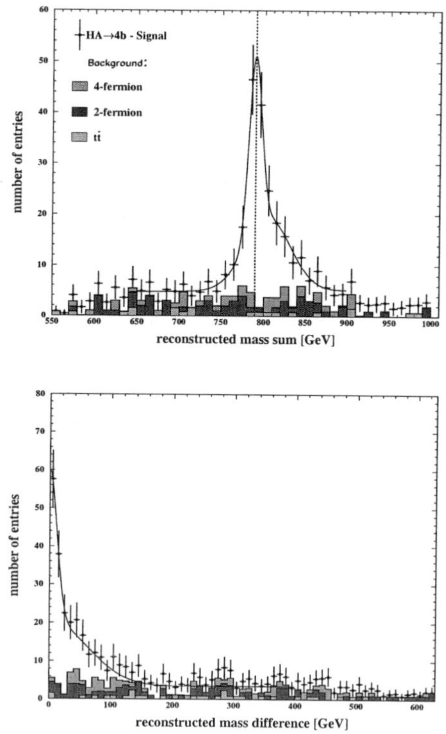

Fig. 4. Discovery significance as a function of the Higgs boson mass (assuming $m_H = m_A$) in the HA $\to$ b$\bar{\text{b}}$b$\bar{\text{b}}$ channel at $\sqrt{s} = 800$ GeV (upper figure). η^2 as a function of the Higgs boson mass for the 5σ discovery limit (lower figure).

Fig. 5. Upper figure: distribution of the di-jet mass sum after selection cuts, kinematic fit and cut on di-jet mass difference. Lower figure: distribution of the di-jet mass difference after selection cuts, kinematic fit and cut on di-jet mass sum. Both distributions are in the HA $\to$ b$\bar{\text{b}}$b$\bar{\text{b}}$ channel for the SPS 1a benchmark point with Higgs boson mass hypothesis $(m_{\text{H}}, m_{\text{A}}) = (394.90, 394.65)$ GeV at $\sqrt{s} = 1$ TeV.

0.785 GeV (1.251 GeV).

Figure 5 presents results for the mass sum and the mass difference after selection cuts, kinematic fit and final cuts for the mass difference and the mass sum, respectively. The masses can be measured with precision of 1.3 GeV. The signal efficiency is 29% after selection cuts and 24% after cuts on di-jet mass sum and difference. The cross section can be measured with the relative uncertainty of 9%.

1.6 Measurement of parameters of CP-violating MSSM

CP violation in the MSSM Higgs sector is predominantly mediated by the CP-odd phase of the soft SUSY-breaking trilinear coupling of the Higgs boson to the top and bottom squarks $\arg(A_{t,b})$ at the

one-loop level, with $\arg(m_{\tilde{g}})$ entering at the two-loop level. The CP-violating benchmark scenario is defined as follows [22]:
$\widetilde{M}_Q = \widetilde{M}_t = \widetilde{M}_b = M_{SUSY}$, $\mu = 4M_{SUSY}$,
$|A_t| = |A_b| = 2M_{SUSY}$, $\arg(A_t) = 90°$,
$|m_{\tilde{g}}| = 1\text{TeV}$, $\arg(m_{\tilde{g}}) = 90°$.
Assuming the realization of the CP-violating scenario of MSSM, we maintain in the present analysis that $\tan\beta$ and μ could be measured in the chargino sector while m_{H+} could be measured directly. The remaining free parameters are $\mathrm{Im}(A_{t,b})$ and $\mathrm{Re}(A_{t,b})$. We choose $\tan\beta = 3$ and $m_{H+} =$ 200 GeV, $\mu = 2000$. For this parameter set it is possible to generate model points in the plane $(\mathrm{Im}(A_{t,b}), \mathrm{Re}(A_{t,b}))$ and presuppose that any point chosen is in fact realized in nature. In this case the chosen point has the parameters presented in Table 3. The analysis was done for the process

Parameter	Value
m_{H_1}	95 GeV
m_{H_2}	180 GeV
m_{H_3}	200 GeV
$\tan\beta$	3
m_{H+}	200 GeV
N_{events}	360

Table 3. Parameter values for the chosen point $(\mathrm{Im}(A_{t,b}), \mathrm{Re}(A_{t,b}))$.

$e^+e^- \to H_1H_2 \to b\bar{b}b\bar{b}$, with $\sqrt{s} = 500$ GeV, integrated luminosity of 500 fb^{-1} and for the condition that the number of events exceeds 50. The 1σ deviations for the values $\sigma \times \mathrm{BR}$ and m_{H_1} in $\mathrm{Im}(A_{t,b}), \mathrm{Re}(A_{t,b})$ parameter space are presented in Fig. 6. The precision of TESLA measurements for m_{H_1} and m_{H_2} is assumed to be 1 GeV, for $\sigma \times \mathrm{BR}$ it is 10% ([6] or see 1.4). The 1σ region for the combination of $\sigma \times \mathrm{BR}$ and m_{H_1} and m_{H_2} is presented in Fig. 7. As one can see, it is possible to measure $\arg(A_{t,b})$ for this case. The result of the measurement is $\mathrm{Im}(A_{t,b}) = 700 \pm 50$ and $\mathrm{Re}(A_{t,b}) = 800 \pm 300$. The measurement of m_{H_2} in addition to m_{H_1} does not give any new information about the region of the interest.

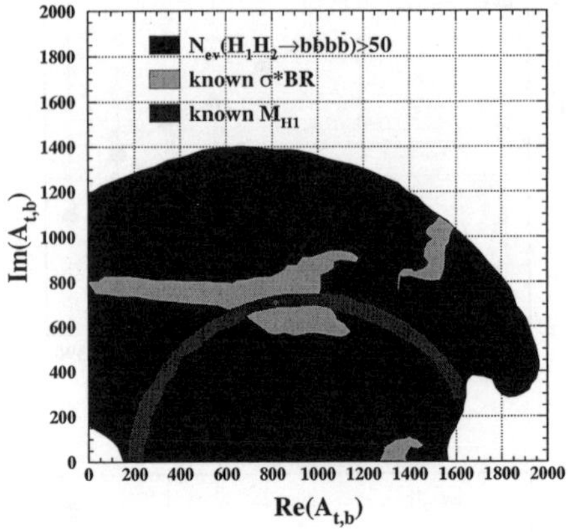

Fig. 6. 1σ deviation for $\sigma \times \mathrm{BR}$ for the process $e^+e^- \to H_1H_2 \to b\bar{b}b\bar{b}$ and m_{H_1} in the case of two variables $\mathrm{Im}(A_{t,b})$ and $\mathrm{Re}(A_{t,b})$.

2 Heavy Quark Production at HERA using the H1 Detector

One of the most important processes to study in quantum chromodynamics (QCD) is heavy quark production. Measurements of the open charm (c) and beauty (b) cross sections in low Q^2 Deep Inelastic Scattering (DIS) at HERA have mainly been done using exclusive methods that are statistically limited [23]. The present analysis is of inclusive c and b cross sections in e^+p scattering at HERA in the range $Q^2 > 12$ GeV^2. Events containing heavy quarks can be distinguished from light quark events by the long lifetimes of c and b hadrons, which lead to displacements of tracks from the primary vertex. The distance of a track to the primary vertex is reconstructed using precise spatial information from the H1 vertex detector. A similar analysis has already been performed for high Q^2 DIS events [24].

At values of $Q^2 \sim M^2$ the production of heavy quarks is described by the 'massive' scheme [25], in which the heavy flavour partons are treated as massive quarks. The heavy quarks are produced perturbatively with their mass providing the hard

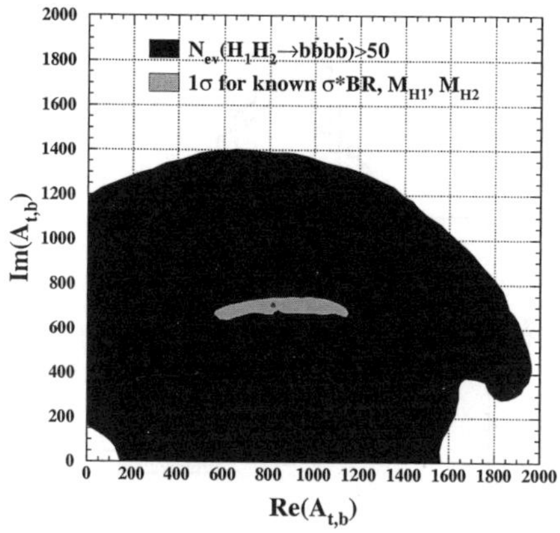

Fig. 7. 1σ region for the combination of $\sigma \times$ BR for the process $e^+e^- \to H_1H_2 \to b\bar{b}b\bar{b}$, m_{H_1} and m_{H_2} in the case of two variables $Im(A_{t,b})$ and $Re(A_{t,b})$.

erated for the c and b and inclusive processes are passed through a detailed simulation of the detector response based on the GEANT3 program [28], and through the same reconstruction software as is used for the data.

Charged particles are measured in the central tracking detector (CTD). The CTD tracks are linked to hits in the vertex detector (central silicon tracker CST) to provide precise spatial track reconstruction. In this paper, the CST-improved CTD tracks are called 'CST tracks'. The different quark flavours are distinguished on the basis of the different lifetimes of the produced hadrons. Due to the relatively low cross sections and limited CST track reconstruction efficiency, the decay length of the heavy hadrons is not reconstructed directly. Instead, the transverse distance of closest approach (DCA) of the track to the primary vertex point is used (Fig. 9). Tracks from the decays of long lived particles will mainly have a positive true DCA, but those produced at the primary vertex will have zero true DCA. Reconstructed tracks with negative DCA values result mainly from detector resolution.

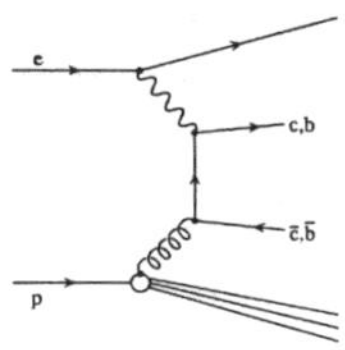

Fig. 8. The process of boson gluon fusion.

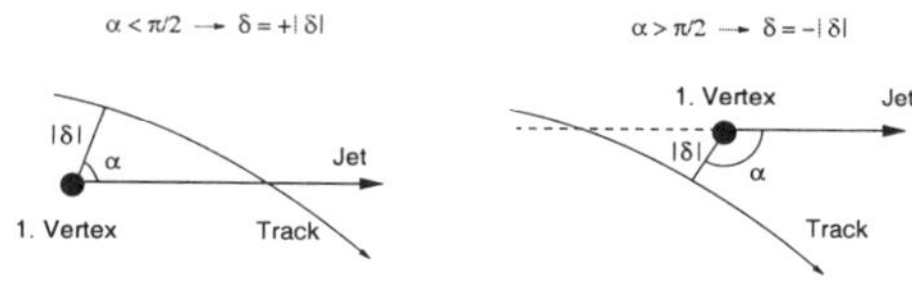

Fig. 9. The DCA is defined as positive if the angle between the jet axis and the line joining the vertex to the point of DCA is less than 90°; otherwise, it is defined as negative.

scale. The dominant leading order (LO) process is boson gluon fusion (BGF) (Fig. 8). The scheme is often referred to as the fixed flavour number scheme (FFNS).

The analysis is based on the low Q^2 data of neutral current (NC) events, collected in the years 1999-2000, when HERA was operated in unpolarised e^+p mode, with an ep centre of mass energy $\sqrt{s} = 318$ GeV. The Monte Carlo program RAPGAP [26] is used to generate low Q^2 NC DIS events for the processes $ep \to eb\bar{b}X$, $ep \to ec\bar{c}X$, and an inclusive sample is generated using the DJANGO [27] Monte Carlo program. The samples of events gen-

The DCA distribution of CST tracks associated to the jet axis is presented in Fig. 10. The data appear asymmetric, with positive values exceeding negative values, thus indicating the presence of long lived particles. The data is reasonably well described by the simulation. The light-quark component of the simulation is almost symmetric at low |DCA|. The asymmetry at DCA$\geq$0.1 cm is mostly attributable to long-lived strange particles, such as K_s. The significance, defined as the ratio of the DCA to its error, is used for better separation of the quark

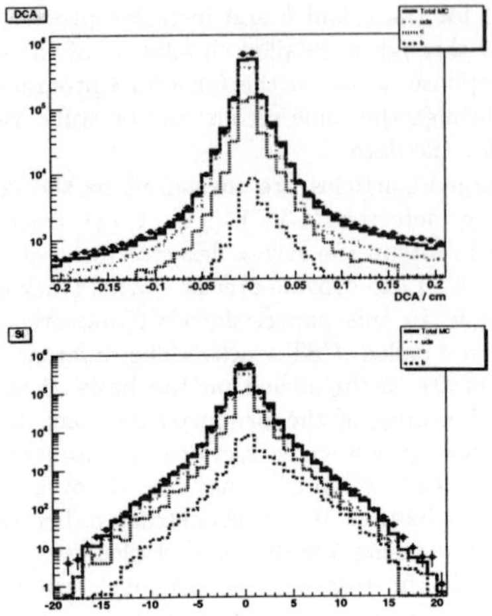

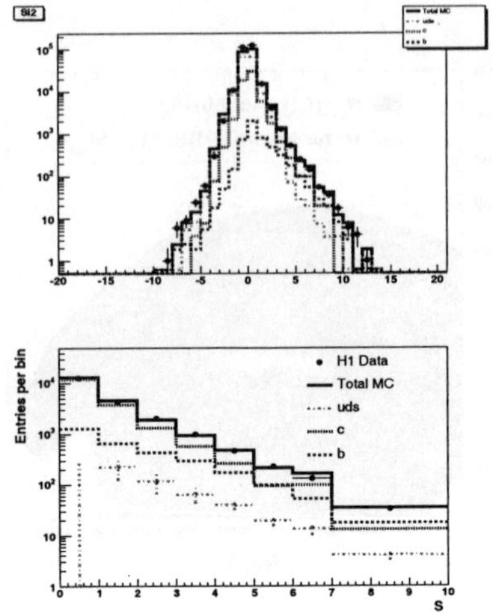

Fig. 10. The distance of closest approach of a track to the vertex (DCA) in the $x-y$ plane (upper plot) and the significance (lower plot) for all CST tracks. Included in the figure is the RAPGAP Monte Carlo [26] after adjusting the contributions from the various quark flavours to the data.

Fig. 11. The significance S_2 distribution per event (upper plot) of the track with the second highest absolute significance for events with ≥ 2 reconstructed CST tracks matched to the jet. The distribution S_2 (lower plot) after subtracting the bins with equal magnitude but negative sign from the positive ones.

flavours (Fig. 10).

Furthermore, the second significance distribution S_2 (Fig. 11) is defined for events with two or more tracks associated with the jet, and represents the significance of the track with the second highest absolute significance. The track with the second highest significance is chosen because heavy quarks with more than two tracks are usually produced with high significance, whereas it is highly improbable for light quarks to produce two tracks with large significance due to resolution effects. Therefore S_2 gives a better separation power of light from heavy quarks.

For the reduction of the uncertainty due to DCA resolution and light quark normalisation, the negative bins in the S_2 distributions are subtracted from the positive ones. The subtracted S_2 is shown in Fig. 11. As one can see, the resulting distribution is dominated by c quark events, with an increasing b fraction with increasing significance. The contribution of the light quarks is a small fraction for all values of significance.

In the final step, the fractions of c, b and light quarks of the data will be extracted for several $x-Q^2$ intervals by a simultaneous least squares fit to the subtracted S_2 distributions and the total number of inclusive events before track selection, using the Monte Carlo samples as templates.

References

1. P.W. Higgs, Phys. Lett. **12** (1964) 132, Phys. Rev. Lett. **13** (1964) 508 and Phys. Rev. **145** (1966) 1156;
   F. Englert and R. Brout, Phys. Rev. Lett. **13** (1964) 321;
   G.S. Guralnik, C.R. Hagen and T.W.B. Kibble, Phys. Rev. Lett. **13** (1964) 585.
2. A. Pilaftsis and C.E.M. Wagner, Higgs bosons in the minimal supersymmetric standard model with

explicit CP violation, Nucl. Phys. **B553** (1999) 3, hep-ph/9902371.

3. M. Carena, J.R. Ellis, A. Pilaftsis and C.E.M. Wagner, Renormalization-group-improved effective potential for the MSSM Higgs sector with explicit CP-violation, Nucl. Phys. **B586** (2000) 92, hep-ph/0003180.
4. M. Carena, J.M. Moreno, M. Quiros, M. Seco and C.E.M. Wagner, Supersymmetric CP-violating currents and electroweak baryogenesis, Nucl. Phys. **B599** (2001) 158, hep-ph/0011055.
5. K. Desch, T. Klimkovich, T. Kuhl, A. Raspereza, Study of Higgs Boson Pair Production at Linear Collider, hep-ph/0406229, LC-PHSM-2004-006, 2004.
6. T. Behnke, S. Bertolucci, R.-D. Heuer, R. Settles, TESLA : The Superconducting Electron-Positron Linear Collider with an Integrated X-Ray Laser Laboratory. Technical Design Report, Part IV : A Detector for TESLA, DESY 2001-011 and ECFA 2001-209 (2001).
7. A. Andreazza, C. Troncon, Study of HA Production in e^+e^- Collisions at $\sqrt{s} = 800$ GeV, DESY-123-E, p. 417.
8. S. Kiyoura, S. Kanemura, K. Odagiri, Y. Okada, E. Senaha, S. Yamashita and Y. Yasui, arXiv:hep-ph/0301172.
9. T. Abe et al., Linear Collider Physics Resource Book for Snowmass 2001, BNL-52627, CLNS 01/1729, FERMILAB-Pub-01/058-E, LBNL-47813, SLAC-R-570, UCRL-ID-143810-DR, LC-REV-2001-074-US.
10. SIMDET V4.0, M. Pohl and H.J. Schreiber, SIMDET : a Parametric Monte Carlo for a TESLA Detector, DESY-02-061 LC-DET-2002-005 (2002).
11. CIRCE V6, T. Ohl, Comp. Phys. Comm. **94** (1996) 53.
12. T. Sjöstrand, L. Lonnblad, S. Mrenna, PYTHIA 6.2: Physics and Manual, hep-ph/0108264.
13. J.F. Gunion, H.E. Haber, G. Kane, S. Dawson, The Higgs Hunter's Guide, Addison Wesley, 1990.
14. A. Djouadi, HPROD: A Program for SM and MSSM Higgs Boson Production in e^+e^- Collisions, http://w3.lpm.univ-montp2.fr/~djouadi/GDR/programs/hprod.html
15. D. Jackson, Nucl. Instr. and Meth. **A388** (1997) 247.
16. S. Xella-Hansen, D.J. Jackson, R. Hawkings, C.J.S. Damerell, Flavour Tagging Studies for the TESLA Linear Collider, LC-PHSM-2001-024.
17. N. Kjaer and R. Moller, Reconstruction of invariant masses in multi-jet events, DELPHI Note 91-17 PHYS 88.
18. S. Catani, Yu.L. Dokshitzer, M. Olsson, G. Turnock and B.R. Webber, Phys. Lett. B269 (1991) 432.
19. B.C. Allanach, M. Battaglia, G.A. Blair, M. Carena et al., hep-ph/0202233.
20. A. Djouadi, J. Kalinowski, M. Spira, Hdecay: a program for Higgs boson decays in the Standard Model and its supersymmetric extension, Comput. Phys. Commun. 108:56-74,1998; hep-ph/9704448.
21. S. Heinemeyer, W. Hollik, G. Weiglein, Feynhiggs and Feynhiggsfast: programs for higher order calculations in the neutral CP even Higgs boson sector of the MSSM, LC-TH-2001-065, KA-TP-5-2001, 2001. In 2nd ECFA/DESY Study 1998-2001 2393-2421.
22. M. Carena, J. Ellis, A. Pilaftsis and C.E.M. Wagner, CP-Violating MSSM Higgs Bosons in the Light of LEP 2, hep-ph/0009212, 2000.
23. F. Sefkow, Heavy Quark Production in Deep-Inelastic Scattering, hep-ex/0110036, 2001.
24. Measurement of $F_2^{c\bar{c}}$ and $F_2^{b\bar{b}}$ at high Q^2 using the H1 Vertex Detector at HERA, H1 Preliminary.
25. E. Laenen, S. Riemersma, J. Smith and W.L. van Neerven, Nucl. Phys. **B392** (1993) 162; E. Laenen, S. Riemersma, J. Smith and W.L. van Neerven, Nucl. Phys. **B392** (1993) 229; S. Riemersma, J. Smith and W.L. van Neerven, Phys. Lett. **B347** (1995) 143, hep-ph/9411431.
26. H. Jung, Comput. Phys. Commun., **86** (1995) 147; (see also http://www.desy.de/ jung/rapgap/)
27. H. Charchula, G.A. Schuler and H. Spiesberger, Comput. Phys. Commun., **81** (1994) 381; (see also http://www.desy.de/ hspiesb/django6.html)
28. R. Brun, R. Hagelberg, M. Hansroul and J.C. Lassalle, CERN-DD-78-2-REV.

$\Sigma(1385)$ Results with STAR

Sevil Salur for STAR Collaboration

Physics Department, Yale University,
Sloane Physics Laboratory, P.O. Box 208120,
New Haven, CT, 06520-8120, USA

Abstract. This paper introduces the Solenoidal Tracker experiment at the Relativistic Heavy Ion Collider. The corrected p_T spectra and the yields of the $\Sigma(1385)^{\pm}$ and their antiparticles in the most central Au+Au and elementary p+p collisions are presented. Comparison of the $\Sigma(1385)$ $\langle p_T \rangle$ measurement with the other particles corroborates the fact that the $\langle p_T \rangle$ behavior for hyperons (mass> 1.2 GeV) is similar in p+p and Au+Au collisions. The $\Sigma(1385)/\Lambda$ ratio, along with other resonance-to-stable-particle ratios, suggests that a regeneration mechanism recovers the signal loss due to re-scattering in the final state in Au+Au collisions.

PACS: 01.30.Bb

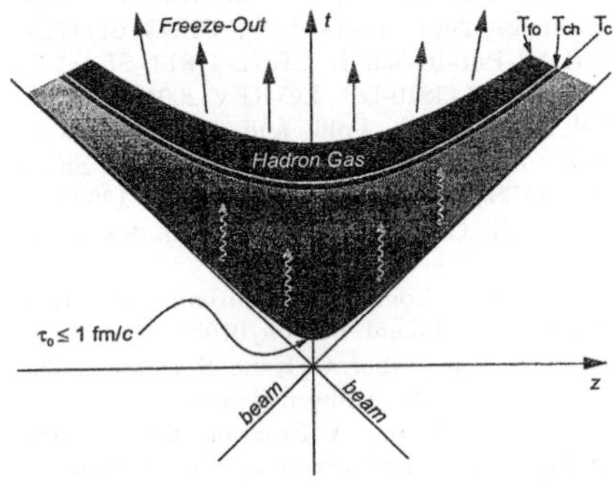

Fig. 1. A lightcone diagram of a collision when QGP is formed. T_{fo} is the abbreviation for thermal freeze-out, while T_{ch} is for chemical freeze-out. The hadronization starts at T_c, the critical temperature.

1 Introduction

The Relativistic Heavy Ion Collider (RHIC) [1] is located at Brookhaven National Laboratory (BNL) on Long Island, NY, USA. The circumference of the two independent accelerator rings is 3.8 km. The first collisions took place in the year 2000 and since then collisions of Au+Au at $\sqrt{s_{NN}}$=19, 62, 130, 200 GeV, and p+p and d+Au at $\sqrt{s_{NN}}$=200 GeV have been performed. There are 6 beam interaction points at RHIC and 4 dedicated heavy-ion experiments. BRAHMS [2] and PHOBOS [3] are the two small experiments and the two large ones are PHENIX [4] and STAR [5].

The Solenoidal Tracker At RHIC (STAR) is a large acceptance detector system that is designed to investigate strongly-interacting matter at high energy densities and search for signatures of Quark-Gluon Plasma (QGP), the hadronic deconfinement phase consisting of "free" partons, and its space-time evolution. The physics program of STAR also includes the study of nucleon spin structure functions with polarized p+p collisions, the study of pomeron and photon interactions from intense electromagnetic fields of the colliding ions at RHIC and the study of initial parton distribution functions of the incident nuclei with p+p and d+Au collisions. The p+p and d+Au collisions are also essential for establishing the reference data for heavy ion collisions.

During the expansion of the hot and dense matter (fireball) that is created in heavy ion collisions, chemical freeze-out is reached when the hadrons stop interacting inelastically (see Fig.1), and elastic interactions continue until thermal freeze-out. Due to the very short lifetime ($\tau < \tau_{fireball} \sim$ 10 fm) of most resonances, a large fraction of their decays oc-

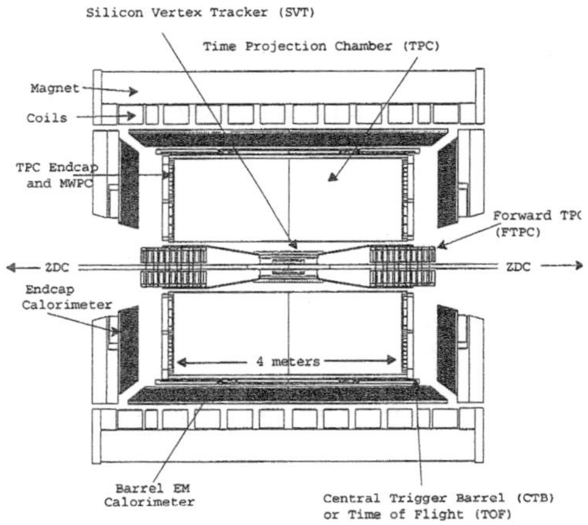

Fig. 2. Side view of the STAR detector configuration. TPC is the main component of the STAR detector used in this analysis.

cur before thermal freeze-out. The decay products interact elastically with the surrounding particles prior to thermal freeze-out, thus resulting in a signal loss for the reconstructed resonances. Secondary interactions (regeneration), however, increase the resonance yield (such as $\Lambda + \pi \to \Sigma(1385)$). The contribution of re-scattering and regeneration to the total observed yields depends on the time span between the chemical and thermal freeze-out, the lifetime of each resonance, and the regeneration and rescattering probabilities [6, 7]. Thus the study of resonances provides an additional tool for determining the hadronic expansion time between chemical and thermal freeze-out through the comparison of resonance-to-stable-particle ratios.

2 STAR Overview and Particle Identification

A schematic diagram of the components of the STAR experiment is presented in Fig. 2 [5]. The STAR detector's main design features include a Time Projection Chamber (TPC) surrounded by a solenoidal magnet, which has a field of 0.5 Tesla, with an acceptance of $|\eta| < 1.5$ for charged particle tracks [8]. The Silicon Vertex Tracker surrounding the beam pipe improves the resolution of the interaction vertex [9]. In addition, Forward-TPCs increase the rapidity coverage of charged particles to $2.5 < |\eta| < 4$, and electromagnetic calorimeters provide a full azimuthal coverage for high p_T photons, electrons and electromagnetically decaying mesons [10, 11, 12]. Glauber model calculations applied to measurements from the zero degree calorimeters (ZDC) in coincidence with the central trigger barrel (CTB) are used to select the most central 5% of Au+Au events. Beam-beam counters and the CTB are used to select minimum bias p+p events.

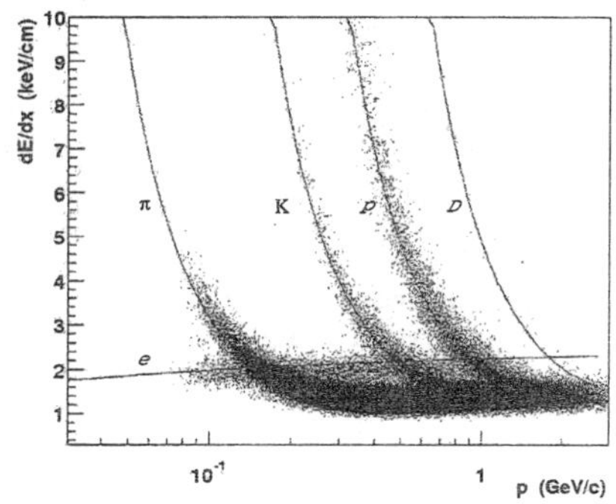

Fig. 3. The energy loss per unit length (de/dx) distribution in the STAR TPC as a function of the momentum of the charged particles.

Charged particles such as π, K, p and e are identified by the energy lost per unit length (dE/dx) and the magnetic field information within the TPC. Fig. 3 presents the dE/dx distribution as a function of the momentum of the charged particles in the STAR TPC detector. These charged particles can be clearly identified with the Bethe-Bloch parametrization (solid lines in Fig. 3) up to 800 MeV.

Due to the large acceptance of STAR's detector components, the broad physics program includes precision measurements of a wide variety of strange and charm particles. Long-lived ($c\tau$ > few cm) neutral strange particles, such as Λ and K_s^0, can be reconstructed in the central TPC from their charged decay products. All of the pairs of oppositely charged tracks that originate in the same secondary vertex are selected to represent the decay topology of the neutral particles. Additional geometrical cuts are used to improve the signal-to-background ratio. Signals of K_s^0 in Fig. 4, Λ and $\overline{\Lambda}$ in Fig. 5 are presented for p+p collisions at $\sqrt{s_{NN}} = 200$ GeV.

The direct measurement of resonances is not possible due to their short lifetimes (e.g. $c\tau_{\Sigma(1385)} = 5$ fm) [13, 14]. The $\Sigma(1385)$ resonances are identified instead through the invariant mass analysis of π and Λ decay particle candidates. A mixed event technique is used to determine the background for uncorrelated pair combinations. The $\Sigma(1385)$ signal is obtained by subtracting this normalized mixed-event background from the invariant mass distribution [15]. The signal before and after the background subtraction is presented in Fig. 6 for p+p collisions at $\sqrt{s_{NN}} = 200$ GeV. Since $\Xi^- \to \Lambda + \pi^-$ shares the same decay channel as $\Sigma^-(1385) \to \Lambda + \pi^-$, we observe both signals in the invariant mass spectrum. The background subtracted signal is fitted with a Gaussion for Ξ and a Breit-Wigner for $\Sigma(1385)$.

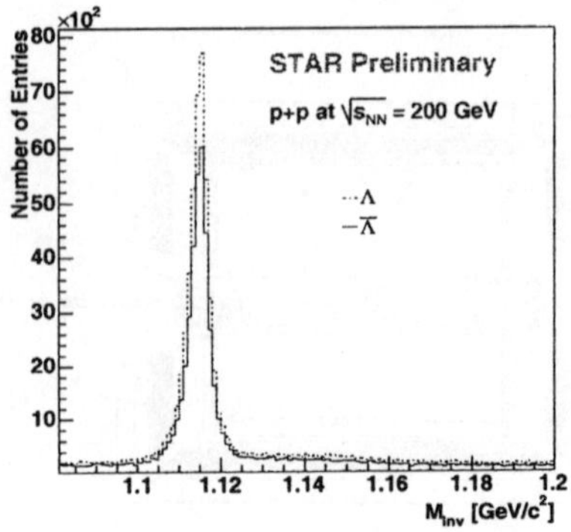

Fig. 5. Invariant mass spectra for $\Lambda \to p + \pi^-$ and $\overline{\Lambda} \to \overline{p} + \pi^+$ with decay topology information in p+p collisions at $\sqrt{s_{NN}} = 200$ GeV.

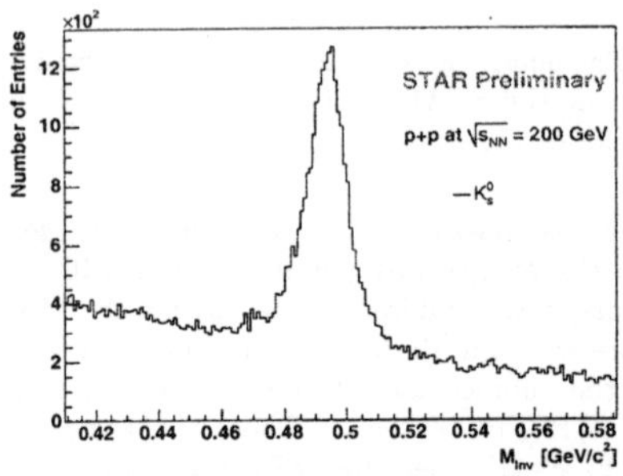

Fig. 4. Invariant mass spectrum for $K_s^0 \to \pi^- + \pi^+$ with decay topology information in p+p collisions at $\sqrt{s_{NN}} = 200$ GeV.

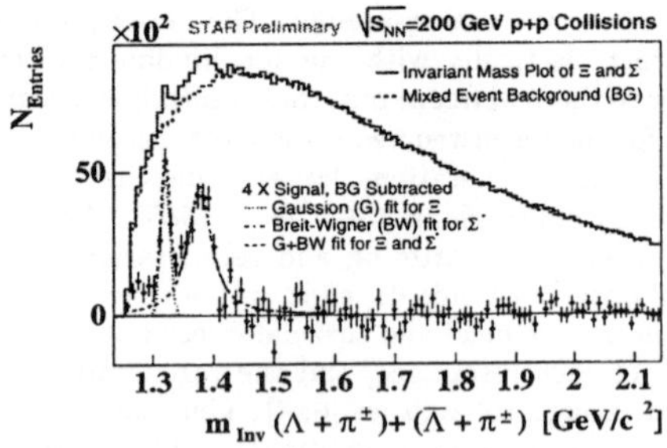

Fig. 6. $\Sigma(1385)$ invariant mass spectra before and after normalized mixed event background subtraction in p+p at $\sqrt{s_{NN}} = 200$ GeV.

3 $\Sigma(1385)$ Results

The transverse mass $(m_T)^1$ spectra of $\Sigma^{\pm}(1385)$ and their antiparticles in p+p (open circles) and

[1]The transverse mass is defined as $m_T = \sqrt{p_T^2 + m^2}$

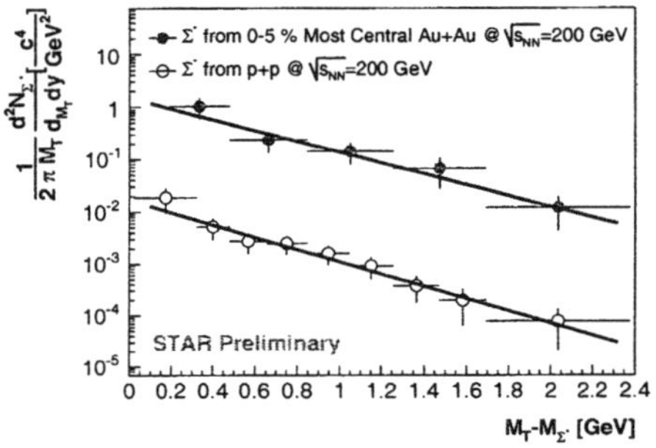

Fig. 7. The transverse mass spectrum for $\Sigma(1385)$ drawn as open circles for p+p and as closed circles for the 5% most central Au+Au collisions at $\sqrt{s_{NN}} = 200$ GeV. The solid lines represent exponential fits to the data.

$\Sigma^{\pm}(1385)$	p+p	Au+Au
T [MeV]	358 ± 47	420 ± 84
$\langle p_T \rangle$ [GeV/c]	1.08 ± 0.15	1.20 ± 0.24
Yields (dN/dy)	$(4.66 \pm 0.98)10^{-3}$	4.72 ± 1.38

Table 1. Temperature T, $\langle p_T \rangle$ and yield obtained from the exponential fits of the p_T spectra in Fig. 7 for elementary p+p and 5% most central collisions. The statistical uncertainties are given and the systematic error, mostly due to normalization and the shape of the background, $\sim 15\%$, must be included in the given values.

Au+Au (closed circles) collisions are shown in Fig. 7. The acceptance and efficiency correction for the spectra is accomplished by embedding Monte Carlo simulated resonances into real p+p and Au+Au events. The solid lines in Fig. 7 represent exponential fits to the data with the function directly proportional to the yield (dN/dy) and inversely proportional to the slope parameter $T^2 + m_0 T$ within the framework of a thermal model. The data coverage is 91% and 85% of the fully integrated yield in p+p and Au+Au collisions, respectively. The $\langle p_T \rangle$ is derived from the full range integration of the corresponding exponential fit. For p+p and Au+Au collisions at $\sqrt{s_{NN}} = 200$ GeV, Table 1 presents inverse slope parameters of the exponential fit functions, the $\langle p_T \rangle$s and the yields (dN/dy) of the summed signal of $\Sigma^{\pm}(1385)$ and their antiparticles.

The $\langle p_T \rangle$ as a function of particle mass for p+p and Au+Au collisions at $\sqrt{s_{NN}} = 200$ GeV is presented in Fig. 8. The behavior of $\langle p_T \rangle$ vs. mass for the various particles in p+p and Au+Au collisions is compared to two parameterizations. The triangles represent the short-lived resonances and the circles indicate long-lived stable particles. The black curve is an empirical fit to the ISR π, K and p data in p+p collisions and the band is a blast wave fit using π, K and p in STAR Au+Au collisions [16, 17]. The empirical parametrization for the ISR data at $\sqrt{s} = 25$ GeV in p+p collisions can describe the behavior of the lower mass particles, such as π, K and p, despite the fact that our collision energy is one order of magnitude higher. However, this empirical parametrization does not represent the behavior of the higher mass particles in p+p collisions. Similarly, the blast wave parametrization, which can describe the lower mass particles ($\sim 98\%$ of all the particles observed) in Au+Au collisions, fails to explain the behavior of higher mass particles.

The heavy particles in p+p and Au+Au collision show a similar behavior of $\langle p_T \rangle$. It is expected that resonances with higher transverse momentum are more likely to be reconstructed because they have longer relative lifetimes due to Lorentz contraction, which means that they are more likely to decay outside of the medium. As a consequence, their daughter particles should interact less with the medium in Au+Au collisions. Any loss at low p_T would increase the T parameter of the p_T spectra for the central Au+Au collisions with respect to p+p collisions. However we do not see any significant increase in the T parameter for $\Sigma(1385)$ from p+p to the most central Au+Au collisions within the statistical and systematic errors. This might be due to our p+p reference data. The higher mass particles might be produced in more violent (mini-jet) p+p collisions than the lower mass particles, so that the $\langle p_T \rangle$ for heavy particles in p+p collisions would be higher. In Au+Au collisions there is evidence that heavier particles flow radially at a lower velocity

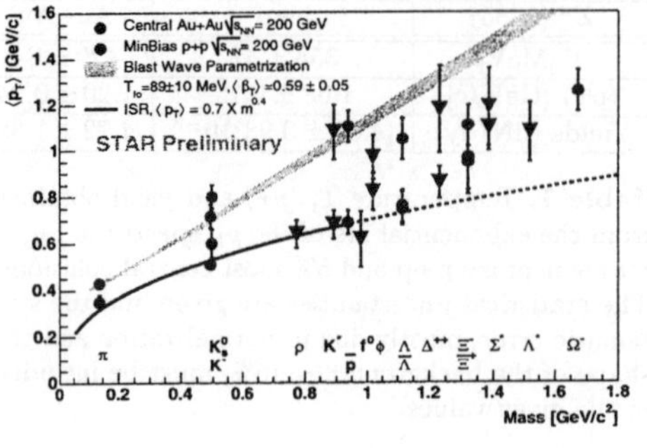

Fig. 8. The $\langle p_T \rangle$ vs particle mass measured in p+p and Au+Au collisions at $\sqrt{s_{NN}} = 200$ GeV. The black curve represents the ISR parametrization from π, K and p for $\sqrt{s_{NN}} = 25$ GeV p+p collisions. The yellow band is the blast wave fit using π, K and p for Au+Au collisions.

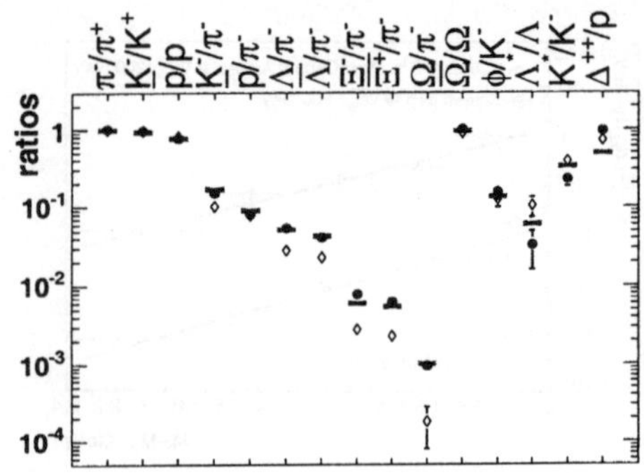

Fig. 9. Particle ratios in $\sqrt{s_{NN}} = 200$ GeV p+p (open diamonds) and Au+Au (solid circles) collisions as compared to thermal model (solid lines) predictions with $T_{ch} = 160$ MeV, $\mu_B = 2.65$, $\mu_s = 1.31$, $\gamma_s = 0.99$ [25].

than the lighter mass particles (such as π mesons). These two independent effects in p+p and Au+Au collisions might cause the apparent merging of the $\langle p_T \rangle$ which is corroborated by the $\Sigma(1385)$ measurement.

The resonance production also presents an important test of thermal production and hydrodynamics models. The ratio of resonances to their stable particles are insensitive to fugacities and phase space occupancies [24]. Fig. 9 presents a comparison between the thermal model predictions and the measured particle ratios [25]. Lines represent the thermal model predictions, the open diamonds indicate p+p collisions and closed circles are for Au+Au collisions. The thermal model accurately describes the stable particle ratios in both p+p and Au+Au collisions. However, the deviations observed for the resonance particles are large and must be investigated. Regeneration and rescattering are excluded in thermal models, which might suggest one explanation for why the experimental values differ from the calculations of the statistical model for most of the resonances.

The measurements of the ratios of resonances to stable particles as a function of charged parti-

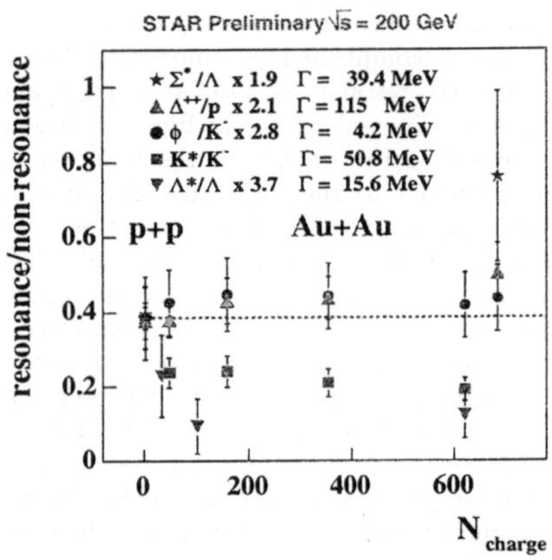

Fig. 10. Resonance to stable particle ratios of ϕ/K^- [19], Δ^{++}/p [20], $\mathrm{K}^*(892)/\mathrm{K}^-$ [21], $\Sigma(1385)/\Lambda$ and $\Lambda(1520)/\Lambda$ [22, 23] for p+p and Au+Au collisions at $\sqrt{s_{\mathrm{NN}}} = 200$ GeV. The ratios are normalized to the ratio of $\mathrm{K}^*(892)/\mathrm{K}^-$ in p+p collisions. Statistical and systematic errors are included.

cle multiplicity for $\Sigma(1385)$ and other resonances are presented in Fig. 10. The observed suppression of $\Lambda(1520)/\Lambda$ and $K*(892)/K$ from peripheral to most central Au+Au collisions compared to p+p collisions is in agreement with a rescattering scenario of the decay particles in the dense medium that causes a signal loss. The $\Sigma(1385)/\Lambda$ ratio does not show this suppression (within errors) for the 5% most central Au+Au collisions in comparison to the p+p collision environment, even though $c\tau_{\Sigma(1385)} < c\tau_{\Lambda(1520)}$, which should cause more rescattering for $\Sigma(1385)$. Therefore a strong regeneration cross section has to be assumed for the $\Lambda + \pi \rightarrow \Sigma(1385)$ in order to recover the signal loss due to rescattering. Resonance-to-stable-particle ratios can also be used to test microscopic (UrQMD) models, which include rescattering and regeneration in their calculations. The UrQMD model prediction in Fig. 11 shows the collision energy dependence of the $\Sigma^{\pm}(1385)/\Lambda$ ratios for Au+Au collisions. The measured $\Sigma^{\pm}(1385)/\Lambda$ ratio is 0.295 ± 0.086 for the 5% most central Au+Au collisions, which is about a factor of 2 below the UrQMD prediction at $\sqrt{s_{NN}} = 200$ GeV. The regeneration cross section assumed in UrQMD appears to be too high and has to be revised in light of these resonance measurements.

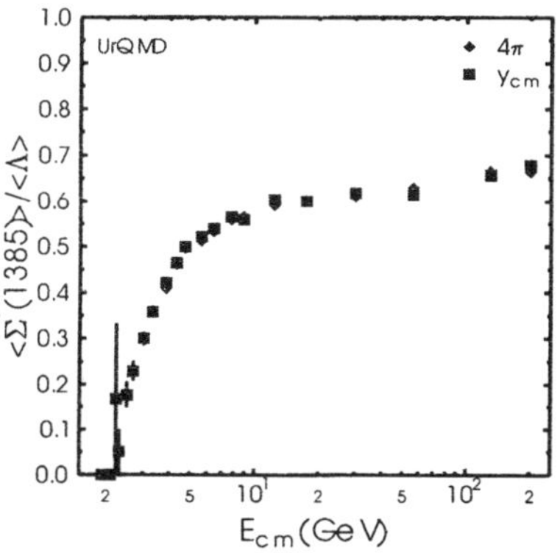

Fig. 11. Collision energy dependence of $\Sigma^{\pm}(1385)/\Lambda$ ratios for Au+Au collisions predicted by the microscopic models (UrQMD). Plot from M. Bleicher [18].

4 Conclusions

$\Sigma^{\pm}(1385)$ resonances and their anti-particles can be reconstructed via event mixing techniques in p+p and Au+Au collisions with the STAR detector. The $\Sigma(1385)$ $\langle p_T \rangle$ in p+p collisions is similar (within the errors) to that measured in Au+Au collisions. This follows the behavior of the other high mass particles ($m > 1.2$ GeV) whose $\langle p_T \rangle$ measurement in p+p approaches the Au+Au value. This behavior suggests a smaller radial flow for heavy particles and/or the preferred production of heavy particles from hard scattering and jet fragmentation in p+p collisions.

No significant suppression is observed in the $\Sigma(1385)/\Lambda$ ratios from p+p to the 5% most central Au+Au collisions. Since the previously measured suppression of K*(892)/K and $\Lambda(1520)/\Lambda$ ratios agrees with a signal loss due to rescattering in the medium, the null suppression of the $\Sigma(1385)$ ratio suggests a significant regeneration mechanism for recovering the signal lost via rescattering [18, 24]. Current statistical and microscopic models need to be modified to account for the resonance measurements coming from heavy ion collisions.

The event mixing technique clearly works well for resonance identification. The same technique is being used in the search for the Θ^+ pentaquark, a five-quark bound system consisting of uudd$\bar{s}$ and other exotic particles [26].

References

1. W. Barletta *et al*, Nuc. Ins. & Meth., **A 499**, (2003).
2. K. H. Adamczyk *et al*, Nuc. Ins. & Meth., **A 499**, 437, (2003).
3. B. B. Back *et al*, Nuc. Ins. & Meth., **A 499**, 603, (2003).
4. K. Adcox *et al*, Nuc. Ins. & Meth., **A 499**, 469, (2003).
5. K. H. Ackermann *et al*, Nuc. Ins. & Meth., **A 499**, 624, (2003).
6. G. Torrieri and J. Rafelski, Phys. Lett. **B509**, 239, (2001).

7. M. Bleicher and J. Aichelin, Phys. Lett. **B530**, 81, (2002).
8. M. Anderson *et al*, Nuc. Ins. & Meth. **A 499**, 659,(2003).
9. R. Bellwied *et al*, Nuc. Ins. & Meth. **A 499**, 640,(2003).
10. K. H. Ackermann *et al*, Nuc. Ins. & Meth. **A 499**, 713, (2003).
11. M. Beddo *et al*, Nuc. Ins. & Meth. **A 499**, 725, (2003).
12. C.E. Allgower *et al*, Nuc. Ins. & Meth. **A 499**, 740, (2003).
13. C. Adler *et al*, STAR Collaboration, Phys. Rev. C **65**, 041901 (2002).
14. C. Adler *et al*, STAR Collaboration, Phys. Rev. C **66**, 061901 (2002).
15. S. Salur, for STAR Collaboration, NATO Science Series II, **166**, 665, (2004) hep-ex/0403016.
16. M. Bourquin, J.-M. Gaillard, Nucl. Phys. **114**, 334 (1976).
17. J. Adams, STAR Collaboration, Phys. Rev. Lett. **92**, 182301, (2004).
18. Marcus Bleicher, Nucl.Phys. A, **715**, 85, (2003).
19. J. Ma, for STAR Collaboration, J. Phys. G, **30**, S543, (2004).
20. Z. Xu, J. Phys. G, **30**, S325, (2004).
21. H. Zhang, for STAR Collaboration, J. Phys. G, **30**, S577, (2004).
22. L. Gaudichet, for STAR Collaboration, J. Phys. G, **30**, S549, (2004).
23. Ch. Markert, for STAR Collaboration, 19th Winter Workshop on Nuclear Dynamics 71 (2003).
24. G. Torrieri and J. Rafelski, Phys. Rev. **C68**, (2003).
25. O. Barannikova, for STAR Collaboration, Private Communication.
26. S. Salur, for STAR Collaboration, nucl-ex/0403009.

Reprinted with permission from the electronic-only supplement "Young Scientists Papers" (DOI: 10.1140/epj/s2004-02053-6) of The European Physical Journal (EPJ) C

Compact stars in the standard model - and beyond

F. Sandin

Department of Physics, Luleå University of Technology, SE-97187 Luleå, Sweden;
e-mail: fredrik.sandin@ltu.se

Abstract. In the context of the standard model of particle physics, there is a definite upper limit to the density of stable compact stars. However, if there is a layer of constituents that is deeper than that of quarks and leptons, stability may be re-established far beyond this limiting density so that a whole new class of compact stars could exist. These objects would cause gravitational lensing of white dwarfs and gamma-ray bursts, that might be observable as a diffraction pattern in the spectrum. Such observations could provide means for obtaining new clues about the fundamental particles and the origin of cold dark matter.

PACS: 12.60.Rc - 04.40.Dg - 95.35.+d

1 Introduction

The different types of compact objects traditionally considered in astrophysics are white dwarfs, neutron stars (including quark and hybrid stars), and black holes. The first two classes are supported by Fermi degeneracy pressure from their constituent particles. For white dwarfs, electrons provide the pressure that counterbalances gravity. In neutron stars, the neutrons play this role. For black holes, the degeneracy pressure is overcome by gravity and the object collapses to a singularity, or at least to the Planck scale ($\rho \sim 10^{93}$ g/cm^3). For a recent review of neutron stars, hybrid stars, and quark stars, see, *e.g.*, [1] and references therein.

The distinct classes of degenerate compact stars originate directly from the properties of gravity, as was made clear by a theorem of Wheeler and collaborators in the mid 1960s [2]. This theorem states that for the solutions to the stellar structure equations, whether Newtonian or relativistic, there is a change in stability of one radial mode of vibration whenever the mass reaches a local maximum or minimum as a function of the central density. The theorem assures us that distinct classes of stars, such as white dwarfs and neutron stars, are separated in central density by a region in which there are no stable configurations.

In the standard model of quarks and leptons (SM), the theory of the strong interaction between quarks and gluons predicts that with increasing energy and density, the coupling between quarks fades away [3, 4] asymptotically. As a consequence of this so-called asymptotic freedom, matter is expected to behave as a gas of free fermions at sufficiently high densities. This puts a definite upper limit on the density of stable compact stars, since the solutions to the stellar equations end up in a never-ending sequence of unstable configurations, with increasing central density. Thus, in light of the standard model, the densest stars likely to exist are neutron stars, quark stars, or the slightly more dense hybrid stars [5, 6, 7]. However, if there is a layer of constituents that is deeper than that of quarks and leptons, the possibility opens up for a new class of compact stars [8].

While being a successful theory quantitatively, the SM consists of a large number of exogenous *ad hoc* rules and parameters, which were introduced solely to fit the experimental data. The SM provides no explanation for the deeper meaning of these rules. At a closer look, however, the SM seems to be full of hints to its deeper background. Considered from a historical point of view, a "simple" and appealing explanation for these rules is *compositeness* [9], *i.e.*, that the quarks, leptons, and gauge bosons are composite particles, built out of more elementary *preons* [10]. Preons provide natural explanations for the particle families of the SM and phenomena such as neutrino oscillations, the mixing of weak gauge bosons, and quarks of different flavour.

Over the last decades, many papers have been

written about preons, but so far no direct evidence has been found for (or against!) their existence. In the late 1970s, a number of consistency conditions were formulated by 't Hooft [11]. In the same work, a vector-like non-Abelian $SU(3)$ gauge group was considered, but no solution to the consistency conditions was found. Later, it was shown that with another choice for the gauge group and the flavour structure of preons, *e.g.*, three different preon flavours, the consistency conditions could be satisfied [12]. For a more detailed discussion of preon models, see [9, 10, 13] and references therein.

Not all clues point to preon models, but the existence of preons is still an open question; as a consequence so is the question of whether or not a new class of compact stars exists. This paper is based on the ideas and results presented in [8]. Assuming that quarks and leptons are composite particles, built out of more elementary preons, I will:

I. Give an estimate for the mass and radius of stars composed of preons.

II. Show that for some particular equations of state, stable solutions to the general relativistic stellar equations do exist, with densities far beyond the maximum density of stars composed of quarks and leptons.

III. Briefly discuss some potential astrophysical consequences and how these objects could be observed. Herein lies the potential importance of this qualitative speculation, since these objects are candidates for cold dark matter and could be found as, *e.g.*, gravitational "femtolenses".

2 Compact stars and the maximum density prophecy

In order to explain why there is a maximum density for stars composed of quarks and leptons, or any other composite particle composed of these two species (*e.g.*, nucleons and ^{56}Fe) some basic knowledge about the theory of compact stars is needed. In the following, I give a short introduction and a summary of the main points.

Due to the high density and large mass of compact stars, a general relativistic treatment of the equilibrium configurations is necessary. This is especially important for the analysis of stability when a star is subject to radial oscillations. Such oscillations are excited, to some extent, and for a star to be stable the amplitude of the oscillations must not grow spontaneously with time. The starting point for a general relativistic consideration of compact stars is the Oppenheimer-Volkoff (OV) equations [14] for hydrostatic, spherically symmetric equilibrium:

$$\frac{dp}{dr} = -\frac{G\left(p+\rho c^2\right)\left(mc^2+4\pi r^3 p\right)}{r\left(rc^4-2Gmc^2\right)}, \quad (1)$$

$$\frac{dm}{dr} = 4\pi r^2 \rho. \quad (2)$$

Here p is the pressure, ρ the density and $m = m(r)$ the mass within the radial coordinate r. The total mass of the star is:

$$M = 4\pi \int_0^R r^2 \rho \, dr, \quad (3)$$

where R is the coordinate radius of the star. Combined with an equation of state (EOS), $p = p(\rho)$, obtained from some microscopic (quantum field) theory, the OV solutions give the possible equilibrium states of spherically symmetric stars.

As an example, I show two sequences of compact star configurations. One is composed of nuclear matter (neutron stars) and the other of a deconfined quark matter core with a nuclear matter crust (hybrid stars), see Fig. 1. These configurations were obtained by solving the OV equations (1)-(2) numerically. The low-density part of the nuclear matter EOS was extracted from [15] and the high-density part comes from [16]. For the deconfined quark matter phase an unpaired massless quark approximation, $\rho c^2 = 3p + 4B$, was used and the "bag pressure", B, was fitted such that the transition from quark matter to nuclear matter occurs at 1% of the nuclear saturation density, $n_0 \sim 0.16$ fm^{-3}. The exact density at which cold nuclear matter decomposes into quark matter is unknown, so the transition density used here serves as an example only.

The composition of matter at neutron star densities remains an open question, and many different

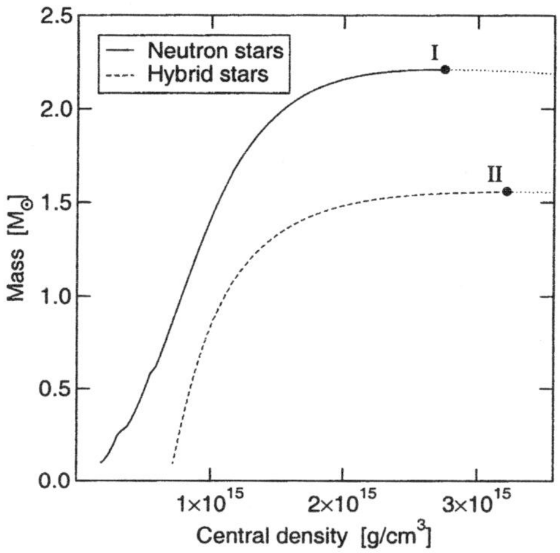

Fig. 1. Two sequences of compact stars obtained by solving the OV equations. The hybrid stars have a core of unpaired quarks and the nuclear matter crust extends down to 1% of the nuclear saturation density. The neutron star sequence is stable up to the maximum mass configuration (I). For this particular equation of state, this configuration has the highest possible (central) density, as stars more massive and dense than this are unstable and collapse into black holes. The stable hybrid star sequence terminates at II. $M_\odot \simeq 2 \times 10^{30}$ kg is the solar mass.

models for the EOS exist, *e.g.*, EOSs for nuclear matter, matter with hyperons, and superconducting quark matter. Regardless of the specific model, the maximum mass and corresponding radius are roughly one solar mass $M_\odot \simeq 2\times10^{30}$ kg and 10 km. Substantially more dense configurations composed of quarks and leptons are impossible. The motivation goes roughly like this: at white dwarf densities, the nucleons occupy nuclei that contribute little to the pressure, and electrons provide the pressure that counterbalances gravity. With increasing density, the pressure rises and the electrons become more energetic. Eventually, the electrons are captured by protons and the pressure drops. As a consequence, the white dwarf sequence becomes unstable and terminates. At a density that is roughly six to seven orders of magnitude higher than in the maximum-mass white dwarf, nuclei dissolve and the Fermi pressure of nucleons (in nuclear matter) and quarks (in quark matter) stabilize the next sequence of stable stars. The maximum mass of this sequence is of the order of one solar mass, for all compositions (nuclear matter, strange matter, hyperons etc.).

The reason why this is the limiting mass of stable compact stars composed of quarks and leptons is simply that there is no particle that can stabilize another sequence of stars. Each quark flavour is accompanied by an extra Fermi sea that relieves the growth of pressure, and quark Fermi pressure is only won at the expense of pressure from other species. Also, the chemical potential in stable hybrid stars is much smaller than the charm mass, so stars composed of quarks heavier than the strange quark do not exist [17, 18].

Hence, beyond the very rich and beautiful landscape of structures composed of quarks and leptons, at $\sim 10^{16}$ g/cm^3, there is again a desert of instability, just as there are no stable stellar configurations in-between white dwarfs and neutron stars. The question now is whether the desert ends before the Planck scale.

3 Compact stars beyond the desert

A definite upper limit to the density of any static spherically symmetric star can be obtained from the Schwarzschild radius,

$$R = 2GM/c^2, \tag{4}$$

since any object more dense than this would collapse into a black hole. By using the expression for the Schwarzschild radius and the relations:

$$M \sim mA, \tag{5}$$
$$R \sim d_0 A^{1/3}, \tag{6}$$

where A is the number of constituent particles, m their mass and d_0 the distance between adjacent particles, an order of magnitude estimate for the maximum mass and radius of the corresponding class of compact stars can be calculated. For an idealized neutron star composed of nucleons of mass $m_n \simeq 939$ MeV/c^2 and size $d_n \simeq 0.5 \times 10^{-15}$ m, equations (4)-(6) give us $A \sim 3 \times 10^{57}$ baryons, $R \sim 7$ km, and $M \sim 5 \times 10^{30}$ kg ~ 2.5 $M_\odot$. In reality, a somewhat larger radius and smaller mass are expected, since the density is non-uniform in the star, say $R \sim 10$ km and $M \sim 2$ $M_\odot$. In any case, the correct order of magnitude for the maximum mass and corresponding radius of a neutron star is obtained. The average density is $\bar{\rho} \simeq 10^{15}$ g/cm^3.

Since the Schwarzschild limit is already almost reached for the most massive neutron stars, it is reasonable to assume that this should be the case also for a more dense class of compact stars. Then, in order to provide similar estimates for the mass and radius of a star composed of preons, something must be known about the mass and "size" of preons. Before trying to achieve this, it should be emphasized that we know nothing about preons, not even whether or not they exist. So whatever method is used, the result remains a speculative estimate of the order of magnitude. As I will show, however, it is still possible to approach some qualitative conclusions.

Guided by the observation that the density of nuclear matter is of roughly the same order of magnitude as for deconfined quark matter, I assume that the density of preon matter is of roughly the same order of magnitude as for a closely spaced lattice of some "fundamental" particle of the SM. In this case the problem is simplified to the finding of a fundamental SM particle, with known mass and maximum spatial extension. The simplest and least ambiguous choice seems to be the electron, since the mass of an electron is well known, and from scattering experiments it is known that electrons do not have any visible substructure down to a scale of $\sim \hbar c$/TeV $\sim 10^{-19}$ m. Using the electron mass $m_e \simeq 511$ keV/c^2 and an upper estimate for its radius $r_e \sim 10^{-19}$ m, the maximum mass and radius of a star composed of preons is found to be of the order $M \sim 10^2$ $M_\oplus$ and $R \sim 1$ m. Here $M_\oplus \simeq 6 \times 10^{24}$ kg is the mass of the Earth. The average density is of the order $\sim 10^{23}$ g/cm^3.

This crude estimate gives metre-sized objects that are a hundred times more massive than the Earth. In particular, to be a bit more specific, it would be interesting to see whether such objects could be stable or not. In order to do this, I extrapolate an effective model for hadrons, the so-called MIT bag model [19]. In its simplest form, the MIT bag is a gas of massless fermions (partons), enclosed in a region (the bag) and subject to an external pressure B (the bag constant). The EOS for a gas of massless fermions is $\rho c^2 = 3p$ and by including B one obtains:

$$\rho c^2 = 3p + 4B. \tag{7}$$

This result does not depend on the degeneracy factor, *i.e.*, the number of fermion species, spin, etc. For a single hadron the pressure is practically zero, so that $\rho c^2 = 4B$ and the total energy E of a hadron is [19]:

$$E = 4B\langle V\rangle, \tag{8}$$

where $\langle V\rangle$ is the time-averaged volume of the

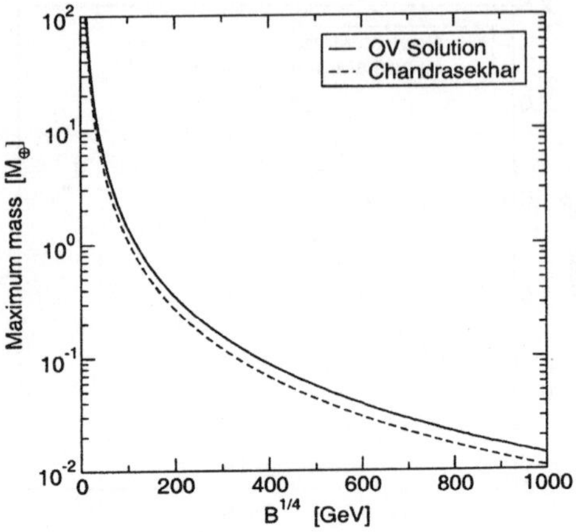

Fig. 2. The maximum mass of preon stars vs. the bag constant B. The solid line represents the general relativistic OV solutions, while the dashed line represents the Newtonian (Chandrasekhar) estimate. Despite the high central density, the mass of these objects is below the Schwarzschild limit, as is always the case for static solutions to the stellar equations. $M_\oplus \simeq 6 \times 10^{24}$ kg is the Earth mass.

bag (hadron). Hence, the bag pressure, B, must be of the order of 1 $\text{GeV/fm}^3 \sim 10^{35}$ Pa for hadrons. This is in agreement with experiments and other independent methods of calculating light-quark hadronic masses; most of the mass-energy is due to the confining interactions, not the "bare mass" of the constituents.

The MIT bag model is frequently used for the description of deconfined quark matter and applications to compact stars. Its usefulness in this regime originates in asymptotic freedom, simplicity and the possibility to include pertubative corrections. The bag pressure, B, is introduced in order to confine partons; it is a phenomenological parametrization of the strong interactions that confine quarks into hadrons. These interactions are present also in deconfined quark matter, so the "bag model" should also be applicable in this regime. However, the value of the bag pressure is different, since the density is higher and the interactions weaker. Thus, the so-called bag constant, B, is not really a constant, but a density dependent parameter. For strange quark matter, the bag constant is roughly $B^{1/4} \sim 150\ \text{MeV}/(\hbar c)^{3/4}$ [20] and the corresponding contribution to the energy density is $4B \sim 260\ \text{MeV/fm}^3$. This means that a considerable fraction of the density in quark matter, roughly 10^{15} g/cm^3, is due to the bag constant, *i.e.*, interactions.

The fundamental assumption here is that preons exist and are fermions. Since preons constitute light particles, such as neutrinos and electrons, the "bare" preon mass should be fairly small. Thus a massless fermion approximation, $\rho c^2 = 3p$, can be used. This EOS does not allow for stable superdense stars, however, so something more is needed. That 'something' is dynamics - the preon interactions that give mass-energy to the particles of the SM. The question is how, since there is not yet any quantitative model for preon interactions. Indeed, a fundamental problem in preon models is to find a suitable dynamics, capable of binding preons into fermions with masses that are essentially negligible with respect to their inverse radius. With this in mind, the principle of parsimony ("Occam's razor") seems to be the only guidance.

A simple solution is to include the dynamics in the form of a bag constant [8], which roughly reproduces the minimum energy density of an electron,

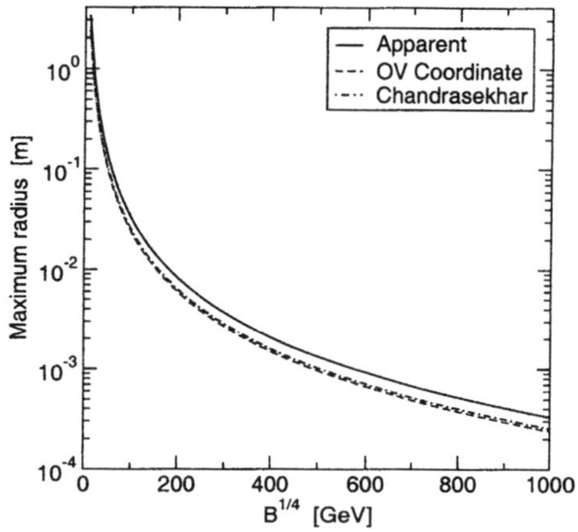

Fig. 3. The maximum radius of preon stars vs. the bag constant. The solid line is the "apparent" radius, $R^\infty = R/\sqrt{1-2GM/Rc^2}$, as seen by a distant observer. The dashed line represents the general relativistic coordinate radius obtained from the OV solution. The dotted line represents the Newtonian (Chandrasekhar) estimate.

$$
\begin{aligned}
B &= \frac{E}{4\langle V\rangle} \sim \frac{3\times 511\,\text{keV}}{16\pi(10^{-19}\,\text{m})^3} \sim 10^4\,\text{TeV/fm}^3 \\
&\implies B^{1/4} \sim 10\,\text{GeV}/(\hbar c)^{3/4}. \qquad (9)
\end{aligned}
$$

The very high density contribution from the bag constant $4B/c^2 \sim 10^5\ \text{TeV}\,c^{-2} fm^{-3} \sim 10^{23}$ g/cm^3 might seem a bit peculiar. It should, however, be kept in mind that the density contribution from the bag constant in deconfined quark matter is $\sim 10^{15}$ g/cm^3, which is a large fraction of the maximum density in any type of star composed of deconfined quark matter. So the high density is not that peculiar. On the contrary, should anything at all be expected, is that B would be much higher for preon matter than for quark matter, since the typical "preon bag" is much smaller and more dense than a hadron. In the following, for simplicity, I put $\hbar c = 1$ for the bag constant and express $B^{1/4}$ in eV.

The density introduced by the bag constant is of

the same order of magnitude as the density used in the mass-radius estimate above. The improvement here is the transition to a proper EOS for fermions; the possibility to apply the EOS in a general relativistic framework, for the analysis of mass-radius relations and stability. In addition to the full general relativistic analysis, the mass and radius can be estimated from first principles as a function of the bag constant [21]. The result is somewhat similar to the original Chandrasekhar limit, but the role of the fermion mass is replaced by the bag constant, B,

$$M = \frac{16\pi}{3c^2} B R^3, \tag{10}$$

$$R = \frac{3c^2}{16\sqrt{\pi G B}}. \tag{11}$$

Inserting $B^{1/4} \sim 10$ GeV in (10)-(11), an estimate for the (maximum) mass $M \simeq 10^2$ $M_\oplus$ and radius $R \sim 1$ m is obtained. This is consistent with the somewhat simpler mass-radius estimate given above.

Since $B^{1/4} \sim 10$ GeV is only an order of magnitude estimate for the lower limit, the bag constant is considered as a free parameter of the model, constrained by a lower limit of $B^{1/4} = 10$ GeV and an upper limit chosen as $B^{1/4} = 1$ TeV. The latter value corresponds to a spatial extension of the electron of the order $\sim \hbar c/10^3$ TeV $\sim 10^{-22}$ m. In Figs. 2-3 the maximum mass and radius of a preon star are plotted as a function of the bag constant.

A necessary (but not sufficient) condition for the stability of a compact star is that the total mass is an increasing function of the central density, $dM/d\rho_c > 0$. This condition is a consequence of a generic microscopic relation known as Le Chatelier's principle. Roughly speaking, this condition implies that a slight compression or expansion of a star will result in a less favourable state, with higher total energy. Obviously, this is a necessary condition for a stable equilibrium configuration. Equally important, a star must be stable when subject to (small) radial oscillations, in the sense that the amplitude of the oscillations must not grow spontaneously with time. Otherwise a small perturbation would bring about a collapse of the star.

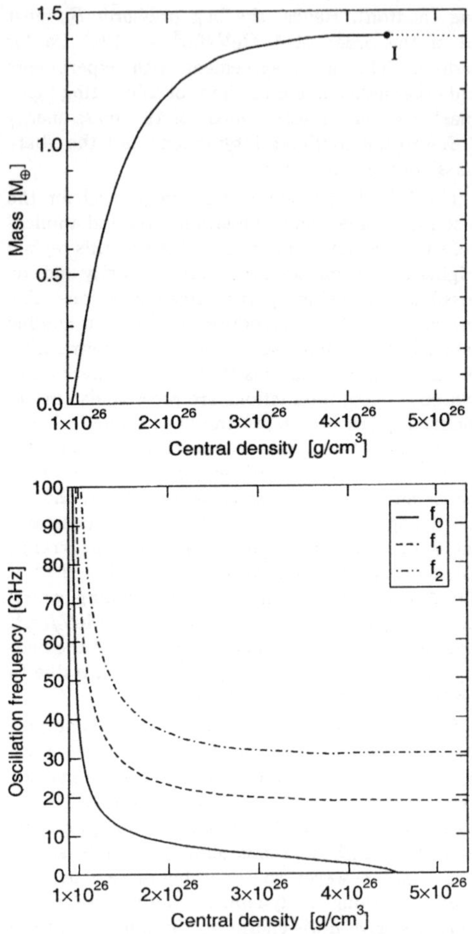

Fig. 4. The mass and the first three eigenmode oscillation frequencies (f_0, f_1, f_2) vs. the central density ρ_c of preon stars. Here, a fixed value of $B^{1/4} = 100$ GeV has been used. For the maximum mass configuration (I) the fundamental (nodeless) mode, f_0, has zero frequency, indicating the onset of instability. Preon stars with densities less than the density of the maximum mass configuration are stable.

The equation for the analysis of such radial modes of oscillation is due to Chandrasekhar [22]. An overview of the theory, and some applications, can be found in [23]. A catalogue of various numerical methods for solving the original set of equations can be found in [24]. However, a far more practical form of the oscillation equations has been derived by Gondek *et al.* [25]. The details of the stability analysis can be found in [8]. Here I summarise only the main points.

Assuming a time dependence of the radial displacement of fluid elements of the form $e^{i\omega t}$, the equation governing the radial oscillations is found to be a Sturm-Liouville eigenvalue equation for ω^2. A necessary and sufficient condition for stability is that all ω_i^2 are positive, since imaginary frequencies give exponentially increasing amplitudes. Furthermore, since ω_i^2 are eigenvalues of a Sturm-Liouville equation, it turns out that it is sufficient to prove that the fundamental (nodeless) mode, ω_0^2, is positive for a star to be stable. In Fig. 4, the first three oscillation frequencies, $f_i = \omega_i/2\pi$, for various stellar configurations with $B^{1/4} = 100$ GeV are plotted. In agreement with the theorem of Wheeler *et al.* [2] the onset of instability is the point of maximum mass, as ω_0^2 becomes negative for higher central densities. Thus, for $B^{1/4} = 100$ GeV, preon stars are stable up to the maximum mass configuration. The same is true for other values of B [8].

Despite the large uncertainty regarding preon interactions, manifested here as a large uncertainty in the bag constant, preon stars should have central densities beyond $\sim 10^{23}$ g/cm^3. This makes preon stars fundamentally different from the traditional types of compact stars, since such high densities imply that the stars must be very small and light in order to be stable, see Fig. 5.

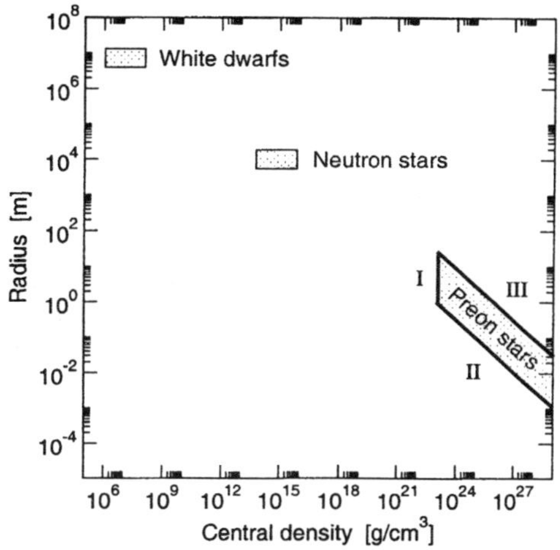

Fig. 5. The different types of compact stars traditionally considered in astrophysics are white dwarfs and neutron stars (including quark and hybrid stars). In white dwarfs, electrons provide the Fermi pressure counterbalancing gravity. In neutron stars, the neutrons (quarks, hyperons etc.) play this role. If quarks and leptons are composite particles, a new class of compact stars (preon stars) could exist. The minimum density (I) of preon stars is roughly given by the minimum density of leptons and quarks. The minimum size (II) for a given central density is due to the Schwarzschild radius (actually 4/3 of it) and a maximum size (III) exists due to instability.

4 Formation, detection and astrophysical consequences

The list of possible connections between the properties of the fundamental particles and the large scale structures in the universe is long. However, not much has been proposed beyond a density of $\sim 10^{23}$ g/cm^3 because there are strong arguments against the existence of stable objects beyond $\rho \sim 10^{16}$ g/cm^3. That is, if quarks and leptons are fundamental entities.

If preons exist, and if objects composed of preon matter are as small and light as suggested here and are stable, density fluctuations in the early universe might have produced primordial preon stars (or "nuggets"). As this material did not take part in the ensuing nucleosynthesis, the abundance of preon nuggets is not constrained by the hot big bang model bounds on baryonic matter. Also, preon nuggets are immune to Hawking radiation [26], which rapidly evaporates small primordial black holes, thus making it possible for preon nuggets to survive to our epoch. They can therefore serve as the mysterious dark matter needed in many dynamical contexts in astrophysics and cosmology [27, 28]. The idea that preons could be connected to dark matter is already recognized in the literature [29, 30], but the picture presented here is rather different.

The Friedmann equation for the early universe is $H^2(t) = 8\pi G\rho/3$ and in the radiation-dominated epoch $\rho c^2 \sim T^4$ (Boltzmann's law). When including the number of internal degrees of freedom, g_{eff}, an expression for the Hubble parameter, H, in units where $\hbar = c = k_B = 1$, is [31]:

$$H \simeq 1.66\sqrt{g_{eff}}\frac{T^2}{m_{pl}}. \qquad (12)$$

Here T is the temperature in eV, g_{eff} the effective number of degrees of freedom and $m_{pl} \simeq 1.2 \times 10^{19}$ GeV/c^2 the Planck mass. For the SM, the fermions and the gauge and Higgs bosons give $g_{eff}(T = 1\,\text{TeV}) = 106.75$. In the preon phase, this number should be smaller, say $g_{eff} \sim 10$ for simplicity. Then the Hubble radius at a temperature of 1 TeV is $H^{-1} \sim 1$ mm and the mass within the Horizon (a causally connected region) is $\rho H^{-3} \sim 10^{-1}$ $M_\oplus$. This is the maximum mass of any structure that could have formed in this early epoch. Hence, the maximum mass within causally connected regions, at the minimum temperature when deconfined preon matter might have formed preon nuggets (and the particles of the SM), is of the correct order of magnitude for stable configurations.

A potential problem is that the Jeans length, which defines the minimum length scale of regions that can contract gravitationally, was roughly of the same order of magnitude as the Hubble radius at that temperature. The Jeans length, λ_J, is [31]:

$$\lambda_J = v_s\sqrt{\frac{\pi}{G\rho_0}}, \qquad (13)$$

where v_s is the speed of sound and ρ_0 the average background density. For a relativistic fluid with EOS $\rho c^2 = 3p + 4B$ the speed of sound is $v_s = c/\sqrt{3}$ and $\lambda_J \sim 1$ mm $\sim H^{-1}$. However, considering the high level of approximation used here, this is not yet a serious problem. It merely shows that the numbers are in the correct intervals.

But perhaps it will be the other way around. After all, Popper's idea that we make progress by falsifying theories is not always true. By utilizing gamma-ray bursts (GRB) or white dwarfs in the large magellanic cloud as light sources, gravitational lenses with very small masses might be observable as a diffraction signal in the spectrum [32, 33, 34, 35]. For a lens with a mass within the range 10^{-16} $M_\odot \leq M \leq 10^{-11}$ $M_\odot$, the angular separation of images would be in the femto-arcsec range (femtolensing) and for more massive lenses, $M \leq 10^{-7}$ $M_\odot$, the angular separation is in the pico-arcsec range (picolensing). The mass within the Hubble radius at $T = 1$ TeV is $\sim 10^{-1}$ $M_\oplus \sim 10^{-7}$ $M_\odot$. This roughly defines the maximum mass of preon nuggets that could be abundant enough to be observed as gravitational lenses. Hence, preon nuggets fall in the correct mass range for picolensing and femtolensing.

In Fig. 6 the magnification of a distant light source due to gravitational lensing by an intermediate preon star is plotted as a function of the dimensionless frequency:

$$\nu = \frac{\tilde{\nu}(1 + z_L)2GM}{c^3}. \qquad (14)$$

Here, M is the mass of the lens, z_L the redshift of (distance to) the lens and $\tilde{\nu}$ the frequency of light. This result was calculated with a physical-optics model, as described in [34]. In essence, the time dependent amplitude due to a single light pulse from the source was calculated, and then the power spectrum was obtained by a Fourier transform of the amplitude. The magnification is normalized to a unit flux in the absence of a lens, *i.e.*, the signal equals the flux from the source times the magnification factor. The exact shape of the magnification curve depends on the relative position of the source and lens. Here the source is slightly off-axis, corresponding to $\theta = 0.2$ in [34].

As mentioned in [8], preon stars could also form in the collapse of normal massive stars, if the collapse is slightly too powerful for the core to stabilize as a neutron star, but not sufficiently violent for the formation of a black hole. Due to the potentially very large magnetic field and rapid rotation

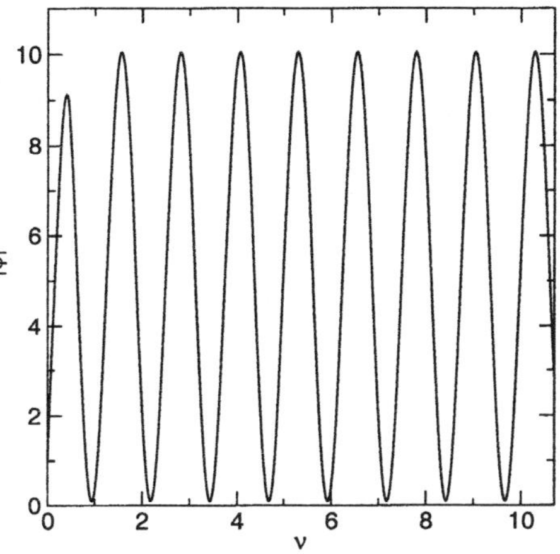

Fig. 6. $|\psi|^2$, the magnification of a distant light source vs. the dimensionless frequency, $\nu = \tilde{\nu}(1 + z_L)2GM/c^3$, due to gravitational lensing by an intermediate preon star. The flux entering the detector is obtained by multiplying the magnification, $|\psi|^2$, with the flux in the absence of a lens. For a 10^{-6} $M_\oplus$ preon star located in the halo of our galaxy, $\nu = 1$ corresponds to a photon energy of 0.14 keV. See the text for further details.

of preon stars formed in this way, the astrophysical consequences could be important, *e.g.*, for acceleration of ultra-high energy (UHE) cosmic rays. However, even though the idea might seem appealing at first sight, the possibility of expelling such a large fraction of the mass of the progenitor star needs to be better understood. What should be noted here is merely a potential connection to UHE cosmic rays, which might provide a second means for locating and observing preon stars.

5 Conclusions

If there is a layer of fermionic constituents (preons) deeper than that of quarks and leptons, a new class of stable compact stars could exist. By fitting a simple equation of state for fermions to the minimum energy density of an electron, the maximum mass for stars composed of preons can be estimated to $\sim 10^2$ Earth masses and the maximum radius to ~ 1 m. The minimum central density is of the order of $\sim 10^{23}$ g/cm^3. Preon stars (or "nuggets") with a maximum mass of $\sim 10^{-1}$ Earth masses and radius $\sim 10^{-3}$ m could have been formed by the primordial density fluctuations in the early universe. By utilizing gamma-ray bursts, or white dwarfs in the large magellanic cloud as light sources, an intermediate preon star (nugget) would produce a diffraction signal in the spectrum, which might be observable. Due to the need for observational clues in the cold dark matter sector, this could prove compositeness to be plausible, without much dedicated effort. This approach might complement direct tests of preon models at particle accelerators, especially at high energies, since preon stars fall in the observable range even if the energy scale of preon interactions is far beyond the range of any existing or planned particle accelerator.

6 Acknowledgements

I acknowledge support from the Swedish National Graduate School of Space Technology. I thank S. Fredriksson and J. Hansson for several useful discussions and for reading the manuscript. I thank M. Alford for providing me with the data tables of the high density nuclear matter equations of state, J. Goodman for useful discussions regarding the femtolensing signature, and J. Bourjaily, D. Casadei, A. Geiser, A. Giazotto, and J. Kamenik for interesting discussions about the astrophysical consequences of preon stars. Finally, I thank G. 't Hooft and A. Zichichi for organizing an excellent 42nd course of the international school of subnuclear physics, and for the award that this work was designated.

References

1. F. Weber, *Strange Quark Matter and Compact Stars*, astro-ph/0407155 (submitted).
2. B.K. Harrison, K.S. Thorne, M. Wakano & J.A. Wheeler, *Gravitation Theory and Gravitational Collapse* (University of Chicago Press, Chicago, 1965).

3. D.J. Gross & F. Wilczek, Phys. Rev. Lett. **30**, 1323 (1973).
4. H.D. Politzer, Phys. Rev. Lett. **30**, 1346 (1973).
5. U.H. Gerlach, Phys. Rev. **172**, 1325 (1968), U.H. Gerlach, Ph.D. thesis, Princeton University, 1968.
6. N.K. Glendenning & C. Kettner, Astron. Astrophys. **353**, L9 (2000).
7. K. Schertler *et al.*, Nucl. Phys. **A 677**, 463 (2000), astro-ph/0001467.
8. J. Hansson & F. Sandin, *Preon stars: a new class of cosmic compact objects*, 2004, astro-ph/0410417.
9. S. Fredriksson, in *Proc. of the Fourth Tegernsee Int. Conf. on Particle Physics Beyond the Standard Model, 2004*, ed. by H.-V. Klapdor-Kleingrothaus (Springer-Verlag, Heidelberg, 2004), p. 211, hep-ph/0309213.
10. I.A. D'Souza & C.S. Kalman, *Preons* (World Scientific, Singapore, 1992).
11. G. 't Hooft, Cargèse Lecture Notes, 1979.
12. R. Barbieri, L. Maiani & R. Petronzio, Phys. Lett. **B 96**, 63 (1980).
13. J.-J. Dugne, S. Fredriksson & J. Hansson, Europhys. Lett. **57**, 188 (2002).
14. J.R. Oppenheimer & G. Volkoff, Phys. Rev. **55**, 374 (1939).
15. J.W. Negele & D. Vautherin, Nucl. Phys. **A 207**, 298 (1973).
16. A. Akmal, V.R. Pandharipande & D.G. Ravenhall, Phys. Rev. **C 58**, 1804 (1998), nucl-th/9804027.
17. C. Kettner *et al.*, Phys. Rev. **D 51**, 1440 (1995).
18. M. Prisznyak, B. Lukacs & P. Levai, *Are There Top Quarks in Superdense Hybrid Stars?*, astro-ph/9412052.
19. A. Chodos *et al.*, Phys. Rev. **D 9**, 3471 (1974).
20. N.K. Glendenning, *Compact Stars* (Springer-Verlag, New York, 1997).
21. S. Banerjee, S.K. Ghosh & S. Raha, J. Phys. **G 26**, L1 (2000), astro-ph/0001246.
22. S. Chandrasekhar, Phys. Rev. Lett. **12**, 114 (1964).
23. C.W. Misner, K.S. Thorne & J.A. Wheeler, *Gravitation* (Freeman and Co., San Francisco, 1973).
24. J.M. Bardeen, K.S. Thorne & D.W. Meltzer, Astrophys. J. **145**, 505 (1966).
25. D. Gondek, P. Haensel & J.L. Zdunik, Astron. Astrophys. **325**, 217 (1997), astro-ph/9705157.
26. S.W. Hawking, Commun. Math. Phys. **43**, 199 (1975).
27. M.S. Turner, Phys. Rep. **333**, 619 (2000).
28. L. Bergström, Rep. Progr. Phys. **63**, 793 (2000).
29. V. Burdyuzha *et al.*, in *Proc. of the Second Int. Workshop on Particle Physics and the Early Universe, 1999*, ed. by D.O. Caldwell (Springer-Verlag, 1999), p. 392, astro-ph/9912555.
30. O. Lalakulich & G. Vereshkov, in *Proc. of the Second Int. Conf. on Dark Matter in Astro and Particle Physics, 1999*, ed. by H.-V. Klapdor-Kleingrothaus and L. Baudis (IOP Publishing, Great Yarmouth, 1999), p. 668.
31. L. Bergström & A. Goobar, *Cosmology and Particle Astrophysics*, 2nd ed. (Springer-Verlag, Germany, 2004).
32. A. Gould, ApJ **386**, L5 (1992).
33. K.Z. Stanek, B. Paczyński & J. Goodman, ApJ **413**, L7 (1993).
34. A. Ulmer & J. Goodman, ApJ **442**, 67 (1995), astro-ph/9406042.
35. R.J. Nemiroff & A. Gould, *Probing for MACHOs of Mass $10^{-15}M_\odot - 10^{-7}M_\odot$ with Gamma-Ray Burst Parallax Spacecraft*, astro-ph/9505019 (submitted).

DIPLOMAS

Thirteen Diplomas were open for competition among the participants. They have been awarded as follows:

- John S. BELL Diploma to:

  Giovanni IMPONENTE
  Queen Mary University of London, London, UK

- Patrik M. S. BLACKETT Diploma to:

  Sevil SALUR
  Yale, New Haven, USA

- Paul A. M. DIRAC Diploma to:

  Jacob BOURJAILY
  CERN, Geneva, Switzerland

- Vladimir N. GRIBOV Diploma to:

  Wolfgang SCHLEIFENBAUM
  Technische Universität, Darmstadt, Germany

- Robert HOFSTADTER Diploma to:

  Iwona GRABOWSKA-BOLD
  DESY, Hamburg, Germany

- Gunnar KÄLLEN Diploma to:

  Angelo Raffaele FAZIO
  JINR, Dubna, Russia

- Oreste PICCIONI Diploma to:

  Jernej KAMENIK
  Institute Jozef Stefan, Ljubljana, Slovenia

- Isidor I. RABI Diploma to:

  Oleg LOKTIK
  Columbia University, New York, USA

- Giulio RACAH Diploma to:

  Christian KLEIN-BÖSING
  Westfälische Wilhelms Universität, Münster, Germany

- Victor F. WEISSKOPF Diploma to:

  Johan ALWALL
  Uppsala University, Uppsala, Sweden

- Eugene P. WIGNER Diploma to:

  Jakub WAGNER
  Warsaw University, Warsaw, Poland

- Bjorn H. WIIK Diploma to:

  Robert TWEEDIE
  University of Edinburgh, Edinburgh, UK

- Chien Shiung WU Diploma to:

  Christine A. AIDALA
  Brookhaven National Laboratory, Upton, USA

AWARDS

The Award for the best "New Talents" – "Speaker" is divided into four parts:

i) for an original work in theoretical physics, to:

Jan Fredrik SANDIN
Lulea University of Technology, Lulea Sweden

ii) for an original work in experimental physics, to:

Tatsiana KLIMKOVICH
DESY, Hamburg, Germany

iii) for the best overview of a theoretical subject, to:

Jacob BOURJAILY
CERN, Geneva, Switzerland

iv) for the best overview of an experimental subject, to:

Sevil SALUR
Yale, New Haven, USA

- **Award for <u>BEST STUDENT</u> given to:**

Jacob BOURJAILY
CERN, Geneva, Switzerland

- **Award for <u>BEST SCIENTIFIC SECRETARY TEAM</u> given to:**

a) **Christine A. AIDALA**
Brookhaven National Laboratory, Upton, USA

Christian KLEIN-BÖSING
Westfälische Wilhelms Universität, Münster, Germany

b) **Dafne BALDASSARI**
University of Bologna, Bologna, Italy

Veronica BINDI
University of Bologna, Bologna, Italy

PARTICIPANTS

Stefano ACTIS Italy	Dipartimento di Fisica Teorica – Università di Torino Via Pietro Giuria 1 – 10125 TORINO, Italy
Christine A. AIDALA USA	Brookhaven National Laboratory Bldg. 510C UPTON, NY 11973, USA
Konstantin ALKALAEV Russia	Theory Department Lebedev Physical Institute Leninsky prospect 53 – MOSCOW 119991, Russia
Johan ALWALL Sweden	High Energy Physics Uppsala University Box 535 – S-75121 UPPSALA, Sweden
Silvia ARCELLI Italy	Dipartimento di Fisica Università degli Studi Via Irnerio 46 – 40126 BOLOGNA, Italy
Andriy Victorovich BADIN Ukraine	Department of Physics and Technology Kharkov National University 4 Svobody sq. – KHARKOV, 61077 Ukraine
Dafne BALDASSARI Italy	Dipartimento di Fisica Università degli Studi Viale Berti Pichat 6 – 40126 BOLOGNA, Italy
Veronica BINDI Italy	Dipartimento di Fisica Università degli Studi Viale Berti Pichat 6 – 40126 BOLOGNA, Italy
Jacob BOURJAILY USA	CERN CH-1211 GENEVA 23, Switzerland
Elena BRUNA Italy	Dipartimento di Fisica Sperimentale, Università di Torino Via Pietro Giuria 1 – 10125 TORINO, Italy

Diego CASADEI
Italy

Dipartimento di Fisica
Università degli Studi
Via Irnerio 46 – 40126 BOLOGNA, Italy

Luisa CIFARELLI
Italy

Dipartimento di Fisica
Università degli Studi
Via Irnerio 46 – 40126 BOLOGNA, Italy

Michael CLARK
UK

Department of Physics – Columbia University
538 West 120th Street – Mail Code 5272
NEW YORK, NY 10027, USA

Daniele DE GRUTTOLA
Italy

Dipartimento di Fisica "E.R. Caianiello"
Università degli Studi
Via S. Allende – 84081 SALERNO, Italy

Giuseppe DE RISI
Italy

Dipartimento di Fisica – Università di Bari
Via Amendola 173 - 70126 BARI, Italy

Sergiy DUBYNSKIY
Ukraine

Dept. n. 55-00 - Institute for Theoretical Physics
Kharkov Physics and Technology Institute
Akademicheskaya st. 1 - KHARKOV 61108, Ukraine

Angelo Raffaele FAZIO
Italy

JINR - Joint Institute for Nuclear Research
Joliot Curie 6 – DUBNA 141980, Moscow Region, Russia

Sergio FERRARA
Italy

CERN - TH Division
CH-1211 GENEVE 23, Switzerland

Achim GEISER
Germany

DESY
Notkestrasse 85 - D-22607 HAMBURG, Germany

Adalberto GIAZOTTO
Italy

VIRGO - INFN Pisa, Via E. Amaldi
56021 Santo Stefano Di Macerata - Cascina, PISA, Italy

Michel GOURDIN
France

Laboratoire de Physique Théorique et Hautes Energies
LPTHE - Université Pierre et Marie Curie
4 Place Jussieu - F-75252 PARIS Cedex 05, France

Miklos GYULASSY
USA

Dept of Physics Pupin Laboratory, Columbia University
538 West 120th Street - NEW YORK, NY 10027, USA

Iwona GRABOWSKA-BOLD
Poland

H1-Zeuthen Group, DESY Deutsches Elektronen-Synchroton
Notkestrasse 85 – D 22607 HAMBURG, Germany

Giovanni IMPONENTE
Italy

School of Mathematical Sciences
Queen Mary, University of London
Mile End Road – E1 4NS LONDON, UK

Jernej KAMENIK
Slovenia

Institute Jozef Stefan
Jamova 39 - 1000 LJUBLJANA, Slovenia

Richard D. KENWAY
UK

School of Physics and Astronomy, The University of Edinburgh
The King's Buildings - EDINBURGH, EH9 3JZ, Scotland, UK

Christian KLEIN-BÖSING
Germany

Institut für Kernphysik
Westfälische Wilhelms-Universität
Wilhelm-Klemm Str. 9 - D-48149 MÜNSTER, Germany

Tatsiana KLIMKOVICH
Belarus

DESY – FLC Group
Notkestrasse 85, 1d/39b – D-22607 HAMBURG, Germany

Christian P. KORTHALS-ALTES
Germany

Centre de Physique Théorique
CNRS - Luminy
163 avenue de Luminy, Case 907
F-13288 MARSEILLE Cedex 9, France

Benjamin LABONNE
France

Laboratoire de Physique Corpusculaire, CNRS-IN2P3
Université Blaise Pascal
24 Avenue des Landais
F-63177 AUBIÉRE CEDEX, France

Tiehui Ted LIU
USA

FERMI LAB - MS 318 Fermilab
P.O. Box - BATAVIA, IL 60510, USA

Oleg LOKTIK
Belarus

Columbia University
195 Claremont Avenue Apt 51
NEW YORK, NY 10027, USA

Luciano MAIANI San Marino	Dipartimento di Fisica, Università La Sapienza Piazzale Aldo Moro 2 – 00185 ROMA, Italy
Dmitry NESTEROV Russia	Theory Department, Lebedev Physical Institute Leninsky prospect 53 – MOSCOW 119991, Russia
Fulvio PICCININI Italy	INFN Sezione di Pavia Via Bassi 6 - 27100 PAVIA, Italy
Antonio POLOSA Italy	LAPTH 9 Chemin de Bellevue, BP 110 F-74941 ANNECY-LE-VIEUX, France
Lucio QUADRANI Italy	Dipartimento di Fisica Università degli Studi Viale Berti Pichat 6 – 40126 BOLOGNA, Italy
Verónica RIQUER Mexico	Theory Division – CERN CH-1211 GENEVE 23, Switzerland
Sevil SALUR Turkey	A.W. Wright Nuclear Structure Laboratory Yale University P.O. Box 208124 – 272 Whitney Avenue NEW HAVEN, CT 06520-8124, USA
Jan Fredrik SANDIN Sweden	Department of Physics Lulea University of Technology SE-97187 LULEA, Sweden
Gabriella SARTORELLI Italy	Dipartimento di Fisica Università degli Studi Via Irnerio 46 – 40126 BOLOGNA, Italy
Wolfgang SCHLEIFENBAUM Germany	Institut für Kernphysik Technical University Darmstadt Schlossgartenstrasse 9 - D-64289 DARMSTADT, Germany

Gilda SCIOLI
Italy

Dipartimento di Fisica
Università degli Studi
Via Irnerio 46 – 40126 BOLOGNA, Italy

Marco SELVI
Italy

Dipartimento di Fisica
Università degli Studi
Viale Berti Pichat 6 – 40126 BOLOGNA, Italy

Rosetta SILVESTRI
Italy

Dipartimento di Fisica "E.R. Caianiello"
Università degli Studi
Via S. Allende – 84081 SALERNO, Italy

Reinhard SIMON
Germany

Nuclear Physics KP1
GSI
Planckstrasse 1 – D-64291 DARMSTADT, Germany

Gerardus 't HOOFT
The Netherlands

Department of Theoretical Physics
Spinoza Institute - Utrecht University
NL-3508 TD UTRECHT, The Netherlands

Robert TWEEDIE
UK

School of Physics
University of Edinburgh
Room 4414 JCMB – King's Buildings
EDINBURGH EH93JL, Scotland, UK

Gabriele VENEZIANO
Italy

CERN - TH Division
CH-1211 GENEVA 23, Switzerland

Jakub WAGNER
Poland

Institute of Theoretical Physics
Warsaw University
Hoia 69 – 00-681 WARSAW, Poland

Antonino ZICHICHI
Italy

CERN
CH-1211 GENEVA 23, Switzerland
Dipartimento di Fisica - Università di Bologna
40126 BOLOGNA, Italy